CARNEGIE LIBRARY
LIVINGSTONE COLLEGE
SALISBURY, NC 28144

128928

# TECHNOLOGY AS A HUMAN AFFAIR

# TECHNOLOGY AS A HUMAN AFFAIR

Larry A. Hickman
Texas A&M University

**McGRAW-HILL PUBLISHING COMPANY**

New York St. Louis San Francisco Auckland Bogotá Caracas
Hamburg Lisbon London Madrid Mexico Milan Montreal New Delhi
Oklahoma City Paris San Juan São Paulo Singapore Sydney Tokyo Toronto

This book was set in Times Roman by the College Composition Unit
in cooperation with Monotype Composition Company.
The editors were Cynthia Ward and Bernadette Boylan;
the production supervisor was Friederich W. Schulte.
The cover was designed by Joan E. O'Connor.
Cover photo courtesy of DPS.
R. R. Donnelley & Sons Company was printer and binder.

Acknowledgments appear on pages 493-495 and on this page by reference.

**Technology As a Human Affair**

Copyright © 1990 by McGraw-Hill, Inc. All rights reserved. Printed in the United States of America. Except as permitted under the United States Copyright Act of 1976, no part of this publication may be reproduced or distributed in any form or by any means, or stored in a data base or retrieval system, without the prior written permission of the publisher.

1 2 3 4 5 6 7 8 9 0 DOC DOC 9 5 4 3 2 1 0

ISBN 0-07-028688-4

Library of Congress Cataloging-in-Publication Data

Technology as a human affair. Edited and with introductions by Larry A. Hickman.
p. cm.

ISBN 0-07-028688-4
1. Technology—Philosophy. I. Hickman, Larry.
T14.T394 1990
601—dc20 89-14517

# CONTENTS

# LIST OF AUTHORS

C. E. AYRES (1891–1972) was Professor of Economics at the University of Texas at Austin. His books include *Science: The False Messiah* (1927), *The Theory of Economic Progress* (1944), and *Toward a Reasonable Society* (1961).

DANIEL BOORSTIN is the Librarian of Congress Emeritus and a Pulitzer Prize–winning historian. His books include *The Americans: The Democratic Experience* (1973), *The Republic of Technology* (1978), and *The Discoverers* (1983).

DOUGLAS BROWNING is Professor of Philosophy at the University of Texas at Austin. His publications reflect his concern with ethics, metaphysics, and philosophical anthropology. In addition to a book of poetry, his publications include *Act and Agent* (1964) and *Philosophers of Process* (1965).

EDMUND S. CARPENTER is a noted anthropologist whose specialty is the study of Eskimo cultures. Carpenter collaborated with Marshall McLuhan to produce *Explorations in Communication* (1960), from which the essay in this volume is taken. Among his other books are *Eskimo Realities* (1973) and *Oh, What a Blow that Phantom Gave Me* (1973).

RUTH SCHWARTZ COWAN teaches history at the State University of New York at Stony Brook. Her published work includes studies of Francis Galton and of household technology. In 1983 she published *More Work for Mother: The Ironies of Household Technology from the Open Hearth to the Microwave.*

JOHN DEWEY (1859–1952) is widely regarded as the most original and influential American philosopher of the twentieth century. His books include *Democracy and Education* (1916), *Experience and Nature* (1925), *The Quest for Certainty* (1929), *and Art as Experience* (1935).

ALAN R. DRENGSON teaches philosophy at the University of Victoria, British Columbia. He is the editor (with Rodger Beehler) of *The Philosophy of Society* (1978) and the editor in chief of *The Trumpeter: Journal of Ecosophy.*

D. O. EDGE is Director of the Science Studies Unit of the University of Edinburgh. His work on technological myths and metaphors includes *Meaning and Control* (1973), edited with J. N. Wolfe.

JACQUES ELLUL is Professor of Law and History at the University of Bordeaux. His books include the groundbreaking *The Technological Society,* published in French in 1954 and in English in 1964, and *The Technological System,* published in French in 1977 and in English in 1980.

GEORGE GERBNER is Dean of the Annenberg School of Communications at the University of Pennsylvania. He is the editor of the *Journal of Communication* and the author of numerous books and articles that assess the social impacts of television.

GODFREY GUNATILLEKE is a director of the Marga Institute, the Sri Lanka Center for Development Studies, Colombo. He has written widely on the technological problems of developing countries. He is the editor, together with Neelan Tiruchelvam and Radhika Coomaraswamy, of *Ethical Dilemmas of Development in Asia* (1983), from which the essay that appears in this volume was excerpted.

JÜRGEN HABERMAS is Professor of Philosophy at the University of Frankfurt. His books include *Toward a Rational Society* (1970), *Legitimation Crisis* (1976), and *Communication and the Evolution of Society* (1979).

LARRY HICKMAN is Professor of Philosophy at Texas A&M University. His research interests include the history of logic, American pragmatism, and the philosophy of technology. His books include *Modern Theories of Higher Level Predicates* (1980), *Technology and Human Affairs* (edited with A. al-Hibri, 1981), and *John Dewey's Pragmatic Technology* (1990).

DON IHDE is Professor of Philosophy at the State University of New York at Stony Brook. His numerous books on the philosophy of technology include *Technics and Praxis* (1979) and *Existential Technics* (1983).

GLEN JEANSONNE teaches history at the University of Wisconsin, Milwaukee. His interests include political history and biography. He has published two books on the political history of Louisiana.

HANS JONAS is Alvin Johnson Professor of Philosophy Emeritus at the New School for Social Research. In addition to his work in the history of philosophy and religion, he has published numerous essays on technology and ethics. His book *The Imperative of Responsibility: In Search of an Ethics for the Technological Age* was published in 1984.

PAUL LEVINSON is a member of the Senior Faculty of Graduate Media Studies at the New School for Social Research. He is also president of Connected Education, Inc., a not-for-profit organization that offers university courses entirely via computer teleconferencing. Among his publications is *Mind at Large: Knowing in the Technological Age* (1988).

ROBERT LINHART was Professor of Economics at Vincennes University in 1968 when he left academia to join and assess the French labor force. *The Assembly Line,* from which the selection in this volume is taken, is an account of his experiences in the Citroën factory at Choisy.

W. H. LOWRANCE is Senior Fellow and Director of the Life Sciences and Public Policy Program at Rockefeller University. His research interests include national and international science policy, technological ethics, and nuclear proliferation.

JOHN J. McDERMOTT is Distinguished Professor of Philosophy at Texas A&M University. His many scholarly interests include American pragmatism, urban studies, and the philosophy of medicine. He is the author of *The Culture of Experience: Philosophical Essays in the American Grain* (1976) and *Streams of Experience: Reflections on the History and Philosophy of American Culture* (1986).

ROBERT E. McGINN, although trained as a philosopher, is Professor of Industrial Engineering and Associate Chair of the Stanford Values, Technology, Science, and Society Program. Among his many published assessments of the human dimensions of technology are "The Anatomy of Modern Technology" (with N. Bruce Hannay), "The Problem of Scale in Human Life: A Framework for Analysis," and *Science, Technology, and Society: An Introduction to the Field.*

HERBERT MARCUSE (d. 1973) was a political philosopher who deeply influenced the New Left movement in the United States during the 1960s. Along with Theodor Adorno, Max Horkheimer, and Walter Benjamin, he was a member of the "Frankfurt School" of critical theory. He taught philosophy at Brandeis University and at the University of California at San Diego. His books include *Eros and Civilization: A Philosophical Inquiry into Freud* (1955), *One Dimensional Man: Studies in the Ideology of Advanced Industrial Society* (1964), and *Studies in Critical Philosophy* (1972).

MAURICE MERLEAU-PONTY, an influential French philosopher and psychologist, was until his death in 1961 Professor at the Collège de France. Among his many books are *Phenomenology of Perception* (1962), The Structure of Behavior (1963), and *The Visible and the Invisible* (1968).

LEWIS MUMFORD has written extensively on the history of technology and the history of architecture. Among his many influential books are *Technics and Civilization* (1934), *The Highway and the City* (1964), and *The Myth of the Machine: Technics and Human Development* (1967).

JOSÉ ORTEGA Y GASSET (d. 1955) was a Spanish philosopher and journalist. During the brief life of the Spanish Republic, he was a member of its parliament. Among his numerous books are *The Revolt of the Masses* (1932) and *The Mission of the University* (1944).

ARNOLD PACEY, who has a degree in physics, has written and lectured on the history and public policy dimensions of technology. He is particularly interested in the development of technologies that are appropriate for the needs of developing countries. His publications include *The Maze of Ingenuity* (1974) and *The Culture of Technology* (1983), from which the selection in this volume was taken.

RICHARD POWERS is an award-winning novelist. He is the author of *Three Farmers on Their Way to a Dance* (1985) and *The Prisoner's Dilemma* (1988).

HERBERT I. SCHILLER is Professor of Communications at the University of California at San Diego. His works include *Mass Communications and American Empire* (1969) and *Information and the Crisis Economy* (1986).

AUTUMN STANLEY is a free-lance scholar and writer. In addition to her poetry and fiction, she has published a cookbook.

PAUL B. THOMPSON is Associate Professor of Philosophy and Agricultural Economics at Texas A&M University. He is the author of numerous articles on risk assessment and public policy.

LYNN WHITE, JR. (d. 1987) was an influential historian of the medieval and Renaissance periods of technology. He taught history at Sanford, Princeton, and UCLA. His many publications include *Medieval Technology and Social Change* (1962) and *Medieval Religion and Technology* (1978).

LANGDON WINNER is Associate Professor in the Department of Science and Technological Studies at Rensselaer Polytechnic Institute. His works include the highly regarded *Autonomous Technology: Technics-Out-of-Control as a Theme in Political Thought* (1977) and *The Whale and the Reactor: A Search for Limits in an Age of High Technology* (1986).

SHOSHANA ZUBOFF is Associate Professor of Organizational Behavior and Human Resource Management at Harvard University. Among her publications are "New Worlds of Computer-Mediated Work," "Technologies that Informate," and *In the Age of the Smart Machine* (1988), from which the essay reprinted in this volume was excerpted.

# PREFACE

This book brings together essays by humanists and social scientists whose methods differ significantly and who appraise technology from widely divergent points of view. Their disagreements are profound. They share no common definition of technology, nor do they agree about what counts as a technological problem. Some exhibit deep pessimism about the future of technology. Others offer reasons why we should confront the future with energy and hope. Each brings to technology studies his or her unique insights as a Marxist, a feminist, a pragmatist, an analytic philosopher, a phenomenologist, or an analyst of third world cultures. It has been my aim in selecting these essays to exhibit the diversity, the rich complexity, the profound problems, and the myriad possibilities of our technological milieu.

Beneath their diversity, however, these essays have much in common. Each finds its place within a set of debates that has been a part of philosophical activity since the time of Plato. The relation of individual men and women to the social groups of which they are a part; the meaning of embodiment; the ramifications of aesthetic experience; the nature and consequences of human freedom; the significances of quotidian or everyday artifacts; the methods and limits of human knowledge: these basic issues have occupied philosophers for over two and a half centuries, and they now form the basis for the humanities and social sciences as those disciplines are currently understood.

Approximately half of the essays collected here were written by philosophers. The rest, written by historians, sociologists, communication theorists, economists, political scientists, and others, approach their topics in terms that are thoroughly philosophical.

This book may be used in a variety of ways. Some of its essays are general in nature, others are more technical; some represent classical positions, others advance novel proposals; some are playful and evocative, others are somber and hortatory. Since I have attempted to design a book for undergraduate or graduate study, as well as for general reading, these essays vary in terms of their length and complexity. I have preceded each of the seven sections with an essay that introduces its problems, indicates various possibilities of resolution, and relates its material to other relevant essays and books.

The literature of technology studies constitutes a vast territory, and it is the purpose of this book to provide a map of that territory. It is not the function of a map to cover a territory point for point, but rather to indicate important landmarks, points of interest, and routes of travel. For the reader who wishes to examine one or more of these features in a more detailed way, I have provided a section entitled "For Further Reading."

Some of the essays included here, and a portion of the material that introduces them, were parts of two earlier readers. The first, *Technology and Human Affairs*, which I coedited with Azizah al-Hibri, was published in 1981. The second, *Philosophy, Technology and Human Affairs*, of which I was the sole editor, was published in 1985.

I would be remiss if I failed to express my gratitude for the work of those who have helped to define the field we now call the philosophy of technology. Conversations and correspondence with Carl Mitcham, Paul Durbin, Paul Thompson, Joe Pitt, Langdon Winner, Stanley Carpenter, Ed Byrne, Fritz Rapp, Kristin Schrader-Frechette, Paul Levinson, Alan Drengson, Don Ihde, Albert Borgmann, Frederick Ferré, John Schumacher, and many others have been an invaluable aid in helping me to clarify my own ideas and to render them more coherent. I would also like to thank my assistant Ed Gabrielsen for the careful attention he gave to the details of manuscript preparation.

*Larry A. Hickman*

# TECHNOLOGY AS A HUMAN AFFAIR

PART **ONE**

# TOWARD A PHILOSOPHY OF TECHNOLOGY

## INTRODUCTION

The task of determining the place of technology within human affairs is as difficult as it is important. A measure of its importance is the ubiquity of technology: technology conditions and permeates virtually every human experience. But this ubiquity is also the source of a great difficulty: because technology is everywhere we look, we are not always able to see it. Because we use hundreds of artifacts—tools, gadgets, implements, appliances, machines, and methods—each day of our lives, these things have become invisible to most of us. We tend just to take them for granted. We tend to think about artifacts only when they break down, and we tend to think about methods only when they fail to serve us in the ways to which we have become accustomed. But this kind of tardy response to our technological environment has not served us very well. If we are to understand and control technology, instead of merely drifting wherever its constraints and facilities take us, we must learn to anticipate its problems and develop coherent plans of action.

Another difficulty is that despite the fact that we often refer to technology as "it," as I did in the last paragraph, the term "technology" doesn't refer to any one thing. "Technology" has come to have so many meanings that those who speak and write about "it" often seem to have no common vocabulary. This collection of essays has been designed to render our technological experiences more visible, to sharpen the conceptual tools with which we can explore the many meanings of technology, and to advance the development of a common technological vocabulary.

In the first essay in this section, Robert E. McGinn helps us get a foothold in this treacherous territory by asking, "What Is Technology?" He characterizes technology quite broadly as one form of human activity, alongside such others as science, art, religion, and sports. He focuses upon eight aspects of technology.

First, technology is concerned with *material,* as opposed to ideational, outcomes. His view is thus much narrower, for example, than that of Emmanuel Mesthene, who once defined technology as "the organization of knowledge for the achievement of practical purposes."[1] Unlike McGinn, Mesthene and others would view such nonmaterial objects as scientific theories, the square root of minus one, laws, regulations, and institutions as technological objects because they are human artifacts.[2] This difference of viewpoint regarding the very nature of technology offers an important key to understanding the debates presented in the essays collected in this volume. In emphasizing the material aspects of technological activity, however, McGinn is careful to include not only "material product-making" activities, such as making a pair of shoes out of leather, but also "object-transforming" activities, such as plastic surgery and genetic engineering.

Second, technology is, in McGinn's view, *fabricative*. This means that when human beings engage in technological activities they are *making* something artificial rather than just facilitating something that nature ordinarily does. What this means is that McGinn rejects agriculture or cloud seeding (to induce rain) as legitimate instances of technological activity.

Third, technology involves *purpose*. This is more than just the adaptation of methods to some preconceived end. Instead of working with a well-defined stock of ends for which methods are invented, technological activity constantly enlarges the domain of ends; it expands the number of possibilities. In other words, technology is far from being "value free": it is instead a rich source of values.

Fourth, technology is *resource-based* and *resource-expending*. Some of these resources, such as iron ore, are raw materials; others, such as plastics, are the results of previous technological activity.

Fifth, technology is *knowledge*. It is knowledge of how to do certain things, knowledge of resources, and, on a more abstract level, knowledge of methods. It is important to note that McGinn thinks that science is an "integral aspect of technology." This means that he rejects the common view that technology is "applied science." But this is also an important modification of his first point, that technology is concerned only or primarily with material objects.

Sixth, technology involves *method*. The primary method utilized during the early history of technology was trial and error. But technological methods now include complex experimental techniques, some of which are so specialized that they are applicable to only one type of material.

Seventh, technology generates, and operates within, a *sociocultural-environmental context*. McGinn calls this an *enabling* and *informing* activity rather than a *constitutive* one. This means that technological activity both reflects and alters its context in any given stage of its development. Economic, political, and ideological considerations enter into technological decisions, and they are in turn conditioned by technological change. Some forms of technological activity, for example, are possible only within agriculture-based cultures; others are possible only within cultures that are heavily industrial.

Finally, technology has profound connections to the *practitioner's mental set*. This involves not just the purposes of the practitioner but the affective context in which he or she operates. The mental set of the producers of ceremonial pottery among pre-Columbian meso-American Indians was, for example, quite different from the mental set of the producers of the space shuttle.

Because technology is often described as "value free," it is particularly important to examine McGinn's controversial claim that it is in fact a rich source of values. He reminds us that technological activity is almost always undertaken because it is of *benefit* to someone and that technological activity is one of the ways in which human beings render their world more rational and more meaningful. To some technologists, the exhilaration of problem solving is valued and valuable, even when their solutions are not immediately translatable into improvements in the everyday lives of men and women. Moreover, technological activity generates particular institutions which then set the goals or values for further technological activity. And finally, technological activity allows for the expression of the individual values of designers and producers and of those for whom they work. One of the premises of McGinn's essay is that technology is not so much a "thing" as it is an overlapping family of processes and activities. "Technology" is a term that has no single meaning.

"Philosophy" is also a term that has many different senses and that refers to a wide variety of activities. In a general and popular sense, for example, most people profess to have a "philosophy of life." But the essays in this section employ the term in ways that are much more precise. Following the lead of the ancient Greek thinkers Plato and Aristotle, "philosophy" is here used to denote a concern with certain perennial human questions: How is knowledge possible and what can I know? What kinds of things are there? Are some of my actions better than others, and how can I know them to be so? What is the best type of society, and how can I know my place in it? How can I explain and defend my judgments about what I take to be beautiful? What counts as evidence in any given situation?

Far from being made obsolete by contemporary technology, these traditional questions have become more relevant than ever in our age of automobiles, nuclear weapons, and televisions. The philosophical essays collected in this volume invite us to consider the metaphysical meanings of automobiles, the ethical dimensions of nuclear weapons, and the aesthetics of television.

Alan R. Drengson identifies and analyzes several competing senses of "philosophy" in his essay "Four Philosophies of Technology." The first is a kind of generic or nonexplicit attitude toward life—a philosophy of life. The second is an "explicit elaboration of a particular position which spells out assumptions, axioms, etc., and argues for its conclusions." The third is what he regards as the mature sense of the word, "a kind of creative activity of conceptual inquiry which frees us of attachment to specific models and doctrines in order to develop more appropriate cultural practices." For Drengson, philosophy at its best is like a "non-position and an activity. It is a sort of jazz played with concepts."

Drengson writes in a similar tone when characterizing what he takes to be four developmental stages of technology. He characterizes *technological anarchy* as a youthful and exuberant attitude which seeks to do anything that can be done to facilitate the total technological exploitation of the natural world. Technological activity during this stage is like the activity of the young child who is intent on seeing what he or she can do but who is innocently oblivious of consequences. *Technophilia* is like an adolescent love affair in which there is total identification with the object of one's affections and in which caution is thrown to the wind. This is not the heedless domination of nature that is produced by technological anarchy but instead a love of artifacts as toys. Drengson compares his third stage, *technophobia,* to the disenchantments of early adulthood, in which old models for actions have broken down and new ones emerge only at the cost of considerable pain. He argues that each of these stages is severely limited and immature in comparison to the pluralism and possibilities of the last stage, which he calls *appropriate technology*.

Unlike its three predecessors, appropriate technology is self-critical. It is interested in the promotion of community vitality and seeks to build links to every area of human concern. If we are to have a mature relationship to our technology, Drengson argues, our choices must tend to preserve diversity; to promote benign interactions between humans, their machines, and the biosphere; to develop prudential methods of generating and using energy; to balance all costs; and to promote human development.

In addition to serving as an excellent introduction to the philosophy of technology, Drengson's essay also constitutes an important supplement to the essays in Part Five by José Ortega y Gasset, Lynn White, Jr., and Paul Levinson. Each of these authors considers the developmental stages of technology from his own unique point of view. Drengson's presentation makes a particularly interesting companion piece to Levinson's essay, in which technological artifacts are said to go through the stages of toy, mirror (of reality), and art form.

The task that Drengson here undertakes so masterfully, namely, the assessment of the influence on one another of technological and philosophical metaphors, has been an essential component of philosophical discussions since at least the fifth century. But philosophers have not always exhibited as positive an attitude toward methods and artifacts as does Drengson. Nor have they always been as aware as is he of the interpenetration of their metaphors of choice with those of the technology of their time.

Plato and Aristotle, each in his own distinctive way, demeaned that sphere of human interaction with changeable materials which we today call technological. During one phase of his long career, Plato described the world as the product of a "divine craftsman,"[3] outside and above the realm of nature. Quotidian, or everyday, instrumental artifacts—the products of human artisans, such as tables, chairs, and beds—were regarded by Plato as inferior to the ideal, eternal, unchanging, perfect, and transcendent exemplars that he called the Forms. It was thus the task of a human artisan to create instrumental likenesses of the Forms. But those likenesses were treated as inferior to the Forms, and the work of the human artisan was said to be inferior to the work of the divine artisan.

Plato regarded the products of the visual, literary, and dramatic artists, which he called "semblances," as inferior even to the likenesses produced by carpenters and sculptors. Consequently, the poets fared even worse than the carpenters in Plato's ideal state. He tells us in Book X of the *Republic,* for example, that the work of mimetic art is twice removed from reality, since such products are just copies of real things, which are in turn copies of the Forms, that is, of the rational and perfect objects. Plato not only had little use for artists but even harbored plans to banish certain of them from his Republic.[4] Late in his life, in *The Laws,* he proposed that citizens be prohibited from practicing the crafts. It was his view that it is the duty of the citizen to engage in statecraft or political affairs in a single-minded way: Statecraft became for him the supreme technological enterprise to which he subordinated all others.[5]

Plato's treatment of technology is thus deeply ironic. The truest forms of production are divine and transcendent, and the highest form of human production is statecraft. Nevertheless, he had obviously taken the methods and insights of the artisans as the model for his divine artisan, and he had taken the products of the human artisan as the model for what the divine artisan produces in the world of nature. It is therefore remarkable that Plato wished to deny the artisans access to citizenship on the grounds that they were occupied with the material world and thus incapable of contemplating the world of the Forms.

The only type of human production that Plato thought of any real worth was the statecraft of the philosopher-kings—and the success of such production was in his view to be measured by the extent to which the philosopher-kings were able to contemplate and replicate the Forms. The part of technology that manipulates and alters the physical world—the concrete activity of the artisan—was thus for Plato subordinated to a kind of pseudo- or proto-science whose method was to possess through contemplation a realm of Forms which was already complete, finished, and beyond the domain of everyday affairs. His method was pseudo- or proto-science in comparison to the experimental methods of the scientific technology that arose in the seventeenth century and dominates our own era. Modern science replaced the method of contemplation with active and ongoing experimentation whose goal is the transformation of problematic features of concrete environments.

Aristotle, in his turn, moved the locus of production from the supernatural to the natural. In rejecting Plato's transcendent Forms, he made nature itself the grand artisan, depicting nature as the author of the fixed natural ends of objects and organisms in the world. But he did manage to retain an important aspect of Plato's philosophy of technology. Like Plato, he thought that natural ends are fit to be contemplated as ends in themselves, not as means for the new and different goals which human beings might develop as a consequence of their technological activities. Aristotle's work thus marks a great advance over that of Plato, for he at least moved the source and ground of technology out of the supernatural realm and into nature. For Aristotle, however, as for Plato before him (and for the mainstream of Western Christian theology, which owes so much to the two of them), the universe was still considered a finished work of art.

It was not until the seventeenth century that the leading metaphors of the philosophy of technology underwent a crucial change. There is a growing body of detailed and informative accounts of the actual practices of technology during the period from the decay of the Roman Empire to the Renaissance of the fifteenth and sixteenth centuries. One of the foremost historians of medieval technology, Lynn White, Jr., convincingly argues that there was during that period a more or less steady increase of technological activity and that the accomplishments of that period laid the basis without which experimental science could not have developed during the seventeenth century. But it is necessary to distinguish the actual practice of technology from the account given it by philosophers. The philosophy of technology during the medieval period, such as it was, was still mostly dominated by Aristotle. Further, as White reminds us, medieval society "remained hierarchical, and outside monastic circles the contempt for labor of the hands that probably had slowed technological change in antiquity remained normal."[6]

For much of philosophy between René Descartes at the beginning of the seventeenth century and the middle of the nineteenth, a period usually called the "modern" period, nature, human beings, and human societies were regarded as machines, manipulable in a way that the organic but finished and complete nature described by Aristotle could never have been.

In the seventeenth century, Descartes characterized the human body as a kind of machine in which there dwells a very un-machinelike "spirit," which he called "thinking stuff." Jean Jacques Rousseau and Thomas Jefferson, in the eighteenth century, utilized machine metaphors to describe and prescribe political and social organization. (The idea of "checks and balances," which has been such an important political metaphor in the United States, is one such technological metaphor; it draws on the vocabulary of the pendulum clock.) Jeremy Bentham, in the nineteenth century, went so far as to characterize human beings as "reckoning machines."

As primitive as this characterization of the universe and its contents as a machine may sound to us today, it was nevertheless in its time a significant advance in human thinking. As David Edge reminds us in his essay in Part Two, the metaphor of the machine represented a real advantage over the dominant metaphors of the classical and medieval periods, in which technology was grounded either in transcendent supernature or in an immanent nature regarded as finished and complete. Philosophers and scientists in the seventeenth, eighteenth, and nineteenth centuries effectively treated nature as open to manipulation and alteration, even though at the level of theory they still paid homage to the older metaphor of nature as the handiwork (a word which itself possesses rich technological connotations) of a transcendent deity.

It is difficult for those of us alive at the end of the twentieth century to imagine how liberating these new machine metaphors must have been at the time they were introduced. Not only can machines do work, thereby eliminating human drudgery, they (and by extension, nature, society, and human beings themselves) can be manipulated and repaired. This new spirit was particularly important to medical practice. The method of passive observation advanced by Galen (A.D. 130–200), based upon the doctrine that the human body is perfectly designed and constructed and that it is the role of the physician to be the watchful servant of a protecting and curing nature, gave way to active experimentation: new surgical techniques, new pharmaceutical agents, and new forms of therapy were actively developed. Rousseau, Leibniz, Jefferson, Bentham, and Newton all operated with and advanced these new technological metaphors.

But the new metaphors also had a dark side. In his essay in Part Five; Lynn White, Jr., charges that they contributed to an exploitation and debasement of nature and of human beings in ways theretofore unknown. It is not that the Greeks and Romans did not pollute their rivers and deforest their hills: they certainly did so. But after the seventeenth century, nature was accepted as the proper object for domination in a sense that Plato and Aristotle had never dreamed of. What had been unfortunate and unthinking practice became, in the modern period, a program sanctioned by official doctrine.

The new metaphors also contained a vestige of the old metaphysics of Plato and Aristotle that became more dysfunctional and dangerous than it had ever been in the hands of the Greeks and medievals. If nature was for the Greeks an object, it became for the post-Cartesians an object-to-be-used-and-depleted, or, to use the phrase of Martin Heidegger, it became a ''standing reserve'' or ''storehouse'' to be used up. Nature became so completely separated from the interests of the newly articulated human subject that influences were thought to operate in one direction only, namely, from human agency to a dumb and pliable natural world.

During our own time, technological metaphors have once again undergone a dramatic shift. John Dewey, whose essay ''Science and Society'' appears in Part Seven, not only censured the metaphysics of contemplation that was an important part of the philosophical systems of Plato and Aristotle but was equally critical of the metaphysics of domination that characterized many of the major texts of the modern period.

Like the philosophers of the modern period, Dewey depicted the world as open to change by human agency, especially by means of technological activity, which includes inquiry into materials of all types. But unlike Plato and his modern intellectual heirs, Dewey regarded technology as grounded entirely in the here and now instead of in a supernatural world different from (and a measure of) our own. Dewey rejected the metaphysics of objects whose meanings are fixed for all time by nature, by God, or even by human activity. He regarded objects, events, and even individual human beings as in continual flux and as participating in ongoing transactions with their environments. In Dewey's view, these transactions result in a continuing alteration of all parties.

Dewey was quite uncomfortable with the tendency of many modern philosophers to divide the world into two radically separate entities: a thinking subject and an objective world. Against this view he argued that technology must take its place in an organic realm in which the perceived ends of objects, individuals, and events may be utilized as the means to further development. Like Ortega y Gasset, who argues in a similar vein in his essay reprinted in Part Five of this volume, Dewey held that human beings are in large measure responsible for their experiences because they have created the situations which give rise to those experiences, and they continue to interact with and re-create those situations. These situations, of which human beings are a part and in which they find themselves, are, for Dewey, artifacts; they are constructed by intelligent transaction with the worlds of the past, that is, worlds that have been previously created and altered. Both Dewey and Ortega saw human beings as definable in terms of their transactions with the worlds of their experience, and both regarded such transactions as productive of an ongoing re-creation of human beings as well as their experienced worlds. Ortega labeled this phenomenon human "autofabrication," and Dewey's terms for it were "experimentation" and "technology."

It was Dewey's view that technology is *historically* prior to science: the crafts were in evidence long before the rise of experimental science. But it was also his view that technology is *functionally* inclusive of science: scientific theories and concepts are themselves technological artifacts.

This brief and in some ways oversimplified historical sketch provides the background for understanding the position taken by Hans Jonas in his essay "Toward a Philosophy of Technology." He suggests that we consider technology under three aspects: the "formal dynamics" of technology, its material content, and its overarching moral problems. In the first of these three tasks one would attempt to understand technology as an abstract whole, in much the same sense that Jacques Ellul seeks an understanding of what he calls "technique." What are its overarching properties? What are its internal laws? Ellul assimilates most of what he has to say about technology to this aspect of it, identifying technique as "the ensemble of practices by which one uses available resources in order to achieve certain valued ends." It is of technology taken in this sense that one may ask, as Ellul does in his essay included in this section and in his essays in Part Six, "Is it autonomous, or are human beings still free?" This overarching sense of technology is also under discussion in the essays brought together in Part Five, which present alternative patterns of technological history, and in those collected in Part Seven, which attempt to locate technology with respect to human social and political institutions.

By its "material content" Jonas understands the concrete practices and uses of technology. Also included here are attempts to construct taxonomies of technological artifacts, analyses of technological novelties, and work in the ontology of technology. The phenomenology of everyday affairs, including inquiry into the philosophical meanings of automobiles, television, clocks, nuclear missiles, contraceptives, and computers, is also a part of a study of technology's material content.

Finally, Jonas invites us to consider the place of valuation in a philosophy of technology. He reminds us that if we are to think critically about technology, we must ask hard questions about the source of our values, about the image we have of human beings, about the duties of the makers and users of technological systems and artifacts, and about how we can balance long-range goals with those that are short-range.

In "The Technological Order," the final essay in this section, Jacques Ellul suggests that there are several issues of which we make too much. He calls these issues the "fake" problems of technology. Such disagreeable conditions as overcrowding, increased nervous tension, air and water pollution, the perception that morals are in decay and that art is becoming sterilized, and the fear that technology will ultimately eliminate instinctive human values and powers: Ellul suggests that each of these worries can be put to rest by an examination of what is actually the case. He thinks that most of the disagreeable features of technology are open to solution by technology itself. He is unconvinced by those who argue that morality in the age of technology is inferior to that of previous eras. Even a cursory examination of modern cinema, music, painting, and poetry provides testimony to the vitality of art in our time. Further, compensatory mechanisms such as the increasing appreciation of aesthetic eroticism (he mentions with approval the works of Henry Miller) mark an enhancement of what he calls "the vital forces."

But if these concerns fail to take root in Ellul's characterization of technology, there are others that do. Ellul is concerned that technology has come to include all other aspects of contemporary civilization in a way that both limits human freedom and fails to provide the counterweight of an alternative value system by which technology may be judged.

These introductory essays provide a rich diversity of suggestions concerning how we may approach the problems of coming to terms with our immersion in a technological environment. Together, they serve to establish a framework by means of which the problems and issues articulated in the remaining essays in this volume may be profitably explored.

## NOTES

1 Emmanuel G. Mesthene, *Technological Change: Its Impact on Man and Society* (New York: New American Library, 1970), 25.
2 There has been no more persistent advocate of the view that ideas are artifacts—therefore technological—than John Dewey. In *Experience and Nature,* Dewey writes that: "The idea is, in short, art and a work of art. As a work of art, it directly liberates subsequent action and makes it more fruitful in a creation of more meanings and more perceptions." John Dewey, *The Later Works, 1925–1953,* Vol. I, ed. Jo Ann Boydston (Carbondale and Edwardsville: Southern Illinois University Press, 1981), 278.
3 *Theatetus,* 265c
4 *Republic,* 398a.
5 *The Laws,* 846d–e.
6 Lynn White, Jr., *Medieval Religion and Technology* (Berkeley, Los Angeles, and London: University of California Press, 1978), xx.

## What Is Technology?

**Robert E. McGinn**

With the increasing recognition of technology as a pivotal force in modern society, philosophers and other scholars have begun to consider problems arising out of the availability and use of new technologies. However, perusal of the emerging body of literature reveals an underlying divergence of opinion on the basic question, "What *is* technology?" A similar lack of consensus characterizes (and vitiates) much of the public discussion of technology that has occurred in recent years. While supporting the elaboration of a comprehensive social philosophy of technology and analyses of ethical and human values issues raised by modern technologies, I believe it important to improve our understanding of the nature of technology per se. What follows is intended as a contribution to this goal.

I propose to answer the question "What is technology?" in two parts; first, by inquiring into the structure of technology, i.e., by answering the question "What kind of activity-form is technology?" and second, by specifying in general terms what constitutes the content of technology. After offering a substantive example of technological activity to illustrate the presence of each of the previously identified structural aspects, I will conclude by showing how the resultant more differentiated concept of technology may be of use in clarifying the controversy over whether technology is "value-laden" or "value-free."

Definitions of technology have tended to fall into two general categories. Either it is defined in terms of one or two salient characteristics thought to be central to it, or as a concatenation of a larger, more disparate set of such characteristics, none of which, it is suggested, can be omitted without failing to capture something essential to technology. Illustrating the first tendency, Singer, *et al.*, in their *History of Technology,* define technology as "what things are done or made" and "how things are done or made,"[1] while Mesthene defines technology as "the organization of knowledge for the achievement of practical purposes."[2] Alternatively, Donald Schon asserts that technology means "any tool or technique, any product or process, any physical equipment or method of doing or making, by which human capability is extended."[3] In my view, neither definitional strategy succeeds in conveying an adequate notion of what technology is. My approach here is not definitional but characterological. I propose to treat technology as a *form of human activity,* others of which include science, art, religion, and sport. (It should be added that recent work of Jane Goodall and associates has shown that humans are not the sole practitioners of technology. Not only do chimpanzees use tools—e.g., compressing leaves, inserting them in water-filled tree trunks as sponges—but they also engage in rudimentary tool manufacture: stripping leaves off branches and inserting them into tree trunks to extract termites.) After making an important preliminary distinction, I will identify and discuss eight important aspects of the activity-form technology, some more "internal" to it than others; each, however, worth including in an attempt to understand better what kind of activity technology is. Taken individually, each aspect in its specific content constitutes a necessary condition for an activity to be properly termed an instance of the activity-form technology.

Before proceeding, a preliminary distinction is necessary. On those rare occasions when "technology" *is* used to refer to a kind of human activity, this usage tends to lump together two distinguishable albeit related kinds of activity. Consider the following quote: "...a technologist...must before long turn back towards the real world, so to speak, and design his motor car, his factory, his computer, [and] his transport system in the real world rather than in the laboratory."[4] Notice that the things the technologist is said to design fall into two different categories. Earlier the author calls them "device[s]" or "machines" and "systems." Here I shall be concerned to characterize not the activity of the person—usually called the "systems design engineer"—who produces designs of various types of socio-technical systems (e.g., factories and transport systems), but rather the activity (taken as a whole) of those individuals, engineers and others, whose efforts issue in material objects of one sort or another. In the sequel, unless otherwise indicated, when speaking of technology as a form of human activity and analyzing its various aspects, "technology" should be understood as referring only to the latter activity, not the former. The activity of making material objects should not be assimilated to the activity of designing and constructing the systems in which those material objects will operate or be used. This, of course, is not to suggest that these two activities ought to be isolated in practice. On the contrary, the criteria governing the making of material objects ought to take more seriously into account those governing the operations of the socio-technical systems in which the objects will be deployed.

The first aspect of technology pertains to the nature of the results or "outputs" of the activity-form: its characteristic outcomes are *material* as opposed to ideational in nature, as, e.g., tends to be the case with religious activity. Thus prayer, although sometimes involving technique, is not an instance of technology. This is not to say, as will shortly become clear, that ideational phenomena play no role in technology. As technology is widely thought to involve the making of things, it might seem adequate to say that technology is a material product-making activity. This, however, would fail to cover some recent and other imaginable activities which would likely be termed instances of technology but which are not accurately subsumed under the material product-making rubric; e.g., plastic surgery, genetic engineering, and a technical process which made a person invisible. I suggest the rubric be broadened to "material product-making or object-transforming activity," where the transformation in the material object is from an initial state or condition to a subsequent one. Evidently one could subsume both elements in this revised rubric under the latter, i.e., material object-transforming, with material product-making viewed as a transformation from the raw materials stage to the final product stage. However, I shall retain the dual rubric.

Some technological activity results in the production of what might be termed "stuff": "raw" materials (e.g., in a certain size or shape) or partially or wholly "cooked" materials (e.g., synthetic fibers or plastics) out of or with the aid of which the final products (objects) of subsequent technological activity will be made (transformed). Setting aside this subset of the material outcomes or outputs of technological activity, we shall call the elements that remain *technics* (if they are material products) and *quasi-technics* (if they are transformed material objects, e.g., people who have undergone plastic surgery).[5] Items of technics thus include not only tools and the usual cars, planes, and computers, but also, e.g., contact lenses, musical instruments, clothes, certain kinds of foods and furniture (viz., those that have undergone some significant transformational process), as well as certain works of art (e.g., pieces of sculpture). Not included under the category of technics as here characterized would be the trimmed tree trunk or block of marble on which the sculptor begins to work, the plastic out of which the contact lens is to be made, or the synthetic crystals with which semiconductors are made.

It is well to point out that "technology" is perhaps most often used to mean the complex of technics—more specifically, modern technics—or a subset thereof: viz., machines, tools, and weapons. In my view this usage is undesirable, for not only does it cast its net too narrowly as regards the physical objects it embraces, but it also clouds the important fact that technology is as much a multifaceted human activity as any other, and is not just "those machines out there," mistakenly viewed as somehow possessing lives and possibly demonic powers of their own.

The second aspect of the activity-form technology concerns a characteristic of the overall process by means of which technics are made; viz., such processes are *fabricative* in nature. By this is meant that at least some of the significant features of the technic are due primarily to the technologist's working of his or her will on the constituent ingredients or parts, rather than their all being primarily the result of the operation of chemical, biological, or physical laws. A practitioner may, of course, *facilitate* the occurrence of interactions among ingredients. Thus the fact that the material outcome possesses properties resulting from the operation of such laws may in a sense be said to be due to the volition of the practitioner, but not *primarily* so as here understood and intended. Not every stage of the technological process involved in the production of a given kind of technics is fabricative, particularly so in the complex processes of modern technology. Consider, e.g., what engineers call the "preliminary design stage." However, taken as a whole, it is still the case that all technological processes producing technics are fabricative. Technologists and ancillary staff fashion, shape, construct, form the technic (or prototype or model of same). On this account I contend that neither rain-making by seeding clouds nor agriculture are tokens of the activity-type technology. Although they both, when successful, generate material output, neither is fabricative. Whether such activities are based on scientific understanding or not, both involve facilitation of the occurrence of chemical and biological processes. Of course, each may be called technological in the weak and here uninteresting sense that their practice involves various technics, the products of prior full-blooded technological activity. Essentially, however, both are in their most sophisticated forms specimens of applied science. In sum, the fabricative character of technology pertains to the nature of the processes in which its issue is brought into being.

The third aspect of the activity-form technology is that of *purpose*. Technology is an explicitly and highly purposive form of human activity. But is there any specific purpose which can be said to be characteristic of technology in general? Edwin Layton claims that "the central purpose of technology" is "design": "adaptation of means to some preconceived end."[6] Not so: rather than being technology's general purpose, design is rather a vitally important *means* to the realization of that purpose, viz., expansion of the realm of the humanly possible; by this is meant the realm of possibilities which may be utilized by human beings, not mere theoretical possibilities.[7] For technology is a quintessentially practical form of human activity. Particular expansions of the realm of the humanly possible are often undertaken in the name of increasing the power of their beneficiaries; either directly, through increasing their capacities to realize goals in the face of obstacles to doing so, or indirectly, through the expansions' provision of vehicles for practitioners' self-realization. The expansion of the humanly possible through technology occurs in at least six different modes.

**1** direct extension: providing a direct extension of some already existing human function or capacity, as in the cases of the telescope, the megaphone, and the computer;

**2** qualitative innovation: offering a qualitatively new addition to the repertoire of human capacities, as do, e.g., the airplane and the submarine;

**3** risk reduction or elimination: enabling one to do something, previously done, but only with the risk of incurring certain costs, without or with significantly reduced risks, as, e.g., in the case of the birth-control pill and the asbestos fire-fighting suit;

**4** improvement of performance: offering the ability to do something easier or more efficiently than it would be done previously, as, e.g., in the cases of snowshoes, chainsaws, and solar-energy cells;

**5** substitution: enabling one to do Y, previously precluded by doing X, by X's being done technologically, e.g., reading while one's lawn is watered by an automatic sprinkler;

**6** increasing the means for expression of the inner life: providing for the aesthetically or otherwise motivated representation of emotions, beliefs, perceptions or other states or conditions of consciousness in external, tangible forms, as, e.g., in the cases of musical instruments, sculpture, perfumes, etc. (This might be viewed as the provision of additional means whereby humans can express their capacity for the representation of consciousness or experience. Thus, this mode of expansion differs from that of direct extension described in (1) above.)

It should be noted that more than one of these modes may pertain to a given technic.

These modes of expansion suggest the following two more general purposes of technology: (1) facilitating successful human adaptation to a possibly threatening environment (natural or human) by augmenting one or another dimension of human power with respect to that environment (e.g., sunglasses used in driving, a well operating in an arid land, or weapons used for obtaining food or self-protection); and (2) aggrandizement of human power even in a context or environment devoid of threat; put differently, in the name of thrival rather than survival (e.g., the development and use of technology in order to increase one's economic power far beyond that required for survival).

There is a claim lurking in the background which should be brought to the foreground. Although difficult or impossible to confirm, it may be that at a deeper level, the most general purpose of technology is to increase human power via the above-mentioned expansion of the realm of the humanly possible. Sometimes the increase may serve human survival interests, on other occasions it may serve human aspirations to self-aggrandizement or domination. On a somewhat more metaphysical and speculative level, recall the early Sartre's claim that human life in all its aspects, including, e.g., sexuality, is a fruitless attempt to escape from the anguish of ontological contingency by becoming God, the being at once *pour-soi* and *en-soi*. I think it worth conjecturing whether technology itself might finally be one of man's attempts to deny and transcend his finitude and its associated vulnerability, and to become Godlike by acquiring more and more power in all its dimensions. While urging consideration of this conjecture, I do not mean to deny that one or more of the relatively more mundane levels of purpose or modes of expansion might not be operative in the development of particular technologies. The question remains however: is technology's purpose to be understood, following Marx, as a human means of successfully adapting to a recalcitrant natural or social environment, or, alternatively, in the spirit of Nietzsche, as motivated by survival *only in special cases,* but more generally and profoundly as a vehicle for the expression of the alleged insatiable "will to power"?[8] It would appear then that a full understanding of the most general or deepest level of the purpose of technology, if such there be, must await further understanding of human nature itself.

Now we may ask: Is the activity of, say, the organic chemist an instance of technology? After all, much of his/her work involves the synthesis of new compounds and seems, intuitively speaking, more fabricative than either agricultural or cloud-seeding activity. However, even if the synthesizing activity of the organic chemist could not be ruled out as technology by either of the first two aspects of technology, there is a sense in which it might still be disqualified on the basis of the third. When synthesizing compounds, the organic chemist, *qua* organic chemist, is interested in the structure, composition, properties, and reactions of the compounds made, not in the fact that the compounds synthesized may be put to some mode of technological purpose, to aid in expanding the realm of the humanly possible. Only insofar as the production of a material substance (chemical or otherwise) for one or another technological purpose is a goal informing the practitioner's activity, rather than simply the achievement of scientific understanding through the discovery of knowledge, does the activity remain eligible for the appellation "technology." To the extent that a technological purpose is operative, the practitioner is not acting *qua* organic chemist but *qua* technologist. Thus one and the same specific activity may be either an instance of the activity-form of science or one of technology, depending on the nature of the purpose with which the activity is pursued. But can one and the same specific activity be simultaneously an instance of both activity-forms? Consider the following: even the synthesis of new compounds by the organic chemist in order, say, to develop a new contraceptive pill, fails to be fabricative in the sense in which technology is. The activity of compound-synthesis is sometimes fabricative, but only in a weaker, second-order, indirect sense: besides facilitating the interaction of materials, the organic chemist may be said to manipulate or control, the *environment* within which the laws of chemical composition and reaction, as it were, operate to determine the synthesized compounds. But in technology the material outcome itself is fabricated by the practitioner(s).

Thus, if one reserves "technology" for activities which are fabricative in the strong sense and which are not disqualified by any of the other aspects discussed, then it follows that the work of our organic chemist is, although akin to, not a bona fide instance of the activity-form technology. To the extent that this synthesizing activity is felt to be an instance of technology, this may reflect the family resemblance that obtains between the sets of aspects or features of the activities of our chemist and those of paradigmatic technological activities. If, however, one has no quarrel with the weak, second-order, indirect sense of "fabricative," then, other necessary conditions being satisfied, one is liable to maintain that it is possible for a specific activity to be legitimately viewed as being *simultaneously* a token of the activity-types science and technology. This possibility, however, does not in any way negate the important differences between these two activity-forms.

The fourth aspect of the activity-form technology is *resources*. Technology is a resource-based, resource-expending form of human activity. In the course of making its products or transforming its objects, technological activity utilizes a variety of resources, some of which are precisely its prior products: technics and extracted raw and processed materials. In addition, technological resources include information, available energy, nonprocessed and, on occasion, nonextracted materials, people (witness the construction of the Egyptian pyramids and the Alaskan pipeline), and, more recently, capital, a resource increasingly needed to acquire other resources.

The fifth aspect of technology is *knowledge*. Technology is based upon, utilizes, and generates a complex body of knowledge, part of which may reasonably be called specifically technological knowledge. Evidently, knowledge might be viewed as of a piece with the technological resources noted in the preceding section. However, since technology is often thought to differ from science precisely on the ground that it, unlike science, has essentially nothing to do with knowledge worthy of the name, it seems worthwhile to individuate it as an integral aspect of technology.

The knowledge in technology is not wholly reducible to nonintellectual (e.g., neuromuscular) skills or to independently developed scientific knowledge. There seem to be three components to the body of technological knowledge. First is the knowledge of how to do certain things by making or utilizing certain material products or by transforming certain material objects. Consider the following: there was once a time when it was not part of this component of technological knowledge that bodies located in different planes could have their movements coordinated by a system of linkages or interlocking gears. The same is true of the making, controlling, or extinguishing of fire.

Granted, the discovery and application of such possibilities may well have involved certain skills, mental, physical, or mental-physical (e.g., perception). However, it seems likely that the processes of coming to "know how to do" that have occurred in these cases were bound up with grasping certain generalizations of the form "if P (is done), then Q (happens or can be done)." Moreover, such acquisition of knowledge may be coupled with insight into how such results can best be achieved and into the designs,[9] however rudimentary, for achieving them. If it is seen that throwing sand or water on fire can, under certain circumstances, be effective in controlling or extinguishing it, this knowledge, although until relatively recently wholly nonscientific (in the sense of being grounded in the understanding of underlying scientific laws), is not simply an instance of nonintellectual "know-how." Increases or improvements in "knowing how to do" are not only compatible with the achievement of new intellectual knowledge but are often undergirded by such knowledge.

The second component is the knowledge of the resources, especially energy and materials, used in technological activity, both of their general properties and those of the resources used on a particular occasion, the more so if those selected (or rejected) have been tested in some way or other. If so, the resultant test data—part of the activity's information resources—may also be viewed as part of this component of technological knowledge. Indeed, such data can be correlated and used for predictions at what might be termed a pretheoretical level. The engineer's prediction of matters as apparently simple as the pressure drop in turbulent flow through a pipe is based on such a procedure even today. There is, in fact, a body of "theory" on how to construct such correlations.[10]

This brings us to the third component of technological knowledge, consisting of knowledge of the methods used in reaching the desired result of the activity. It is artificial to maintain that in all cases this knowledge is nothing but a species of nonintellectual know-how, particularly for modern engineering, many of whose practitioners have intellectual knowledge and understanding of a host of sophisticated methodologies, e.g., that of "parameter variation."[11]

In sum, there is a broad spectrum of kinds of knowledge integral to technology, one which has evolved over historical time: from inchoate, atheoretical praxis to skilled, purposive correlation and planning methods for taking and organizing data in the name of some form of improvement or optimization. That much technological knowledge of the kinds distinguished above has traditionally[12] arisen from praxis or inductively from acquaintance and testing, rather than deductively from scientific laws, does not negate its status as bona fide knowledge.

The sixth aspect of technology is *method*. Technology incorporates and proceeds in accordance with certain methods. Method would seem to have always been an element of technology in several ways. First, method has been present in the process of determining whether the production of a technic in one way should yield to a different, possibly better way of making it. The traditional method for doing so has been that of trial and error, while in modern times there are different, more formal methods used in making such determinations. On a more microlevel, method has always been present in technology in the more or less careful selection of materials and in the development and employment of techniques used to acquire, prepare, and formally or informally test them as well as to do likewise to the result of the technological activity. Method is involved in technology in a third sense, in that technological activity generally proceeds through its various stages in a nonarbitrary order, one not dictated primarily by the law of nature, but by the exigencies of fabrication and the desire to realize a successful outcome. From initial conception of desired goal to final testing of material outcome, there is a multileveled methodicity in the procession of technological activity; whether in the manufacture of stone axes with its rudimentary methodicity, or in the development of windmills, water wheels, propellers, and space capsules with the complex of more sophisticated methods involved in making those technics.[13]

The seventh aspect of technology is the *sociocultural-environmental context* within which the activity unfolds. This aspect might be called an informing and enabling rather than a constitutive one, although it is inseparable from technology and no less important than the preceding aspects for understanding it. With one kind of possible exception, technology always unfolds in a sociocultural context. The possible exceptions involve early practitioners, some of whom may have made their tools in isolation and from scratch, i.e., such that their methods, knowledge, resources, and purposes were not provided for them by a social unit or cultural tradition. In all other cases, such provision makes the sociocultural context a condition for the possibility of technological activity. In the exceptional case, it is evidently the environmental context that shapes technology, affecting the products produced, purposes pursued, resources utilized, and methods employed.

Technology depends on sociocultural context in a second way: the economic, political, ideological, and social interests of various parties may have directive effects on technological practice.[14] Consider the effects of economic conditions on research into energy and automative technologies, of international politics on weapons technology during the Cold War, of claims to ideological superiority and fears of losing prestige on the development of aerospace technology needed to reach the moon, and of social unrest on research into optical recognition devices used to keep nonresidents from entering apartment buildings. More strictly cultural factors can exert similar directive pressures. The effort invested in sophisticated biomedical technology is not unrelated to Western attitudes toward death and identity. If we awoke tomorrow imbued with Schopenhauerian world views, would we remain as supportive of and interested in, say, research in prosthetic and cryogenic technology aimed at increasing human longevity?[15]

A third connection between technology and its sociocultural context lies in the fact that the making and operation and use of technics frequently and increasingly take place in settings which may reasonably be termed sociotechnical systems—bounded complexes of interacting social and technical elements—e.g., research and development laboratories, armies, factories, hospitals, communications and transportation systems, etc. Indeed, much of modern technology, both its practice and the use of its products, can occur only within the context of such large-scale sociotechnical support systems.

The eighth and final aspect of the activity-form technology is, like the seventh, more informing than constitutive, but is equally inseparable from technology: that part of the *practitioner's mental set* which relates to any of the preceding aspects, as well as to the meaning and significance of his or her activity. In what relations do practitioners of technology stand to its results, purposes, knowledge, resources, methods, and sociocultural-environmental contexts? Of concern here are both cognitive and affective phenomena: those of perception (including observation and prehension), thought, and reasoning, as well as those of feeling and emotion. It would be surprising indeed if the appropriate mental sets of Yir Yoront makers of stone axes (a totem in their society, indeed one present in their archetypal world view), Samurai warrior makers of swords, the craftsmen in George Sturt's wheelwright shop in Farnham, England, in the 1880s and 1890s, and today's General Motors automotive engineers were not significantly and interestingly different.

Here are two examples of technology-related mental sets which are, it would seem, worlds apart. The following rich passage is from Marx's essay "On James Mill" and presents the author's romantic conception of the complex, self-conscious mental set of the unalienated practitioner of technology:

> Supposing that we had produced in a human manner; each of us would in his production have doubly affirmed himself and his fellow men. I would have (1) objectified in my production my individuality and its peculiarity and thus in my activity both enjoyed an individual expression of my life and also in looking at the object have had the individual pleasure of realizing that my personality was objective, visible to the senses and thus a power raised beyond all doubt. (2) In your enjoyment or use of my product I would have had the direct enjoyment of realizing that I had both satisfied a human need by my work and also objectified the human essence and therefore fashioned for another human being the object that met his need. (3) I would have been for you the mediator between you and the species and thus been acknowledged and felt by you as a completion of your own essence and a necessary part of yourself and have thus realized that I am confirmed both in your thought and in your love. In my expression of my life I would have fashioned your expression of your life, and thus in my own activity have realized my own essence, my human, my communal essence. In that case our products would be like so many mirrors, out of which our essence shone.[16]

Compare the mental set evoked by Marx with that of the individual quoted in the middle of the following passage. Speaking of the Italian passion for embellishment, Luigi Barzini recalls:

> I remember examining an Isotta Fraschini motor many years ago in New York, with an eminent American automotive engineer. He looked at everything for a very long time, then shook his head, looking very puzzled. He asked me: "Can you tell me why parts that do not work are as polished and well finished as those that work? This seems to me to be an irresponsible waste of labor and money." I did not explain to him that one could not restrain Italian engineers, designers, and workers from beautifying everything they made. In

fact, most Italians thought that a pleasant appearance was as important as (and often more important than) usefulness. "*L'occhio,*" they always said, "*vuole la sua parte.*"[17]

Evidently there are myriad technology-related mental sets. Among their many elements the following would seem worth investigating: What is the practitioner's attitude toward aesthetic considerations in technological activity, both in regard to the environmental effects of resource acquisition as well as vis-à-vis the final product? What is the nature of his conception of the beautiful? Is the practitioner a perfectionist in technological activity? Are personal identity and self-esteem intimately bound up with his work? If so, in what ways? Does the practitioner perceive work as unfolding in a larger social context, as contributing to the strengthening or weakening of a way of life? Does he feel in any way personally responsible for what is done with his products? What are his conceptions of efficiency, of human progress, of the good life—if he has such notions? What are his animating values? Finally, what is the practitioner's concept of nature and of relationships between man and nature? Some of these elements will be at least in part reflections of or responses to cultural milieus. Others, perhaps even more difficult to identify, will be substantially the expressions of individual idiosyncrasy. Investigation of practitioners' technology-related mental sets may contribute to a more complete elaboration of the thought component of technology. If ignorance or neglect of this component has contributed to the mistaken view of technology as less of a human activity than, say, religion or art, as less intellectual an enterprise than, say, science, perhaps attention to its affective subcomponent will help combat the equally mistaken view of technology as essentially a matter of bloodless precision.

To summarize our characterization of technology: it is a form of activity that is fabricative, material product-making or object-transforming, purposive (with the general purpose of expanding the realm of the humanly possible), knowledge-based, resource-employing, methodical, embedded in a sociocultural-environmental influence field, and informed by its practitioners' mental sets. Technology can be differentiated from, as well as related to, other forms of human activity (e.g., science) by comparing it with them regarding differences in substance in the general aspects identified and discussed above, particularly the initial six: its "outputs," the adverbial quality of its processes, its purpose(s), knowledge, resources, and methods.

Before turning to the second part of my answer to the question "What is technology?", here is a fascinating example of a person actually engaged in the activity of technology in one of its fields: canoe technology. I have selected from a detailed account of this man's work,[18] passages which reveal the presence of the above-mentioned aspects of technology, each of the aspects taking on particular substance in the example.

The person in question is one Henri Vaillancourt, of Greenville, New Hampshire, one of the four or five remaining professional makers of bark canoes. To save time and avoid belaboring the obvious, I shall minimize comments on the connections between the passages and the aspects:

### Technics

Henry makes several different kinds and styles of bark canoes, making all but one of their parts from basic raw materials obtained from the woods.

### Purpose

He became passionately interested in Indian life, and "the aspect of it that most attracted him was the means by which the Indians had moved so easily on lakes and streams through otherwise detentive forests. He wanted to feel—if only approximately—what that had been like. His desire to do so became a preoccupation....He formed an ambition, which he still has, to make a perfect bark canoe, and he says he will not rest until he has done so." He has visited almost all the other living bark canoe-makers, and has learned from the Indians. He has returned home believing, though, that he is the most skillful of them all.

### Resources

Virtually all the material resources he uses are obtained directly from nature: from white birch, cedar, black spruce, and white pine trees, to porcupine quills. He has, however, taken to substituting plain asphalt roofing cement for Indian gum made from the pitch of white and black spruce. The only resources of technics he uses are an ax, a froe, an awl, and a crooked knife—no power tools, nails, screws, or rivets. The energy used is that of his body. The only capital he requires is that needed to purchase his tools. He obtains that by selling his canoes (as well as paddles and snowshoes). Apart from the companies that make those tools, Henri's technological activity does not rely in any crucial way upon institutional resources.

### Knowledge

Henri's mastery of canoe technology is based on a substantial body of knowledge of several kinds. "In 1965 he laid out a building bed, went out and cut bark and saplings, and began to grope his way into a technology that had evolved in the forest under anonymous hands and—as he would learn—was too complex merely to be called ingenious." When he finished it he destroyed it with an ax. "It was a piece of junk," he explained. After visiting the few remaining Indian practitioners in Canada he found his master through the printed sketch and word: Edwin Tappen Adney. "Adney was an American who went to New Brunswick in the 1880s and built a bark canoe under the guidance of a Malecite [Indian]...he recorded everything the Malecite taught him. For the next six decades, he continued to collect data on the making and use of bark canoes. He compiled boxes and boxes of notes and sketches, and...made models of more than a hundred canoes, illustrating different tribal styles, differences within tribes, and differences of design purpose. A short, low-ended canoe was the kindest to portage and the best to paddle among the overhanging branches of a small stream. A canoe with a curving, rocker bottom could turn with quick response in white water. A canoe with a narrow bow and stern and a somewhat V-sided straight bottom could hold its course across a strong lake wind. A canoe with a narrow beam moved faster than any other and was therefore the choice for war. Adney so thoroughly dedicated himself to the preservation of knowledge of the bark canoe that he was still doing research, still getting ready to write the definitive book on the subject, when, having reached the age of eighty-one, he died." The book was compiled and published by the Smithsonian Institution in 1964. Vaillancourt has also acquired extensive knowledge of his *materials:* "You split cedar parallel to the bark," he commented. "Hickory you can split both ways. There are very few woods you can do that with." He makes his canoes in the shade because "in direct sunlight the bark becomes less flexible, and it needs all the elasticity it can retain during the building process." "If the month is June or early July, the bark will almost pop off by itself....Winter bark is really tight. I'm telling you. You need a special tool, a type of spud, to get it off." Like the Indians, Vaillancourt will spend long periods of time roving birch-filled woods, carefully looking for a perfect tree. Moreover, Henri has intimate knowledge of the Indian *methods* and has mastered the skill of using the crooked knife, his finishing tool. His knowledge relative to canoe-making is thus in significant part but by no means wholly nonintellectual skill, even if all of that knowledge is ancillary to a practical technological goal.

### Method

There is a definite general order in which the bark canoe is assembled which differs substantially from that of the canvas canoe. "In the latter, the canvas comes last. It is stretched across an essentially completed and rigid frame. The Indian, on the other hand, began the assembly with bark. He rolled it right out on the building bed, white side up, and built it from there." There is also method in the sense of technique in many of the stages: when putting the ribs in with a mallet as the canoe moves close to completion, Henri says, "you work from the ends toward the middle, putting the ribs in. You pour very hot water on the bark to keep it flexible at this stage." Method is involved in testing samples of bark in the woods: "[After unloading his canoe last night, Henri] went straight to one of the big birches and took a sample of its bark. Along the grain he bent the sample, steadily applying pressure with his fingers as the bark formed the shape of a U. Gradually he increased the pressure until the bark cracked. It was pretty good bark, he said. It had good relative elasticity. Its layers did not tend to separate." He is presently experimenting with abrasive bone tools of the Indians which had been replaced by the steel tools of the white man.

### Sociocultural-Environmental Context

On the environmental context side, there is the obvious connection between the nature of the Indian's forest-cum-lake environment and the character of the canoe technics they built. In Henri's case, the initial choice of bark as his basic material was a product of two facts: he was too poor to afford processed materials like canvas, aluminum, or fiberglass, and the woods around his town were packed with white birch. On the sociocultural context side, two points are noteworthy: Originally in paying the seams of their canoe hulls, the Indians used the gum of the pitch of white and black spruce which they obtained by the culturally transmitted method of tapping trees, analogous to the collection of maple sap. "They boiled the pitch and strained it, and at tribal campgrounds they had communal pitchpots where anyone could go for sealant to touch up a canoe." Secondly, Henri, as noted above, although eschewing power tools, has switched to asphalt roofing cement for his sealant, because collecting and boiling spruce pitch involves unnecessary tedium, not clearly an Indian cultural attitude.

### Mental Set of the Practitioner

Elements of Henri's mental set in relation to each of the foregoing aspects are implicit in the above. He is self-consciously a perfectionist, a sentinel for the tradition of bark-canoe-making, aesthetically inclined, and deeply attached to each of his canoes. When he travels he visits his canoes and loves to get them back to "the yard" for repairs. He decorates them with Indian emblems and symbols using porcupine quills. "I've never seen a bark canoe that wasn't graceful," he says. William Morris, take heart!

Turning to the second part of my answer to the question "What is technology?", having dealt with the general structure of the activity-form, I should like to examine what I call "the content of technology." To do so, I need to make clear what is meant by *a technology* (as opposed to, say, several technologies) rather than by speaking of technology as a form of human activity or of "technology as a whole," whatever that may mean. What are some examples of technologies in this singular/plural sense? Canoe technology, aerospace technology, contraceptive technology, pollution control technology, and tunnel technology. But what kind of a thing is "a technology"? Although they are related, there is a subtle but important difference between a technology of a certain kind and the *process* of manufacturing the associated kind of technic. For example, to speak of what canoe technology is, per se, is not to speak of the process of making canoes, whereas to speak of the practice of canoe technology does entail talk about the process of making canoes. By "canoe technology," at a given point in time, we mean the complex of knowledge, resources, and methods used in making canoes. In general, by a (particular kind of) technology, we shall mean the complex of knowledge, resources, and methods involved in the production (transformation) of the associated technics (quasi-technics) of that kind. We may now indicate what we mean by "the content of technology." As a first approximation we shall say that the content of technology is the set-theoretical union or sum of all particular (kinds of) technologies, each of which is a complex of the kind just discussed. However, since technologies change, new ones coming into existence, old ones lapsing into desuetude, the content of technology is evidently a function of time. Thus we should speak of the content of technology at a particular point in time. Moreover, a given kind of technic may be made differently in different societies at the same time. Think of contemporary bicycle technology or of the fact that the Bic Company currently manufactures and sells a nineteen-cent disposable razor only in Greece. Thus the dependence of the content of technology on a societal variable should also be made explicit. One may then speak of, say, medieval Chinese technology or of British technology in 1800. Finally, the content of technology at a particular point in time in a given society is the union of all technologies practiced in that society at that time.

What is the relationship between technology as a form of human activity and technologies? Technology, as a form of human activity, is variously incarnated or embodied in the practice of each technology. Technologies may be individuated or related by characterizing the substantial differences or similarities in technics, specific purposes, methods, bodies of knowledge, resource bases, contexts, and practitioners' mental sets associated with the different incarnations or embodiments of the activity-form technology in the practice of the technologies under consideration.

We conclude by making use of this more differentiated concept of technology in clarifying the controversial issue of whether technology is "value-laden" or "value-neutral."

## TECHNOLOGY: "VALUE-LADEN" OR "VALUE-NEUTRAL"?

There are at least five ways in which technology is a value-laden enterprise.

1. If, following Kurt Baier,[19] the value of a thing consists in its ability to confer a benefit on someone (by whom or in whose behalf it is used), then to the extent that it is successfully practiced, technology is inherently value-laden. For, as our discussion of the general and specific purposes of technology indicated, technics, the primary products of technological activity, are made precisely in order to fulfill value-laden purposes, whether of practitioners or of those in whose service they labor. Thus the (noneconomic) *value* of a technic generally reflects the *values* of those for whose purposes it is made and used.

2. Technology is also value-laden in a sense similar to that in which, according to Neitzsche's analysis, *Wissenschaft* is. For Nietzsche, science, a different form of human activity, constitutes an essentially Socratic optimistic and would-be triumphant approach to dealing with the problematic nature of the human condition, as opposed to, say, the consolatory approach implicit in the tragic world view.[20] Science, Nietzsche claims, places something akin to an absolute value on truth and assumes that life can be made meaningful by understanding it through rational, intellectual means.[21] As a different way of approaching the world, technology, with its animating Promethean *Geist*, is also optimistic and assigns a kind of categorical value to "technological progress." The predominant spirit informing post-Renaissance technological activity, a spirit liberally fueled by its remarkable successes, assumes that, *ceteris paribus*, the human condition can only be ameliorated and rendered more meaningful by ongoing technological progress (i.e., improvements in the knowledge or resource or methodological sectors of various technologies).

3. Technology is value-laden insofar as use of resources to advance its "cutting-edge" may preclude their use in the more prosaic work of ameliorating people's everyday lives. There is, of course, nothing in the nature of the activity of technology per se which necessitates such value-laden trade-offs, although the

categorical value widely ascribed to technological progress makes such conflicts more likely to arise in the modern world of Spaceship Earth.

4. Another extrinsic dimension of technology's value-laden character arises from the institutionalization of modern technology, i.e., from the institutional locus of its practice in modern society. Technology's products and the values they serve are increasingly determined not by practitioners' values and specific goals but externally, by those directing the institutions controlling the resources requisite for modern technological practice.

5. Finally, technology is value-laden in the young Marxian sense that its products are generally expressions, objectifications, or crystallizations of at least some of the individual preferences, tastes, idiosyncrasies, world views and cultural values of their designers or makers (or those for whom they labor). For example, preferences for or against ornamentation, unalloyed functionality, maximum efficiency, durability, or convenience, character, organicity or simplicity, may be and often are reflected in technics and make them expressive of a range of evaluative attitudes at best tangentially related to the central specific purpose the technic is intended to serve. The quotation concerning the Isotta Fraschini motor illustrates this point nicely.

To sum up, I have tried to answer the question "What is technology?" by distinguishing between the structure and the content of technology; i.e., between the various aspects of its structure as a form of human activity and the various technologies, the practice of each of which embodies the activity-form technology, the union of which makes up what I called the content of technology. After illustrating the presence of these structural aspects in a particular technology, that of the bark canoe, I made use of this more differentiated concept of technology in elucidating the value-laden character of technology. It remains a task for another occasion to elaborate and analyze the structure of the other branch of technological activity to which I referred at the outset: the design of socio-technical systems in which technics are used. It is appropriate that this task be deferred temporarily since, in response to environmental, consumer, technology assessment, and accountability movements, the designers and operators of socio-technical systems are beginning to incorporate steps which, if widely adopted, will lead to significant changes in this branch of technological activity.[22]

## FOOTNOTES

1 C. Singer, E. J. Holmyard, and A. R. Hall, eds., *A History of Technology* (Oxford University Press: Oxford, 1954), Vol. I, p. vii.
2 Emmanuel G. Mesthene, *Technological Change* (Mentor: New York, 1970), p. 25.
3 Donald A. Schon, *Technology and Change* (Delacorte: New York, 1967), p. 1.
4 John J. Sparkes, "Technology and General Education," *Prospects*, Vol. 4, 1974, p. 94.
5 Below, when a statement is made about technics, the parallel statement about quasi-technics will not ordinarily be affixed.
6 Edwin T. Layton, Jr., "Technology as Knowledge," *Technology and Culture*, Vol. 15, Number 1, January, 1974, p. 37.
7 I am indebted to Professor Robert J. Baum for this refinement.
8 This expression is not used here in a pejorative sense.

**9** To some readers it may seem that the notion of *design* has been given short shrift in this essay. Perhaps, it may be argued, design should have been accorded the status of one of the individual aspects of technology. I prefer, however, to think of design as a stage in the process of making (many) technics or their prototypes in which knowledge, methods, and resources are drawn upon and utilized to formulate plans for the fabrication of technics. Design has emerged as a vital stage of most processes for making technics only in the modern era. To claim that design has always been part and parcel of technology, including, say, ancient technology, is to reduce "design" to a point at which it amounts to little more than a conception of a specific technological purpose and an image of a desired technical outcome.

**10** See, e.g., P. W. Bridgman, *Dimensional Analysis* (Harvard: Cambridge, 1922), and S. J. Kline, *Similitude and Approximation Theory* (McGraw-Hill: New York, 1964).

**11** Walter G. Vincenti, "The Air Propeller Tests of W. E. Durand and E. P. Lesley: An Example of Engineering Research by Parameter Variation" (unpublished paper).

**12** In modern times an increasing portion of the resource component of technological knowledge, particularly as regards material resources, has become more specifically scientific in nature.

**13** Methods used in the practice of modern technology include a variety of qualitatively distinct procedures and techniques: e.g., descriptive geometry, design procedures, comparative systems studies, testing, similitude techniques, order of machining and assembly, and parameter variation.

**14** See, e.g., Nathan Rosenberg, "The Direction of Technological Change: Inducement Mechanisms and Focusing Devices," *Economic Development and Cultural Change*, Vol. 18, No. 1, Part 1, October 1969, pp. 1–24.

**15** One would like to see serious research aimed at articulating various aspects of the cultural milieu which have affected the development of specific technologies, research complementing that devoted to the more familiar politico-economic directive factors.

**16** Karl Marx, "On James Mill," in *Early Texts*, David McLellan, Transl. and ed. (Blackwell: Oxford, 1971), p. 202.

**17** Luigi Barzini, *From Caesar to the Mafia* (Bantam: New York, 1972), p. 269.

**18** John McPhee, "The Survival of the Bark Canoe," *The New Yorker*, February 24, 1975, pp. 49–94.

**19** Kurt Baier, "What Is Value? An Analysis of the Concept," *Values and the Future*, K. Baier and N. Rescher, eds. (Free Press: New York, 1971), pp. 33–67.

**20** Friedrich Nietzsche, *The Birth of Tragedy* (Vintage: New York, 1967), pp. 95–96.

**21** Nietzsche, *On the Genealogy of Morals* (Vintage: New York, 1969), p. 151.

**22** I am indebted to my colleagues S. J. Kline and W. G. Vincenti for detailed critiques of an earlier draft of this essay.

## Four Philosophies of Technology

**Alan R. Drengson**

### ABSTRACT

Technological maturity requires mastery of technological processes. This mastery is the last of four developmental stages, which are characterized by four different views of technology. These four philosophies are: (1) technological anarchy; (2) technophilia; (3) technophobia; and (4) appropriate technology. These philosophies are discussed, the features of appropriate technology described. The philosophy of appropriate technology involves mastery of activities which introduce moral and ecological values into the design and application of ecologically sound technologies.

## PHILOSOPHY AND CREATIVE INQUIRY

The aims of this essay are threefold: First, to describe four main philosophies of technology manifest in our culture; second, to engage in a process of creative inquiry that will make it progressively more obvious the extent to which an unwitting adherence to some of these philosophies can affect perceptions of technological possibilities; third, to outline the interconnection between conception, action, and social process with the aim of clarifying the role of conceptual design in intentional technological innovation.

In order to advance the aims of this essay, it is first necessary to explain what is meant by "philosophy" in this context. There are three levels to the term here: At the lowest level, a philosophy can be nonexplicit; at an intermediate level, it is an explicit elaboration of a particular position which spells out assumptions, axioms, etc., and argues for its conclusions; in the final and mature sense, philosophy is a creative activity of conceptual inquiry which frees us of attachment to specific models and doctrines in order to develop more appropriate cultural practices.

In the title of this essay, then, I speak of philosophy in the sense that one can express and live by a philosophy which is neither explicit nor clear, but which forms the structure and quality of one's experience. By "philosophy," then, is meant a way of life formed by attitudes and assumptions which, taken together, constitute a systematic way of conceptualizing actions and experiences by means of an implicit process of unquestioned judgments and conditioned emotional responses. In some dimensions these are cultural, in others they are familial or personal. Together these responses and judgments, constituted by both assumptions and evaluations, and an articulation of them in word and deed, make up one's philosophy of life. Most of the four philosophies of technology analyzed in this essay are culturally at the first level. The aim of this essay is to raise them to the second level, and then to move them to the third level by engaging in creative philosophizing about technological innovation and appropriate design. "Appropriate" here refers to right and artful fit between technique, tool, and human, moral, and environmental limits.

A caveat needs to be made at this point. The four philosophies of technology described here each occupies a given range on the continuum of responses to current technological development. The precise boundaries between each are difficult to mark. Moreover, each of these "philosophies" has certain specific adaptive and economic advantages. For example, a technophobic reaction to modern technology involves in part an attempt to revive and preserve simple, "primitive" technologies which, in the event of disaster, could serve survival and preservation of certain culture values. Technological change is highly dynamic in terms of its material manifestations, and the four philosophies described herein represent dominant views associated with technologically advanced societies. Nonetheless, the attitudes these philosophies represent tend to be primary human responses to change. A specific person may go through stages of development that pass through each of these philosophies. The creative philosopher recognizes the usefulness and limitations of each within this whole developmental process. He or she also recognizes the importance of a balance between each (as represented by different groups within a society) and within the dynamics of healthy social change.

Creative philosophy, as a form of inquiry, aims to free us of an attachment to doctrines and views, but enables us to use such doctrines and views to facilitate positive change and growth in understanding. In order to achieve this end, various metaphors and models are used as part of the activity of creative reflection on the four philosophies of technology. The use of such devices has certain risks. As has been observed by numerous sages, philosophers, insightful psychologists, novelists, Zen masters, and others, human thought tends to become fixated on various stereotypes, metaphors, models, paradigms and belief systems. Creative philosophizing recognizes their inherent limitations, but uses these various models, paradigms, etc., as a way of freeing understanding of their dominance. Initially one uses such models and the like as a way of conceptualizing the world in order to gain understanding and to serve practical aims. However, when these paradigms and their accompanying ideas, ideals, beliefs, and so on, become part of a belief system, it is easy to invest one's identity in them. When we invest our identities in beliefs we resist reflecting on them, and we resist their change, for this can seem a threat to one's self-identity and sense of reality. Thus belief systems tend to become static. Since life is a dynamic process, flexibility and creative adaptation suffer, when cultural processes involving dynamic factors such as science (as inquiry) and technology (as creative technique) get out of harmony with these more static belief systems.

The four philosophies sketched here are offered as provisional models to facilitate insight into the patterns of philosophy of technological development inherent in our culture. The creative philosopher recognizes the limitations in these patterns of thought and approaches them with a serious, but playful attitude so that distinctions can be recast through a continuous process of conceptual adjustment, readjustment, and improvisation. In creative philosophy, concepts become tools, paradigms heuristic devices, clarity and insight products of philosophical activity. In creative philosophy the aim is not *a* philosophy, but the activity of philosophizing as a way of continuously clarifying human intelligence by freeing it from its conceptual constraints. The fully sound human understanding is one that sees the world as it is, while it also realizes that cultural adaptation (of which technology is a part) is a creative affair and has a range of possible options, given the nature of the world.

A final word of caution. Creative philosophizing in its mature form is a nonposition and an activity. It is a sort of jazz played with concepts. It is a creative art that one acquires through long practice. It is classically illustrated in many of Plato's Socratic dialogues. As was observed in *The Republic*, ultimate reality lies beyond all of our forms of thought. The contemporary creative philosopher realizes that as long as we do not identify with these forms, they can be adjusted to better fit reality as revealed through fully aware immediate experience. By approaching philosophy creatively, as a process of dialogue and interaction, of give and take, playfully adopting a variety of perspectives, we free our capacity for creative thought and insight. Insight involves (in part) a direct grasp of networks of relationships and a seeing of the world that reveals its significance and value intensity, which are part of a common ground in the unity of being.

In contemporary Western industrial culture there is wide disagreement about how we should develop resources, whether or how to exploit animal species, whether and which new technologies to develop, and how to manage our collective activities in relation to individual rights and to the biosphere. Thus, the four philosophies discussed here represent the kind of broad, pluralistic mix that one would expect in modern Western democracy. This is particularly evident if we think of this matrix as a dynamic process that displays dialectical features. Within democratic society as a whole, complete consensus is not possible, especially since different people are at different stages of development. The four philosophies to be discussed could be said to represent the stages of maturation of an industrial society, and its gradual transformation into a mature, postindustrial culture characterized by human-scaled, ecologically sound, appropriate technologies, consciously designed to achieve compatibility with fundamental moral values. These matters will be explored now in greater detail.

## FOUR PHILOSOPHIES

There are four fundamental attitudes toward technology that can be discerned in current cultural processes in the industrial West. These attitudes form a continuum from an extreme faith in, to a complete distrust of, technology. The degree to which the various possibilities in between are held varies from person to person and between various subcultural groups. They do not readily correspond to any particular economic philosophy. These four philosophies can be conceived of as nodal points or as dense nexes of social attitudes which are centered on constellations of paradigms and beliefs. Within the whole continuum of social response their features can be described. Since the culture as a whole is in process, and since individuals within the culture are also changing at varying rates, depending on their particular circumstances, these nodal points are not static. They do not define all or nothing positions for the culture as a whole. If Western culture were to become either too static, or too dynamic, these views could become polarized, and then precipitate unresolvable conflicts and statements. As it is, they now appear to represent developmental stages of a continuous growth in which each successively becomes emphasized, as persons and the culture evolve.

For the purposes of this discussion I will designate the four philosophies under consideration as the following: (1) technological anarchy, (2) technophilia, (3) technophobia, and (4) technological appropriateness.[1] I shall now discuss the essential characteristics of each position and the interrelationships between them.

*Technological anarchy* was a dominant philosophy throughout much of the nineteenth-century industrial development of the West. In brief, technological anarchy is the philosophy that technology and technical knowledge are good as instruments and should be pursued in order to realize wealth, power, and the taming of nature. Whatever can be done to serve these ends should be done. The fewer government regulations over technology and the marketplace, the better. Ideally, there should be none, but this is impossible, since some basic order is necessary to further private ends. The market alone will determine which technologies will prevail. Technological anarchy is a philosophy of exuberant, youthful curiosity and self-centeredness. It is an expression of optimistic self-assertion and individual opportunism.

Technological anarchy helped to stimulate rapid technological development. It tends to encourage technological diversity. As industrial development matures, technological anarchy (within a given culture) tends to become less dominant. Technology becomes a more powerful directing force in the whole social process. Technology begins to take on certain autonomous features on a large scale. Technology, which was originally pursued as an instrument to satisfy desires and needs, tends in such a context to become an end in itself. As this process completes itself, technological anarchy loses its dominant position, even though it rarely completely disappears. It then gives way to technophilia, which in turn develops into a structure with technocratic features. (At the international level, technological anarchy still seems a dominant force.)

*Technophilia,* as the word implies, is the love of technology. It is like the love of adolescence. Humans become enamored with their own mechanical cleverness, with their techniques and tricks, their technical devices and processes. The products of our technology become not only productive instruments but also our toys. Technology becomes our life game. This is like the adolescent affair in which we identify with the objects of our love. As a result they tend to control us, for our unconscious identification with them invests these objects with our person. This identification becomes a form of control over us, since we are unable to disassociate ourselves from our technology. We cannot see it objectively. This can be illustrated by our love affair with the automobile. We can become so infatuated with automobiles that they become extensions of our selves. "Insults" to them become personal affronts, and can be felt as threats to self-esteem. This represents a loss of an objective understanding of the positive and the negative features of the technology of the auto—which includes the whole infrastructure of factories, gas stations, parking lots, roads, freeways, legal structures, supported, of course, by a whole complex of human routines and skills. Thus, although the automobile was first a means to an end, viz., transportation, it and its supporting infrastructure eventually became a dominant feature of the culture as a whole. Cities, land use, and even economic well-being have become entangled with the technology of the auto. What began as an instrumental value, as a means to the end which was transport, becomes an end in itself. Paradoxically this works to frustrate the original human values involved. Finally, the technology of the automobile can become a threat to life, health, economy, the environment, and even to our way of life.

Technophilia, as the love of technology, turns the pursuit of technology into the main end of life. It eventually aims to apply technology to everything: To education, government, trade, office work, health care, personal psychology, sex, etc. In this way it becomes technocracy, for technology is now a governing force. This represents the overwhelming of spontaneity by technique. In its most complete form, as technocratic, it represents the rule by and for technological processes. At this point humans are technologized by their own love of the technical and of techniques. Life becomes mere mechanism. However, this is only the implied logical terminus of technophilia. It is unlikely that it could achieve a complete technocracy because the social process is a stream with diverse elements. The application of technology to nearly everything stirs counterforces, and the imagined logical end of this pursuit is unacceptable to many. The love affair with technology cools as the process of maturation leads many people to realize that technology is becoming an autonomous force endangering human and nonhuman values. Even the biosphere as a whole becomes threatened by the products and processes of human technological activity. The initial reaction to these imagined and perceived threats is first to attempt to control technology and its hazards by means of technique and the technological fix. But these are both only extensions of the technophilia which furthers the development of technocracy.

*Technophobia* emerges when it is realized that only human and humane values can curb the threats of a technology running out of human control. As an extreme reaction technophobia attempts to detechnologize human life, for to many persons the idea of applying engineering techniques and technocratic control to all aspects of human culture is repugnant. It is seen as a mechanization of the human, leading to the loss of the sensitive, spontaneous and vital organism. There is a natural desire to return to human autonomy, which was originally one of the motives in pursuing technology, but it is now seen as frustrated by the technostructure. This autonomy is perceived to reside in the revitalization of crafts and arts, of simpler, "neoprimitive" technologies. A do-it-yourself attitude characterizes it. The aim is self-sufficiency; a distrust of complex technologies is one of its features. Even while this reaction is developing the forces of technocracy are consolidating their control of extensive industrial technologies, which in turn, by their own inner dynamics, are evolving toward post-industrial maturity through smaller scaled, flexible systems of production. Ultimately, technophobia aims to bring the large-scale technologies to an end, and to bring technology once more under local human control. It helps prepare the ground for evolution to appropriate technological design.

Technophobia can be compared to the disenchantments of early adulthood. One learns that attachments which are centered in romantic and erotic identification can frustrate growth and can generate suffering, pain, grief, and fear of loss. Such loss is felt initially as a severe threat to one's self-image. Unable to accept full responsibility for oneself (in every dimension), because one does not understand the exact nature of the situation, one inevitably suffers disappointment and may attempt to avoid such relationships in the future. This is usually not possible, although it is probably necessary to take this "pledge" as a step toward more mature relationships with others. In a similar way, perceiving the dangerous character of the technological panoply can at first be very disorienting, especially since it was originally thought that building such a technostructure would make life easier and safer. However, direct planning and innovation has often been done by persons who were not able to be fully responsible because they lacked sufficient understanding of the nature and implications of powerful technologies, or because they were caught in structures that made responsibility difficult. When human imagination is harnessed to technophilia in order to create and to proliferate technologies (as in the chemicals industry, e.g.), and when competition becomes an important force (whether national or international), it then becomes very difficult to control these technological forces. Fearing that this technological power will ultimately lead to total control of humans, or even to ecocide, finally brings disenchantment with the whole process. The romantic entanglement with technology (technophilia) is now perceived as threatening human integrity and survival.

Technophobia rejects technological autonomy and asserts human autonomy over it. This accomplishes two important things. First, it brings renewed commitment to humane values. Second, as already noted, it leads to the revitalization and preservation of arts, crafts, techniques, and skills that *emphasize* personal and interpersonal development as more important than technological supremacy over humans and nature. This not only preserves simpler technologies, but it insures that the process of maturation will continue, since it is necessary to psychologically distance ourselves from these activities, if we are to understand them. This understanding is necessary, if we are to perceive the possibilities for new forms of technology that are under our control and that are more appropriate to human and to natural values.

It is realized at this stage that it is the relationship between technology and ourselves that we must understand. This means understanding the relationship between nature and technology as well, for humans are born as nature and through technē and other cultural activities they modify themselves. The tendency is to see this cultural process as fixed, rather than the stochastic process that it is. In this case, jazz is good paradigm for the art of self-creation as a stochastic process, for here there is the possibility of both control *and* spontaneity. The culture provides different roles for us to play. Thus there are patterns through which our activities can cohere and gain meaningful harmony. We could compare this process to the capacity of learning how to learn. Becoming aware of the possibility of knowing how to learn sets the stage for a continuous, consciously ordered transformation. If one becomes adept at learning, then one is adept at adjustment to ongoing changes in the world. One then becomes sensitively attuned to these changes, and can stay with them. When one learns how to bring one's full attention to a subject, and becomes capable of learning all there is to learn about it, then one becomes a master learner. From this vantage point technophobia can be seen as one of the stages of growth that involves becoming aware of the use of technology in a consciously reflective, critical way. We have the chance to see it from a meta-level.

*Appropriate technology* represents the fourth stage of technological development we have been describing in terms of the evolution of philosophy of technology and technological design. The fourth stage involves a maturing of the reciprocal relationships between technology, person, and world. Appropriate technology requires that we reflect on our ends and values, before we commit ourselves to the development of new technologies, or even to the continuation and use of certain older ones. As in mature love, one becomes capable of compassion and helping others to attain their ends (this is the very essence of the compassionate person), so in this stage we become capable of mastering our technology as instrumental to ends about which we become progressively more clear.

In the philosophy of appropriate technology, technologies should be designed so that they meet the following requirements. First, they should preserve diversity; second, they should promote benign interactions between humans, their machines, and the biosphere; third, they should be thermodynamically sound in the generation and use of energy; fourth, they should dynamically balance all costs; fifth, they should promote human development through their use. Let us reflect upon these points. Diversity is one of the features of both stable ecosystems and stable economies. Diverse technologies provide a large range of options to individuals and to further social development. Benign, symbiotic interactions between technology and the biosphere are necessary features of future technologies, if we are to develop sustainable economies. Compatibility with ecosystem principles is a minimal requirement. This is emphasized in sound thermodynamic design, for ecological compatibility and thermodynamic soundness work together to balance social, economic, and environmental costs. Finally, when technologies evolve to the level of appropriateness, they can be designed in such a way as to facilitate human development. Such technologies are designed to allow humans to master whole processes as arts, which stimulate the development of the complete human person. Thus the maturation of appropriate technology involves the transformation of the technological process into an art. The technological processes then become a life-enhancing part of a significant set of values. Labor thus becomes meaningful work. Comparing the stage of maturity of appropriateness to the capacity to love, we can say that it corresponds to the capacity for compassion. The compassionate person loves in order to enhance the other. Here technology is designed to enhance individual persons, ecological integrity, and cultural health.

From what has been said so far, we can see that technology cannot be separated from the selves that create and perpetuate it. If its creators and its perpetuators are immature selves, then the technical process will reflect their characteristics in various ways. Some of the bad consequences of technology are the result of design unduly influenced by immaturity, ignorance, confusion of ends, impatience, and too narrow values.

Appropriate technology is the most complete philosophy of the four outlined, since it addresses more of the relevant values, and since it also brings subject and object together in a responsible, reciprocal interaction. Furthermore, it recognizes the useful roles that the other philosophies can play. At this stage there is the possibility for continuing technological development in ways that resolve the negative consequences of the technological imperative of modern human history.

At present we seem to be moving toward the emergence of the philosophy of appropriate technology as a major force in our society. The period of the maximum influence of technophobia might be waning, but this is by no means certain, for there remain powerful forces of technocratic intent which are supported by vast resources with great institutional momentum. This tends to increase political and environmental opposition to technocratic policies. Technophobia could wax, particularly if there are large-scale failures of major technological projects which fully reveal all of their hazardous dimensions, such as pollution, debt, tyranny, and their displacement of human workers.

## APPROPRIATE TECHNOLOGY, INNOVATION AND MASTERY

Each of the philosophies outlined has played a role throughout the process of industrialization in the West. The technological anarchy that was dominant earlier was important in exploring and developing options that led to the industrial revolution. The forces of technophilia and the technocratic mind-set helped to create large-scale processes and infrastructures of continental and global extent that have importance and value. Without the function of these four philosophies expressed in individual lives and in collective social activities, we would lack many positive things we have today.

It would seem that the revolution in modern electronics, the miniaturization of technologies, the emerging solar technologies, improved organic agriculture, and various forms of personal and spiritual growth taken together point toward the possible emergence of the philosophy of appropriate technology as a major cultural force. It is a philosophy conducive to, and compatible with, these postindustrial technologies. From the perspective of appropriate technology, we have the opportunity to create new benign technologies with a clear intent of purpose.

An important feature of appropriate technology is that it forces us to ask central questions for the philosophy of technology: What shall be our relationship to technology? How should we define it? These and other fundamental questions receive our conscious attention. We are able to articulate assumptions of current policy and evaluate them in terms of the human context. Ultimately, appropriate technology aims to transform our relationship with technology in such a way that it becomes a means to the realization of abiding values we fully understand and freely choose. This means that the limits of technology are clearly perceived, and the values of simplicity realized in reduced dependence on heavy technologies.

Appropriate technology will also help to promote the re-creation of community vitality. Many of our current systems are too centralized. It is now necessary to shift to community revitalization through the development of decentralized, human-scaled technologies that preserve the values of the places in which communities have their being. Some of the large processes that are built into the system have now generated spin-off technologies that make such down-scaling and decentralization possible. The transition to such appropriate technologies can be aided by government policies, such as tax incentives and facilitating citizen participation in planning, but ultimately it can only be fully realized as the result of community and personal commitments which grow out of a mature understanding of the values at stake. As Plato saw so clearly, beyond all ideas, at the very center of existence is the Good. Putting this in twentieth-century terms, we can say that appropriate technology leads us to reflect deeply on life as a whole, and mature reflection leads us to realize that life has value at its center. The division between fact and value is only a logical division of concepts with limited usefulness. The practice of science and technology is a value-laden activity. Understanding life requires cognizance and appreciation for its many dimensions of value.

The philosophy of appropriate technology can be further illuminated by considering the four levels of innovation it recognizes. With respect to technological innovation, appropriate technology recognizes four fundamental forms: (1) technological modification, (2) technological hybridization, (3) technological mutation, and (4) technological mastery and creation. Technological modification involves improvement of a technology by means of gradual modification. This process relies heavily on trial and error. In the case of hybridization, we have the merging of two or more technologies to form a new technology or a new technological solution to an existing problem. An example of this would be the design of hybrid vehicles such as a propane-electric automobile. Technological mutation is the transformation of a technology to some other form, or for some radically different purpose. For example, the Chinese used gunpowder for fireworks entertainment, but not to do work or to fight battles. The Mongols and then the Europeans transformed this technology and applied it not only to armaments and warfare, but also for use in the construction of roads, tunnels, dams, and other things. In a reverse direction, atomic bomb technology has been transformed to nonwarfare applications in medicine, the generation of electricity, and the propulsion of ships. These forms of innovation are recognized by the other philosophies of technology discussed here. But appropriate technology emphasizes technological mastery and creation, which involves the capacity to transcend technology and much of human dependence on it. At the same time it opens endless possibilities for the creation of new appropriate technologies. One masters an art by transcending one's fascination with techniques; for the master there is fluency and freedom in the art.[2] Rules and a breakdown of techniques are useful for instructing learners. Mastery transcends these since it leads to spontaneous, creative activity. We often depend on rules and techniques because we have not achieved complete mastery or fluency in the art. Technological mastery in the context of appropriate technology leads to the possibility of transcending technology as a force in human life that lies beyond our control.[3]

The philosophy of appropriate technology encompasses the possibility of mastery and creativity. In its mature form this can be seen as the possibility for a self-mastery that transcends self-manipulation and the desire to control others. In short, the end of domination by technology is seen to lie beyond technology in the realization of human possibilities for mastery of technology in a way that emphasizes the value of persons, develops creative community, and promotes communion with nature. This is the ultimate raison d'être of a fully mature, appropriate technology.

## TECHNOPHILIA AND APPROPRIATE TECHNOLOGY COMPARED: FOUR EXAMPLES

In this section we shall explore more fully the contrast between the approach of appropriate technology (which involves the self-mastery necessary for the wise use of technology) and the approach of technophilia (which uses technology as a means to provide the power to control nature and other humans). The philosophy of appropriate technology applies technology to the natural world in a way respectful of its intrinsic values, whereas technophilia seeks to impose technology upon a nature seen only as resources having instrumental values. The appropriate technologist is respectful of the values *in* the world, whereas the technocratic mind of technophilia attempts to impose patterns of its own devising *on* the world. The aim of appropriate technology is to understand the world and appreciate it, so that humans can interact with it to realize a maximum of reciprocal benefits and also of such values as wonder, delight, and compassion. Technophilia (in contrast) does not seek to know the other, to experience the other, but only to manipulate and control the other, to possess the other. It sees the other as object, not as subject. For the appropriate technologist, however, the living world is filled with subjects. Its dynamic, untamed, organic processes are interdependent. It cannot be approached in a fragmentary way, as a collection of objects to be subdued. It must be approached as a subject-other.

Appropriate technology is a philosophy that includes the human self as part of nature's selves. Questions of ends are primary, and ends depend upon knowing the kinds of beings that we are and can be. This finally leads beyond all techniques and tools, beyond their limits to our own limits. These limits are known through self-knowledge and self-mastery. Self-mastery leads to a mastery of technology that is appropriate to ends worthy of human pursuit. In order to illustrate this important aspect of the philosophy of appropriate technology let us now consider four examples. These examples will help to illustrate the difference between appropriate technology and technophilia in the use and design of technology. The four examples we will discuss are interpersonal conflict, alpine hiking, exercise, and energy generation and use.

Consider, then, some of the levels of technology available for resolving interpersonal conflicts. We shall use examples of warfare and a specific martial art to illustrate the practical difference in philosophy between technophilia and appropriate technology. Suppose that two tribes, two countries, or two treaty groups have a disagreement that seems unresolvable and tending toward violence. Naturally, in these situations tempers can become inflamed. Tension builds while the conflict simmers on. Under these conditions fears arise. These fears magnify the perception of what are interpreted as threats. At a certain stage one or both of the antagonists will think of resorting to force in order to remove the tension. If they apply the full range of modern technology to this conflict, the forces involved could destroy one or even both sides. If they think in terms of winning and losing, then an all-out technological response would seem irrational, given nuclear arms. Hence, they are forced to consider other options. Negotiation and willingness to compromise could be buttressed by this powerful technology, but only if the parties know that there is no armed technological solution to their conflict. It becomes clear at this point that the total use of this vast technological power negates its practicality. It is no longer useful for its originally designed purpose, for technological power has undermined the rationale of war. The pursuit of a technological solution to the conflict could then lead beyond a focus on technology, as a result of the very logic of technophilia's total technological response. At this point it can be seen that the appropriate response to human conflict is not technological warfare. The application of technology leads us to realize that ignorance, immaturity, and lack of self-mastery underlie much interpersonal and international conflict. This can be brought out more clearly at the level of an interpersonal conflict restricted to two persons.

Let us consider the range of options open to two persons, assuming a high level of conflict between them. At the technical level they could resort to bombs, guns, swords, knives, clubs, stones, fists, and feet. If they are martial artists they might use karate, judo, or boxing. Now the technocratic approach is to try to control the other person through the use of technology and techniques. The philosophy of appropriate technology can be illustrated by the martial art of aikido. The aikido martial artist practices the martial way but uses the energy that would be spent on fighting to transcend fighting. The master aikidoist is the ultimate martial artist, since there can be no aggression and competition. Aikido is such a complete art that it resolves conflicts before they can progress to fighting. It is highly subtle, since it masters the impulse to fight by transcending the small self that would fight. It leads one to understand others and the reasons for our impulses toward aggression. This is an art that has its origin in the techniques of fighting, but ultimately it transcends fighting and techniques by means of a practice which leads toward self-mastery. Instead of attempting to manipulate and control others through techniques and fighting technology, aikido resolves conflicts through self-mastery, self-correction, and understanding.

Consider as our second example alpine hiking. Let us compare two hikers: One is loaded with every conceivable camping device modern technology has produced. He is also involved in learning all available techniques. His weekend pack weighs at least 100 pounds. When he camps he employs these various techniques and technology to make a well-organized, "comfortable" camp. He engages in lots of wood craft, lots of "wild-river Jim," nailing, chopping, and building. He loads up his gear in the morning, after spending two hours flipping pancakes on a fancy griddle.

In contrast, the appropriate technology hiker travels light. She is not a "live off the wilderness" wildperson, digging up roots, rooting out berries, and eating the flowers. She is there to celebrate the joy of being alive, and the joy of being able to know nature in an intimate way. She is there to listen to the softer voices of the world and to the deeper voices within herself. Her equipment is carefully designed to be simple, light, durable, minimally polluting, and harmless to the world in its production and use. She is comfortable, but not isolated from the elements of nature she would know. The rain is not an enemy, nor is the sunshine the only pleasure. She eats simple food, such as a breakfast of homemade granola, that requires no or minimal cooking, but nonetheless is optimally nutritious and aesthetically satisfying.

These two hikers illustrate the differences in philosophy between the technophiliac and the appropriate technologist. For the former, the equipment becomes a burden that isolates him from the natural world. For the latter, the equipment is a minimal intrusion which is efficient and enhances her enjoyment of the natural world. It is not a burden, but a joy to use.

As our third example let us consider the range of possibilities open to us with respect to technology and exercise. Ideally, the aims of exercise are self-discipline, fun, and a strong, healthy, flexible, and aesthetically balanced body. Technology can be used to assist in this process. However, the ultimate end of applying technology to exercise undermines many of these aims, as is seen in the exercise machines that do all the moving for you. There is no interaction. You become the manipulated. The other contrasting attitude approaches exercise as a form of self-discipline to be enjoyed also for its own sake.

In jogging, one needs only running shoes, nothing else. Aikido can be done with soft clothes, a padded floor, and one other person. Isometrics and calisthenics require no equipment or helpers. For the philosophy of appropriate technology, the approach to exercise is an integrated and elegant one that uses technology minimally, and would emphasize self-mastery instead of some "easy" technological solution to overweight and lack of sound conditioning. In the technocratic approach, machines become a substitute for this self-discipline and tend to alienate one from one's own body.

Finally, for our fourth example consider the generation and use of energy to illustrate the contrast between the technocratic thrust of technophilia and the approach of appropriate technology. The epitome of the technocratic approach is represented by nuclear power. The use of nuclear fission to boil water to generate steam to power electric generators involves the use of highly capitalized and centralized technology. In the form of electricity this power is distributed through complex grids to distant end users. Electricity is applied to a variety of uses, such as cooling, cooking, and space heating. Nuclear power is highly complex and requires vast subsidies in the form of publicly financed insurance and storage of dangerous wastes. It presents difficult problems of security and increases the probability of the spread of nuclear weapons. In terms of energy use it employs high-temperature processes to accomplish many practical ends which are of low thermodynamic quality. It adds thermopollution to rivers. For these and many other reasons, nuclear power is environmentally, economically, and thermodynamically unsound. It raises serious moral questions. Nonetheless, to the technocrat it is a "logical" way to go.

In contrast, for the appropriate technologist the aim is to diversify and decentralize the use and production of energy. Instead of relying on vast power systems (although some may be developed), the aim is to develop a large variety of smaller scale technologies such as photovoltaic, hydroelectric, and solar. Such approaches as cogeneration and conservation within communities create local systems that use generated power and heat over several times. It gives to local communities greater control over their future, lower costs and debt, and broader public participation, in contrast to many of the large-scale projects which promote complex bureaucratic management structures, increased environmental hazards, and large debt. Appropriate technology emphasizes thermodynamic soundness, doing more with less, conservation, and keeping open a large variety of options. It is rich in understanding of natural processes and takes advantage of the rhythms of natural sources of energy that are readily available on site. It relies on a mastery of design that blends technology and ecological processes, rather than imposing powerful technologies upon nature. In contrast, technocratic forces strive to master nature by controlling and overwhelming rather than working with it.

We can see from these examples, and from earlier comments in this essay, that attempting to resolve the problems caused by technology without first appreciating the human elements involved leads nowhere. The problems of technology that have social and personal implications are not just problems of technology. If we do not appreciate the influence of the particular philosophy of technology that underlies our own individual approach, and see its contrast with other views within our culture, then we will lack a perspective that enables us to move beyond the search for technical solutions to nontechnical problems. In philosophizing about these philosophies of technology I have attempted to sketch how their conceptions of technology affect self, society, and nature. If through this activity we are better able to attend to these attitudes directly, then the chances for a flexible, creative adjustment of our interactions with one another and the world will be increased.

## CONCLUSION

The problems of technology do not all have technical solutions, for the root of some problems of technology lies in the problems of human life itself. Our attitudes toward technology define us, and they bind us to the creation of processes that magnify our initial failure to understand life as the interrelated, holistic process that it is. Powerful modern technologies express in their material forms problems for human life precisely because these technologies reflect the nonresolution of underlying uncertainties about existence and value. Martin Heidegger was one twentieth-century thinker who realized this. He saw that much modern technology grows out of a confused metaphysics that manifests itself in our material and other cultural processes. This confused metaphysics, he observed, is essentially the result of a failure to understand Being and what it means to dwell in the world. Our failure is not that we have linked our industrial technology to profit; it is rather that our pursuits and their technology fail to understand what it is to be in the world in the full openness (the mystery) of Being. Modern industrial technology, as often applied, is an example of a lack of comprehension of Being, a lack of care for the world, and a failure to perceive the fundamental essence of things. It lacks an understanding of the sense of life and of values. With this failure goes the inability to let others be. It begins with confused, calculative thinking, but once this thinking is expressed in the material of technology, that technology then carries it across political and economic boundaries. This is why in the contemporary world industrial technologies and their negative features are transpolitical. The philosophy of appropriate technology recognizes these failings and is open to new possibilities. Because of this it can help us to free our minds of narrower technological concerns, and the sense of being overwhelmed by the "inevitability" of the domination of humans by their own technology. Technology need not be an alien power that overrides responsible human choice. We are better able to solve problems because we better understand their source. The dialogue of creative philosophy frees our minds, the philosophy of appropriate technology frees our practical work of technical and technological tyranny. Together they blend science and art in creative adaptation to a natural world that embodies values to which humans contribute.[4]

## NOTES

1 "Appropriate technology" is a term sometimes used for intermediate technologies (Dunn, 1978). Intermediate technologies are designed for application in developing economies. As we use the term, "appropriate technology" refers to the philosophy we have here described. It is capable of guiding technological designs for many levels of development. Dunn's definition of appropriate technology in his first chapter is not incompatible with the one used here. For a more detailed discussion of the philosophy of appropriate technology, see my article, "Toward a Philosophy of Appropriate Technology," *Humboldt Journal of Social Relations,* Spring/Summer, 1982, vol. 9, no. 2, pp. 161–176. This issue of the journal is devoted entirely to appropriate technology.

2 On the mastery of arts as a form of self-development and self-transcendence, see my paper, "Masters and Mastery," *Philosophy Today,* Fall, 1983, vol. 27, no. 3/4, pp. 230–246. On the relationship between art, imagination, and technology, see my paper, "Art and Imagination in Technological Society," *Research in Philosophy and Technology,* Fall 1983, vol. 6, pp. 77–91.

3 One example of the creation of a completely new technology would be learning how to directly influence the informational forms that underlie matter, and which direct energy to create specific material forms. Gene splicing would be another example (perhaps just a different application of the former). Such new technologies depend on a deep understanding of natural processes, which could work with them, rather than attempting to subdue or overwhelm them. Many earlier (and present) industrial technologies are less subtle, poorer in understanding, and are often crudely overpowerful. However, biotechnologies carry some profound risks. There are also inherent limits to the pursuit of a technological fix. For an exploration of some of these issues, see my paper, "The Sacred and the Limits of the Technological Fix," *Zygon,* September, 1984, vol. 19, no. 3, pp. 259–275. This issue of *Zygon* contains other articles relevant to new biotechnologies.

4 Thanks to Dr. Ingrid Leman and to Kirke Wolfe for their helpful comments and suggestions, which improved the quality of this essay. Thanks also to Robert Lechner, the editor of *Philosophy Today,* for his encouragement and positive support.

## BIBLIOGRAPHY

Barrett, William. *The Illusion of Technique: A Search for Meaning in a Technological Civilization.* New York: Anchor/Doubleday, 1978.

Boulding, Kenneth E. *The Meaning of the 20th Century: The Great Transition.* New York: Harper & Row, 1965.

Commoner, Barry. *The Closing Circle: Nature, Man and Technology.* New York: Bantam, 1972.

———. *The Poverty of Power*. New York: Knopf, 1976.
Dunn, P. D. *Appropriate Technology: Technology with a Human Face*. New York, Schocken Books, 1978.
Durban, P. T., ed. *Research in Philosophy and Technology*. 6 vols. Greenwich, Conn.: JAI Press, 1978–1983.
Ellul, Jacques. *The Technological Society*. New York: Vintage, 1964.
Fromm, Eric. *The Revolution of Hope: Toward a Humanized Technology*. New York: Bantam, 1968.
Galbraith, John K. *The New Industrial State*. New York: Signet, 1967.
Heidegger, Martin. *The Question Concerning Technology and Other Essays*. New York: Harper & Row, 1977.
Illich, Ivan. *Tools for Conviviality*. New York: Harper & Row, 1973.
Jantsche, Erich. *Design for Evolution*. New York: George Braziller, 1975.
Lovins, Amory B. *Soft Energy Paths: Toward a Durable Peace*. Cambridge, Mass.: Ballinger, 1977.
Mitcham, C., and R. Mackey, eds. *Philosophy and Technology*. New York: Free Press, 1972.
Mumford, Lewis. *The Myth of the Machine*. Vol. 1, *Technics and Human Development*. New York: Harcourt Brace, 1967.
———. *The Myth of the Machine*. Vol. 2, *The Pentagon of Power*. New York: Harcourt Brace, 1970.
Odum, Eugene. *Fundamentals of Ecology*. Philadelphia: Saunders, 1971.
Papanek, Victor. *Design for the Real World*. New York: Bantam, 1973.
Roszak, Theodore. *Personal Planet*. New York: Anchor/Doubleday, 1978.
Schumcher, E. F. *Small Is Beautiful: Economics As If People Mattered*. New York: Harper & Row, 1973.
Shepard, Paul, and Daniel McKinley, eds. *The Subversive Science: Essays Towards an Ecology of Man*. New York: Houghton Mifflin, 1969.
Stavrianos, L.S. *The Promise of the Coming Dark Age*. San Francisco: W. H. Freeman, 1976.
Tawney, R. H. *The Acquisitive Society*. New York: Harvest Books, 1948.
Watt, James. *The Titanic Effect*. Stanford: Senaur and Associates, 1974.
Weizenbaum, Joseph. *Computer Power and Human Reason*. San Francisco: W. H. Freeman, 1976.
Wilber, Ken. *The Spectrum of Consciousness*. Wheaton, Ill.: Quest, 1977.
Young, Arthur M. *The Reflexive Universe: Evolution of Consciousness*. San Francisco: Delacorte, 1976.

## Toward a Philosophy of Technology

**Hans Jonas**

Are there philosophical aspects to technology? Of course there are, as there are to all things of importance in human endeavor and destiny. Modern technology touches on almost everything vital to man's existence—material, mental, and spiritual. Indeed, what of man is *not* involved? The way he lives his life and looks at objects, his intercourse with the world and with his peers, his powers and modes of action, kinds of goals, states and changes of society, objectives and forms of politics (including warfare no less than welfare), the sense and quality of life, even man's fate and that of his environment: all these are involved in the technological enterprise as it extends in magnitude and depth. The mere enumeration suggests a staggering host of potentially philosophic themes.

To put it bluntly: if there is a philosophy of science, language, history, and art; if there is social, political, and moral philosophy; philosophy of thought and of action, of reason and passion, of decision and value—all facets of the inclusive philosophy of man—how then could there not be a philosophy of technology, the focal fact of modern life? And at that a philosophy so spacious that it can house portions from all the other branches of philosophy? It is almost a truism, but at the same time so immense a proposition that its challenge staggers the mind. Economy and modesty require that we select, for a beginning, the most obvious from the multitude of aspects that invite philosophical attention.

The old but useful distinction of "form" and "matter" allows us to distinguish between these two major themes: (1) the *formal dynamics* of technology as a continuing collective enterprise, which advances by its own "laws of motion"; and (2) the *substantive content* of technology in terms of the things it puts into human use, the powers it confers, the novel objectives it opens up or dictates, and the altered manner of human action by which these objectives are realized.

The first theme considers technology as an abstract whole of movement; the second considers its concrete uses and their impact on our world and our lives. The formal approach will try to grasp the pervasive "process properties" by which modern technology propels itself—through our agency, to be sure—into ever-succeeding and superseding novelty. The material approach will look at the species of novelties themselves, their taxonomy, as it were, and try to make out how the world furnished with them looks. A third, overarching theme is the *moral* side of technology as a burden on human responsibility, especially its long-term effects on the global condition of man and environment. This—my own main preoccupation over the past years—will only be touched upon.

## THE FORMAL DYNAMICS OF TECHNOLOGY

First some observations about technology's form as an abstract whole of movement. We are concerned with characteristics of *modern* technology and therefore ask first what distinguishes it *formally* from all previous technology. One major distinction is that modern technology is an enterprise and process, whereas earlier technology was a possession and a state. If we roughly describe technology as comprising the use of artificial implements for the business of life, together with their original invention, improvement, and occasional additions, such a tranquil description will do for most of technology through mankind's career (with which it is coeval), but not for modern technology. In the past, generally speaking, a given inventory of tools and procedures used to be fairly constant, tending toward a mutually adjusting, stable equilibrium of ends and means, which—once established—represented for lengthy periods an unchallenged optimum of technical competence.

To be sure, revolutions occurred, but more by accident than by design. The agricultural revolution, the metallurgical revolution that led from the neolithic to the iron age, the rise of cities, and such developments, *happened* rather than were consciously created. Their pace was so slow that only in the time-contraction of historical retrospect do they appear to be "revolutions" (with the misleading connotation that their contemporaries experienced them as such). Even where the change was sudden, as with the introduction first of the chariot, then of armed horsemen into warfare—a violent, if short-lived, revolution indeed—the innovation did not originate from within the military art of the advanced societies that it affected, but was thrust on it from outside by the (much less civilized) peoples of Central Asia. Instead of spreading through the technological universe of their time, other technical breakthroughs, like Phoenician purple-dying, Byzantine "greek fire," Chinese porcelain and silk, and Damascene steel-tempering, remained jealously guarded monopolies of the inventor communities. Still others, like the hydraulic and steam playthings of Alexandrian mechanics, or compass and gunpowder of the Chinese, passed unnoticed in their serious technological potentials.[1]

On the whole (not counting rare upheavals), the great classical civilizations had comparatively early reached a point of technological saturation—the aforementioned "optimum" in equilibrium of means with acknowledged needs and goals—and had little cause later to go beyond it. From there on, convention reigned supreme. From pottery to monumental architecture, from food growing to shipbuilding, from textiles to engines of war, from time measuring to stargazing: tools, techniques, and objectives remained essentially the same over long times; improvements were sporadic and unplanned. Progress therefore—if it occurred at all[2]—was by inconspicuous increments to a universally high level that still excites our admiration and, in historical fact, was more liable to regression than to surpassing. The former at least was the more noted phenomenon, deplored by the epigones with a nostalgic remembrance of a better past (as in the declining Roman world). More important, there was, even in the best and most vigorous times, no proclaimed *idea* of a future of *constant progress* in the arts. Most important, there was never a deliberate method of going about it like "research," the willingness to undergo the risks of trying unorthodox paths, exchanging information widely about the experience, and so on. Least of all was there a "natural science" as a growing body of theory to guide such semitheoretical, prepractical activities, plus their social institutionalization. In routines as well as panoply of instruments, accomplished as they were for the purposes they served, the "arts" seemed as settled as those purposes themselves.[3]

### Traits of Modern Technology

The exact opposite of this picture holds for modern technology, and this is its first philosophical aspect. Let us begin with some manifest traits.

1. Every new step in whatever direction of whatever technological field tends *not* to approach an equilibrium or saturation point in the process of fitting means

to ends (nor is it meant to), but, on the contrary, to give rise, if successful, to further steps in all kinds of direction and with a fluidity of the ends themselves. "Tends to" becomes a compelling "is bound to" with any major or important step (this almost being its criterion); and the innovators themselves expect, beyond the accomplishment, each time, of their immediate task, the constant future repetition of their inventive activity.

2. Every technical innovation is sure to spread quickly through the technological world community, as also do theoretical discoveries in the sciences. The spreading is in terms of knowledge and of practical adoption, the first (and its speed) guaranteed by the universal intercommunication that is itself part of the technological complex, the second enforced by the pressure of competition.

3. The relation of means to ends is not unilinear but circular. Familiar ends of long standing may find better satisfaction by new technologies whose genesis they had inspired. But equally—and increasingly typical—new technologies may suggest, create, even impose new ends, never before conceived, simply by offering their feasibility. (Who had ever wished to have in his living room the Philharmonic orchestra, or open heart surgery, or a helicopter defoliating a Vietnam forest? or to drink his coffee from a disposable plastic cup? or to have artificial insemination, test-tube babies, and host pregnancies? or to see clones of himself and others walking about?) Technology thus adds to the very objectives of human desires, including objectives for technology itself. The last point indicates the dialectics or circularity of the case: once incorporated into the socioeconomic demand diet, ends first gratuitously (perhaps accidentally) generated by technological invention become necessities of life and set technology the task of further perfecting the means of realizing them.

4. Progress, therefore, is not just an ideological gloss on modern technology, and not at all a mere option offered by it, but an inherent drive which acts willy-nilly in the formal automatics of its *modus operandi* as it interacts with society. "Progress" is here not a value term but purely descriptive. We may resent the fact and despise its fruits and yet must go along with it, for—short of a stop by the fiat of total political power, or by a sustained general strike of its clients or some internal collapse of their societies, or by self-destruction through its works (the last, alas, the least unlikely of these)—the juggernaut moves on relentlessly, spawning its always mutated progeny by coping with the challenges and lures of the now. But while not a value term, "progress" here is not a neutral term either, for which we could simply substitute "change." For it is in the nature of the case, or a law of the series, that a later stage is always, in terms of technology itself, *superior* to the preceding *stage*.[4] Thus we have here a case of the entropy-defying sort (organic evolution is another), where the internal motion of a system, left to itself and not interfered with, leads to ever "higher," not "lower" states of itself. Such at least is the present evidence.[5] If Napoleon once said, "Politics is destiny," we may well say today, "Technology is destiny."

These points go some way to explicate the initial statement that modern technology, unlike traditional, is an enterprise and not a possession, a process and not a state, a dynamic thrust and not a set of implements and skills. And they already adumbrate certain "laws of motion" for this restless phenomenon. What we have described, let us remember, were formal traits which as yet say little about the contents of the enterprise. We ask two questions of this descriptive picture: *why* is this so, that is, what *causes* the restlessness of modern technology; what is the nature of the thrust? And, what is the philosophical import of the facts so explained?

### The Nature of Restless Technology

As we would expect in such a complex phenomenon, the motive forces are many, and some causal hints appeared already in the descriptive account. We have mentioned *pressure of competition*—for profit, but also for power, security, and so forth—as one perpetual mover in the universal appropriation of technical improvements. It is equally operative in their origination, that is, in the process of invention itself, nowadays dependent on constant outside subsidy and even goal-setting: potent interests see to both. War, or the threat of it, has proved an especially powerful agent. The less dramatic, but no less compelling, everyday agents are legion. To keep one's head above the water is their common principle (somewhat paradoxical, in view of an abundance already far surpassing what former ages would have lived with happily ever after). Of pressures other than the competitive ones, we must mention those of population growth and of impending exhaustion of natural resources. Since both phenomena are themselves already by-products of technology (the first by way of medical improvements, the second by the voracity of industry), they offer a good example of the more general truth that to a considerable extent technology itself begets the problems which it is then called upon to overcome by a new forward jump. (The Green Revolution and the development of synthetic substitute materials or of alternate sources of energy come under this heading.) These compulsive pressures for progress, then, would operate even for a technology in a noncompetitive, for example, a socialist setting.

A motive force more autonomous and spontaneous than these almost mechanical pushes with their "sink or swim" imperative would be the pull of the quasi-utopian *vision* of an ever better life, whether vulgarly conceived or nobly, one technology had proved the open-ended capacity for procuring the conditions for it: perceived possibility whetting the appetite ("the American dream," "the revolution of rising expectations"). This less palpable factor is more difficult to appraise, but its playing a role is undeniable. Its deliberate fostering and manipulation by the dream merchants of the industrial-mercantile complex is yet another matter and somewhat taints the spontaneity of the motive, as it also degrades the quality of the dream. It is also moot to what extent the vision itself is *post hoc* rather than *ante hoc*, that is, instilled by the dazzling feats of a technological progress already underway and thus more a response to than a motor of it.

Groping in these obscure regions of motivation, one may as well descend, for an explanation of the dynamism as such, into the Spenglerian mystery of a "Faustian soul" innate in Western culture, that drives it, nonrationally, to infinite novelty and unplumbed possibilities for their own sake; or into the Heideggerian depths of a fateful, metaphysical decision of the will for boundless power over the world of things—a decision equally peculiar to the Western mind: speculative intuitions which do strike a resonance in us, but are beyond proof and disproof.

Surfacing once more, we may also look at the very sober, functional facts of industrialism as such, of production and distribution, output maximization, managerial and labor aspects, which even apart from competitive pressure provide their own incentives for technical progress. Similar observations apply to the requirements of *rule* or control in the vast and populous states of our time, those giant territorial superorganisms which for their very cohesion depend on advanced technology (for example, in information, communication, and transportation, not to speak of weaponry) and thus have a stake in its promotion: the more so, the more centralized they are. This holds for socialist systems no less than for free-market societies. May we conclude from this that even a communist world state, freed from external rivals as well as from internal free-market competition, might still have to push technology ahead for purposes of control on this colossal scale? Marxism, in any case, has its own inbuilt commitment to technological progress beyond necessity. But even disregarding all dynamics of these conjectural kinds, the most monolithic case imaginable would, at any rate, still be exposed to those noncompetitive, natural pressures like population growth and dwindling resources that beset industrialism as such. Thus, it seems, the compulsive element of technological progress may not be bound to its original breeding ground, the capitalist system. Perhaps the odds for an eventual stabilization look somewhat better in a socialist system, provided it is worldwide—and possibly totalitarian in the bargain. As it is, the pluralism we are thankful for ensures the constancy of compulsive advance.

We could go on unravelling the causal skein and would be sure to find many more strands. But none nor all of them, much as they explain, would go to the heart of the matter. For all of them have one premise in common without which they could not operate for long: the premise that there *can* be indefinite progress because there *is* always something new and better to find. The, by no means obvious, giveness of this objective condition is also the pragmatic conviction of the performers in the technological drama; but without its being true, the conviction would help as little as the dream of the alchemists. Unlike theirs, it is backed up by an impressive record of past successes, and for many this is sufficient ground for their belief. (Perhaps holding or not holding it does not even greatly matter.) What makes it more than a sanguine belief, however, is an underlying and well-grounded, theoretical view of the nature of things and of human cognition, according to which they do not set a limit to novelty of discovery and invention, indeed, that they of themselves will at each point offer another opening for the as yet unknown and undone. The corollary conviction, then, is that a technology tailored to a nature and to a knowledge of this indefinite potential ensures its indefinitely continued conversion into the practical powers, each step of it begetting the next, with never a cutoff from internal exhaustion of possibilities.

Only habituation dulls our wonder at this wholly unprecedented belief in virtual "infinity." And by all our present comprehension of reality, the belief is most likely true—at least enough of it to keep the road for innovative technology in the wake of advancing science open for a long time ahead. Unless we understand this ontologic-epistomological premise, we have not understood the inmost agent of technological dynamics, on which the working of all the adventitious causal factors is contingent in the long run.

Let us remember that the virtual infinitude of advance we here seek to explain is in essence different from the always avowed perfectibility of every human accomplishment. Even the undisputed master of his craft always had to admit as possible that he might be surpassed in skill or tools or materials; and no excellence of product ever foreclosed that it might still be bettered, just as today's champion runner must know that his time may one day be beaten. But these are improvements within a given genus, not different in kind from what went before, and they must accrue in diminishing fractions. Clearly, the phenomenon of an exponentially growing *generic* innovation is qualitatively different.

### Science as a Source of Restlessness

The answer lies in the interaction of *science* and *technology* that is the hallmark of modern progress, and thus ultimately in the kind of nature which modern science progressively discloses. For it is here, in the movement of *knowledge,* where relevant novelty first and constantly occurs. This is itself a novelty. To Newtonian physics, nature appeared simple, almost crude, running its show with a few kinds of basic entities and forces by a few universal laws, and the application of those well-known laws to an ever greater variety of composite phenomena promised ever widening knowledge indeed, but no real surprises. Since the mid-nineteenth century, this minimalistic and somehow finished picture of nature has changed with breathtaking acceleration. In a reciprocal interplay with the growing subtlety of exploration (instrumental and conceptual), nature itself stands forth as ever more subtle. The progress of probing makes the object grow richer in modes of operation, not sparer as classical mechanics had expected. And instead of narrowing the margin of the still-undiscovered, science now surprises itself with unlocking dimension after dimension of new depths. The very essence of matter has turned from a blunt, irreducible ultimate to an always reopened challenge for further penetration. No one can say whether this will go on forever, but a suspicion of intrinsic infinity in the very being of things obtrudes itself and therewith an anticipation of unending inquiry of the sort where succeeding steps will not find the same old story again (Descartes's "matter in motion"), but always add new twists to it. If then the art of technology is correlative to the knowledge of nature, technology too acquires from this source that potential of infinity for its innovative advance.

But it is not just that indefinite scientific progress offers the *option* of indefinite technological progress, to be exercised or not as other interests see fit. Rather the cognitive process itself moves by interaction with the technological, and in the most internally vital sense: for its own *theoretical* purpose, science must generate an increasingly sophisticated and physically formidable technology as its tool. What it finds with this help initiates new departures in the practical sphere, and the latter as a whole, that is, technology at work provides with its experiences a large-scale laboratory for science again, a breeding ground for new questions, and so on in an unending cycle. In brief, a mutual feedback operates between science and technology; each requires and propels the other; and as matters now stand, they can only live together or must die together. For the dynamics of technology, with which we are here concerned, this means that (all external promptings apart) an agent of restlessness is implanted in it by its functionally integral bond with science. As long, therefore, as the cognitive impulse lasts, technology is sure to move ahead with it. The cognitive impulse, in its turn, culturally vulnerable in itself, liable to lag or to grow conservative with a treasured canon—that theoretical eros itself no longer lives on the delicate appetite for truth alone, but is spurred on by its hardier offspring, technology, which communicates to it impulsions from the broadest arena of struggling, insistent life. Intellectual curiosity is seconded by interminably self-renewing practical aim.

I am conscious of the conjectural character of some of these thoughts. The revolutions in science over the last fifty years or so are a fact, and so are the revolutionary style they imparted to technology and the reciprocity between the two concurrent streams (nuclear physics is a good example). But whether those scientific revolutions, which hold primacy in the whole syndrome, will be typical for science henceforth—something like a law of motion for its future—or represent only a singular phase in its longer run, is unsure. To the extent, then, that our forecast of incessant novelty for technology was predicated on a guess concerning the future of science, even concerning the nature of things, it is hypothetical, as such extrapolations are bound to be. But even if the recent past did not usher in a state of permanent revolution for science, and the life of theory settles down again to a more sedate pace, the scope for technological innovation will not easily shrink; and what may no longer be a revolution in science, may still revolutionize our lives in its practical impact through technology. "Infinity" being too large a word anyway, let us say that present signs of potential and of incentives point to an indefinite perpetuation and fertility of the technological momentum.

### The Philosophical Implications

It remains to draw philosophical conclusions from our findings, at least to pinpoint aspects of philosophical interest. Some preceding remarks have already been straying into philosophy of science in the technical sense. Of broader issues, two will be ample to provide food for further thought beyond the limitations of this paper. One concerns the status of knowledge in the human scheme, the other the status of technology itself as a human goal, or its tendency to become that from being a means, in a dialectical inversion of the means-end order itself.

Concerning knowledge, it is obvious that the time-honored division of theory and practice has vanished for both sides. The thirst for pure knowledge may persist undiminished, but the involvement of knowing at the heights with doing in the lowlands of life, mediated by technology, has become inextricable; and the aristocratic self-sufficiency of knowing for its own (and the knower's) sake has gone. Nobility has been exchanged for utility. With the possible exception of philosophy, which still can do with paper and pen and tossing thoughts around among peers, all knowledge has become thus tainted, or elevated if you will, whether utility is intended or not. The technological syndrome, in other words, has brought about a thorough *socializing* of the theoretical realm, enlisting it in the service of common need. What used to be the freest of human choices, an extravagance snatched from the pressure of the world—the esoteric life of thought—has become part of the great public play of necessities and a prime necessity in the action of the play.[6] Remotest abstraction has become enmeshed with nearest concreteness. What this pragmatic functionalization of the once highest indulgence in impractical pursuits portends for the image of man, for the restructuring of a hallowed hierarchy of values, for the idea of "wisdom," and so on, is surely a subject for philosophical pondering.

Concerning technology itself, its actual role in modern life (as distinct from the purely instrumental definition of technology as such) has made the relation of means and ends equivocal all the way up from the daily living to the very vocation of man. There could be no question in former technology that its role was that of humble servant—pride of workmanship and esthetic embellishment of the useful notwithstanding. The Promethean enterprise of modern technology speaks a different language. The word "enterprise" gives the clue, and its unendingness another. We have mentioned that the effect of its innovations is disequilibrating rather than equilibrating with respect to the balance of wants and supply, always breeding its own new wants. This in itself compels the constant attention of the best minds, engaging the full capital of human ingenuity for meeting challenge after challenge and seizing the new chances. It is psychologically natural for that degree of engagement to be invested with the dignity of dominant purpose. Not only does technology dominate our lives in fact, it nourishes also a belief in its being of predominant worth. The sheer grandeur of the enterprise and its seeming infinity inspire enthusiasm and fire ambition. Thus, in addition to spawning new ends (worthy or frivolous) from the mere invention of means, technology as a grand venture tends to establish *itself* as the transcendent end. At least the suggestion is there and casts its spell on the modern mind. At its most modest, it means elevating *homo faber* to the essential aspect of man; at its most extravagant, it means elevating *power* to the position of his dominant and interminable goal. To become ever more masters of the world, to advance from power to power, even if only collectively and perhaps no longer by choice, can now be seen to be the chief vocation of mankind. Surely, this again poses philosophical questions that may well lead unto the uncertain grounds of metaphysics or of faith.

I here break off, arbitrarily, the formal account of the technological movement in general, which as yet has told us little of what the enterprise is about. To this subject I now turn, that is, to the new kinds of powers and objectives that technology opens to modern man and the consequently altered quality of human action itself.

## THE MATERIAL WORKS OF TECHNOLOGY

Technology is a species of power, and we can ask questions about how and on what object any power is exercised. Adopting Aristotle's rule in *de anima* that for understanding a faculty one should begin with its objects, we start from them too—"objects" meaning both the visible *things* technology generates and puts into human use, and the *objectives* they serve. The objects of modern technology are first everything that had always been an object of human artifice and labor: food, clothing, shelter, implements, transportation—all the material necessities and comforts of life. The technological intervention changed at first not the product but its production, in speed, ease, and quantity. However, this is true only of the very first stage of the industrial revolution with which large-scale scientific technology began. For example, the cloth for the steam-driven looms of Lancashire remained the same. Even then, one significant new product was added to the traditional list—the machines themselves, which required an entire new industry with further subsidiary industries to build them. These novel entities, machines—at first capital goods only, not consumer goods—had from the beginning their own impact on man's symbiosis with nature by being consumers themselves. For example: steam-powered water pumps facilitated coal mining, required in turn extra coal for firing their boilers, more coal for the foundries and forges that made those boilers, more for the mining of the requisite iron ore, more for its transportation to the foundries, more—both coal and iron—for the rails and locomotives made in these same foundries, more for the conveyance of the foundries' product to the pitheads and return, and finally more for the distribution of the more abundant coal to the users outside this cycle, among which were increasingly still more machines spawned by the increased availability of coal. Lest it be forgotten over this long chain, we have been speaking of James Watt's modest steam engine for pumping water out of mine shafts. This syndrome of self-proliferation—by no means a linear chain but an intricate web of reciprocity—has been part of modern technology ever since. To generalize, technology exponentially increases man's drain on nature's resources (of substances and of energy), not only through the multiplication of the final goods for consumption, but also, and perhaps more so, through the production and operation of its own mechanical means. And with these means—machines—it introduced a new category of goods, not for consumption, added to the furniture of our world. That is, among the objects of technology a prominent class is that of technological apparatus itself.

Soon other features also changed the initial picture of a merely mechanized production of familiar commodities. The final products reaching the consumer ceased to be the same, even if still serving the same age-old needs; new needs, or desires, were added by commodities of entirely new kinds which changed the habits of life. Of such commodities, machines themselves became increasingly part of the consumer's daily life to be used directly by himself, as an article not of production but of consumption. My survey can be brief as the facts are familiar.

### New Kinds of Commodities

When I said that the cloth of the mechanized looms of Lancashire remained the same, everyone will have thought of today's synthetic fibre textiles for which the statement surely no longer holds. This is fairly recent, but the general phenomenon starts much earlier, in the synthetic dyes and fertilizers with which the chemical industry—the first to be wholly a fruit of science—began. The original rationale of these technological feats was substitution of artificial for natural materials (for reasons of scarcity or cost), with as nearly as possible the same properties for effective use. But we need only think of plastics to realize that art progressed from substitutes to the creation of really new substances with properties not so found in any natural one, raw or processed, thereby also initiating uses not thought of before and giving rise to new classes of objects to serve them. In chemical (molecular) engineering, man does more than in mechanical (molar) engineering which constructs machinery from natural materials; his intervention is deeper, redesigning the infra-patterns of nature, making substances to specification by arbitrary disposition of molecules. And this, be it noted, is done deductively from the bottom, from the thoroughly analyzed last elements, that is, in a real *via compositiva* after the completed *via resolutiva,* very different from the long-known empirical practice of coaxing substances into new properties, as in metal alloys from the bronze age on. Artificiality or creative engineering with abstract construction invades the heart of matter. This, in molecular biology, points to further, awesome potentialities.

With the sophistication of molecular alchemy we are ahead of our story. Even in straightforward hardware engineering, right in the first blush of the mechanical revolution, the objects of use that came out of the factories did not really remain the same, even where the objectives did. Take the old objective of travel. Railroads and ocean liners are relevantly different from the stage coach and from the sailing ship, not merely in construction and efficiency but in the very feel of the user, making travel a different experience altogether, something one may do for its own sake. Airplanes, finally, leave behind any similarity with former conveyances, except the purpose of getting from here to there, with no experience of what lies in between. And these instrumental objects occupy a prominent, even obtrusive place in our world, far beyond anything wagons and boats ever did. Also they are constantly subject to improvement of design, with obsolescence rather than wear determining their life span.

Or take the oldest, most static of artifacts: human habitation. The multistoried office building of steel, concrete, and glass is a qualitatively different entity from the wood, brick, and stone structures of old. With all that goes into it besides the structures as such—the plumbing and wiring, the elevators, the lighting, heating, and cooling systems—it embodies the end products of a whole spectrum of technologies and far-flung industries, where only at the remote sources human hands still meet with primary materials, no longer recognizable in the final result. The ultimate customer inhabiting the product is ensconced in a shell of thoroughly derivative artifacts (perhaps relieved by a nice piece of driftwood). This transformation into utter artificiality is generally, and increasingly, the effect of technology on the human environment, down to the items of daily use. Only in agriculture has the product so far escaped this transformation by the changed modes of its production. We still eat the meat and rice of our ancestors.[7]

Then, speaking of the commodities that technology injects into private use, there are machines themselves, those very devices of its own running, originally confined to the economic sphere. This unprecedented novum in the records of individual living started late in the nineteenth century and has since grown to a pervading mass phenomenon in the Western world. The prime example, of course, is the automobile, but we must add to it the whole gamut of household appliances—refrigerators, washers, dryers, vacuum cleaners—by now more common in the lifestyle of the general population than running water or central heating were one hundred years ago. Add lawn mowers and other power tools for home and garden: we are mechanized in our daily chores and recreations (including the toys of our children) with every expectation that new gadgets will continue to arrive.

These paraphernalia are machines in the precise sense that they perform work and consume energy, and their moving parts are of the familiar magnitudes of our perceptual world. But an additional and profoundly different category of technical apparatus was dropped into the lap of the private citizen, not labor-saving and work-performing, partly not even utilitarian, but—with minimal energy input—catering to the senses and the mind: telephone, radio, television, tape recorders, calculators, record players—all the domestic terminals of the electronics industry, the latest arrival on the technological scene. Not only by their insubstantial, mind-addressed output, also by the subvisible, not literally "mechanical" physics of their functioning do these devices differ in kind from all the macroscopic, bodily moving machinery of the classical type. Before inspecting this momentous turn from power engineering, the hallmark of the first industrial revolution, to communication engineering, which almost amounts to a second industrial-technological revolution, we must take a look at its natural base: electricity.

In the march of technology to ever greater artificiality, abstraction, and subtlety, the unlocking of electricity marks a decisive step. Here is a universal force of nature which yet does not naturally appear to man (except in lightning). It is not a datum of uncontrived experience. Its very "appearance" had to wait for science, which contrived the experience for it. Here, then, a technology depended on science for the mere providing of its "object," the entity itself it would deal with—the first case where theory alone, not ordinary experience, wholly preceded practice (repeated later in the case of nuclear energy). And what sort of entity! Heat and steam are familiar objects of sensuous experience, their force bodily displayed in nature; the matter of chemistry is still the concrete, corporeal stuff mankind had always known. But electricity is an abstract object, disembodied, immaterial, unseen; in its usable form, it is entirely an artifact, generated in a subtle transformation from grosser forms of energy (ultimately from heat via motion). Its theory indeed had to be essentially complete before utilization could begin.

Revolutionary as electrical technology was in itself, its purpose was at first the by now conventional one of the industrial revolution in general: to supply motive power for the propulsion of machines. Its advantages lay in the unique versatility of the new force, the ease of its transmission, transformation, and distribution—an unsubstantial commodity, no bulk, no weight, instantaneously delivered at the point of consumption. Nothing like it had ever existed before in man's traffic with matter, space, and time. It made possible the spread of mechanization to every

the latitude of spontaneity smaller; and whether man has not actually been weakened in his decision-making capacity by his accretion of collective strength.

However, in speaking, as I have just done, of "his" decision-making capacity, I have been guilty of the same abstraction I had earlier criticized in the use of the term "man." Actually, the subject of the statement was no real or representative individual but Hobbes' "Artificiall Man," "that great Leviathan, called a Common-Wealth," or the "large horse" to which Socrates likened the city, "which because of its great size tends to be sluggish and needs stirring by a gadfly." Now, the chances of there being such gadflies among the numbers of the commonwealth are today no worse nor better than they have ever been, and in fact they are around and stinging in our field of concern. In that respect, the free spontaneity of personal insight, judgment, and responsible action by speech can be trusted as an ineradicable (if also incalculable) endowment of humanity, and smallness of number is in itself no impediment to shaking public complacency. The problem, however, is not so much complacency or apathy as the counterforces of active, and anything but complacent, interests and the complicity with them of all of us in our daily consumer existence. These interests themselves are factors in the determinism which technology has set up in the space of its sway. The question, then, is that of the possible chances of unselfish insight in the arena of (by nature) selfish *power,* and more particularly: of one long-range, interloping insight against the short-range goals of many incumbent powers. Is there hope that wisdom itself can become power? This renews the thorny old subject of Plato's philosopher-king and—with that inclusion of realism which the utopian Plato did not lack—of the role of myth, not knowledge, in the education of the guardians. Applied to our topic: the *knowledge* of objective dangers and of values endangered, as well as of the technical remedies, is beginning to be there and to be disseminated; but to make it prevail in the marketplace is a matter less of the rational dissemination of truth than of public relations techniques, persuasion, indoctrination, and manipulation, also of unholy alliances, perhaps even conspiracy. The philosopher's descent into the cave may well have to go all the way to "if you can't lick them, join them."

That is so not merely because of the active resistance of special interests but because of the optical illusion of the near and the far which condemns the long-range view to impotence against the enticement and threats of the nearby: it is this incurable shortsightedness of animal-human nature more than ill will that makes it difficult to move even those who have no special axe to grind, but still are in countless ways, as we all are, beneficiaries of the untamed system and so have something dear in the present to lose with the inevitable cost of its taming. The taskmaster, I fear, will have to be actual pain beginning to strike, when the far has moved close to the skin and has vulgar optics on its side. Even then, one may resort to palliatives of the hour. In any event, one should try as much as one can to forestall the advent of emergency with its high tax of suffering or, at the least, prepare for it. This is where the scientist can redeem his role in the technological estate.

The incipient knowledge about technological danger trends must be developed, coordinated, systematized, and the full force of computer-aided projection techniques be deployed to determine priorities of action, so as to inform preventive efforts wherever they can be elicited, to minimize the necessary sacrifices, and at the worst to preplan the saving measures which the terror of beginning calamity will eventually make people willing to accept. Even now, hardly a decade after the first stirrings of "environmental" consciousness, much of the requisite knowledge, plus the rational persuasion, is available inside and outside academia for any well-meaning powerholder to draw upon. To this, we—the growing band of concerned intellectuals—ought persistently to contribute our bit of competence and passion.

But the real problem is to get the well-meaning into power and have that power as little as possible beholden to the interests which the technological colossus generates on its path. It is the problem of the philosopher-king compounded by the greater magnitude and complexity (also sophistication) of the forces to contend with. Ethically, it becomes a problem of playing the game by its impure rules. For the servant of truth to join in it means to sacrifice some of his time-honored role: he may have to turn apostle or agitator or political operator. This raises moral questions beyond those which technology itself poses, that of sanctioning immoral means for a surpassing end, of giving unto Caesar so as to promote what is not Caesar's. It is the grave question of moral casuistry, or of Dostoevsky's Grand Inquisitor, or of regarding cherished liberties as no longer affordable luxuries (which may well bring the anxious friend of mankind into odious political company)—questions one excusably hesitates to touch but in the further tide of things may not be permitted to evade.

What is, prior to joining the fray, the role of philosophy, that is, of a philosophically grounded ethical knowledge, in all this? The somber note of the last remarks responded to the quasi-apocalyptic prospects of the technological tide, where stark issues of planetary survival loom ahead. There, no philosophical ethics is needed to tell us that disaster must be averted. Mainly, this is the case of the ecological dangers. But there are other, noncatastrophic things afoot in technology where not the existence but the image of man is at stake. They are with us now and will accompany us and be joined by others at every new turn technology may take. Mainly, they are in the biomedical, behavioral, and social fields. They lack the stark simplicity of the survival issue, and there is none of the (at least declaratory) unanimity on them which the spectre of extreme crisis commands. It is here where a philosophical ethics or theory of values has its task. Whether its voice will be listened to in the dispute on policies is not for it to ask; perhaps it cannot even muster an authoritative voice with which to speak—a house divided, as philosophy is. But the philosopher must try for normative knowledge, and if his labors fall predictably short of producing a compelling axiomatics, at least his clarifications can counteract rashness and make people pause for a thoughtful view.

Where not existence but "quality" of life is in question, there is room for honest dissent on goals, time for theory to ponder them, and freedom from the tyranny of the lifeboat situation. Here, philosophy can have its try and its say. Not so on the extremity of the survival issue. The philosopher, to be sure, will also strive for a theoretical grounding of the very proposition that there ought to be men on earth, and that present generations are obligated to the existence of future ones. But such esoteric, ultimate validation of the perpetuity imperative for the species—whether obtainable or not to the satisfaction of reason—is happily not needed for consensus in the face of ultimate threat. Agreement in favor of life is pretheoretical, instinctive, and universal. Averting disaster takes precedence over everything else, including pursuit of the good, and suspends otherwise inviolable prohibitions and rules. All moral standards for individual or group behavior, even demands for individual sacrifice of life, are premised on the continued existence of human life. As I have said elsewhere,[11] "No rules can be devised for the waiving of rules in extremities. As with the famous shipwreck examples of ethical theory, the less said about it, the better."

Never before was there cause for considering the contingency that all mankind may find itself in a lifeboat, but this is exactly what we face when the viability of the planet is at stake. Once the situation becomes desperate, then what there is to do for salvaging it must be done, so that there be life—which "then," after the storm has been weathered, can again be adorned by ethical conduct. The moral inference to be drawn from this lurid eventuality of a moral pause is that we must never allow a lifeboat situation for humanity to arise.[12] One part of the ethics of technology is precisely to guard the space in which any ethics can operate. For the rest, it must grapple with the cross-currents of value in the complexity of life.

A final word on the question of determinism versus freedom which our presentation of the technological syndrome has raised. The best hope of man rests in his most troublesome gift: the spontaneity of human acting which confounds all prediction. As the late Hannah Arendt never tired of stressing: the continuing arrival of newborn individuals in the world assures ever-new beginnings. We should expect to be surprised and to see our predictions come to naught. But those predictions themselves, with their warning voice, can have a vital share in provoking and informing the spontaneity that is going to confound them.

## REFERENCES

1 But as serious an actuality as the Chinese plough "wandered" slowly westward with little traces of its route and finally caused a major, highly beneficial revolution in medieval European agriculture, which almost no one deemed worth recording when it happened (cf. Paul Leser, *Entstehung und Verbreitung des Pfluges,* Münster, 1931; reprint: The International Secretariate for Research on the History of Agricultural Implements, Brede-Lingby, Denmark, 1971).

2 Progress did, in fact, occur even at the heights of classical civilizations. The Roman arch and vault, for example, were distinct engineering advances over the horizontal entablature and flat ceiling of Greek (and Egyptian) architecture, permitting spanning feats and thereby construction objectives not contemplated before (stone bridges, aqueducts, the vast baths and other public halls of Imperial Rome). But materials, tools, and techniques were still the same, the role of human labor and crafts remained unaltered, stonecutting and brickbaking went on as before. An existing technology was enlarged in its scope of performance, but none of its means or even goals made obsolete.

3 One meaning of "classical" is that those civilizations had somehow implicitly "defined" themselves and neither encouraged nor even allowed to pass beyond their innate terms. The—more or less—achieved "equilibrium" was their very pride.

4 This only seems to be but is not a value statement, as the reflection on, for example, an ever more destructive atom bomb shows.

piness as is to be achieved through techniques. Modern man in choosing is already incorporated within the technical process and modified in his nature by it. He is no longer in his traditional state of freedom with respect to judgment and choice.

To understand the problem posed to us, it is first of all requisite to disembarrass ourselves of certain fake problems.

1. We make too much of the disagreeable features of technical development, for example urban overcrowding, nervous tension, air pollution, and so forth. I am convinced that such inconveniences will be done away with by the ongoing evolution of Technique itself, and indeed, that it is only by means of such evolution that this can happen. The inconveniences we emphasize are always dependent on technical solutions, and it is only by means of techniques that they can be solved. This fact leads to the following two considerations:

Every solution to some technical inconvenience is able only to reinforce the system of techniques *in their ensemble;*

Enmeshed in a process of technical development like our own, the possibilities of human survival are better served by more technique than less, a fact which contributes nothing, however, to the resolution of the basic problem.

2. We hear too often that morals are being threatened by the growth of our techniques. For example, we hear of greater moral decadence in those environments most directly affected technically, say, in working-class or urbanized milieux. We hear, too, of familial disintegration as a function of techniques. The falseness of this problem consists in contrasting the technological environment with the moral values inculcated by society itself.[2] The presumed opposition between ethical problematics and technological systematics probably at the present is, and certainly in the long run will be, false. The traditional ethical milieu and the traditional moral values are admittedly in process of disappearing, and we are witnessing the creation of a *new* technological ethics with its own values. We are witnessing the evolution of a morally consistent system of imperatives and virtues, which tends to replace the traditional system. But man is not necessarily left thereby on a morally inferior level, although a moral relativism is indeed implied—an attitude according to which everything is well, *provided* that the individual obeys some ethic or other. We could contest the value of this development *if* we had a clear and adequate concept of what good-in-itself is. But such judgments are impossible on the basis of our general morality. On that level, what we are getting is merely a substitution of a new technological morality for a traditional one which Technique has rendered obsolete.

3. We dread the "sterilization" of art through technique. We hear the artist's lack of freedom, calm, and the impossibility of meditation in the technological society. This problem is no more real than the two preceding. On the contrary, the best artistic production of the present is a result of a close connection between art and Technique. Naturally, new artistic form, expression, and ethic are implied, but this fact does not make art less art than what we traditionally called such. What assuredly is *not* art is a fixation in congealed forms, and a rejection of technical evolution as exemplified, say, in the neoclassicism of the nineteenth

century or in present day "socialist realism." The modern cinema furnishes an artistic response comparable to the Greek theater at its best; and modern music, painting, and poetry express, not a canker, but an authentic aesthetic expression of mankind plunged into a new technical milieu.

4. One last example of a false problem is our fear that the technological society is completely *eliminating* instinctive human values and powers. It is held that systematization, organization, "rationalized" conditions of labor, overly hygienic living conditions, and the like have a tendency to repress the forces of instinct. For some people the phenomenon of "beatniks," "*blousons noirs,*"[3] and "hooligans" is explained by youth's violent reaction and the protestation of youth's vital force to a society that is overorganized, overordered, overregulated, in short, technicized.[4] But here too, even if the facts are established beyond question, it is very likely that a superior conception of the technological society will result in the integration of these instinctive, creative, and vital forces. Compensatory mechanisms are already coming into play; the increasing appreciation of the aesthetic eroticism of authors like Henry Miller and the rehabilitation of the Marquis de Sade are good examples. The same holds for music like the new jazz forms which are "escapist" and exaltative of instinct; *item,* the latest dances. All these things represent a process of "*défoulement*,"[5] which is finding its place in the technological society. In the same way, we are beginning to understand that it is impossible indefinitely to repress or expel religious tendencies and to bring the human race to a perfect rationality. Our fears for our instincts *are* justified to the degree that Technique, instead of provoking conflict, tends rather to *absorb* it, and to *integrate* instinctive and religious forces by giving them a place within its structure, whether it be by an adaptation of Christianity[6] or by the creation of new religious expressions like myths and mystiques which are in full compatibility with the technological society.[7] The Russians have gone farthest in creating a "religion" compatible with Technique by means of their transformation of Communism into a religion.

What, then, is the real problem posed to men by the development of the technological society? It comprises two parts: (1) Is man able to remain master[8] in a world of means? (2) Can a new civilization appear inclusive of Technique?

1. The answer to the first question, and the one most often encountered, seems obvious: Man, who exploits the ensemble of means, *is* the master of them. Unfortunately, this manner of viewing matters is purely theoretical and superficial. We must remember the autonomous character of Technique. We must likewise not lose sight of the fact that the human individual himself is to be an ever greater degree the *object* of certain techniques and their procedures. He is the object of pedagogical techniques, psychotechniques, vocational guidance testing, personality and intelligence testing, industrial and group aptitude testing, and so on. In these cases (and in countless others) most men are treated as a collection of objects. But, it might be objected, these techniques are exploited by other men, and the exploiters at least remain masters. In a certain sense this is true; the exploiters *are* masters of the particular techniques they exploit. But, they, too, are subjected to the action of yet other techniques, as, for example, propaganda.

Above all, they are spiritually taken over by the technological society; they believe in what they do; they are the most fervent adepts of that society. They themselves have been profoundly technicized. They never in any way affect to despise Technique, which to them is a thing good in itself. They never pretend to assign values to Technique, which to them is in itself an entity working out its own ends. They never claim to subordinate it to any value because for them Technique *is* value.

It may be objected that these individual techniques have as their end the best adaptation of the individual, the best utilization of his abilities, and, in the long run, his happiness. This, in effect, is the objective and the justification of all techniques. (One ought not, of course, to confound man's "happiness" with capacity for mastery with, say, freedom.) If the first of all values is happiness, it is likely that man, thanks to his techniques, will be in a position to attain to a certain state of this good. But happiness does not contain everything it is thought to contain, and the *absolute disparity between happiness and freedom* remains an ever real theme for our reflections. To say that man should remain *subject* rather than *object* in the technological society means two things, viz., that he be capable of giving direction and orientation to Technique, and that, to this end, he be able to master it.

Up to the present he has been able to do neither. As to the first, he is content passively to participate in technical progress, to accept whatever direction it takes automatically, and to admit its autonomous meaning. In the circumstances he can either proclaim this life is an absurdity without meaning or value; *or,* he can predicate a number of indefinitely sophisticated values. But neither attitude accords with the fact of the technical phenomenon any more than it does with the other. Modern declarations of the absurdity of life are not based on modern technological efflorescence, which none (least of all the existentialists) think an absurdity. And the predication of values is a purely theoretical matter, since these values are not equipped with any means for putting them into practice. It is easy to reach agreement on what they are, but it is quite another matter to make them have any effect whatever on the technological society, or to cause them to be accepted in such a way that techniques must evolve in order to realize them. The values spoken of in the technological society are simply there to justify what is; *or,* they are generalities without consequence; *or,* technical progress realizes them automatically as a matter of course. Put otherwise, neither of the above alternatives is to be taken seriously.

The second condition *that man be subject rather than object,* i.e., the imperative that he exercise mastery over technical development, is facilely accepted by everyone. But factually it simply does not hold. Even more embarrassing than the question "How?" is the question "Who?" We must ask ourselves realistically and concretely just who is in a position to choose the values which give Technique its justification and to exert mastery over it. If such a person or persons are to be found, it must be in the Western world (inclusive of Russia). They certainly are not to be discovered in the bulk of the world's population which inhabits Africa and Asia, who are, as yet, scarcely confronted by technical problems, and who, in any case, are even less aware of the questions involved than we are.

Is the arbiter we seek to be found among the *philosophers,* those thinking specialists? We well know the small influence these gentry exert upon our society, and how the technicians of every order distrust them and rightly refuse to take their reveries seriously. Even if the philosopher could make his voice heard, he would still have to contrive means of mass education so as to communicate an effective message to the masses.

Can the *technician* himself assume mastery over Technique? The trouble here is that the technician is *always* a specialist and cannot make the slightest claim to have mastered any technique but his own. Those for whom Technique bears its meaning in itself will scarcely discover the values that lend meaning to what they are doing. They will not even look for them. The only thing they can do is to apply their technical specialty and assist in its refinement. They cannot *in principle* dominate the totality of the technical problem or envisage it in its global dimensions. Ergo, they are completely incapable of mastering it.

Can the *scientist* do it? There, if anywhere, is the great hope. Does not the scientist dominate our techniques? Is he not an intellectual inclined and fit to put basic questions? Unfortunately, we are obliged to reexamine our hopes here when we look at things as they are. We see quickly enough that the scientist is as specialized as the technician, as incapable of general ideas, and as much out of commission as the philosopher. Think of the scientists who, on one tack or another, have addressed themselves to the technical phenomenon: Einstein, Oppenheimer, Carrel. It is only too clear that the ideas these gentlemen have advanced in the sphere of the philosophic or the spiritual are vague, superficial, and contradictory in extremis. They really ought to stick to warnings and proclamations, for as soon as they assay anything else, the other scientists and the technicians rightly refuse to take them seriously, and they even run the risk of losing their reputations as scientists.

Can the *politician* bring it off? In the democracies the politicians are subject to the wishes of their constituents, who are primarily concerned with the happiness and well-being which they think Technique assures them. Moreover, the further we get on, the more a conflict shapes up between the politicians and the technicians. We cannot here go into the matter, which is just beginning to be the object of serious study.[9] But it would appear that the power of the politician is being (and will continue to be) outclassed by the power of the technician in modern states. Only dictatorships can impose their will on technical evolution. But, on the one hand, human freedom would gain nothing thereby, and, on the other, a dictatorship thirsty for power has no recourse at all but to push toward an excessive development of various techniques at its disposal.

*Any* of us? An individual can doubtless seek the soundest attitude to dominate the techniques at his disposal. He can inquire after the values to impose on techniques in his use of them, and search out the way to follow in order to remain a man in the fullest sense of the word within a technological society. All this is extremely difficult, but it is far from being useless, since it is apparently the only solution presently possible. But the individual's efforts are powerless to resolve in any way the technical problem in its universality; to accomplish this would mean that *all* men adopt the same values and the same behavior.

2. The second real problem posed by the technological society is whether or not a new civilization can appear that is inclusive of Technique. The elements of this question are as difficult as those of the first. It would obviously be vain to deny all the things that can contribute something useful to a new civilization: security, ease of living, social solidarity, shortening of the work week, social security, and so forth. But a civilization in the strictest sense of the term is not brought into being by all these things.

A threefold contradiction resides between civilization and Technique of which we must be aware if we are to approach the problem correctly:

The technical world is the world of material things; it is put together out of material things and with respect to them. When Technique displays any interest in man, it does so by converting him into a material object. The supreme and final authority in the technological society is fact, at once ground and evidence. And when we think of man as he exists in this society it can only be as a being immersed in a universe of objects, machines, and innumerable material things. Technique indeed guarantees him such material happiness as material objects can. But the technical society is not, and cannot be, a genuinely humanist society since it puts in first place not man but material things. It can only act on man by lessening him and putting him in the way of the quantitative. The radical contradiction referred to exists between technical perfection and human development because such perfection is only to be achieved through quantitative development and necessarily aims exclusively at what is measurable. Human excellence, on the contrary, is of the domain of the qualitative and aims at what is not measurable. Space is lacking here to argue the point that spiritual values cannot evolve as a function of material improvement. The transition from the technically quantitative to the humanly qualitative is an impossible one. In our times, technical growth monopolizes all human forces, passions, intelligences, and virtues in such a way that it is in practice nigh impossible to seek and find anywhere any distinctively human excellence. And if this search is impossible, there cannot be any civilization in the proper sense of the term.

Technical growth leads to a growth of power in the sense of technical means incomparably more effective than anything ever before invented, power which has as its object only power, in the widest sense of the word. The possibility of action becomes limitless and absolute. For example, we are confronted for the first time with the possibility of the annihilation of all life on earth, since we have the means to accomplish it. In *every* sphere of action we are faced with just such absolute possibilities. Again, by way of example, governmental techniques, which amalgamate organizational, psychological, and police techniques, tend to lend to government absolute powers. And here I must emphasize a great law which I believe to be essential to the comprehension of the world in which we live, viz., that when power becomes absolute, values disappear. When man is able to accomplish anything at all, there is no value that can be proposed to him; when the means of action are absolute, no goal of action is imaginable. Power eliminates, in proportion to its growth, the boundary between good and evil, between the just and the unjust. We are familiar enough with this phenomenon in totalitarian societies. The distinction between good and evil disappears beginning with the moment that the ground of action (for example, the *raison d'état,* or the instinct of the proletariat) claims to have absolute power and thus to incorporate ipso facto all value. Thus it is that the growth of technical means tending to absolutism forbids the appearance of values, and condemns to sterility our search for the ethical and the spiritual. Again, where Technique has place, there is the implication of the impossibility of the evolution of civilization.

The third and final contradiction is that Technique can never engender freedom. Of course, Technique frees mankind from a whole collection of ancient constraints. It is evident, for example, that it liberates him from the limits imposed on him by time and space; that man, through its agency, is free (or at least tending to become free) from famine, excessive heat and cold, the rhythms of the seasons, and from the gloom of night; that the race is freed from certain social constraints through its commerce with the universe, and from its intellectual limitations through its accumulation of information. But is this what it means really to be free? Other constraints as oppressive and rigorous as the traditional ones are imposed on the human being in today's technological society through the agency of Technique. New limits and technical oppressions have taken the place of the older, natural constraints, and we certainly cannot aver that much has been gained. The problem is deeper—the operation of Technique is the contrary of freedom, an operation of determinism and necessity. Technique is an ensemble of rational and efficient practices; a collection of orders, schemas, and mechanisms. All of this expresses very well a necessary order and a determinate process, but one into which freedom, unorthodoxy, and the sphere of the gratuitous and spontaneous cannot penetrate. All that these last could possibly introduce is discord and disorder. The more technical actions increase in society, the more human autonomy and initiative diminish. The more the human being comes to exist in a world of ever-increasing demands (fortified with technical apparatus possessing its own laws to meet these demands), the more he loses any possibility of free choice and individuality in action. This loss is greatly magnified by Technique's character of self-determination, which makes its appearance among us as a kind of fatality and as a species of perpetually exaggerated necessity. But where freedom is excluded in this way, an authentic civilization has little chance. Confronted in this way by the problem, it is clear to us that no solution can exist, in spite of the writings of all the authors who have concerned themselves with it. They all make an unacceptable premise, viz., rejection of Technique and return to a pretechnical society. One may well regret that some value or other of the past, some social or moral form, has disappeared; but, when one attacks the problem of the technical society, one can scarcely make the serious claim to be able to revive the past, a procedure which, in any case, scarcely seems to have been, globally speaking, much of an improvement over the human situation of today. All we know with certainty is that it was different, that the human being confronted other dangers, errors, difficulties, and temptations. Our duty is to occupy ourselves with the dangers, errors, difficulties, and tempta-

> tions of modern man in the modern world. All regret for the past is vain; every desire to revert to a former social stage is unreal. There is no possibility of turning back, of annulling, or even of arresting technical progress. What is done is done. It is our duty to find our place in our present situation and in no other. Nostalgia has no survival value in the modern world and can only be considered a flight into dreamland.

We shall insist no further on this point. Beyond it, we can divide into two great categories the authors who search for a solution to the problem posed by Technique: The first class is that of those who hold that the problem will solve itself; the second, of those who hold that the problem demands a great effort or even a great modification of the whole man. We shall indicate a number of examples drawn from each class and beg to be excused for choosing to cite principally French authors.

Politicians, scientists, and technicians are to be found in the first class. In general, they consider the problem in a very concrete and practical way. Their general notion seems to be that technical progress resolves all difficulties pari passu with their appearance, and that it contains within itself the solution to everything. The sufficient condition for them, therefore, is that technical progress be not arrested; everything that plagues us today will disappear tomorrow.

The primary example of these people is furnished by the Marxists, for whom technical progress is the solution to the plight of the proletariat and all its miseries, and to the problem posed by the exploitation of man by man in the capitalistic world. Technical progress, which is for Marx the motive force of history, *necessarily* increases the forces of production, and simultaneously produces a progressive conflict between forward moving factors and stationary social factors like the state, law, ideology, and morality, a conflict occasioning the periodic disappearance of the outmoded factors. Specifically, in the world of the present, conflict necessitates the disappearance of the structures of capitalism, which are so constituted as to be completely unable to absorb the economic results of technical progress, and are hence obliged to vanish. When they do vanish, they of necessity make room for a socialist structure of society corresponding perfectly to the sound and normal utilization of Technique. The Marxist solution to the technical problems is therefore an automatic one since the transition to socialism is *in itself* the solution. Everything is *ex hypothesi* resolved in the socialist society, and humankind finds therein its maturation. Technique, integrated into the socialist society "changes sign": from being destructive it becomes constructive; from being a means of human exploitation it becomes humane; the contradiction between the infrastructures and the suprastructures disappears. In other words, all the admittedly difficult problems raised in the modern world belong to the structure of capitalism and not to that of Technique. On the one hand, it *suffices* that social structures become socialist for social problems to disappear; and on the other, society *must necessarily* become socialist by the very movement of Technique. Technique, therefore, carries in itself the response to all the difficulties it raises.

A second example of this kind of solution is given by a certain number of technicians, for example, Frisch. All difficulties, according to Frisch, will inevitably be resolved by the technical growth that will bring the technicians to power. Technique admittedly raises certain conflicts and problems, but their cause is that the human race remains attached to certain political ideologies and moralities and loyal to certain outmoded and antiquated humanists whose sole visible function is to provoke discord of heart and head, thereby preventing men from adapting themselves and from entering resolutely into the path of technical progress. Ergo, men are subject to distortions of life and consciousness which have their origin, not in Technique, but in the conflict between Technique and the false values to which men remain attached. These fake values, decrepit sentiments, and outmoded notions must inevitably be eliminated by the invincible progress of Technique. In particular, in the political domain, the majority of crises arise from the fact that men are still wedded to certain antique political forms and ideas, for example, democracy. All problems will be resolved if power is delivered into the hands of the technicians who alone are capable of directing Technique in its entirety and making of it a positive instrument for human service. This is all the more true in that, thanks to the so-called human techniques (for example, propaganda) they will be in a position to take account of the human factor in the technical context. The technocrats will be able to use the totality of Technique without destroying the human being, but rather by treating him as he should be treated so as to become simultaneously useful and happy. General power accorded to the technicians become technocrats is the only way out for Frisch, since they are the only ones possessing the necessary competence; and, in any case, they are being carried to power by the current of history, the fact which alone offers a quick enough solution to technical problems. It is impossible to rely on the general improvement of the human species, a process which would take too long and would be too chancy. For the generality of men, it is necessary to take into account that Technique establishes an inevitable discipline, which, on the one hand, they must accept, and, on the other, the technocrats will humanize.

The third and last example (it is possible that there are many more) is furnished by the economists, who, in very different ways, affirm the thesis of the automatic solution. Fourastié is a good example of such economists. For him, the first thing to do is to draw up a balance between that which Technique is able to deliver and that which it may destroy. In his eyes there is no real problem: What Technique can bring to man is incomparably superior to that which it threatens. Moreover, if difficulties *do* exist, they are only temporary ones which will be resolved beneficially, as was the case with the similar difficulties of the last century. Nothing decisive is at stake; man is in no mortal danger. The contrary is the case: Technique produces the foundation, infrastructure, and suprastructure which will enable man really to become man. What we have known up to now can only be called the *prehistory* of a human race so overwhelmed by material cares, famine, and danger, that the truly human never had an opportunity to develop into a civilization worthy of the name. Human intellectual, spiritual, and moral life will, according to Fourastié never mature except when life is able to start from a complete satisfaction of its material needs, complete security, including security, from famine and disease. The growth of Technique, therefore, initiates the genuinely human history of the whole man. This new type of human being will clearly be different from what we have hitherto known; but this fact should occasion no complaint or fear. The new type cannot help being superior to the old in every way, *after* all the traditional (and exclusively material) obstacles to his development have vanished. Thus, progress occurs automatically, and the inevitable role of Technique will be that of guaranteeing such material development as allows the intellectual and spiritual maturation of what has been up to now only potentially present in human nature.

The orientation of the other group of doctrines affirms, on the contrary, that man is dangerously imperiled by technical progress; and that human will, personality, and organization must be set again to rights if society is to be able to guard against the imminent danger. Unfortunately, these doctrines share with their opposites the quality of being too optimistic, in that they affirm that their thesis is even feasible and that man is really capable of the rectifications proposed. I will give three very different examples of this, noting that the attitude in question is generally due to philosophers and theologians.

The orientation of Einstein, and the closely related one of Jules Romains, are well known, viz., that the human being must get technical progress back again into his own hands, admitting that the situation is so complicated and the data so overwhelming that only some kind of "superstate" can possibly accomplish the task. A sort of spiritual power integrated into a world government in possession of indisputable moral authority might be able to master the progression of techniques and to direct human evolution. Einstein's suggestion is the convocation of certain philosopher-scientists, whereas Romains's idea is the establishment of a "Supreme Court of Humanity." Both of these bodies would be organs of meditation, of moral quest, before which temporal powers would be forced to bow. (One thinks, in this connection, of the role of the papacy in medieval Christianity vis-à-vis the temporal powers.)

A second example of this kind of orientation is given by Bergson, at the end of his work, *The Two Sources of Morality and Religion*. According to Bergson, initiative can only proceed from humanity, since in Technique there is no "*force des choses*." Technique has conferred disproportionate power on the human being, and a disproportionate extension to his organism. But, "in this disproportionately magnified body, the soul remains what it was, i.e., too small to fill it and too feeble to direct it. Hence the void between the two." Bergson goes on to say that "this enlarged body awaits a supplement of soul, the mechanical demands the mystical," and "that Technique will never render service proportionate to its powers unless humanity, which has bent it earthwards, succeeds by its means in reforming itself and looking heavenwards." This means that humanity has a task to perform, and that man must grow proportionately to his techniques, but that he must *will* it and *force* himself to make the experiment. This experiment is, in Bergson's view, a possibility, and is even favored by that technical growth which allows more material resources to men than ever before. The required "supplement of soul" is therefore of the order of the possible and will suffice for humans to establish mastery over technique. The same position, it may be added, has in great part been picked up by E. Mounier.

A third example is afforded by a whole group of theologians, most of them Roman Catholic. Man, in his actions in the domain of the technical, is but obeying the vocation assigned him by his Creator. Man, in continuing his work of technical creation, is pursuing the work of his creator. Thanks to Technique, this man, who was originally created "insufficient," is becoming "adolescent." He is summoned to new responsibilities in this world which do not transcend his powers since they correspond exactly to what God expects of him. Moreover, it is God Himself who through man is the Creator of Technique, which is something not to be taken in itself but in its relation to its Creator. Under such conditions, it is clear that Technique is neither evil nor fraught with evil consequences. On the contrary, it is good and cannot be dangerous to men. It can only become evil to the extent that man turns from God; it is a danger only if its true nature is misapprehended. All the errors and problems visible in today's world result uniquely from the fact that man no longer recognizes his vocation as God's collaborator. If man ceases to adore the "creature" (i.e., Technique) in order to adore the true God; if he turns Technique to God and to His service, the problems must disappear. All of this is considered the more true in that the world transformed by technical activity *must* become the point of departure and the material support of the new creation which is to come at the end of time.

Finally, it is necessary to represent by itself a doctrine which holds at the present a place of some importance in the Western world, i.e., that of Father Teilhard de Chardin, a man who was simultaneously a theologian and a scientist. His doctrine appears as an intermediate between the two tendencies already sketched. For Chardin, evolution in general, since the origin of the universe, has represented a constant progression. First of all, there was a motion toward a diversification of matter and of beings; then, there supervened a motion toward Unity, i.e., a higher Unity. In the biological world, every step forward has been effected when man has passed from a stage of "dispersion" to a stage of "concentration." At the present, technical human progress and the spontaneous movement of life are in agreement and in mutual continuity. They are evolving together toward a higher degree of organization, and this movement manifests the influence of Spirit. Matter, left to itself, is characterized by a necessary and continuous degradation. But on the contrary, we note that progress, advancement, improvement do exist, and, hence, a power of contradicting the spontaneous movement of matter, a power of creation and progress exists which is the opposite of matter, i.e., it is Spirit. Spirit has contrived Technique as a means of organizing dispersed matter, in order simultaneously to express progress and to combat the degradation of matter. Technique is producing at the same time a prodigious demographic explosion, i.e., a greater density of human population. By all these means it is bringing forth "communion" among men; and likewise creating from inanimate matter a higher and more organized form of matter which is taking part in the ascension of the cosmos toward God. Granting that it is true that every progression in the physical and biological order is brought about by a condensation of the elements of the preceding period, what we are witnessing today, according to Chardin, is a condensation, a concentration of the whole human species. Technique, in producing this, possesses a function of unification *in-*

side humanity, so that humanity becomes able thereby to have access to a sort of unity. Technical progress is therefore synonymous with "socialization," this latter being but the political and economic sign of communion among men, the temporary expression of the "condensation" of the human species into a whole. Technique is the irreversible agent of this condensation; it prepares the new step forward which humanity must make. When men cease to be individual and separate units, and all together form a total and indissoluble communion, then humanity will be a single body. This material concentration is always accompanied by a spiritual, i.e., a maturation of the spirit, the commencement of a new species of life. Thanks to Technique, there is "socialization," the progressive concentration on a planetary scale of disseminated spiritual personalities into a supra-personal unity. This mutation leads to another Man, spiritual and unique, and means that humanity in its ensemble and in its unity, has attained the supreme goal, i.e., its fusion with that glorious Christ who must appear at the end of time. Thus Chardin holds that in technical progress man is "Christified," and that technical evolution tends inevitably to the "edification" of the cosmic Christ.

It is clear that in Chardin's grandiose perspective, the individual problems, difficulties, and mishaps of Technique are negligible. It is likewise clear how Chardin's doctrine lies midway between the two preceding ones: On the one hand, it affirms a natural and involuntary ascension of man, a process inclusive of biology, history, and the like, evolving as a kind of will of God in which Technique has its proper place; and, on the other, it affirms that the evolution in question implies consciousness, and an intense *involvement* on the part of man who is proceeding to socialization and thus *committing* himself to this mutation.

We shall not proceed to a critique of these different theories, but content ourselves with noting that all of them appear to repose on a too superficial view of the technical phenomeon; and that they are *practically* inapplicable because they presuppose a certain number of *necessary* conditions which are not given. None of these theories, therefore, can be deemed satisfactory.

## NOTES

**1** In his book *La Technique,* Jacques Ellul states he is "in substantial agreement" with H. D. Lasswell's definition of technique: "the ensemble of practices by which one uses available resources in order to achieve certain valued ends." Commenting on Lasswell's definition, Ellul says: "In the examples which Lasswell gives, one discovers that he conceives the terms of his definition in an extremely wide manner. He gives a list of values and the corresponding techniques. For example, he indicates as values riches, power, well-being, affection, and so on, with the techniques of government, production, medicine, the family. This notion of value may seem somewhat novel. The expression is manifestly improper. But this indicates that Lasswell gives to techniques their full scope. Besides, he makes it quite clear that it is necessary to bring into the account not only the ways in which one influences things, but also the ways one influences persons." "Technique," as it is used by Ellul, is most nearly equivalent to what we commonly think of as "the technological order" or "the technological society." (Trans.)

**2** Cf. K. Horney.

3 A kind of French beatnik. (Trans.)
4 The psychoanalyst Carl Jung has much to say along this line.
5 An untranslatable French play on words. *Défoulement* is an invented word which presumably expresses the opposite of *refoulement,* i.e., repression. (Trans.)
6 Teilhard de Chardin represents, in his works, the best example of this.
7 Examples of such myths are: ''Happiness,'' ''Progress,'' ''The Golden Age,'' etc.
8 French *sujet* . The usual rendering, ''subject,'' would indicate exactly the contrary of what is meant here, viz., the opposite of ''object.'' The present sense of ''subject'' is that in virtue of which it governs a grammatical object. (Trans.)
9 See, for example, the reports of the International Congress for Political Science, October 1961.

# PART TWO

# FURTHER TECHNOLOGICAL METAPHORS

## INTRODUCTION

Aesthetics is the study of beauty wherever it is encountered: it is therefore bound to consider natural objects and events such as waterfalls and sunsets as well as artificial ones such as paintings, sculptures, musical performances, and novels. The philosophy of art, by contrast and by definition, considers only artificial objects of a particular type, namely, those which someone has claimed to have aesthetic value. Consequently, the philosophy of art, itself a large part of aesthetics, exhibits a thoroughgoing concern with the ways in which technology manifests itself in the materials utilized in the visual, plastic, auditory, and literary arts.

In addition to aesthetics and the philosophy of art, there is art criticism.[1] Among art critics there are those who specialize in what might be called "technological art," whose content, as well as the materials used to bring it to fruition, is patently technological.[2] In some forms of art, however, such as in certain types of video, computer, and laser art, medium and content are almost impossible to separate. In other forms of art, such separation can be made quite easily: a book or a painting may be about technology (something artificial) or about something natural.

The four essays in this section deal with technological aesthetics from very different points of view. At the same time, however, they have something in common which ties them together and relates them to most of the other selections in this book: each author exhibits an intense interest in the manner in which metaphors function in our technological life-worlds.

David Edge has reminded us that a metaphor, by definition, is something that is literally absurd.[3] It is literally absurd, for example, to speak of technology as a runaway railway train or of a person as "letting off steam." In addition, metaphors function in contexts that provide some connection or link between elements: not just any absurdity is a metaphor. Even though metaphors are not images, strictly speaking, many of them nevertheless have strong imagistic content. In one of the examples just given, it is quite easy to imagine a person with steam coming out of her ears. Steam technology offers a fund of relatively common and straightforward images that are easily called up for use.

Metaphors create new perceptions because they are constructed from radical juxtapositions of elements not previously related to one another. This happens in the visual arts as well as in the complex of arts we call linguistic and literary. Max Ernst's famous painting "Murdering Airplane," made shortly after World War I, juxtaposes radically different elements—the cold metal of the aircraft and the soft sensuality of human arms and hands—to create a unique and memorable visual metaphor. Of course, the title of the painting is yet another metaphor—a juxtaposition of words not normally used together but whose linkage works in this context.

Metaphors also create tension. When novel, they disturb the status quo. Once used, metaphors that are effective continue to seize and hold us. One helpful way of thinking about metaphors is to imagine them coming together into complexes, associating and working together to form myths. Further, when those myths function as explanatory devices (for example, the electron wave theory), they form complexes called paradigms.

As Edge suggests, there is a sense in which we do not choose our metaphors, myths, or paradigms; quite often, it is as if they choose us. In his essay "Nuclear Weapons and Everyday Life" in Part Four of this book, for example, Paul Thompson explores some of the ways in which the metaphors of nuclear weaponry have seized us. One way of putting this is that metaphors are "negentropic": they reverse the usual entropic tendency to disorganization and exhaustion of resources by providing a new purchase on old problems.

Edge argues that the metaphors that become the most influential are essentially ambiguous. One of his examples is the patently technological metaphor of the railway: "It concerns the tension between the throbbing, hissing, fiery primal energy of the engine, vigorous and eager to be off the leash, and the exact, geometrical, purposeful discipline of the rails. The powerful imagery generated by this tension gives the metaphor much of its symbolic energy."

He goes on to discuss the ways in which railroad metaphors work in American folk songs. The fireman (who is associated with the power of the engine) is invariably depicted as an undisciplined, lustful drunkard; the engineer (who is associated with the guidance of the train) is invariably a sober, happily married man.

A second aspect of metaphor, according to Edge, is its potential "to alter feelings and attitudes towards oneself and others, and the natural world." He cites as an example the Cartesian metaphor of the body as a machine. Whereas the older Galenist medical metaphors had supported the healing power of nature, depicting the physician as a watchful attendant to natural processes, the new mechanistic metaphors envisioned the role of the scientist in general, and the physician in particular, as interacting with and improving upon nature. A machine, after all, may break down and require intervention for its repair.

But the Cartesian metaphors, in their turn, have had their detractors. Herbert Marcuse, in his essay in Part Six, argues against the kind of "rationality" that treats nature as object and therefore as something that is of value only insofar as it functions as a means to human ends. In his essay in Part Seven, John Dewey finds a place between these extremes, that is, between Cartesian mechanism and Marcuse's Marxist idealism. He argues that nature exhibits ends that must be taken into account if human beings are to continue to live on the earth. At the same time, he contends, neither natural ends nor any of the artificial ones that humankind has devised are beyond human ability to improve by means of cultivated and intelligent interaction with them.

Edge's third point about metaphors is that even though they are at base aesthetic responses to our experience, they function in ways that add authority to moral and social injunctions. He quotes in this regard a remark by Ralph Lapp, that "we are aboard a train which is gathering speed, racing down a track on which there are an unknown number of switches leading to unknown destinations. No single scientist is in the engine cab, and there may be demons at the switch." For those who are seized by this metaphor, and some have suggested that Jacques Ellul is one of them, more authoritarian forms of moral and social control might seem desirable.

A second example to support Edge's point is the ways in which the presidential speechwriters of the Reagan administration capitalized upon the Manichean metaphors of light and dark, good and evil, that are a part of the larger store of Hollywood metaphors and myths. These ancient religious metaphors were given a new high-tech twist. Reagan's famous references to an "evil empire" and his "Star Wars" program of space-based weapons draw on the simplistic, futuristic, high-tech movie scenario in which disaster can be averted only by the imposition of the will of a strong leader upon cultural forces that are otherwise pluralistic, diverse, and (in the leader's view) decadent.

Yet another example of Edge's point is the way in which the complex of metaphors that form the Frankenstein myth works to provide the basis for exhortations to curtail certain types of research in genetic engineering. The popular fear, articulated and exacerbated in the work of Jeremy Rifkin, for example, is that we, like Dr. Frankenstein, will be unable to control what we have created and that we will be destroyed by it. It should be noted in passing that Langdon Winner (whose essay "Technē and Politeia" appears in Part Seven) has done a great service to those who have gotten so caught up in the Hollywood Frankenstein scenario and have forgotten the original story by Mary Wollstonecraft Shelley. In the last chapter of his book *Autonomous Technology,* Winner has reminded us that, far from being malicious, Dr. Frankenstein's creature sought to learn the ways of human beings and to be of service to them. The villain of the original story is not Frankenstein's creature but Frankenstein himself, and this precisely because he abandoned his handiwork, refused to accept his responsibility, and fled from it.

Perhaps no analyst of technology had more to say about its metaphors than did Marshall McLuhan. His books abound with descriptions of the shifts of metaphor that accompany the transitions between the several stages of technological history. He invited us, for example, to consider the "horseless carriage" metaphor used in the early days of the automobile. This is a verbal construction that involves the radical juxtaposition of an assertion and a denial, since carriages up to that time had been horse-drawn. This metaphor, he proposed, is not unlike the term "artificial writing," used at the beginning of the sixteenth century to describe the new process of printing.

McLuhan warned us that new technologies cannibalize old ones for the basic stuff of their metaphors, so that the metaphor is always, as it were, in arrears. He likened our construction and use of technological metaphors to "walking backwards into the future" or to driving a car while looking in the rearview mirror. During its early days, for example, electricity was spoken of in terms of the flow of a river and the discharge of a gun. Susan Sontag has made a similar point with respect to the metaphors of photography. Photography took up the language of firearms: one aims, one shoots, and the product is a "snapshot." But Sontag took these metaphors so literally that she was led to describe photography as a predatory activity, one which constitutes, at its very basis, the violation of another person. Some manufacturers of photographic equipment have been so sensitive to the kind of criticism advanced by Sontag that they have sought to replace inherited firearms metaphors with friendlier ones. An advertisement for Olympus cameras once entreated its readers to buy one for the occasions "when you have more to say than just 'smile.'"

In "Glass Without Feet," John J. McDermott begins an exploration of the aesthetic content of the human technological lifescape that he continues in his essay "The Aesthetic Drama of the Ordinary," in Part Three, and in his essay "Urban Time," in Part Four. The title of his essay is itself a rich metaphor of the ways in which the architects of the urban experience have limited human interaction by constructing our cities and their principal buildings to a scale other than the human one.

Approaching the problems of the cities from the point of view of the American pragmatists William James and John Dewey, McDermott utilizes and enlarges their metaphors of the human body as permeable with respect to its environment. Since coming to consciousness is an affair of perceiving, constructing, and securing relationships to and within one's environment, the flat and impermeable slabs of glass that confront the pedestrian in the modern cityscape are for McDermott the occasion for the atrophying of many of our most fundamental and important responses to our world. He begins the rich detailing of his characterization of the contrasts between nature and city by wondering: "If nature is the environment we inherit and city is the environment we build, what then is the ideal relationship between them? What does nature teach us that we need to maintain if we are to remain human as we make an abode, a human place in which many, many of us live together?"

It might be objected that McDermott's prescription for our cities is altogether too ethereal, that it does not come to terms with the real problems of the city, such as crack houses, shootouts between rival gangs, and massive disruptions of functioning neighborhoods by ill-conceived development. But McDermott has some interesting empirical studies on his side. In *The City: Rediscovering the Center*, the sociologist William H. Whyte contends that some of our most tightly held ideas about cities are fallacious.[4] He argues, for example, that the presence of "street people" is a sign of health wherever they are, since if conditions become unsafe, they inevitably move to another location. Moreover, contrary to the opinions of some urban developers, Whyte has found that sheer walls of concrete and glass at sidewalk level render streets not more but less safe. Even more than laws or police action, the vitality of an active community is what is required to displace crime and prevent decay. A central problem of the cities, both Whyte and McDermott tell us, is the creation and maintenance of spaces in which people can feel at home.

McDermott concludes that cities must be planned to excite and allow development of the many aesthetic and intellectual possibilities open to us as human beings. This means that the structures of the city must be rendered "affectionate, sentimental, warm, and symbolically pregnant." He calls for the revitalization of the inner city as a place for people to reside, to stroll, to eat, and to play. "Let us give the glass towers some feet," he pleads, "so that we may walk about in a genuinely human home, our city."

In his essay "The Cheap and Accessible Print," Richard Powers explores the photograph as a source of technological metaphors. Powers's essay is a chapter from his novel *Three Farmers on Their Way to a Dance,* whose title alludes to a well-known photograph by August Sander. The persuasive power of photographs, he contends, as well as of motion pictures and video, lies not so much in their ability to mimic reality as it does in what he calls "selective accuracy wedded to selective distortion." The photograph thus serves as a rich metaphor for the wider technology in the context of which it has come to exist: it conceals as it reveals, and it is both expository of the real and an invitation to participation within a realm of ideals. The hybrid nature of the photograph also serves as a metaphor for the centaurlike lives of men and women in a technological milieu—half in the world of nature and half in the world of artifacts and institutions which they themselves have created.

Powers also pays homage to the work of Walter Benjamin, an art critic and social philosopher who died in 1940 while fleeing the Nazis. Benjamin's essay "The Work of Art in the Age of Mechanical Reproduction" remains one of the seminal contributions to its field.[5] Powers and Benjamin argue that the photo has been the Model T of the arts. The photo has severed the bond between beauty and scarcity, since anyone can now make and own photos. It has been a truly popularizing and democratizing influence with respect to artistic values. But even more important, because it "slices a cross-section through time, presenting an unchanging porthole on a changed event," the photo enlarges the viewer's understanding of the photographer's act, and it engages him or her to an act of complicity with the photographer. As Benjamin had put it, "The distinction between author and public is about to lose its basic character. The difference becomes merely functional."

Finally, "Autonomous Technology in Literature" provides a sampler of treatments of technology by novelists as diverse as John Updike, Robert Pirsig, E. M. Forster, Robert Penn Warren, and Thomas Pynchon. This essay seeks to test the claim made by Langdon Winner in his book *Autonomous Technology* that if we are to be successful in our attempt to evaluate the claim made by Jacques Ellul and others that technology has become autonomous in our lives, we must do two things: we must pinpoint precisely what we mean by "technology," and we must measure claims of its autonomy against the ways in which metaphors, myths, and paradigms actually operate upon us. Winner suggested that "technology" might mean its software, which is to say its techniques, skills, methods, procedures, and routines; its apparatus, that is, its tools, machines, instruments, appliances, weapons, and gadgets; or its system, that is, its rational-productive social arrangements.

Even though only one of the five novelists considered by this essay set out to write about technology (Robert Pirsig's *Zen and the Art of Motorcycle Maintenance* has as its central theme its protagonist's attempts to come to terms with the demands of quantification and technical efficiency), each in his turn demonstrates the usefulness of Winner's distinction and generates striking and fruitful metaphors to aid our understanding of our technologically conditioned lives.

The subject of technological aesthetics is, of course, not exhausted by the brief treatment it has been given in this section. It is a subject that includes the study of music, video art, advertising, and almost every other area of popular culture. It also treats painting, sculpture, and the aesthetics of the printed word. But these four essays propose categories for approaching its broad and complex subject matter which may be termed "heuristic": they are not intended to be exhaustive and exclusive classifications but instead tools to be used to shape useful interpretations of the most concrete of our experiences.

## NOTES

1 The work of the philosopher of art and the work of the art critic often overlap. It may be helpful to think of the philosopher of art as concerned with the aesthetic, ethical, epistemological, and ontological aspects of works of art, whereas the art critic is more often concerned with the history and cultural content of such works, or even with the personalities involved in their production, distribution, and consumption.

2 An excellent example is K. G. Pontus Hultén, *The Machine as Seen at the End of the Mechanical Age,* Museum of Modern Art, New York, 1968.

3 D. O. Edge, "Technological Metaphor," in D. O. Edge and J. N. Wolfe (eds.), *Meaning and Control,* Tavistock Publications, Ltd., London, 1973, pp. 31–59.

4 William H. Whyte, *The City: Rediscovering the Center,* Doubleday, New York, 1988.

5 Walter Benjamin, *Illuminations,* Hannah Arendt (ed.), Schocken Books, New York, 1969, pp. 217–251.

## Technological Metaphor and Social Control

**D. O. Edge**

In 1969, during the big student demonstrations in Washington, the BBC showed a television news sequence (I think it originated on CBS). The sequence consisted mainly of edited clips of "vox pop" interviews with middle-aged bystanders, who talked about the demonstrations as a healthy "letting off of steam" by the young, a "safety valve to relieve society's pressures," and so on. But the final clip was of one of the student leaders, addressing the crowd: "You've heard them all say we're the safety valve—but I say they're dead wrong. We're the explosion!"

In recent years, there has been a notable increase in interest among philosophers of science in the cognitive functions of metaphor.[1] The process whereby we construe an uncertain, obscure, or puzzling area of experience in terms of one both familiar and apparently (in at least some respects) similar, the displaced pattern acting as a metaphorical redescription of the unfamiliar, has been shown to be central to many key scientific innovations.[2] Technological devices are significant components of our familiar, everyday world: they therefore share in this process, forming the literal basis of metaphors which give implicit, tacit structures to our thought and feeling—that "fill our consciousness." Indeed, many of the most influential of the theories of modern science have an explicit origin in such "technological metaphor."[3] (The pervasive metaphors of cybernetics, reconceptualizing the brain and society in terms of the behavior of computers and other electrical networks, offer one striking example: how often do you hear the term "feedback"?) We would do well to explore the extent and the dynamics of this process by which our imagination comes to be dominated by those very devices which we devise in order to dominate and control our environment and human society.

Writers on the cognitive functions of metaphor tend to stress two aspects of the process. First, the successful metaphor does not merely provide answers to pre-existing questions: rather, by radically restructuring our perception of the situation, *it creates new questions, and, in so doing, largely determines the nature of the answers*. Judith Schlanger, for instance, in discussing the technical origins of metaphors which are central to recent advances in molecular biology, writes that "the concept of cell regulation establishes the field for which it sets the boundaries and is the coordinator."[4] That concept, like the cybernetic metaphor of brain function, derives from the "hardware" of control technology. And it is hardly surprising that, when authors come to discuss the inadequacies of such metaphor, they do so in terms derived from the metaphor itself—talking, for instance, of the brain's impenetrable "black boxes," or of "the mysterious nature of human encoding and decoding."[5] This is not a matter of human weakness. We cannot think about the matter otherwise, because otherwise there is no "matter" to think about. As Schlanger comments, "The cybernetic analogue provokes and instigates its own theoretical elaboration."

Second, there is a realization that, in some important sense, we do not "choose" a successful metaphor—rather, *it chooses us*. We are "seized by it."

The metaphor acts to eliminate *confusion,* to structure chaos: its action involves, as Schon notes, "a transition from helplessness to power." What is more, those on whom the metaphor acts in this way are driven to take it literally. Not for them the detachment implicit in the realization that their activity is essentially metaphorical: they "take it seriously," and proceed to "spell out" their new perception in direct detail, imposing the categories determined by the metaphor. In Douglas Berggren's terms, they have lapsed from metaphor to "myth."

These two features are strongly reminiscent of the action of a "paradigm," as T. S. Kuhn characterizes it.[6] And, just as Kuhn's later work is moving in the direction of a sociological analysis of the basis of scientific activity, so we might profitably inquire as to the social forces which may determine the use of metaphor, predisposing us to assimilate it into our imagination, and to "take it seriously." I have three suggestions to make as to how we might approach this problem.

## I

The first starts from the simple observation that successful metaphors tend to have ambiguous associations. (Religious symbols, of course, are the classic instance.) We've already noted one such metaphor: the "safety valve." It is only activated when the pressure reaches dangerous proportions: but the noisy outrush of steam signifies that the danger is *under control.* The metaphor can sometimes reassure the Silent Majority, and sometimes alarm them! (In this it is like another notorious technological metaphor, the Swinging Pendulum.) These confused associations arise from an ambivalence in our experience of the technology itself. The most deeply rooted form of this metaphor, which equates a human skull with a steam boiler, reflects a critically ambivalent attitude to the violent expenditure of human energy (especially by the young), and to enthusiasm in general. Such behavior seems to be both necessary and dangerous. If the connotations of danger can be attributed to the prevalence of spectacular boiler explosions in the mid-nineteenth century,[7] and hence to the ignorance at that time of the second law of thermodynamics and of basic principles of metallurgy, would this alter our feelings?

There is a related ambivalence at the heart of the railway metaphor, which can account for much exploration of it in modern literature.[8] It concerns the tension between the throbbing, hissing, fiery primal energy of the engine, vigorous and eager to be off the leash, and the exact, geometrical, purposeful discipline of the rails. The powerful imagery generated by this tension gives the metaphor much of its symbolic energy. (It has some interesting corollaries: in American folk songs, for instance, the fireman is usually depicted as a randy, undisciplined, hedonistic, drunken character, with a girl at every stopover, while the engineer is "straight," happily married and sober. One stokes the fire, while the other keeps to the rails!) One recalls the uneasy truce that Lovejoy has called "the ethic of the middle link"—the view of the human predicament so widespread and influential throughout the late eighteenth century, which saw men as forced to accept "an unsatisfactory but nevertheless unavoidable compromise between their animal nature and their rational ideals." The railway steam engine symbolizes this "middle link ethic'' to perfection. Life is, indeed, like a railroad, and man like a railway engine: in the face of this competition, buses haven't a hope. What technology has provided us with, then, one might say, in this relatively brief cultural interlude, is merely a fresh way of sharpening our apprehension of both poles of this essential tension. Alternatively, you could argue that this tension, if it really exists, is not all that important. The visions and ideals of the Enlightenment thinkers may have been delusory, and related only to their own, unique social experience; the problem they posed is of no general consequence. In focusing our attention on this tension, and in acting as the vehicle by which an eighteenth-century dilemma has been vividly transported into the twentieth century, technology may have led us all off the rails.

The use of railway metaphor within the emerging class structure of America illustrates another ambiguity.[9] To the middle class, and putative middle class, the railway symbolized the endless possibilities of exploitation, the journey of adventure to those far frontiers of rich potentiality, where they would find their rewards. It had a romantic quality, positive associations. To the underprivileged, as those familiar with traditional Negro blues know, the metaphor had negative associations. The railway station was a place of separation from your lover; the railway recurs as a symbol of desolation and despair; it appears as an impersonal fate, inexorably tearing apart the securities and transitory comforts of personal existence; if it offers any hope, it is the rather attenuated one of escape from a living hell on earth to somewhere (Chicago?) where life might be slightly more tolerable. Only in the Gospel song does it offer, metaphorically displaced, the Endless Frontier of an escape to Jesus.

Two widely differing class (and racial) groups, two radically different perceptions—but each supported by the ambiguous connotations of one central metaphor. And what is true of the particular case of the railway is true, a fortiori, of the general manifestation of the paradigm technological metaphor—Society as Machine. To those in power, it expresses liberation and hope: to those exploited, repression and despair. The ambivalence lends life to the metaphor, even today. It is one that a wide spectrum of class interests can still take seriously.

And this centrality of an ambiguous metaphor can be supported (often inappropriately) by *conflict*. To take a related example: there is a lively debate these days in Britain (and I gather elsewhere, too) over the content of the mass media. Those in control of the media justify their policies in grand, liberal terms—they are enlightening, educating, fostering humane values, and so on: and a vigorous pressure group maintains a continuous crossfire—the media are corrupting, a major cause of increasing crime and violence, morally degenerating, and so forth. Both groups have a vested interest in believing that the media are *influential*. In the midst of this debate, the view (supported by the bulk of empirical research) that the media have relatively little effect on people's beliefs, values, attitudes and behavior doesn't get much air time! And when it does surface—as in the recent United States Presidential Commission Report on Pornography—it is found to be politically embarrassing, and has to be shouted down.

Central to this debate, and lurking within the Society/Machine metaphor, is the notion that the action of society is centrally controlled, coherent, and manipulative. Controversy can keep this alive, too. Those in power like to think that the course of events has been coherent, and guided by humane, rational policy, and they justify their actions in these terms. Their critics also make the "coherence" assumption, but see the policies of those in power as mistaken, foolish, or actively evil.[10] In such a climate, the *lack* of control in society is not acknowledged. Inappropriate images and metaphors can be perpetuated.

## II

The second approach to the social analysis of metaphor recognizes the potentiality of metaphor to alter feelings and attitudes towards oneself and others, and the natural world. Here, for instance, is Owsei Temkin's account of the development of the metaphor of the body as a machine:

> ...Descartes' metaphor of the body machine proved most fruitful in many respects. In the first place, it made room for a more active attitude toward the body. Galen had imagined the human organism to be so perfectly constructed that an improvement was not even thinkable. Besides, nature was constantly at work to protect and cure.... But a machine has only a certain number of regulations, which in many cases may prove insufficient to restore the damage. One of the consequences of the Cartesian concept... was a difference in the evaluation of the healing power of nature and of medical interference. The Galenists upheld the healing power of nature whereas many Cartesians tended to stress its limitations. Boyle, for instance, who followed Descartes in the metaphor of the human machine, argued elaborately that many natural reactions in disease were not beneficial but harmful and that the physician, therefore, had to combat rather than encourage them—this, in spite of Boyle's belief that the human body had been fashioned by God with infinite wisdom. Once this belief weakened it could be asked whether the body was a good or a bad machine. Thus, Helmholtz, in considering the eye as an optical instrument, found it so full of defects that he for one would have felt justified in returning it to the optician who had dared to sell it to him. And perhaps it is not by chance that the period of the nineteenth century which made the most fruitful applications of the metaphor of the body machine also became interested in "dysteleology." By this theory Haeckel designated organs which were useless, and dysteleology found its practical culmination in the removal of the healthy appendix as an altogether useless and dangerous part. In the days of Galen this would have been rank heresy.[11]

This is an account of the slow, steady working out of the affective implications of a metaphor. As the metaphor begins to "bite," cognitively, it brings with it attitudes appropriate to its literal referent. A similar account could be given of the development of the metaphor of the Universe as Machine (specifically, the Universe as Clock, and its attendant Deist theology of God as Clockmaker). And both these accounts run in parallel with the socioeconomic history of Western Europe, the development of technological means to harness energy efficiently and to increase agricultural production, and the associated growth in population and wealth. As Lynn White has argued, in several controversial essays,[12] this development involved a radical change in attitudes towards nature. Cooperation gave way to exploitation. He cites the diffusion of the heavy plough in North West Europe as one technological innovation which made such dominance possible, and traces the shift in attitudes via, for instance, a succession of calendar motifs—which steadily abandon impersonal, natural symbols for the seasons, and substitute pictures of human activities—ploughing, threshing, and so forth. As White puts it: "Man and Nature are now two things, and Man is the master." The idea of the Universe as Clock offers a pervasive metaphor, with impeccable cognitive credentials, within which such attitudes and feelings come to be seen as "perfectly natural," and are hence given social validation and support. The ambiguity of the symbolism and myths of the Judeo-Christian religious tradition allows an appropriate reinterpretation of the sacred texts.

We are, of course, now talking about religious perspectives. James Fernandez has recently offered an analysis, in similar terms, of certain tribal systems of religious symbols and rituals.[13] He talks of a "metaphoric strategy," by which we can alter our position in "quality space." Metaphoric strategies "involve the placing of self and other pronouns on continua." The religious believer, for instance, enters a ritual in a depressed relationship to his social and physical environment. The ritual starts by metaphorically equating him, say, with a worm, with its earthbound, submissive connotations. Thus identified, the ritual then transposes him metaphorically to, say, an eagle—dominant, free. Fernandez comments: "We need to become objects to ourselves, and others need to become objects to us as well....The shift in feeling tone—of adornment and disparagement—may be the dominant impulse to metaphor....People undertake religious experiences because they desire to change the way they feel about themselves and the world in which they live....We come to understand these operations only if we study metaphoric predications upon pronouns as they appear in persuasion and performance. The strategy of emotional movement in religion lies in them."

## III

The third approach to the social basis for the diffusion of metaphor includes aspects of the first two, and is, to my mind, the most powerful and suggestive. It emphasizes the role of metaphor (and of institutionalized forms of knowledge in general) in establishing and reinforcing moral and social control.

Charles Rosenberg has recently illustrated the way in which scientific and technological metaphors can be used to add authority to moral injunctions.[14] He discusses the popularity, in America in the late nineteenth century, of the metaphor of the human nervous system as an electrical (telegraph) network, around which flows a (limited) amount of a fluidlike "nervous energy." This metaphor, when allied with thermodynamic notions (notably the conservation of energy), produced a potent source of moral homilies on the virtues of moderation, adding the force of "science" to the controlling moral notions of the time. It was popular precisely because it did so. I would guess that much of the popularity of the Skull/Boiler metaphor, and of the Energy/Restraint tension exhibited by the railway engine, could be analyzed in similar terms. (Rosenberg also shows how the electrical metaphor for the nervous system "helped to express the ambivalence of many Americans toward progress, toward urbanization, toward the treacherous fluidity of American life"—and, again, the analysis could serve equally well for aspects of the railway metaphor.)

Metaphors of this kind, with similar moral implications, are, of course, still with us. Konrad Lorenz, for instance, writes in his classic work, *On Aggression:* "I believe that present-day civilized man suffers from insufficient discharge of this aggressive drive." And Lorenz has been vigorously attacked by many of his scientific colleagues for just this kind of "metaphorical sloppiness," which encourages the reader to draw out moral and social implications where they do not "properly" exist, and to call on Lorenz's authority to support "controversial" political and ethical positions.[15] But, in the perspective I am suggesting, the authority of the scientist, in society at large, rests on what he says: if people find it congenial, he will be believed; if not, his authority vanishes. Of course, attempts to reverse this trend, and to derive a new ethic from science, are not new[16]—attempts, that is, to take the scientific community as an overriding authority, and to wield that authority in order to "persuade" people to alter their moral beliefs and behavior—but the verdict on all such attempts seems to echo Stephen Toulmin's verdict on Evolutionary Ethics: "The support given by Evolution to ethics serves as a source of confidence in our moral ideas, rather than as an intellectual justification for them." (Still less, of course, as a reason to *change* them!)

The most comprehensive statements of the position I am describing here can be found in the tradition in social anthropology that stems from the work of Emile Durkheim.[17] Mary Douglas' article, "Environments at Risk," is particularly apposite. She discusses the problem faced by those ecologists associated with the environmental conservation movement in "being believed." She notes that they portray the Earth as a unified system—the Eco-System. (In our terms, this is to say that they are propounding a fresh metaphor for the world, and man's place within it—with all the appropriate cognitive and affective implications.) They argue that the integrity of that system—and ultimately the fate of the human race—is at risk, due to the activities of the human race itself—or, at least, of identifiable portions of it—the exploiters and polluters. This picture, of course, is presented by these scientists with clear moral intent: the object is to persuade us to modify our desires and values, and to act so as to *restrain* our inclinations towards material advancement. But, as Mary Douglas argues, drawing on a range of anthropological material, "no one can impose a moral view of nature on another person who does not share the same moral assumptions." So, to put it crudely, if the moral consensus is towards restraint, then these ecologists are likely to be listened to, and their advice acted on; but if there is no such consensus, they will not be generally believed, and their metaphorical world view will not gain any wide popularity.

In her recent book, *Natural Symbols,* Mary Douglas elaborates the notion that "the view of the universe, and a particular kind of society holding this view, are closely interdependent. They are a single system. Neither can exist without the other." She suggests a typology with which to classify four relatively distinct kinds of social experience, and, with a wealth of anthropological detail, she shows that these can be plausibly related to four distinct kinds of cosmological belief. The details need not concern us here, except to note two points. First, that modern Western society contains people whose social experience (in Mary Douglas' terms) are of distinct kinds. (She is herself centrally concerned with the plight of the orthodox, Friday-fasting, working-class "Bog Irish" immigrants in the Catholic Church in London, ministered upon by articulate, mobile, middle-class, symbol-blind, and uncomprehending priests—and on the rationally unbridgeable cosmological chasm between them.) Second, that the related cosmologies differ markedly in their metaphysical and moral coherence (and hence in the extent to which they lend themselves to a unified system/organism/mechanism metaphor), and also in the degree to which the Universe, and the powers (if any) which control it, are seen as benign, or neutral, or malevolent. This scheme, when combined with the other aspects I have mentioned, seems to me to offer a powerful potential source of illumination on both the optimistic, "rational" assurance of those who sponsor and manage the modern technological enterprise, and also on the pessimistic, "irrational" passion of those whose social experience and cosmology force them to view that enterprise as more impersonal than humane, more sinister than benign.

## IV

One final, rather gloomy, meditation. Mary Douglas writes: "Credibility depends so much on the consensus of a moral community that it is hardly exaggerated to say that a given community lays on for itself the sum of the physical conditions which it experiences."[18] Technology now has a central role in determining "the sum of the physical conditions which our society experiences," and it is itself expressly designed and promulgated in order to accommodate consensus views of how things should rightly proceed. If it lends itself to ambiguous symbolic reference, carries affective implications, reflects moral patterns, and evokes cosmological echoes, it will therefore tend to generate its own metaphorical force. And that force—embracing, as it does, critical perceptions of ourselves, our society, and our environment—will be essentially conservative.

With that thought in mind, consider the kind of cartoon image with which we are all now familiar, in the pages of *The Ecologist* and elsewhere: "Spaceship Earth" is plunging through smoke and grime, its structure slowly disintegrating and no one at the controls, but carrying a complacent complement of passengers. Or this remark by Ralph Lapp, in his book *The New Priesthood:* "We are aboard a train which is gathering speed, racing down a track on which there are an unknown number of switches leading to unknown destinations. No single scientist is in the engine cab, and there may be demons at the switch." The sense of "crisis" is conceived of in terms of a technology which enshrines an extreme form of centralized control.[19] Society (and the world) is "out of control" like a crashing, disintegrating aircraft without a pilot. And this brings with it the corollary that order, sense, purpose will be restored by mending the plane, and by putting in a competent, well-meaning pilot (or, as Lapp would have it, "a scientist in the engine cab"). One common reaction to our present social problems, such as environmental pollution, is for people to say that it demonstrates that our existing centralized institutions of political control are defective, and then to proceed to attempt to shore up and strengthen those centralized "controls." Technological metaphor tends to confirm this conservative reaction, since we devise technologies specifically to "fit," to serve, and to extend our preconceived notions of "control," as we have institutionalized them. The very form of our technology necessarily spells out this assumption, refining it, in sophisticated forms, to a high art. It may be that the dialogue between the conservation movement and the government agencies is unwittingly perpetuating inappropriate metaphors.[20]

A common view of the relationship of man and technology is contained in this remark by Jurgen Schmandt: "Technology is seen as an all-consuming monster which has so enslaved man that the normal relationship between man and his tools—that of master and servant—has been turned upside down." This notion of the "normal relationship" seems closely tied to metaphors deriving from simple "tool users," such as carpenters. A carpenter uses tools selectively for a preconceived purpose; the tools lie passively until used; their use affects neither the carpenter nor his preconceived goal. Here, indeed, is a "master/servant" metaphor: the tools are "commanded." When such a metaphor fails to do justice to the situation, the only possible modification is to stand it on its head (in electronic parlance, it defines a "flip-flop," with only two stable states); hence the power of the Sorcerer's Apprentice myth. But how "normal" is this kind of tool using? Perhaps the experience of the artist is (or could be) more typical: as he struggles with his materials, he is forced to reconceive his goals (and himself); he neither dominates, nor is dominated by, his materials. The model of artistic creation, with its dialectic symbiosis, seems more congruent with the data of cultural anthropology, the experience of great scientists—and, indeed, with the interaction of men with computers. The popularity of the simpler, "cleaner," more authoritarian and dualist alternative may reflect some of modern society's hidden structure, and its deeper fears. But this is no reason for taking it as a model with which to shape the way in which we attempt to "control technology." Rather, it can be seen as a challenge so to change society that people will be predisposed to adopt more humane and creative (and less fear-ridden) alternatives. Technology itself, by epitomizing other styles of "control," can provide bases for such alternative metaphors: it can also change society in such a way as to make those alternatives seem "appropriate."

Mary Douglas senses parallel possibilities and dangers. In *Purity and Danger,* she argues that all pollution taboos in primitive societies can be interpreted as ways of preventing the fundamental schemes of classification in those societies from being infringed: the taboos "keep the categories pure," and protect them from danger. Ritual cleansing, and the expulsion of polluters, represent the essentially conservative reenactment and reinforcement of traditional categories and principles. In "Environments at Risk," she cites the example of the Eskimo girl who, by persistently eating caribou meat after winter had begun, broke a fundamental taboo of her tribe. She was, by unanimous decision, banished, to freeze to death. Thinking of the conservation movement, Mary Douglas comments: "Are we going to react, as doom draws near, with rigid applications of the principles out of which our intellectual system has been spun? It is horribly likely that, along with the Eskimo, we will concentrate on eliminating and controlling the polluters. But for us there is the recourse of thinking afresh about our environment in a way which was not possible for the Eskimo level of scientific advance. Nor is it only scientific advance which lies in our grasp. We have the chance of understanding our own behavior."

That "our own behavior" must include the metaphorical activity I have been describing, both within science and in society at large, is an appropriate point with which to close. I have suggested that the essential ambiguity of successful metaphors, and their ability to express and develop fundamental feelings and attitudes, conspire to consolidate our tendency so to conceptualize the world that its supposed categories impose upon us the order and control which we seem unable to achieve by self-conscious activity. But if the action of metaphor, like that of the technology from which it so often derives, can seem, at times, like a tyranny, it can also be a source of freedom. For what is freedom, if not the ability to choose by which metaphors we will be seized?[21]

## NOTES

1 See, for instance, Max Black, *Models and Metaphors* (Ithaca, 1962); D. A. Schon, *Displacement of Concepts* (London, 1963), rpt. as *Invention and the Evolution of Ideas* (London, 1967); D. Berggren, "The Use and Abuse of Metaphor," *Review of Metaphysics,* 16 (1962–63), 237–58, 450–72; Berggren, "From Myth to Metaphor," *Monist* 50 (1966), 530–32; Mary Hesse, "The Explanatory Function of Metaphor," *Logic, Methodology, and Philosophy of Science,* ed. Y. Bar-Hillel (Dordrecht, Holland, 1965), pp. 249–59; Hesse, *Models and Analogies in Science* (New York, 1966). The latter contains a useful bibliography. For an extensive, annotated bibliography on metaphor, see Warren A. Shibles, *Metaphor* (Whitewater, Wis., 1971).

2 See, for instance, Owsei Temkin, "Metaphors of Human Biology," *Science and Civilization,* ed. R. C. Stauffer (Madison, Wis., 1949); Eduard Faber, "Chemical Discoveries by Means of Analogies," *Isis,* 41 (1958), 20–26; Harvey Nash, "The Role of Metaphor in Psychological Theory," *Behavioral Science,* 8 (1963), 336–45; Robert M. Young, "Darwin's Metaphor: Does Nature Select?" *Monist,* 55 (1971), 442–503. An interesting early reference in D. Fraser Harris, "The Metaphor in Science," *Science,* 36 (1912), 263–69. Also relevant are D. C. Bloor, "The Dialectics of Metaphor," *Inquiry,* 14 (1971), 430–44; Bloor, "Are Philosophers Averse to Science?" *Meaning and Control,* ed. D. O. Edge and J. N. Wolfe (London, 1973), pp. 1–30; Martin Landau, "On the Use of Metaphor in Political Analysis," *Social Research,* 28 (1961), 331–53. Schon, *Displacement,* discusses both scientific and technological innovation.

3 See esp. Karl W. Deutsch, "Mechanism, Organism, and Society: Some Models in Natural and Social Science," *Philosophy of Science,* 18·(1951), 230–52. This paper is particularly interesting since, after an elegant and concise introduction on the historical role of technological metaphor in the development of scientific theory. Deutsch proceeds to expound another (the cybernetic metaphor). See also my own "Technological Metaphor," *Meaning and Control,* pp. 31–59. Other examples are discussed in the works cited above, n. 2.

4 Judith Schlanger, "Metaphor and Invention," *Diogenes,* 69 (1970), 21.

5 See, for instance, C. C. Anderson, "The Latest Metaphor in Psychology," *Dalhousie Review,* 38 (1958), 176–88. Deutsch's paper closes with a short discussion (pp. 250–52)

in which he analyzes the differences between his metaphorical referents in terms of the metaphor itself.

**6** In *The Structure of Scientific Revolutions* (Chicago, 1970). See also his "Reflections on My Critics," *Criticism and the Growth of Knowledge*, ed. I. Lakatos and A. Musgrave (Cambridge, 1970), and "Second Thoughts on Paradigms," *The Structure of Scientific Theories*, ed. F. Suppe (Urbana, Ill., 1972).

**7** See B. Sinclair, *Early Research at the Franklin Institute: The Investigation into the Causes of Steam Boiler Explosions: 1830–1837* (Philadelphia, 1966); J. G. Burke, "Bursting Boilers and Federal Power," *Technology and Culture*, 7 (1966), 1–23; and review of Burke by Sinclair, *Technology and Culture*, 9 (1968), 230–32.

**8** See T. R. West, *Flesh of Steel* (Nashville, Tenn., 1967); H. L. Sussman, *Victorians and the Machine* (Cambridge, Mass., 1968). The classic and historic commentary on the debilitating effects of the spread of the machine metaphor is, of course, Thomas Carlyle's essay, "Signs of the Times" (1829); I am grateful to David Bloor for drawing my attention to this work.

**9** Leo Marx, *The Machine in the Garden: Technology and the Pastoral Ideal in America* (New York, 1964), is relevant here.

**10** Alasdair MacIntyre accuses Herbert Marcuse of inappropriately perpetuating just such a notion in this way, in *Marcuse* (London, 1970), esp. pp. 71–72. MacIntyre uses the example of the debate over the Vietnam war.

**11** Temkin, "Metaphors of Human Biology," 180–82, n. 2.

**12** Notably, *Medieval Technology and Social Change* (New York, 1962) and *Machina ex Deo* (Cambridge, Mass., 1968). See also Richard A. Underwood, "Toward a Poetics of Ecology: A Science in Search of Radical Metaphors," *Ecology, Crisis and New Vision*, ed. Richard E. Sherrell (Richmond, Va., 1971).

**13** "Persuasions and Performances: Of the Beast in Everybody ... and the Metaphors of Everyman," *Daedalus* (Winter 1972), 39–60. See also Robert P. Armstrong, *The Affecting Presence* (Urbana, Ill., 1971).

**14** "Science and American Social Thought," *Science and Society in the United States*, ed. D. Van Tassel and M. G. Hall (Homewood, Ill., 1966), pp. 137–84. An excerpt can be found in *Sociology of Science*, ed. Barry Barnes (London, 1972), pp. 292–305.

**15** For a particularly vigorous attack, on these lines, on Lorenz's book, see the review by S. A. Barnett, *Scientific American*, Feb. 1967, pp. 135–38. Some similar ideological touches can be found in Deutsch, "Mechanism."

**16** The attempts to derive an ethic from evolutionary theory are particularly well documented and scrutinized. See A. G. N. Flew, *Evolutionary Ethics* (London, 1967); Stephen Toulmin, "Contemporary Scientific Mythology," *Metaphysical Beliefs*, ed. A. MacIntyre (London, 1970); A. Quinton, "Ethics and the Theory of Evolution," *Biology and Personality*, ed. I. Ramsay (Oxford, 1965).

**17** In Britain, two major exponents are Mary Douglas and Basil Bernstein. For Douglas, see her *Purity and Danger* (London, 1966) and *Natural Symbols* (London, 1970) and "Environments at Risk," *Times Literary Supplement*, 30 Oct. 1970, 1273–75, rpt. *Ecology, the Shaping Enquiry*, ed. J. Benthall (London, 1972). For Bernstein, see first his essay in *Knowledge and Control*, ed. Michael F. D. Young (London, 1971), and other bibliographies in that volume. See also S. B. Barnes, "On the Reception of Scientific Beliefs," *Sociology of Science*, ed. Barry Barnes, pp. 269–91.

**18** Douglas, "Environments," 1274.

**19** There is a fairly extensive psychoanalytical literature on this topic. See, particularly, Robert W. Daly, "The Specters of Technicism," *Psychiatry*, 33 (1970), 417–31; and Bruno Bettelheim, "Joey: A 'Mechanical Boy,'" *Scientific American*, Mar. 1959, pp.

116–27. This literature is, I find, somewhat unhelpful in discussing the topic with those involved, since it tends to encourage people to "distance" themselves from the issue, as something that "only affects the abnormal." For most people, the reaction to Daly's remark that "given contemporary symbols of power, efficacy, and heroic human action, can one wonder that troubled persons employ these symbols of power in their neurotic constructions" is an immediate "Not me, Lord! I'm not troubled or neurotic!"

**20** In the same way, and for similar reasons, as the debate over the mass media. See also MacIntyre's remarks on Marcuse.

**21** This paper was prepared during tenure of a joint appointment as Senior Fellow of the Society for the Humanities and Senior Research Associate of the Science, Technology and Society Program at Cornell University. I am indebted to my colleagues in the Science Studies Unit, Edinburgh University, for introducing me to many of the ideas expressed, and many of the sources cited, in this paper.

## Glass Without Feet

**John J. McDermott**

Some time last year I made my way to the area adjoining the Galleria in the imposing new outer city of Houston. Entering one of the more formidable glass towers, so as to experience its inside, I was stopped by a security guard, who asked me if I had an appointment with anyone in the building. I replied that my appointment was with the building, in an effort to see if it had a soul and was open to the presence of personal space. He tossed me out. I went looking for a newspaper, a sandwich, a personal opening to these jutting edifices of technological supremacy. No luck! No kiosk! No paper! No sandwich! No body! No place! Just a marvel of impersonal use of space. Consequently, no person-cityspace.

### COMING TO CONSCIOUSNESS

*Life is coexistent and coextensive with the external natural environment in which the body is submerged. The body's dependence upon this external environment is absolute—in the fullest sense of the word, uterine.*[1]

James Marston Fitch

The human body is neither a container nor a box in a world of boxes. To the contrary, our bodies are present in the world as diaphanous and permeable. The world, in its activity as the affairs of nature and the affairs of things, penetrates us by flooding our consciousness, our skin and our liver with the press of the environment. We, in turn, respond with our marvelous capacity to arrange, relate, reject and above all, symbolize these transactions. In effect, we as humans, and only we, give the world its meaning. Even as embryos in the comparatively closed environment of our mother's womb, we are open to the experience of the other. The quality of the "other" and how we transact with the "other" constitutes our very being as human. Let us diagnose some of these transactions in general terms and then specify our relationship to the world as natural and to the world as artifact, human-made, that is, as urban.

For the human organism, coming to full consciousness is an exquisitely subtle phenomenon. We are always enveloped by the climate known in America as the "weather." We weather the world and the world weathers us. Except for the occasional outrageous performances of nature, such as typhoons, hurricanes, earthquakes and tornadoes, our experience of the weather is largely inchoate, that is subliminally conscious. Unfortunately, inchoate also is our experience of most things, doings, happenings, events and even creatures. I say that this is unfortunate because our tendency to categorize, name and box the affairs of the world, signifies a drastic decline from the alert consciousness as to environment, which characterized our young childhood.

The most telling way to realize how little we actually experience that which takes place within the range of our bodies, is to focus on the very young child. For children, the world comes as philosopher William James suggests, as a buzzing and blooming continuity. The nefarious dualisms between color and shape, past and present, even between organic and inorganic are foreign to the child, who has an extraordinary capacity to experience relations, that webbing which holds the multiple messages of the environment in an organic whole.[2] Despite the ongoing richness of the child's experience of the world, adults quickly strike back, for they have long been conditioned to see the world as disparate, each thing and event complete unto itself, with an appropriate name tag. The children soon yield and like the rest of us, sink into the oblivion of the obvious. The poets rise above this sameness of description, but their special place only reinforces the empty character of how we feel and describe the world in which we find ourselves.

Allow me to offer a way out of this Platonic cave, in which we so often languish, cut off from the teeming richness so potentially present in our experience. Returning to the uterine analogy, I prefer us to experience the world as if we were in a permeable sheath rather than trapped in a linguistic condom. If we listen to the wisdom of the Greek and Roman stoic philosophers, then we know that the world appeals to us from the very depths of being. The *logos,* the word, the utterance is meant for us, if we but listen, feel, touch, see and allow ourselves to be open to the novelty, originality and freshness that is endemic to experience. Openness to experience is crucial, by which I mean that we allow our perceptions a full run before we slap on the defining, naming, locating tags of place, thing and object. Coming to consciousness, in this way, allows us to be bathed by a myriad of impressions, each of them too rich for the task of organization, a task which should follow far behind, gathering only the husks, the leftover from what we have undergone in a personal and symbolic way.

In the above scenario, all experience counts and counts to the end of our days. To wit: the salamander who lives on my porch and dazzles me with its change of skin color; the forgotten rake which leaps up in the melting of an April snow; those lovely Georgian doors that front the row houses in Dublin city; Brahman bulls with human visage; the Irish setter on its haunches, nose poised to spring in response to the advance of an impinging world; the great Gretzky streaking the ice and the soaring of the marvelous Dr. J.; or, more profoundly and demanding, the crippled seventy-year-old mother who upon learning of the impending death of her incontinent, cystic fibrosis, totally retarded forty-eight-year-old daughter, cries out in anguish, not my baby, not my baby!

Boredom and ennui are signs of a living death. In that the only time we have is the time we have, they are inexcusable faults. Sadness and even alienation, given our personal and collective miseries, are proper responses, especially as attenuated, as they can be by occasional joy and celebration. Coming to consciousness has nothing to do with the traditional pursuit of happiness, an attempt illusory and self-deceiving for a human organism whose denouement is the inevitability of death without redemption. Rather, the signal importance of how we come to consciousness is precisely that such a process is all we are, all we have and all we shall ever be.

The world appears to us as given, yet we knead it to our own image, for better and for worse. As we slowly awake to our environment, to our cosmic and planetary womb, we soon find that the more local arrangements are the ones which formulate how we come to know and feel ourselves. In American culture, one of the most prepossessing of these local contexts for consciousness is the dialectic between nature, on the one hand, and things, artifacts, especially in their supreme formulation, cities, on the other hand. Two hands, but one consciousness. Two hands, but one nation. For Americans, the experience of nature and city has been both actual and mythic, depending on the site in which a person came to consciousness. The difference between the experience of nature as primary and the experience of city as primary is not one between good and bad or positive and negative. It is, however a difference which has important ramifications for our evaluation of the kind of environment we should seek to build and for a host of allied personal concerns, values, preferences and attitudes. If nature is the environment we inherit and city is the environment we build, what then is the ideal relationship between them? What does nature teach us that we need to maintain if we are to remain human as we make an abode, a human place in which many, many of us live together? Further, what does the history of cities teach us about what we have discovered to be distinctively human and propitious in that which we have made for ourselves, undreamt by nature or by the animals or by the insects?

### Nature and City

As in all such general terms of description, nature and city[3] are herein used with a specific cultural context in mind, that of the America of North America. We have before our consciousness, Boston and the Berkshire Mountains; Chicago and Lake Michigan; Denver and the Rocky Mountains; San Francisco and its Bay; Miami and the Everglades; Missoula and the Lolo pass; Houston and West Texas; the ocean cities; the river cities, the cities of the plains; and, of course, those dots of nostalgia from the storied and event-saturated past as we traversed the land, for example, North Zulch, Snook and Old Dime Box as representative of Texas fables.

In American terms, nature was always writ large. The history of the meaning of nature for America is as old as first settlements and as central to the meaning of America as is our vaunted political history. The stark generalization is that nature took the place to the Bible as the script by which Americans searched for the signs of conversion. How shall we know when we have been saved, was the plaintive and genuine question most prominent on the lips and in the heart of the early American puritans. Faced, as they were, however, with an environment both deeply forbidding and foreboding, as well as lushly promising, they soon began to equate their justification with their ability to bring nature to its knees and celebrate Zion in the wilderness. Witness this revealing text in 1697, from the "Phaenomena" of Samuel Sewall:

> As long as nature shall not grow old and dote; but shall constantly remember to give the rows of Indian Corn their education, by Pairs: So long shall Christians be born there; and being first made meet, shall from thence be Translated, to be made partakers of the Inheritance of the Saints in Light.[4]

As the American eighteenth century opened, a revealing statistic comes to light. The fixed meetinghouses of the Congregationalists and Presbyterians were slowly but inevitably being replaced by the itinerant preaching of the Baptists and the Methodists. America was on the move! Nature was calling; the West was calling. First, the Berkshires, then the Alleghenies, the Appalachians, western Connecticut, western Virginia, Kentucky, the Northwest Territory, the Mississippi, the Pony Express to California, the Mormon Trek, Texas and as late as 1907, the emergence of Oklahoma. This two-hundred-year history of frontier conflict, frontier conquest and frontier failure became the central theme of American consciousness; lived in the "West" in nature, and vicariously lived in the "East" in the cities, which always "trailed" behind, as redoubts, points of departure. It is essential to realize that the public story of the American odyssey was played out in the clutches of nature, at war with nature's children, the Amerindians, and at war, internally, on the land, about the land, for the control of the land. To the West, they wend. To the West, is to be free.

To the West meant, simply, to be free. To the West meant, simply, to leave the city, wherever it might be, and chase the setting of the sun. This is why California, despite its protracted adolescence, symbolizes the end of continental America.[5] And this also is why Texas retains its mythic value, for Texas is endless, large, vast and still provides the scape for disappearance into the land and the chance to start over. Further, this is why Americans are chary of cities, often seeing them as vulgar necessities and as interruptions in the natural flow of person in nature. The public voice of America tells us that cities are places to go, to sin, to buy, to sell, in short, to visit. The utter unreality of this judgment, given that ninety percent of all Americans live on but ten percent of the land, points to the inordinate power of our long-standing romance with nature, especially as it is biblically sanctioned, and artistically recreated in story and film. This tradition confirms our indigenous belief that mobility, exodus, change of place, are necessary for salvation. Is it any wonder, then, that despite the presence of historical monuments and grandiloquent architectural sorties, we build our cities, not as habitats for persons, but rather as pens for those awaiting to leave, to leave for the land where, allegedly, human beings truly belong.

### The Space and Time of Nature and City

The most dramatic and long-lasting influences on a person coming to consciousness is his sense of space and of time. What could be more different for me as a person, than if I came to believe that I was in space as one creature among others, all of us with space to spare, rather than if I came to believe that no space was present for me unless I seized it and made it my own space, that is, my place? What could be more different for me as a person, than if I believed that time was seasonal, repetitive and allowed for reconnoitering over the same experienced ground again and again, rather than if I believed that time was by the clock, measured in hours, minutes and even in rapidly disappearing seconds? The difference between the space and time of nature and the space and time of cities has its most important impact on our perceptions of the environment, sufficient to generate a very different experience of ourselves, our bodies, our memory and our creative life.

The affairs of nature have an inexorability about them. Even the intrusions, such as drought and tornadoes, follow the larger natural forces. Although spawned elsewhere, they occur periodically, loyal to conditions routinely present in nature. Cities, however, have no correlate to the seventeen-year cycle of the appearance of the cicada. Nature space is spacious, with room for recovery, for redemption, for second and third chances. Vision in the space of nature is predominantly horizontal, as though a person were on permanent lookout. The plains and the desert offered us an extraordinary visual reach, a field for our eyes to play upon, as in an endless cavorting through space with nothing definite, no things to obtrude. Still more powerful is the heightened sense of distance given to us from our mountain perch. This horizontality passes over the billions upon billions of individual items present in nature, space, those of the flora, the fauna and the living creatures of the soil and sky. But the very vastness of nature space requires a special attention, a special ability to grasp and relate the presence of the particulars. Thus, persons of nature space are homemade botanists, entomologists, ichthyologists, and animal scientists constantly aware of variations which await the patient observer of what only seems to be stretches of sameness. For city dwellers, this potential richness of nature space is missed, for they have been schooled to confront the obvious in their experience, multiple sounds, objects, things and events, all up front and requiring no search. Actually, persons of city consciousness often find the virtually endless sky of nature space, as in Montana and especially Texas, to be claustrophobic. This situation is obviously paradoxical but it is instructive. When a person experiences a sense of vast distance, a sky so wide that the eye reaches out endlessly, with no end in sight and no grasp of the subtle particulars of shadow, cloud, and terrain which intervene, then that person shrinks back in an introverted, hovering manner, alone and fearful. The claustrophobia of open space for the person of urban consciousness is precisely the opposite of the experience of a person of nature consciousness, whose claustrophobia erupts when they first find themselves adrift in a teeming city street.

The time of nature space is likewise special. I call it fat time, for it is measurable by seasons, even by decades. The shortest time in nature time is that of sun time and moon time, both of which are larger by far than the hurried ticks of clock time, so omnipresent in cities. Nature time is also consonant with body time, a rhythm akin to the natural processes of the philosophy of the human organism. This rhythm is present in the daily transformation of night into day and then into night once more, just as our bodies sleep, wake and sleep once more. These events mirror exactly the nutrition cycle in our bodies and more poignantly, more naturally, they are microcosmic instances of the life and death of crops, insects, the leaves of the trees and on a larger scale, our own life, whose rhythm, like that of nature, turns over in a century.

If we are properly attuned, nature time like nature space affords an abode for a distinctively human marriage between the body and the affairs of the environment. Being in and of nature, then, seems to make sense. Yet, not by matched rhythm alone doth we live. The human organism, on an even deeper personal terrain, historically and culturally has expressed a need to develop an interior life, singular and independent of the external world. For the most part, this possibility is absent in the time and space framework of nature. I see the result of this absence as twofold; first, a quiet, yet a present sense of personal loneliness and even alienation pervades much of the lives of nature-persons and, second, in response to that situation, from the beginning of humankind, we have abandoned the rightness of nature for us and decided to huddle together by building cities. The great American philosopher Ralph Waldo Emerson, no lover of cities and a master delineator of the majesty of nature, writes nonetheless, as follows:

> We learn nothing rightly until we learn the symbolical character of life. Day creeps after day, each full of facts, dull, strange, despised things, that we cannot enough despise—call heavy, prosaic and desert. The time we seek to kill: the attention it is elegant to divert from things around us. And presently the aroused intellect finds gold and gems in one of these scorned facts—then finds that the day of facts is a rock of diamonds; that a fact is an Epiphany of God.[6]

It is this "aroused intellect," which is at work in the building and maintaining of cities, a homemade environment which demands human symbolization for purpose of survival. The space of city space is vertical rather than horizontal. The sky fades into anonymity, replaced by the artifactual materials of wood, concrete, mortar, steel, and glass, each married to the other in an upward spiral of massive presence and extensive height. Yet, the massiveness and the height, even in the great towers of the John Hancock building in Boston, the Sears Roebuck building in Chicago and the twins of the World Trade Center in New York City, are human scale, for by our hands and by the technological extension of our hands, we built them. They belong to us, they are our creation. The mountains, the great lakes, the rivers, the desert, belong to someone, somewhere, somehow, someway, somewhen else. They have mystery and distance from our hands. City space is not found space not space in which we wander. Nor is it space that fulfills the needs of our body, understood physiologically. Those, rather, are the conditions of the space of nature. Quite to the contrary, city space is space seized, made space, wrestled to a place, our place, a human place. A building is a clot, a clamp, a signature in the otherwise awesome, endless reach of nature space. City space as human place is rife with alcoves, alleys, streets, corners, backs and fronts, basements, tunnels, bridges, and millions of windows which look out on other windows. This introverting character of city space is the source of human interiority. The reach of vision is shortcircuited by the obtrusion of objects, impermeable, stolid and in need of symbolic regathering. Bus stations, train stations, cafeterias, urban department stores, pawnbrokers, betting parlors, ticketrons, stadia, bars, courthouses, jails, hospitals and innumerable other "places" gather us in a riot of interpersonal intensity. These spaces, once they become places, colonize space with an intensity and a personal message which defies their comparatively small size, planetwise, for they multiply meaning independent of both physics and geometry. These city places are spiritual places, transcending the body as a thing in space. City space is person-centered and symbolically rich. The accusation of anonymity as lodged against urban life is a myth of nature dwellers. Cities are fabrics, woven neighborhoods, woven covenants, each integral and thick in personal exchange, yet, each siding up to the other, for one reason or another so as to knot a concatenated, webbed, seamed and organic whole. If the truth be told, the capacity of genuine urban dwellers, as distinct from visitors, transients, or those on the make so as to return to the land, constitute a most remarkable phenomenon in human history. The history of the city is a history of human organisms, ostensibly fit for nature, rendering the artifactual affectionate, sentimental, warm, and symbolically pregnant.

The mountains, rivers, streams, lakes, oceans, insects, animals, plants, stars all belong to nature. But the buildings, trolleys, depots, museums, and the Hyatt-Regency belong to us. They are human versions of nature, found, so far as we know, in no other planet, in no other place, for there is no other place which is human place, our place. Emerson praises this human capacity to render the activities and things of the world as more than they be, as simply taken or simply had.

> This power of imagination, the making of some familiar object, as fire or rain, or a bucket, or shovel do new duty as an exponent of some truth or general law, bewitches and delights men. It is a taking of dead sticks, and clothing about with immortality; it is music out of creaking and scouring. All opaque things are transparent, and the light of heaven struggles through.[7]

The activities which make a city out of naturals, converting them to artifacts with distinctive human significance, function and symbol, are characteristic of a millennia long effort at transubstantiation, the effort to render nature an abode for human life in its liturgy, its interpersonal transactions and its artistic creations. The space of nature is aesthetically rich but it is never art unless rendered so by human ways. City space is artifact supreme, the most extensive, detailed multimedia human creation in the history of what we do to nature, rather than our simply being in nature.

The time of city time also reverses that as characteristic of nature. City time is thin time, transparent time, the experience of virtual instantaneity. Ruled by the clock rather than by the sun, city time measures out our experience in short bursts, such that we place ourselves under enormous pressure to make every minute count. City time seems to have no attendant space, no room to move around and idly look ahead. To exist sanely in city time requires that we build an inner life, an introspective fortress in which we can reflectively lounge impervious to the din accompanying rapid time passing. In this way, despite the rush of events and the multiplicity of persons transacting in an infinite number of ways, city persons soon develop a sense of privacy which they carry deep within them, even in the midst of public urban clamor. It is not ironic but rather fitting that America's most distinctive game, baseball, has been played on a sylvan field smack in the middle of the city. It is a nature game, defying time, which has no permanent role to play, for a baseball game can be tied to infinity. Sitting in the afternoon sun in Wrigley Field in Chicago, where night baseball is forbidden, provides the urban dweller a natural cranny, nestled among the tenements and elevated train, a space in which time is defeated, halted and sent scurrying.[8] This contrasts with the clock games, hockey, basketball and football, which are so driven by time that they attempt to cheat its passing by asking for overtime.

The great Hawaiian sailors, who without instruments, charted the Pacific Ocean and the South Seas, seemingly had sundials for souls. City persons hide their souls and chart their way with urban clocks, that is, watches, which they wear everywhere, even to bed. Analogue, digital, bells, songs, faces of every kind; each way a way to tell the time and in so telling, tell ourselves where we are, how we are and how much time we have left. Although for most of the past decade, I have been a resident of that recent category of the amorphous, a small metropolitan statistical area, having a teeny population of only one hundred thousand people, I am by consciousness still an urban person. Consequently, when someone asks if they can see me, I respond seriously and with conscious intent, that "I have a minute." What an extraordinary statement, and still more extraordinary that increasingly, even in an SMSA, people are not offended. Orthodox psychoanalysis, and urban phenomenon to be sure, features a couch with a large clock visible to the patient so that the hour of soul-searching wears down publicly. The urban factory introduced the tradition of punching the clock. If one were late, the clock would punch this out and you would be "docked." Urban time has a fixed and narrow memory. It exacts its price, twenty-four hours a day, seven days a week, year in and year out. This is wearing and it is, therefore, that urban persons hurry by you, without a howdy, for they are carrying on a conversation with themselves, a dialogue that protects them spiritually from the ravages of urban time.

In this contrast between nature and city, one further contrast should prove helpful, that of the presence of nature in the city. I do not refer to the greening of the city, too often a euphemism for bushing around ugly buildings. Rather, I mean those times when nature shouts at the top of its voice and does so in the confines of a city. Some years ago, with family, while traveling across the horizontal state of Kansas, we came upon, in the distance, a magnificent tornado. It soaked up the sky and drew all of the horizon to its center. I subsequently learned that aside from an occasional ranch fence and an errant cow, it played itself out failing to damage. More recently, the people of the city of Wichita Falls in Texas have no such romantic memories. The tornado of their presence visited their city and cut a violent swath, scattering bodies, homes, automobiles, while leaving a calling card of disaster. These comparisons are apt, for they provide insight to the natural forbearance of nature in taming its own as contrasted with the awkward vulnerability of the city when nature strikes. Two further examples will assist us here. A massive snowstorm on the plains is troublesome and causes some disarray, even death here and there. Yet by contrast, the endless snow dumped on the city of Buffalo some years back created such havoc that the competition for scarce resources brought the viability of social cohesion into serious question. Second, a hurricane that dances its wild dance off the Gulf Coast is far different from the hurricane that comes to visit Galveston Island or, in recent memory, has the temerity to blow out those archetypes of high technology, the windows of the Houston skyscrapers.

This contrast, however, need not always be frightening or deleterious to urban life. Fog in a woodland is eerie and moving. It takes a back seat, however, to the late summer fog which cascades in glorious tufts across the boulevards of the city of San Francisco, washing and whitening, along the way, those lovely row houses that face the Pacific Ocean. And few events can match the fall of a gentle and wet snow on a late-night city street, providing a white comforter for the sleepers and a glistening cover for the nocturnal walkers. One could replicate these appearances of nature in cities around the world: the overcast, ice-bound streets of a December Moscow; the gentle Irish rain on a pedestrian overlooking the River Liffey in Dublin; the London fog; and the rising sun over the River Ulna in the Prague of Franz Kafka, heightening the twin spires of the castle and the cathedral, church and state, yin and yang, the glory of medieval Christendom.

When cities intrude upon nature, it is an assault. When nature intrudes upon cities, it is a challenge, a transformation of urban life into still more imaginative forms.

### Personscape: "Building" the Future

It is now a truism that the older cities of America are in serious trouble. The reasons for the decay of the inner city and the rise of anomie are now well documented. Despite the many discussions as to how to ameliorate this situation, the problem remains critical, although glimmers of hope poke through, as for example in the exciting revitalization of the city of Baltimore. The financial recovery of the incomparable city of New York is also a cause for cheer and efforts to transform Detroit, Philadelphia and St. Louis are welcome. In this essay, however, the issue is quite different. What of the "new" cities, those of the Southwest, as for example, Houston, Dallas and Tucson? Have they learned anything from the demise of their older peers? I think not.

The older American cities long ago learned from their European forebears that the most crucial characteristic of a viable city was scale, human scale. I call this the presence of "personscape," by which I mean the urban ambience, which affords the city dweller the same bodily continuity with the environment as that found in nature. Even New York City, despite its overwhelming population and the dominating presence of its huge buildings, has preserved the sense of neighborhood, intimacy and personal accessibility. Until recently, the city of Los Angeles, a comparatively new city, failed in precisely the way that Boston, New York and Chicago succeeded, namely, to be a resource for ordinary and local experience as well as for the glitter of being on the town. Los Angeles, in its reworking of downtown, has attempted to address that lack and become a city in reality as well as in claim.

Traditionally, our great cities have been water cities, either adjacent to oceans, to lakes or to rivers. Water has the capacity to bathe the soul and to provide a stimulus for the symbolization of the environment. Water yields depth, fog, mist, the refraction of sun and moon, and especially the possibility of escape, by boat, by suicide and by reverie. Many of our new cities, however, are land cities—Denver, Tucson, Phoenix, Dallas and even Houston, despite its ship channel and occasional bayou. Land cities as prepossessing and powerful with major building scapes, are a new event for America. The city of Dallas is the locus classicus for this development. Deeply symbolic of the state of Texas, Dallas, like an oil well, erupts from the land rather than slowly oozing from the water. At the top of the hotel in the Dallas–Fort Worth Airport, there is a lounge surrounded by enormous picture windows. One can look out, as one does in many similar settings in this nation. In this case, one sees nothing, nothing. Yet, catching the cityscape of Dallas by air is an unusual experience. No matter how one approaches Dallas, it is a shocking appearance, seemingly as if out of nowhere. Dallas is a land city, an urban bequest to the interior of America. It defied the loneliness of the plains and seeks to celebrate the land in a way different than ever before. Although water dominates the origin of American cities, it is intriguing to contemplate the metaphors which will feed the land cities, for the earth, the dirt, the plains, hold their share of mysteries as well. And the sky, the sky of the land cities is visible. Despite the reach of the new buildings, the Dallas, Houston, and Tucson sky holds its own. The old cities have no sky. Their citizens are introverted, cellar dwellers, looking down and in, or when up, no further than the tops of closely packed buildings. It is novel and exhilarating to be in a major city and still be able to reach for the sky. Still more inviting are the cities of Austin and San Antonio, river cities, yet surrounded by land, vast stretches of openness. In those cities, the river, the sky and the new buildings make for a scape original in America, powerful and rich in potential poetry.

So much for the possibilities. The actuality is troublesome. Only a few years ago, I was staying at a motel adjacent to the University of Arizona. I inquired of the bellman as to the directions and distance for visiting the city of Tucson. He warned me, gravely and sternly, of the dangers in such a trip. Rejecting those warnings as endemic to the long-standing myth about cities, I walked a gentle, suburban three miles into the city of Tucson, arriving at 9 p.m. This city of some half-million people was closed; not a light, not a joint, not a person. Stopping a police patrol car, I asked where was it happening? Answer: no happening, no shopping, no action, no nothing. I returned to the dreaded anonymity of my tacky suburban motel.

I close as I open, on the theme of glass without feet. A city, to be a city, *must* have a downtown. A city, to be a city, must have a residential downtown; a walking place. Pretentious buildings without arcades, accessibility to the wanderer, the stroller, stifles the very meaning of a city. The contrast between the foreboding office buildings, in business only from 8 to 5 and the fancy hotels, with their lineup of limousines and airport taxis, on the one hand, and the urine-stained bus stations on the other, has become an unpleasant reality in the new cities. It is not that we are without historical wisdom. One of the truly great cities of the world, Rome, can teach us how to build a human city. In Rome, the old and the new, the elegant and the proletariat, the monumental and the occasional, are married, day by day as people of every persuasion, of every ability and every desire, mingle in a quest for the good life. The warning is clear, for those American cities which have abandoned their downtown areas for the ubiquitous external mall, have become faceless, unsouled blots on the landscape. Neither city nor nature, they are witnesses to the emptiness of contemporary American life. Heed the message; cities are for people, ordinary people who move through both the day and night in the search for nutrition, spiritual and aesthetic nutrition.

It would be ironic folly if the new land cities were to repeat the disasters of the older urban areas. Annexation of suburban land does not resolve the fundamental question as to how to build a city. The inner city, downtown, is still the irreplaceable soul of urban life. The city of Houston, a masterpiece of transportation madness, seduced by glass towers dotting the sides of its ugly freeways, sprawling, struggling for identity, has recently looked within. Downtown Houston, as pathetic an area as can be found in a major American city, is recently a cause of concern. Houston, alert to its major symbolic role in the new American, now has thoughts about invigorating its inner city. This is encouraging news and it is to be hoped that the potential success of the city of Houston will be a model for new cities throughout the land. We come to consciousness in the grip of space, time and touch. Nowhere is this more crucial than in the quality of urban life, where, it turns out, even in the Southwest, even in Texas, most of us live. Let us give the glass towers some feet, so that we may walk about in a genuinely human home, our city.[9]

## NOTES

1 James Marston Fitch, "Experiential Basis for Aesthetic Decision," *Environmental Psychology*, ed. Harold M. Proshansky, et al. (New York: Holt, Rinehart and Winston, Inc., 1970), p. 76.

2 From countless sources, to experience the child experiencing the world, I suggest Henry Roth, *Call It Sleep* (New York: Avon Books, 1962) (1934).

3 Earlier versions of this relationship can be found in John J. McDermott, "Nature Nostalgia and the City: An American Dilemma," and "Space, Time and Touch: Philosophical Dimensions of Urban Consciousness," *The Culture of Experience* (New York: New York University Press, 1976), pp. 179–231.

**4** Samuel Sewall, "Phaenomena," in Perry Miller and Thomas H. Johnson, *The Puritans,* vol. I (New York: Harper Torchbooks, 1963), p. 377.

**5** Granting statehood to Hawaii, therefore, ranks as one of our most saving decisions, for Hawaii ties us back to the beginning, to the Orient, and enables us to rejoin the new hegira of East to West, just at the time when the original journey, begun more than one thousand years ago, has come to an end.

**6** Ralph Waldo Emerson, "Education," *Works,* vol. 10 (Boston: Houghton Mifflin, 1903–1904), p. 132.

**7** cf. *The Journals of Ralph Waldo Emerson* (Boston: Houghton Mifflin, 1909–1914), 9: 277–278.

**8** I am aware that beginning with the Astrodome in Houston, and continuing on with the Silverdome and the Kingdome, baseball is now frequently played indoors. Despite customer convenience, this is blasphemous and should be stopped. It is as if an Indian rain dance were held inside a wigwam. After all, is not being "rained out" the stuff of life?

**9** The absence of personscape in the new urban architecture is depressingly obvious in a recent article by Bob Schwaller, "Pillars, Pedestals and Porticoes," *Texas Business* (November, 1983), 57–66. Apparently money and the size of buildings are the dominant theme in modern architecture. A second article, by Michael McCullar, "Scanning the Skylines, Citing the Singular," 68–77, discusses monumentality and the turf of architects. Speaking of the new Southwest Center building in Houston, McCullar writes that it "will reaffirm the apparent fact that—because of land values, human egos and an almost primal need for man to be awestruck by his architecture—big buildings are getting more monumental all the time." This is macho, male chauvinist America at its worst. Cities are for people and not for the aggrandizement of architectural egos. We do not live in the sky. We live on the ground, where we walk seeking to be at home.

## The Cheap and Accessible Print

**Richard Powers**

*... Those who came into contact with the machine process found it increasingly difficult to swallow the presumptions of "natural law" and social differentiation which surrounded the leisure class. And so society divided; not poor against rich, but technician versus businessman, mechanic against war lord, scientist opposed to ritualist.*

Robert L. Heilbroner, *The Worldly Philosophers*

In one of those apocryphal stories that make up the official history of Hollywood, a strapped producer tells a director who wants to take a film crew on location to Africa: "A rock is a rock, a tree is a tree. Shoot it in Griffith Park." The budget-conscious fellow knows that locale is created only partly by the on-screen jungle. The rest of the work is done by the million collaborators and African explorers inside the confines of the darkened theater. Reproduced on celluloid and given the lightest excuse for a narrative frame, Griffith Park can beat the deepest heart of darkness, even if the audience themselves have passed the same park on their way to see this same show.

A sophisticated adult might spend ten hours a week in Griffith Park, know every slab of granite and each graffito. But frame the place with a camera, crop it, mask the boom mike and bordering expressway, put two boys in the mud with gray fatigues and shallow bowl helmets, and add the persuasion of an artillery barrage produced by firing pistols into a trashcan, and this same sophisticated viewer will think: "So that's what Verdun looked like. It must have been horrible."

We explain such willingness to believe by pointing to the improving technology of reproduction, the growing accuracy of the mechanical facsimile. According to the theory, our machines have become so good at mimicking the sound, texture, and color of the original that it takes only the smallest suspension of disbelief to fall for the illusion. But this explanation is not enough. Boccaccio reports that Giotto's contemporaries often mistook his frescoes, which don't even employ rectilinear perspective, for the real thing; and viewers reportedly collapsed in dead faints at a scene in *The Great Train Robbery*—a film, in its best print, jerky, silent, black and white, and out of focus—where an actor takes a potshot at the lens.

Early audiences did not fall for clumsy illusions out of mere technical naïveté. For even the hyperproduced photographs and films of the 1980s come no closer to actual verisimilitude than did Giotto. A camera cannot begin to approximate the way an eye sees the same image. It changes the scale, focus, field depth, dimensionality, perspective, field of view, resolution, surface modulation, and luminosity of the image. Film color is restricted to discrete, subtractive values of the three primaries, and only approximates the continuous spectra of human sight. Each color print is unique, its tones not reproducible at will. Technological reproduction of images has certainly improved beyond all expectation. But our most advanced images are closer to Giotto than to the image a three-dimensional object makes in the eye.

Besides, if photographic technique were really powerful enough to make us confuse the image with its source, then we would at once say to ourselves, "Those are actors in Griffith Park," not "Those are soldiers at Verdun." For the park we pass daily would be so perfectly reproduced that it would force our recognition. The true power of photography and motion pictures, the trick that allows us to live imaginatively in the frame, is not the perfection of technique but the selective obscuring of it.

The strange persuasion of photographs rests on selective accuracy wedded to selective distortion. The reproduction must be enough like the original to start a string of associations in the viewer, but enough unlike the original to leave the viewer room to flesh out and furnish the frame with belief. Photography seems particularly suited for this precarious hybrid. It produces a finger painting in light-sensitive salts, but one regulated mechanically—simultaneously the most free and determined of procedures. One's shutter can only be open or closed, yet the resultant image can never be fully previsualized, corrected, or repeated.

Because the lens works so much more quickly and permanently than the eye, the result surprises even the photographer in its particulars: "That sign to the left on the building—I hadn't noticed that." Because the process mixes mechanical control with the surprise of light, and because the product mixes technical exactitude with veiling and distortion, the viewer's response is a cross between essayistic firmness—"this, then, the dossier, the facts"—and the invitation of fiction—"What can we make of it?"

Early photography had to educate its audience to this mixture. Editors of old pictorial journals, attempting to introduce rotogravure prints into their copy, met enormous resistance from readers, who found the old line engravings more realistic and dramatic. But by 1939, the public could see not Queen Elizabeth nor the anemic Bette Davis nor even a silver-gray, flat, underexposed and oversized phantom, but a hybrid of all these, tailored by the individual ticket holder.

Since inception, the medium implied that the only material difference between audience and artist was the possession of a camera. The technology first arose out of the desire of amateurs to record a scene independent of the unreliable hand. The camera obscura, an eighteenth-century tracing box, while reducing the need for special skill, still betrayed whether clever Hans or clumsy Franz did the tracing. One needed a device to place the Swiss Alps or the Grand Tour directly on the glass as it now stood. That need, and not the larger one of "seeing" better or improving on the Old Masters, precipitated the development of photography, although all the mechanisms had been known for some time.

And when, as often happens, several people at once evolved a process whereby light left its own register on the receiving frame, there arose an art not of talent or wizardry but of pure composition, vision, and decision, not of technical execution—the machine managed that—but of conception. Yet composition, vision, and decision are precisely the skills any intelligent viewer uses when standing in front of and appraising a finished picture. The process of making the thing becomes qualitatively indistinguishable from that of appreciating it. As Walter Benjamin puts it in his seminal essay, *The Work of Art in the Age of Mechanical Reproduction:*

> ...the distinction between author and public is about to lose its basic character. The difference becomes merely functional.

With the advent of the cheap, hand-held, self-focusing, automatic camera, even this functional difference became negligible. The burgeoning, moderately privileged classes carried at their hip a machine that made it possible to select, edit, and preserve moments of reality at their own choosing. They did so by the trunkload, until no home was complete without crateloads of photo albums on display or stashed in the attic. Every family became at once the subjects, the photographers, and the editors of these albums, making the small jump from author to authority.

The enormous popularity of photography and movies comes from their having arrived at a gratifying hybrid of the expository and the participatory. The effect of a filmed image depends less on its content than on weighting, pacing, contrast, and other editing. Quick cuts between a man walking down the street and a woman, across town, anxiously looking at her watch affect us less in their content than their composition. We go to the matinee each Saturday less to see what perils the script inflicts this week on Pauline than to see how we and the director will get her out of them. As Benjamin states:

> The audience's identification with the actor is really an identification with the camera.

A thinly disguised Griffith Park cannot, by technology alone, pass for Verdun. Our enjoyment of the film comes partly from our *knowing* it is Griffith but making it serve as the battlefield. It resembles the battlefield enough to trigger us to build a symbol table in the mind, a table indistinguishable from the photographer's table of editorial and technical decisions. We are children, the photo, the germ of a story told us by a parent that we must elaborate, expand, repeat to ourselves to keep from falling asleep in the threatening dark. Listen to two people describe the same scene from a recent movie, complete with gesture, montage, directorial flair, but never one the same as another. Witnesses at a crime, they have forever sullied the facts with their own involvement.

We build our own symbols into the image, encouraged by the autonomy of the machine-made image. Although photography offers the possibility of manipulation—tinting, retouching, solarizing or mis-exposing, collage, assemblage, painting on the negative—such tamperings rarely fool the eye, let alone the mind. But under the auspices of the limiting machine, viewers give credence to their own free editing.

The nineteenth century adored the formal photographic sitting for three reasons. First, it could do, cheaply, the work of oils, which had been the reserve of the wealthy. Second, a print gave a good likeness without a painter's mannerisms. Third, the customer enjoyed the multiple pleasure of being subject, audience, and—by commissioning, posing, and selecting the final work—*auteur*. The photo-buying and soon the photo-making public saw in the studio portrait the perfect accomplice for its lifelong autobiographical projects.

With a shorter life span and a heightened awareness of death, the nineteenth-century middle class also found in the daguerreotype and photograph more persuasive remembrances than the customary swatch of hair or inherited necklace. A newspaper ad for one portraitist's studio urges, "Secure the shadow ere the substance fade," and with this slogan hits upon the true selling point of the portrait: the shadow, infinitely more pliable, can mean more to the survivors than the substance ever did.

The shadow lends itself better to the continual act of biography the viewers weave in understanding the subject. Oval portraits, religious icons, stand on dresser-altars, remembrances rounding out the interpretive biography. The nature of the image takes second place to the associations of those who took, or feel that they took, the photo: "We had this taken on Stephen's tenth birthday, two months before he passed on. The photographer took a dozen, but I chose this one. His face clearly shows what he might have become. I made copies and sent them out to the aunts and uncles."

Running off copies—identical except to the trained eye—rather than decreasing the value, as with stamps and rare coins, multiplies it. There can be only one Ghent altarpiece, but there can be as many photos of Matthew on the back steps as one has the machine make. The monetary value of the Ghent altarpiece lies beyond calculating, while Matthew weighs in at twenty cents a print. But to the audience—the artistic consumer—the singularity of the altarpiece renders it unfit for anything but worship: five minutes at the museum rail until the folks behind begin "hemming"—a distancing proposition at best. On the other hand, one can possess, alter, love, own, and archive a print. When the *Hindenburg* explodes, it's not enough to look over someone's shoulder on the train and see that the thing has gone off. Buy a copy of the newspaper carrying the photo, and make the reality your own.

While the commercial value of a thing varies inversely with its own ability, this century nevertheless has declared that the book price counts for little against the identification of ownership. The autobiographical impulse—the true measure of worth—must stamp the object with the viewer's mark. One cannot interact with the Mona Lisa while standing behind the chain. But the machine has manufactured enough Betty Grables for everyone's personal consumption.

Certain reactionary photographers have tried to preserve the value of their prints at the expense of their consumability, creating limited editions and swearing on a stack of Stieglitzes that they destroyed the negative after fifty prints. But photos have always been the Model T of the arts. The man who wants to be buried with his five-hundred-dollar Ford because it has pulled him out of every hole so far also wants a five-dollar photo of his wife so that when she passes, he can pull her out of her hole and place her firmly on the dresser. A rock is a rock, a tree is a tree, and a profile does fine for the original.

This ability to reproduce limitless, virtually identical images without manual intervention seems either the greatest debasement or the fullest promise of the machine age. Machines strip processes of any value aside from the result. Packing plants, cameras, and motorcars care nothing about the way from A to B; they only want to get there sitting down, in the easiest manner possible. But the most expedient path is never the most delightful: the two are by definition distinct. We must choose between the getting and the going, the journey itself or the material outcome, aesthetic transport or mechanical transportation.

As a result, a large segment of population in each age attacks the machine as dehumanizing, moribund, ruthless, stultifying, uncontrollable, banal, ugly—in short, the worst sense of "mechanical" and "mass-produced." To these people, an object's value is the measure of humanity poured out and lavished on it. Reproducing destroys the unique quality and value of things. The cult of beauty judges by difficulty and effort: the clumsiest painting holds more value than the most striking photo, as the first came about through the more revealing, because more arduous, path.

For these believers, replacing the simple and beautiful with the mechanical and expedient—tractor with plough, horse and carriage with Ford, saber with musket and then carbine—starts a process of self-replicating escalation that ends only when our tools, regardless of our own say, seek an outlet of power and mechanical efficiency in an act of violence. No one can deny that this century's wars have been exercises of mechanical power, nor can one doubt—equivocal theories of deterrence notwithstanding—that the mere existence of fifty thousand nuclear warheads raises the possibility of annihilation above zero. But this camp takes an even stronger antitechnological stand: mass reproduction of photographic images represents and initiates those values that would destroy beauty, singularity, and all that is human and humane.

To others squared off against this line, mass production and reproduction provide a welcome liberation from the tyranny of privileged aesthetics—an "art of five kopeks." This camp, including Benjamin, believes that equating rareness with beauty, worshiping art in museums instead of using it in homes, keeping the market free of imitations to drive up the price of the original have for too long deprived too many people of the natural material necessary for contemplation and betterment.

In this light, photography—mass reproduction and distribution—at last provides a means for popularizing and democratizing artistic value. For the first time in history, copying an image is no more difficult or expensive than enjoying the original. The previously hallowed barrier between maker and appreciater breaks down. The anti-mechanicals lament the debasing of author to the level of mass audience. The pro-mechanicals celebrate the elevating of mass audience to the level of authority.

To the pro-mechanicals, rapid and unchecked proliferation of printed art, rather than negating or diluting value, promises untold practical and aesthetic worth for unchecked, proliferating mankind. As Benjamin suggests:

> ...the newsreel offers everyone the opportunity to rise from passer-by to movie extra. In this way any man might even find himself part of a work of art....

What of the anti-mechanicals' accusation that the machine's concentration on results over process, on ends over means leads to a state of violence, where domination grows naturally out of efficiency? That mechanical technology creates weapons of unthinkably violent potential lies beyond doubt. But pro-mechanicals counter that our spirit of using or denying machines, rather than mechanics itself, causes or avoids conflict. Just as proxy battles and corporate shakeouts mean to solidify—by conflict—antagonists' holdings, so wars between national states, say the pro-mechanicals, mean to preserve and expand the material value—the uniqueness—of the state while asking the constituency to pay in doughboys.

An art of the few rather than the many, equating beauty with commercial rareness, only perpetuates those material motives that underwrite wars. Says Benjamin:

> If the natural utilization of productive forces is impeded by the property system, the increase in technical devices...will press for an unnatural utilization, and this is found in war. The destructiveness of war furnishes proof that society has not been mature enough to incorporate technology as its organ, that technology has not been sufficiently developed to cope with the elemental forces of society....Imperialistic war is a rebellion of technology which collects, in the form of "human material," the claims to which society has denied its natural material.

The choice is clear: shoot snapshots or shoot rifles. Produce and distribute widely available images that answer the material needs of the world, run off universally available, obtainable stock certificates, or enforce the old system of aesthetic ownership with violence. We have the choice of politicizing aesthetics with the aid of cameras or falsely prettifying political reality, which, according to Benjamin, leads only to war. Treating current events as art or reading history as romantic fiction must result in revering conflict, too, as a formal beauty.

Thus both extremes in the debate of mechanical reproduction accuse each other of stances that result in catastrophe. Between these two groups, the vast majority of us go about using cameras without realizing the consequences at stake. We make our albums, take our snapshots, at times out of a love of rare beauty, at times out of a documentary impulse, but mostly as a healing charm against death. We try, with the aid of the lens, understanding neither technical mechanism nor philosophical import, to beat the annihilation of time, shore up against loss, not just loss of the subject matter—our late aunt Sophie, or last summer's vacation in the Balkans—but the death of the instant of vision, the death of the eye, which, without the permanent record made by the machine, gradually loses the quality revealed to it in the moment of seeing.

Every photographic print invites identification with the photographer, forces re-creation of the values implied in preserving the vanished image. Viewing becomes a memento mori, a reminder of the death of the subject matter, landscape or portrait, long since passed away but remade in our owning and involving ourselves with the print. For the viewer, contemplating the lost scene of a photograph lies well outside the aesthetic *or* the political. Hannah Arendt explains this in her treatise on violence:

> Death, whether faced in actual dying or in the inner awareness of one's own mortality, is perhaps the most antipolitical experience there is. It signifies that we shall disappear from the world of appearances and shall leave the company of our fellow men, which are the conditions of all politics.

Every full appreciation of a photo, every alignment of ourselves with the lens creates in us the profound awareness of such a departure. For we have left, we have died away from the conditions of the photograph's moment. Every mechanical landscape, interior, or portrait comes to the viewer over time, a memory posted forward from the instant of the shutter waiting to come into conjunction with the instant of viewing. Noticing the image, observing it at once implicates the viewer as a partner in that memory. Looking at a photo, we act out and replay, to a copied phantom, parallels of the very decisions and criticisms of the photographer. We ask: "Who would I have to be, what would I have to believe in to have wanted to preserve this instant?"

And just as our daily autobiographical revisions resemble collective history and allow it to write itself, so the act of seeing and loving a photographic image calls us to action, but action circumscribed by the image's historical context. Interpretation asks us to involve ourselves in complicity, to open a path between feeling and meaning, between ephemeral subject matter and the obstinate decision to preserve it, between the author of the photograph and ourselves.

The idea that we look *at* the sitter or subject matter begins to lose its credibility. We look *over* it, attempting to locate something else. When a movie editor cuts from a woman turning her head to a midrange shot of the shops across the street, we follow her gaze, having of our own volition turned our attention to look with her. To make sense of a montage means to reassemble the cut, in reverse, using the editor's criteria. We remake the montage dynamically, as an act of looking.

The same applies to still photos: the lens slices a cross-section through time, presenting an unchanging porthole on a changed event. The frame invites us to feel a synchroneity with the photographer the way museum-case glass, slicing through a beehive, invites us to live in the colony. We cannot worry too much over what lies to the right or left of the restricting frame. If the path between sense and significance opens, it will open in those moments when momentarily delighted by some overlooked detail or construed resemblance, we become aware of what lies in front of the plane of the photo.

We scour *over* a photo, asking not "What world is preserved here?" but "How do I differ from the fellow who preserved this, the fellows here preserved?" Understanding another is indistinguishable from revising our own self-image. The two processes swallow one another. Photos interest us mostly because they look back.

Time and again in Sander's unfinished portrait gallery, *Man of the Twentieth Century,* painfully aware of the serious business of the lens, his subjects abandon hope of portraying their individual peculiarities and instead take up the heavier obligation of representing their upbringing, social role, and class. They focus their gaze distantly, well past the photographer, on a farther, more important concern. When we come behind the photographer's shoulders into conjunction with that gaze, we have the macabre feeling of being its object, the sense that the sitters mean to communicate something to us, to all posterity.

Yet this sitters' arrangement is the very opposite of what is commonly called posing. Sander deprives his subjects of the studio pose, the refined posture meant to convey character. With the cataloging urge of a natural scientist, he treats them as specimens, and they can at best buck up and make themselves presentable. Normally, the subject, desiring immortality, makes an appointment and appears at the studio. Sander reversed all that, bicycling to and pinning down his unsuspecting species where they grew. Depriving them of their active desire to be photographed, Sander enforces his subjects' accountability to the camera. They are caught in the act of revising their own biographies under the examining eye of history, which is the lens.

A fish and child on either side of aquarium glass react to and modify the behavior of the other: a finger jab causes fluking causes a delighted squeal causes a surprised flushing of gills. We, on the far side of the glass, adjust ourselves similarly. The subject of *Man of the Twentieth Century* is this constant interplay between the small self and the larger portrait gallery. The subject of *Man of the Twentieth Century* is us.

Sander, in depriving his subjects of the ability to pose heroically, deprives the viewer of the same evasion and gives even the casual museumgoer the sense of being summoned. When just before the turn of the century, the police photographer Jacob A. Riis brought out his volume of images called *How the Other Half Lives,* he removed the subjects one degree more. His sitters were both unwilling and unwitting, not even aware they were being photographed. Riis and two assistants skulked about at night, clandestinely taking flashpan photos of criminals and the extreme poor. Following the blinding discharge, the photographers would have to disappear just as suddenly, often pursued by startled and angry crowds.

Riis meant to expose the squalid conditions of New York's Lower East Side, to show the poverty and degradation of the human flotsam there. He could do so only surreptitiously, without his subjects' consent. In one startling image, robbers, some bearing cudgels, peer threateningly down their narrow alley hideout at the dazzling phosphorus explosion. Decades slide past the ground glass, and they look out on the scholar, museumgoer, or nostalgist, who has taken the part of the original intruder.

Perhaps Riis intuited that a comfortable, propertied class would believe the debilitating conditions he documented only if they saw, as collaborative evidence, their own unwillingness to look reflected in the poor's unwillingness to be looked at. Riis, in showing the local decrepitude of one city, succeeds in showing the true face of humanity when unable to prepare itself for the record.

The removal of the sitter's pose from the finished portrait becomes complete in the work of Eugène Atget, who photographed Paris early in this century. Tens of thousands of plates show deserted streets, shop windows filled with bric-a-brac or headless mannequins, empty doorways and arcades, and barren cafés waiting for a clientele conspicuous in its absence. The scenes catch the hushed, paralyzed aftermath of tragedy. Atget's empty streets are portraits without sitters, calling attention to a foreground surprisingly vacant. They are the clichéd snapshots of relatives in front of public buildings, only this time the viewers must supply their own relations.

The aperture of a camera forms a two-way portal through which both subject and viewer peer into another time. The subject, conscious of the permanence of the document, posts forward a memory. The viewer, aligning with the memory at some later date, works to preserve the sight from disintegration. Both are present at both moments; both experience the revelation of being adrift in time, sampling it laterally.

The moment of recording and the moment of interpretation lose their basic distinction. Somewhere in time, observer and observed reverse roles. Conscious of being watched through the asynchronous screen, both modify their behaviors, presenting their best profiles: interpenetration of looker and participant, audience and authority, aesthetic escape and polemical display, welded together and mechanically propagated through time.

To look at a thing is already to change it. Conversely, acting must begin with the most reverent looking. The sitter's eyes look beyond the photographer's shoulders, beyond the frame, and change, forever, any future looker who catches that gaze. The viewer, the new subject of that gaze, begins the long obligation of rewriting biography to conform to the inverted lens. Every jump cut or soft focus becomes a call to edit. Every cropping, pan, down-stopping receives ratification, becomes one's own.

Consider a print of you and a lover standing by the side of a house. You can shrink or enlarge it to any size. You can print it on matte, glossy, or color stock. You can mask the negative, tint it, print it up as Christmas cards. You can crop it and edit out your mate or yourself as appropriate. Finally, you can take a twelve-dollar camera and repeat the scene with a new lover, as many times as it takes to get it right.

And a new technology, already on us, extends this ability well beyond still photography. Every home is about to be transformed into an editing studio, with books, prints, films, and tapes serving the new-age viewer as little more than rough cuts to be reassembled and expanded into customized narratives. Reproduction will make the creation and appreciation of works truly interactive. These exotic technologies, like the camera before them, will enlarge the viewer's understanding of the maker's act.

We feel the process of looking most powerfully when not distracted by the object of attention. On a busy street, a normally perfunctory businessman stops on his way to the office and cranes his neck to relieve a crick acquired by sleeping with the window open. The woman behind him stops too, looking up with a voyeur's curiosity. Soon a whole crowd gathers, refusing to miss the excitement, looking up into absolutely vacant sky. Some demand, "What is it? What's happening?" Others imagine: "There; I saw it, just behind that building." Still others feel an oceanic gratitude come over them: "So this is me, on a July morning, stopped a moment, looking into the blue." Then they shrug shoulders, straighten shirtsleeves, and remark to themselves, "A curious thing, consciousness," before heading off to their appointments.

Here at last is an explanation of why we can be moved by a scene clearly filmed in Griffith Park. We respond not so much to the events on film as to the thousand reels we concurrently edit in our mind—movies of our own hopes and terrors. Griffith Park, Verdun, the empty streets of Paris, or three men on a muddy road matter less than the mechanical decision to involve ourselves, to retake the composition and extend the story.

## Autonomous Technology in Fiction

**Larry Hickman**

One of the issues that most often accompanies serious discussions of technology is its alleged autonomy. As Langdon Winner informed us in his excellent book on the subject, this is a topic which has been around in one form or another for a very long time, even before it was propelled to the importance it now has by the publication of Jacques Ellul's *The Technological Society*. Ellul's book was published in France in 1954 and in the United States in English translation in 1964.

Following Harold Lasswell, Ellul defined technology, or "technique" as "the ensemble of practices by which one uses available resources in order to achieve certain valued ends."[1] He argued at length that human beings are no longer in control of this rational system. Ellul's critics argued that he had been overly vague in his characterizations of technology, autonomy and rationality. They suggested that his account raised serious problems of verification (or falsification).

Winner argued that Ellul's earlier critics were right in at least one important respect: that he had failed to provide a characterization of technology that was sufficiently sharp to be of value. Then Winner suggested a way of rendering more precise Ellul's broad strokes. Ellul's "technique," he argued, actually confused three important facets of technology. The first he called "technique," but in a sense quite different from that of Ellul. This is technology's software: its skills, methods, procedures and routines. The second aspect of technology he termed its "apparatus." This is the hardware of technology: its tools, machines, instruments, appliances, weapons and gadgets. Winner also called these the "physical devices of technical performance."[2] Finally, there is technology's "organization." Winner uses this term to signify "all varieties of technical (rational-productive) social arrangements."[3]

If the distinctions which Winner made are important, then we would expect them to afford us an improved purchase on the "phenomena" that we call technological. And this is in fact the case: it turns out that Winner has provided us with a very nice tool for categorizing complaints about technology.

Complaints about technique, for example, take the form that no organization or person can any longer control the methods of technology, or more simply, "the way things are done these days." The majority of the complaints registered by Ellul are of this sort. Technique has taken on a life of its own. It is resistant to human input. It obeys laws which no human being has programmed into the system, but which have just been generated by the system itself.

A second and very different kind of complaint is often made about the apparatus. These complaints are much more modest and certainly much more concrete. In general, they do not tend toward the vague pessimism which accompanies complaints about technique. It is claimed, for instance, that certain artifacts or configurations of them function in ways which tend either to frustrate particular human intentions or to change people in ways which would have been unexpected or even impossible before the invention or popularity of the apparatus in question. There is no talk of "the machine" here, since such talk more properly refers to technical procedures or technique. Rather, people talk of "the automobile" or "the computer" or "the airplane."

Finally, there are complaints about "organization." These are not complaints about any particular organization, even though at times a particular organization may serve to focus them. At their profoundest level, these are the sentiments articulated by Kafka, Orwell and even Czeslaw Milosz. More commonly, "the bureaucracy" is the object of dissatisfaction, despair or even rage. What is present in both accounts, the penetrating and the inarticulate, is a sense of loss of agency in the face of a phenomenon which is effective because of its very anonymity, invisibility and inevitability.

But there is another means by which the usefulness of Winner's distinctions may be tested. Fiction, like technology, is an affair of human making and doing. If Winner's distinctions are important, then we would expect to find them articulated in fiction. Though we might not be surprised if they lack the sharpness available to the theoretical work of the sociologist or philosopher of technology, we would nevertheless expect them to be there.

Let's take the broad brush claim about technique first. The novels of John Updike present individuals who are enmeshed in a system of technique which they perceive as both impersonal and autonomous. In *The Coup,* Updike's central character is Hakim Fêlix Ellelloû, the president of the African republic of Kush. (The ironic and well-read Updike has given his protagonist a name which bears a striking resemblance to that of Jacques Ellul.) On one of his tours of the countryside, he encounters a man in a white button-down shirt and seersucker suit standing atop a pile of Kornkurls, Total, Spam, Carnation Instant Milk and "KIX TRIX CHEX POPS." This pyramid of junk food, it is explained to Ellelloû by the American aid officer who stands at its apex, is relief for the starving, drought-ridden Kushites. Never mind that most of the labels read "just add water," precisely the element most lacking in Kush. As Ellelloû pointedly informs him, "In Kush, water is more precious than blood."[4] Before torching the junk food, and, lamentably, the aid officer who refuses to abandon it, Ellelloû explains that the famine in his country has been caused by the ingression of technology. The vaccination of herds has led to their increased size, subsequent overgrazing, and the inevitable depletion of the water sources of the area. There is no drought, he says, simply bad ecology.[5]

Later in his account, Updike has another American aid officer appear in Kush. Speaking his bureaucratic dialect to an official in Ellelloû's government, he says "...you tell the man for me, no problem. Our technical boys can mop up any mess technology creates. All you need here is a little developmental input, some dams in the wadis and some intensive replanting with the high-energy pampas grass the guys in the green revolution have come up with."[6] A bit later he continues: "Miracles are everyday business for our boys."[7] Technique here, personified for Ellelloû by the invading Americans, with their rationality and mindless optimism, certainly fits the Elullian profile. It is irreversible,[8] it progresses not arithmetically but geometrically, it is a closed system,[9] it brings the elimination of human variability and elasticity,[10] it becomes a new mystery,[11] and it generates in humankind a loss of sense of scale and a loss of sense of agency.[12]

Ellelloû seeks to combat the onrush of this special brand of technique with his personal blend of Marxism and mysticism, together with the fundamentalist Islamic ideology in which they are nested. But for Updike there is no mystical or ideological solution to the autonomy of technique. Despite his temporary victory over the mountain of KIX TRIX CHEX POPS and the aid officer atop it, Ellelloû is eventually deposed and supplanted by those more sympathetic to the march of high technology, technique. To make his fall from power even more degrading, he is pensioned off to Paris.

This is of course not an account of technique in its purest sense. Ellelloû struggles against organization as well. But Updike makes it clear that if one organization fails, others are waiting, interchangeable, eager to resume the larger task of bringing technique to Kush. It is not the Americans per se that bring Ellelloû to ruin, but what they bring; and that is technique.

Of course Harry Angstrom, too, or perhaps I should say Harry Angstrom *especially,* since he is character whose life Updike has had monitor our culture for the last twenty years, must deal with the autonomy of technique. Rabbit is of course rich in Updike's latest installment of his life. Nevertheless, he worries about the things outside his control. Now a successful Toyota dealer, he worries about the energy crisis ("A new industry, gas pump shrouds").[13] He frets at the incomprehensible response to technology registered by his own son, and consequently whether his car will survive his son's next date, which of course it does not ("Dad, it's just a *thing;* you're looking like you lost your best friend").[14] He wonders whether Toyotas will continue to sell (he apologizes to a prospective buyer: "'These Japanese for all their virtues have pretty short legs,' he tells her. The way she has to sit, her ass is nearly on the floor and her knees are up in the air, these young luminous knees inches from his face").[15] He suffers over the status of the yen. And so on. He is, of course, powerless before all these menacing features of his technological environment.

What are his solutions to these vague but threatening problems? Well, for one thing, he reads *Consumer Reports.*

> Naked, [his wife] Janice bumps against the doorframe from their bathroom back into their bedroom. Naked, she lurches onto the bed where he is trying to read the July issue of *Consumer Reports* and thrusts her tongue into his mouth. He tastes Gallo, baloney, and toothpaste while his mind is still trying to sort out the virtues and failings of the great range of can openers put to the test over five close pages of print. The Sunbeam units were the most successful at opening rectangular and dented cans and yet pierced coffee cans with such force that grains of coffee spewed out onto the counter. Elsewhere, slivers of metal were dangerously produced, magnets gripped so strongly that the contents of the cans tended to spatter, blades failed to reach deep lips, and one small plastic insert so quickly wore away that the model (Ekco C865K) was judged Not Acceptable. Amid these fine discriminations Janice's tongue like an eyeless eager eel intrudes and angers him.[16]

Of course it is not Janice's advances per se that anger Rabbit, for Rabbit's response to his entire world is sensual, but rather that he has been distracted from his technological task. He hopes to convince himself that he is coming to grips with technique by mastering the minutiae, the sensual richness of the apparatus. Besides reading *Consumer Reports,* he contemplates the "black-backed Polaroid instant photos"[17] he discovers during his clandestine examination of the bedside table of his friends Webb and Cindy Murkett. "That SX-70 Webb was bragging about,"[18] Rabbit muses.

> The light in the room must have been dying that day for the flesh of both the Murketts appears golden and the furniture reflected in the mirror is dim in blue shadow as if underwater. This is the last picture; there were eight and a camera like this takes ten. *Consumer Reports* had a lot to say a while ago about the SX-70 Land Camera but never did explain what the SX stood for. Now Harry knows. His eyes burn.[19]

He toys with the tube his Kruggerands came in. The tube's lid, he remarks, resembles a miniature toilet seat.[20] He wonders at a "Lavomaster" faucet; "what was wrong with the old two faucets that said H and C?"[21] He complains to Janice that they need a larger safety deposit box.

Thus does Updike have Rabbit successfully tread the seas of technique by holding tight to the flotsam and jetsam of apparatus. If this is not a sophisticated or exciting solution to his problem, let us admit that Rabbit is neither sophisticated nor exciting as a person. But like many, he gets by in a world he regards as saturated with vague technical threats, a world in which he perceives himself not as agent, but as patient. Moreover, he, unlike Ellelloû, stays afloat. The key may be that for Ellelloû, technique comes like the flash floods in the wadis of Kush. But for Rabbit's America, technique is a great, pacific ocean. One may drift, but he need not fear being swept away. Ellelloû has attempted to confront technique with theory and has failed. Rabbit meets technique where it is the most vulnerable, in terms of praxis, even though it is a relatively unsophisticated appeal to the sensuality of apparatus.

Robert Pirsig, too, has grappled with the autonomy of technique. Although he sets up the basic problem in a number of ways in *Zen and the Art of Motorcycle Maintenance,*[22] he clearly rejects two proposed solutions. There are those, such as his friends John and Sylvia, who ride their motorcycles into the countryside, without benefit of extra sparkplug or wrench, to "get away from technology." But of course "technology" intrudes nevertheless when their machine fails to start for the trip home. It is as if they feel that they can use the apparatus of technology to save them from technique, and this without the necessity of attending to the details of its apparatus. But they are more like Ortega's new technological mass man, who merely takes technical artifacts for granted in much the same way that the jungle dweller takes as just given the fruit that grows on the trees, than they are like Rabbit Angstrom. Rabbit at least knows that there is some sense in which he must attend to the trees, even if it just means reading *Consumer Reports*. John and Sylvia define technology, both technique and apparatus, in such a way that it will forever be alien and autonomous for them. They have made it foreign to their way of interacting with the world, and they have no adequate substitute. In a way, they are even more unfortunate than Ellelloû. He at least attempted and failed. John and Sylvia know only defeat without the joy of the struggle of the game.

Pirsig warns of a technological dilemma. On the one hand one must not lose control of the details of his or her life as have John and Sylvia. There is for Pirsig always room to establish agency within the cracks of technique. A motorcycle trip on the backroads is interesting and pleasurable even when the main highways are backed up with holiday traffic. With motorcycles and with one's life, with technique and with apparatus, one must just keep tuning the damn thing up. In this sense Pirsig is interested in doing more than just staying afloat, Rabbit Angstrom style. He is interested in prevailing, in flourishing.

The second horn of the technological dilemma is the danger of filling up so much of one's lifespace with "rationality" and efficiency that there is no room for surprise. At this point Pirsig introduces several rather thinly veiled references to some of his former professors at the University of Chicago. They hold seminars in Greek philosophy. They have been co-opted by the "rationality" of technique in the sense in which it appeals to, and manifests itself in the professional life of, the academic "humanist." They have failed to remember and to celebrate the origins of the rational elements of technique in myth, metaphor, quality and rhetoric. If the world is so open as to be unmanageable for John and Sylvia, for these academics it is so closed as to be finished, boring and uninviting to any save those who appreciate formal systems for their own sake.

There are of course many other fictional works to which one can turn for a treatment of this issue. It would have been nice to consider the works of Vonnegut, Graham Greene, Malamud, and a number of others, had the space for essay been unlimited.

The second group of problems I mentioned earlier has to do with the autonomy of the apparatus. This area lacks the grand dimensions of the previous one. Here we find explorations by an author of the consequences of the introduction of a new technological artifact, for example, or explorations of the results of an artifact that has been around for a while, all in terms of a character which he or she is attempting to develop.

In the work of E. M. Forster the products of technology are, in the short term, troublesome. In both his early novel *Howards End* (1910) and his later and better known work *A Passage to India* (1924), we learn much of Forster's response to technology by observing the role played in his narratives by the automobile.

Particularly in *Howards End,* the automobile is said to rob us of a sense of duration and of the rhythm of nature. "Perhaps," he writes, "Hertfordshire is scarcely intended for motorists. Did not a gentleman once motor so quickly through Westmorland that he missed it? And if Westmorland can be missed it will fare ill with a country whose delicate structure particularly needs the attentive eye."[23]

In *A Passage to India*[24] the automobile presents enormous difficulties: the social difficulties between the Indians and the Europeans on which so much of the narrative turns are exacerbated. What shall be the seating arrangement in the automobile? The Indian "character" which Forster takes so much pain to develop is nowhere more clearly exemplified than in the relation of Nawab Bahadur to his "new little car," and the crisis which ensues after his chauffeur has hit an animal in the road.

But beyond all this, the coming of the technology of individual transportation brings an enormous change for the better in that most central of Forster's themes, the relationships between men and women. The violence of this change is exemplified by the violence generated in an incident which Forster creates about the middle of *Howards End.* On a motor outing, the first of two cars strikes an animal. The women, who are in the first car, are hustled to the second before they fully comprehend the circumstances. Margaret demands that they stop, but Charles, the driver of their car, has been instructed to distance them from the scene of the accident so that the servants and the remaining men may deal with the owner of the animal. When Charles fails to respond to her pleas, Margaret jumps from the car. "She fell on her knees, cut her gloves, shook her hat over her ear. Cries of alarm followed her."[25] As Forster tells us a bit later, "Charles had never been in such a position before. It was a woman in revolt who was hobbling away from him, and the sight was too strange to leave any room for anger."[26]

For Forster, then, technology rearranges not only our experience of nature, but the most sacred bonds of society. But he is by no means pessimistic about the eventual outcome of those changes. In the sense in which he treats the automobile as an agent of change, there does indeed seem to be a kind of autonomy or inevitability present in it. But the inevitability is in the final analysis one of accentuation of forces already in motion. Margaret is already a woman of independent thought and action long before we see this characteristic magnified by the jump from the automobile.

The automobile is also a major device in the work of Robert Penn Warren. *All the King's Men* opens with an account of a memorable drive down Highway 58 in the summer of 1936. Sugar-Boy, the Boss's driver, has

> whipped around a hay wagon in the face of an oncoming gasoline truck and [gone] through the rapidly diminishing aperture close enough to give the truck driver heart failure with one rear fender and wipe the snot off a mule's nose with the other. But the Boss loved it. He always sat up front with Sugar-Boy and looked at the speedometer and down the road and grinned to Sugar-Boy after they got through between the mule's nose and the gasoline truck.[27]

For Warren, technological artifacts, whether they be the Boss's Cadillac or Sugar-Boy's 38, are the media we use to play at being ourselves. The Boss is never quite so much at home as in his automobile. Sugar-Boy is a kind of technological centaur, half man, half Cadillac. As for Jack Burden, the narrator, Warren assigns to him a magnificent reflection on the way in which we "personate"[28] ourselves by means of the automobile. I use the verb "to personate" here in the sense of "to play at being oneself." As such it is related to the verb "to impersonate," that is, to play at being another. In this passage Burden "personates" himself in very different terms than does the extroverted Willie Stark or the technologically competent Sugar-Boy. Warren, by the way, establishes a striking disparity between Sugar-Boy's competence within the world of apparatus, on the one hand, and his lack of competence in the world of the spoken word, on the other. It is as if he "speaks" by means of the artifacts which he employs. Burden personates himself by emptying himself of himself and then projecting himself into other places and times. But the mantra for his reflection is the rain on the automobile, the "frail thread of sound" which emanates from its engine.

> There is nothing more alone than being in a car at night in the rain. I was in the car. And I was glad of it. Between one point on the map and another point on the map, there was the being alone in the car in the rain. They say you are not you except in terms of relation to other people. If there weren't any other people there wouldn't be any you because what you do, which is what you are, only has meaning in relation to other people. That is a very comforting thought when you are in the car in the rain at night alone, for then you aren't you, and not being you or anything, you can really lie back and get some rest. It is a vacation from being you. There is only the flow of the motor under your foot spinning that frail thread of sound out of its metal gut like a spider, that filament, that nexus, which isn't really there, between the you which you have just left in one place and the you which you will be when you get to the other place.
>
> You ought to invite those two you's to the same party, some time. Or you might have a family reunion for all the you's with barbecue under the trees. It would be amusing to know what they would say to each other.[29]

The phenomenon of technology's organization has also been treated in fiction, again, not in the pure sense of the sociologist or philosopher distinguishing it from technique and apparatus, but certainly in a way in which it is recognizably separate from those other aspects of technology.

Oedipa Mass, the central character of Thomas Pynchon's *The Crying of Lot 49,*[30] certainly does not suffer from lack of contact with the apparatus. Before the first two paragraphs are complete, we learn that she has been to a Tupperware party, that she has been confronted by the "greenish dead eye of the TV tube," that she reads the book reviews in *Scientific American,* that she goes to the market not only to shop but to listen to the Muzak, and that today she has listened to a portion of a Vivaldi kazoo concerto. By page five we have learned, by means of a rather lengthy catalogue, what her husband, "Mucho" Maas, the used car dealer and disk jockey, finds behind the seats of trade-ins.

Oedipa's problem is that she fears what lurks beyond all this apparatus. She fears that there is "organization"—not just *an* organization, but a kind of archetypical organization suffused with the magic of technique. Pynchon's account is the story of how she seeks to dispel her fear. Her fear is at first formless.

> Having no apparatus except gut fear and female cunning to examine this formless magic, to understand how it works, how to measure its field strength, count its lines of force, she may fall back on superstition, or take up a useful hobby like embroidery, or go mad, or marry a disk jockey. If the tower is everywhere and the knight of deliverance no proof against its magic, what else?[31]

Unable to call upon the tools of Ellelloû or Rabbit, she is faced with the frightening metaphors of faceless organization.

> What the road really was, she fancied, was this hypodermic needle, inserted somewhere ahead into the vein of a freeway, a vein nourishing the mainliner L. A., keeping it happy, coherent, protected from pain, or whatever passes, with a city, for pain.[32]

She is faced with the awful, interlocking intricacies of organization. Metzger, who is both her attorney and her seducer, tries to placate her with his assertion that beauty lies in the extended capacity for convolution. He, a lawyer, becomes an actor in front of a jury. But Raymond Burr is an actor impersonating a lawyer, who thus becomes an actor. Metzger himself, an actor before he became a lawyer, is having a film made of his life. The main part will be played by a friend of his, an erstwhile lawyer who quit his firm to become an actor.[33]

Organization, as she comes to know it, both thwarts certain individual goals, and at the same time occasionally ensures the seemingly fortuitous completion of projects which one has all but abandoned. The more one knows about organization the less predictable it is, even while becoming more pervasive and overarching.

How does Oedipa finally fare in her attempt to come to grips with organization? Long before we have completed our reading of his book, we know that Pynchon will not tell us in terms so simple as disappointment or satisfaction, success or failure, survival or extinction. What he does tell us is that Oedipa is either like an insect in a web, or like a lover in an extended embrace, in either case, page by page the more enmeshed.

Updike, Pirsig, Forster, Warren and Pynchon: they are a very diverse lot. Their works span most of the twentieth century, and they exhibit remarkable differences in terms of style, temperament and focus. Of the five, only Pirsig and Forster set out to write "about technology." Yet each makes an important statement about its autonomy in the lives lived by their characters.

This essay has been a sampler; more suggestive than demonstrative, more casual than rigorous. Let me repeat its modest claim: Winner's categories of technology—technique, apparatus and organization—are richer and more provocative than Ellul's "technique." Without the sharpness which they afford we would have to read Rabbit's plight, for example, as an undifferentiated uneasiness concerning technology. I have suggested that he is instead engaged in a battle against technique in the service of which he uses apparatus. It may just be because Ellelloû fails to make this distinction that he fails to come to terms with technology. The situation is similar for Pirsig, Pynchon, Forster and Warren; Winner's categories are keys that unlock further differentiation within the treatments of autonomous technology in fiction.

## NOTES

**1** Jacques Ellul, *The Technological Society* (New York: Alfred A. Knopf, Inc., 1964), p. 18.
**2** Langdon Winner, *Autonomous Technology* (Cambridge, Mass.: The M.I.T. Press, 1977), p. 11.
**3** Winner, p. 12.
**4** John Updike, *The Coup* (New York: Fawcett Crest Books, 1978), p. 53.
**5** Updike, 1978, pp. 49–50.
**6** Updike, 1978, pp. 246–247.
**7** Updike, 1978, p. 247.
**8** Ellul, p. 88.
**9** Ellul, p. 93, p. 133.
**10** Ellul, pp. 133 ff.
**11** Ellul, p. 143.
**12** Ellul, p. 146.
**13** John Updike, *Rabbit Is Rich* (New York: Fawcett Crest Books, 1981), p. 17.
**14** Updike, 1981, p. 99.
**15** Updike, 1981, p. 17.
**16** Updike, 1981, pp. 45–46.
**17** Updike, 1981, p. 284.
**18** Updike, 1981, p. 285.
**19** Updike, 1981, p. 286.
**20** Updike, 1981, p. 201.
**21** Updike, 1981, p. 282.
**22** Robert M. Pirsig, *Zen and the Art of Motorcycle Maintenance* (New York: Bantam Books, 1975).
**23** E. M. Forster, *Howards End* (New York: Penguin Books, 1979), p. 98.
**24** E. M. Forster, *A Passage to India* (New York: Harcourt Brace and World, 1952).
**25** Forster, 1979, p. 212.
**26** Forster, 1952, pp. 212–213.
**27** Robert Penn Warren, *All the King's Men* (New York: Bantam Books, 1968), p. 3.

**28** A nice treatment of this verb "to personate" occurs in Douglas Browning's essay "Some Meanings of Automobiles," in Larry Hickman and Azizah al-Hibri, *Technology and Human Affairs* (St. Louis: C. V. Mosby, 1981), pp. 13–17.
**29** Warren, pp. 128–129.
**30** Thomas Pynchon, *The Crying of Lot 49* (New York: Bantam Books, 1967).
**31** Pynchon, p. 31.
**32** Pynchon, p. 14.
**33** Pynchon, p. 20.

# PART THREE

## TECHNOLOGY AS EMBODIMENT

### INTRODUCTION

Almost without exception, philosophers since Plato have exhibited a low regard for the human body. Plato regarded the body as the "prison" of the soul. He thought it ontologically suspect because its being is inferior to the being of the soul as well as to the being of the transcendent Forms. He thought it epistemologically suspect because it is associated with the senses, which he held to furnish belief rather than knowledge. He thought it ethically suspect because it prohibits a full intellectual contemplation of the Forms by encumbering the soul and leading it astray.

Augustine, perhaps reacting to his youthful excesses, thought the body loathsome. René Descartes portrayed the body as a machine occupied by a somewhat ghostly intellect. The work of the intellect, in his view, was to produce the cognitive life necessary to drive the body-machine. David Hume both pushed this position to its conclusion and parodied it with his argument that it is impossible to prove that the body, or for that matter any other material object, exists. More recent reductionists, including B. F. Skinner and the logical behaviorists of the 1940s and 1950s, have attempted to put philosophy on a scientific footing by treating the body as a machine, as had Descartes.

Since technology is deeply rooted in practice, that is, in the ways in which human beings manipulate their environments by means of their bodies, it is not surprising that until the nineteenth century there was no serious work identifiable as a "philosophy of technology." The formation and development of this branch of inquiry had to await ways of understanding human embodiment that both rejected traditional metaphysical assumptions and offered promising alternatives to them.

The major work in this area has been done within three contemporary schools of philosophy: American pragmatism, the existentialism and phenomenology of the continental Europeans, and some of the strands of Marxism. The four essays in this section, together with the one by John Dewey in Part Seven, represent important alternatives to the deprecation of the body found in traditional philosophical literature.

In "Technology and Human Self-Conception," Don Ihde argues that what these alternative philosophical approaches—pragmatism, existentialism and phenomenology, and Marxism—have in common is their view that "some form of human action precedes or grounds conception, or that a theory of action is primitive with respect to [a] theory of knowledge." Put in more strictly philosophical terms, Ihde's claim is an ontological one: the *being* of human action is primitive to and necessary for human cognition or thinking.

This claim is, of course, the inverse of traditional philosophical views from Plato in the fifth century B.C. to the logical positivists of the 1940s and 1950s. A central figure within that long tradition was Descartes. His famous "I think, therefore I am," Ihde suggests, could in one sense be said to have invented both the "subject" and the "external world." This is to say that before Descartes, the metaphysical center of the world was usually taken to be God, from which other things were said to emanate: the closer an entity was to God on the great ladder of Being, the more "reality" it was said to possess. But after Descartes, the metaphysical center of the world became the thinking self: the self was accepted as the author and keeper of certainty and as the judge of things less certain than itself and its ideas, namely, "the external world." Descartes argued that the self knows itself directly and transparently and that it knows the external world only by inference.

Although there were significant modifications of and departures from this model, including the work of G. W. Hegel in the early nineteenth century, the Cartesian model of the relation of the self to the world remained a central concept of western philosophy until it began to suffer major erosion in the late nineteenth century as a result of the work of the German philosopher Edmund Husserl and others. Husserl's model of experience, which he called "phenomenology," was radically different from that of Descartes. Since every self is in some world or other, he argued, it makes no sense to speak, as Descartes had, of a self divorced from its world. The Cartesian self is nothing more than a misleading abstraction. Not only is the self not "self-contained," as Descartes had said, but from the standpoint of the way in which they are concretely experienced, *as phenomena,* the self and the world are equally certain. From the standpoint of analysis, the world may even have primacy: it is only as a part of a world that I could ever understand my self. My relation to the world is what Husserl called "intentional." It is also interactive: I project myself into the world, and I find myself reflected in the world.

The implications of all this for a philosophy of technology are prodigious. Ihde furnishes three technological examples which serve as variations of this relationship between self and world, a relationship he calls "projective-reflective." The first is a culture with a relatively low level of technology: a hunting and gathering society. What projections and reflections are at work in this type of culture? The hunting of animals and the gathering of plants and their products are the focal concerns of the society. These concerns determine the types of knowledge possessed by the culture and the ways in which that knowledge is transmitted. Young males are taught by means of ceremonies in which identification with hunted animals is effected. Young females are initiated by analogous means to the similarities between certain plants and the parts of the body that they are said to be capable of healing. The form-of-life of this culture can be characterized as one which involves a relationship of "likeness" or "similarity" between a human being and his or her world.

For his second example, Ihde takes an idealized agricultural society. Since animals and plants have been domesticated, the focal concerns of the society are now the seasonal cycles and the natural rhythms of birth or germination, growth, decay, and death. This form-of-life projects not similarity, but correlation: repetition becomes what is safe, and change is viewed as threat.

Finally, there is a form-of-life that is mechanized: it comprises a vast system of impersonal, machinelike relations. Even the human body is thought of as a machine with a pump for a heart, levers for arms, and an electrical system of nerves. Although Ihde does not do so, we might call this relationship one of "objectification."

Ihde draws three conclusions from the ways in which these three forms-of-life relate to one another. First, though they may be quite different in other ways, each exhibits some form of projection-reflection. Second, each form-of-life is anthropomorphized: the projection-reflection structures and outcomes in each are human products and are expressed as human concerns. Third, our own contemporary high-tech manner of projection-reflection is functionally equivalent to the other two. Regardless of what its theory may indicate, in its concrete activities it nevertheless operates just as do the two cultures that precede it: it puts practice before cognition and regards the self as a part of the world rather than as something against and separate from the world.

It was once fashionable for anthropologists to treat simpler or earlier human cultures as imperfect precursors of our own. Ihde inverts this treatment, finding in our culture the vestiges of those which were earlier or are simpler. He thus provides an important key to understanding contemporary technological culture precisely as a set of transformations of the concrete ritual practices of earlier and simpler societies. Although we no longer celebrate the spirits of the hunt or those of grapes or grain, for example, we nevertheless still animate our automobiles, calling them "jaguars" or "cougars," and our cinema exhibits detailed mythologies, such as that of *Star Wars*.

If we have made progress, it is not that our "self-knowledge" has improved, Ihde argues, but rather that the existential practice which we project upon the world is reflected back to us in ways that allow us to stay looser with respect to our total situation; we are able to take matters less seriously and to be more resilient. For Ihde, the amelioration or improvement of the human situation lies ultimately in remaining practically flexible in relation to experienced constraints.

The selection from Maurice Merleau-Ponty's classic work *Phenomenology of Perception* provides both a background for, and further development of, Ihde's thesis. Merleau-Ponty contends that the metaphysics of the traditional Cartesian *cogito*—the "I think, therefore I am," which sought to establish the awareness of an "inner core" of the self as a repository of meaning—failed to take into account the breadth of actual practical human interaction with the world. Because it attempted to reduce the practical to the cognitive, that is, to reduce active manipulation of the world to thinking about the world, it failed to acknowledge the imposition of meaning which is not "the work of a universal constituting consciousness." What Merleau-Ponty proposes in place of the Cartesian view is that we understand the human organism as "the movement to and fro of existence which at one time allows itself to take corporeal form and at others moves towards personal acts."

This is of enormous importance for the development of a philosophy of technology: according to Merleau-Ponty, our self-experience involves not just one but several "bodies." Among these are the "constituted" body, which is merely one expressive space among others; the "visual" body, or that part of a person's body which he or she can see; and more fundamentally, the body as "anchorage in the world" and as a "meaningful core which behaves like a general function, and which nevertheless exists, and is susceptible to disease." The function of this last body is richly illustrated with examples that are basic to technology: typing, playing musical instruments, and driving an automobile.

Of particular importance is Merleau-Ponty's novel account of the way we form habits. Traditional accounts had gone something like this: there is an act of understanding, by means of which certain elements are organized according to a formula or rule, and then the understanding withdraws to allow the body to act more or less automatically. Merleau-Ponty rewrites this account in the following way: learning a habit is in fact the "grasping of a significance," but it is the grasping of a motor significance. Most of our habits are not learned by applying a rule but instead by practice operating within what he calls a "motor space."

A good example of Merleau-Ponty's account is what happens when I drive my car. If I try to park my car in a space parallel to the curb, I do not first get out of the car, measure the space and the car, and then decide whether I can squeeze into the space. Rather, it is as if the car has become an extension of my own body, incorporated into its bulk. Like a blind person's stick or the movement of my hands across a computer keyboard, the activities involved in parking a car are not so much rule-governed or formulaic as they are a matter of the give and take of practice. "Thinking" about a well-developed practice may even reduce its efficiency. It is well known that one way to put tennis or golf partners at a disadvantage is to compliment them on their swings: by making their motor activities a matter of conscious attention, we often cause their actual performance to suffer. To take another example, if I ask you about the position of the *u* on a standard keyboard, it is unlikely that you will tell me that it is the seventh key from the left on the top row of letters, starting with the *q*. It is more likely that you will gesture with the index finger of your right hand, answering the question in the vocabulary of the "motor space" which constitutes the actual practice of using the *u* key.

In the third essay in this section, excerpted from Shoshana Zuboff's book *In the Age of the Smart Machine,* these themes are fleshed out even further by an examination of actual industrial environments. The author takes us to the bleach plant of a pulp mill. The plant has been redesigned and its operators have been moved away from its dangerous environs into computerized control rooms that are both clean and safe. But the workers are ambivalent about their situation. They have found ways to override the air lock between the control room and the fume-filled plant floor. They are fearful of the chemicals used to bleach the wood pulp and want to be protected from them, but at the same time they resent the new structure "that no longer requires either the strength or the know-how lodged in their bodies."

Zuboff finds in this situation a metaphor for the larger paradox of automated work. New forms of technology amplify and surpass the organic limits of the body. As a result, there is an atrophying of traditional performance skills. Work becomes the manipulation of symbols, the "reconstruction of knowledge of a different sort." Tasks that have traditionally been action-centered are taken over by machines, and a reskilling process is required. Intellective skills are required where action skills once sufficed. Zuboff regards all this as just a further evolution of the civilizing process, which has "increased the distance between behavior and the impulse life of the animal body."

It might be thought that Zuboff's thesis, and the diremption between action and intellection on which it rests, constitutes a counterexample to the thesis of Ihde and Merleau-Ponty, namely, that adaptive skill is irretrievably tied to an embodied life-world. But far from contradicting their thesis, her work reinforces it. The bleach plant operators open up a "motor space," to use Merleau-Ponty's term, each time they use their new computers. The fact that contemporary technology demands an ever more extensive manipulation of symbols in no way provides the grounds for returning to the outworn Cartesian picture of a thinking self opposed to a material world. Instead, the new electronic tools increasingly function, as Marshall McLuhan suggested, as extensions of our limbs and senses.

In the fourth essay in this section, "The Aesthetic Drama of the Ordinary," John J. McDermott picks up the central themes of the essays that precede his and substantially expands them. Like theirs, his primary concern is to supplant the Cartesian worldview, which he regards as having become both embedded in our language and inadequate and misleading. Our Cartesian-based language places us "in the world" as a button in a box, a marble in a hole, or a coin in a pocket. But for McDermott, to be in the world has connotations that are much richer: it is to "world," that is, to make a world, and to be "worlded," that is, to be made by a world.

Following a path blazed by the American pragmatists John Dewey and William James, McDermott contends that human beings are far from being complete and finished objects, as conventional wisdom would have it. We are instead permeable and interactive. We "incorporate" materials, energy waves, ideas, and thousands of other "things" into our selves on a daily, an hourly basis. We alter, build upon, transform, and decompose these things. Some of them we expel, some of them we absorb, and some of them we store temporarily against an anticipated use. Our existence is "uterine": we are in the womb of the world. McDermott quotes the architectural historian James Marston Fitch: "Life is coexistent with the external natural environment in which the body is submerged.

The body's dependence upon this external environment is absolute—in the fullest sense of the word—uterine."

Stretching his metaphors of the body almost to the breaking point, McDermott suggests that the human being needs two livers. The traditional one transforms our blood by filtering out its poisons. The second more vague and symbolic one "eats the sky and the earth, sorts out tones and colors, and provides a filter through which the experienced environment enters our consciousness. It is this spiritual liver which generates our feelings of queasiness, loneliness, surprise, and celebration."

The secular trinity of McDermott's account of embodiment comprises time, space, and things. Anticipating an investigation he will continue in his essay "Urban Time" in Part Four, McDermott suggests that we tend to embody time as if it were a given, a constant. But as post-Einsteinian beings, we should realize that time is a mock-up, that we can bring time to its knees. We can do this by making time over for our own needs and purposes, that is, by making time revolve around us rather than passing through us.

McDermott calls for active embodiments of place that are no less dramatic and far-reaching than those he demands for the embodiment of time. Since human beings have no natural place, place is ours to make over. As a race, we have moved from natural places to artifactual ones, from caves to skyscrapers. On a more personal level, each of us moves from cradle to coffin, from the small boxes that begin our lives to the small boxes that end them. Along the way we inhabit and incorporate the values of larger boxes: apartments, offices, sports arenas. How can we escape these boxes? We can do so only by the symbolic manipulation of the things they contain and of which they are constructed.

"Thing," in McDermott's reconstructed lexicon, means not simply an object but "concern, assembly, and above all, an affair. . . . Thing is what is to be done or its doing." Things are the clots in the flow of time; they are the materials we use to make our world, and they are the means by which the world makes us. Perhaps the ancients were right in burying the dead with their "things." Cartesian language has it wrong. It is not that things just *are;* they also happen. Things are bundles of relations, snipped at the edges by our interests, our likes, our dislikes.

In McDermott's rich account of the ways in which we embody technology and in which it embodies us, "the world is made sacred by our *hand*ling of our things. We are the makers of our world. It is we who praise, lament, and celebrate. Out of the doom of obviousness and repetition shall come the light, a light lit by the fire of our eyes."

Every technology expresses a double concern with embodiment. Every technology is a form of life that embodies human hopes and goals, and every technology comprises tools and instruments by means of which humans incorporate themselves into their environing situations. The central message of the otherwise diverse essays in this section is that if technological embodiment is to be appreciated, then it is necessary to go beyond preoccupation with abstract theories to an involvement with concrete human practice and production, even if such practice and production occurs at the level of the manipulation of the most abstract of symbols.

## Technology and Human Self-Conception

**Don Ihde**

### PREFACE

It will soon be seen that the thesis I shall explore today reverberates positively with the argument Professor Feibleman has made for technology affecting human nature. By way of preface, however, I must make one quibble: I do not believe that the overstatement that philosophers from Nietzsche to Heidegger have neglected the interaction and thus deep-level effect of technology upon humanity is true—although had Feibleman directed his comment to the mainstream of philosophy, in which the neglect of technology is widespread, he would have been more nearly correct.

Philosophy in its "platonist" and "idealist" form has tended to view technology pretty much as the outcome and result of ideas; or, to put it in contemporary form, technology is viewed as "applied science" with science in the form of theory as the root foundation of technology. This view, as I shall show in a moment, not only tempts the philosopher to overlook fundamental aspects of technology but leads to an arrival at questions of technology which is too late and thus misses the most fundamental questions.

There is, however, a set of traditions within philosophy that have already addressed technology at a fundamental level. Those philosophies are what I shall call the *praxis philosophies*. They include diverse families, such as Marxist and Critical Theory schools of thought, existentialism, phenomenology, and certain branches of American pragmatism. What they have in common are a starting point and conviction about human ontology. It is the conviction that some form of action precedes or grounds conception, or that a theory of action is primitive with respect to theory of knowledge.

Marx argues that human beings interact with their environment and with each other within some fundamental set of productive relations and actions—humans *are* what they make. For Sartre, humans project a project into the world and then seek to become that project—humans are what they do. For Dewey, humans are primarily problem-oriented, and their very intelligence is a tool for solving problems posed by their environment—humans are what they do in term of problems.

In each case some form of *praxis* is what grounds the relationship between humans and their world. Now when this thesis is applied to technology, not only can technology be seen to be important, and in a few cases even central, but it is related to the fundamental dimensions of human life itself. Thus philosophers in this tradition have argued that technology affects what Feibleman has called human nature, even if in some cases the effect is seen to be highly negative. Hans Jonas, for example, has specifically argued that technology affects the essence of humanity itself, and Heidegger as early as *Being and Time* argued for the primacy of the ready-to-hand as the origin and base for science itself.

None of this, of course, detracts from the major point of Feibelman's primary thesis. It links what will follow to a specific exploration to one dimension of that interaction between humans and their world by means of technology. What I am after is an understanding of how humans *interpret* themselves within a technological culture, and the thesis which I shall argue is that technology supplies the dominant basis for an understanding both of the world and of ourselves. In setting up the case, I shall link a more general philosophical question: How do humans come to understand themselves? with a more specific thematic question: How does technology affect this process?

## METHOD

Even before a philosophically informed analysis of technology and human self-conception, a minimal reflection can establish that technology at least provides a certain texture to the context of daily life. To make this point in class, I frequently assign students the task of cataloguing the number and kinds of human-technological artifact interactions in an hour or two of ordinary activity. The result is sometimes overwhelming and is a bit like trying to account for time in Proustian fashion. For example, beginning with the first conscious event of the day, it is likely that the ringing of an alarm or the sound of a clock radio is our first awareness. This is followed by a whole series of interactions and uses, which may include turning off the electric blanket or turning up the heat and in either case throwing back the technologically produced bedclothes from the technologically produced bed, engaging the vast plumbing system, and entering a veritable technologically jungle in the modern kitchen with stove, toaster, hot-water system, lighting, and so on. And even the philosopher takes this technological texture for granted in his or her daily use of telephone, Xerox machine, typewriter, automobile, ad infinitum.

All of this is familiar, even if we do not critically reflect upon its meaning for human life. And if Heidegger is right, precisely because it is familiar it is even more difficult to elicit its existential significance. Such a technological texture to life forms a "life-world," and familiarity itself may be a clue or index for what is taken as "true." If humans always interpret the world and themselves in some dominant way, how do they do this in the midst of technology?

The framework into which I shall cast my inquiry is one derived from the insights of twentieth-century phenomenology, although I shall eschew any intensive or extensive use of its complex tribal language. But there are a few essential fundamentals of the model of interpretation which need be noted. I shall do this by contrasting a phenomenological framework for self-conception with what I take to be the more usually taken-for-granted framework that derives from Descartes.

Descartes's *cogito* stands historically for a method of interpreting the world and the self which derives from the formula *Cogito ergo sum,* "I think, therefore I am." His ideal of clear and distinct ideas grounded upon certainty may in one sense be said to have invented both the "subject" and the "external world." This is to say that when the *cogito* "I think" became the basis for certainty it did so at a certain cost. On one side that cost was that the world and existence became "external," and on the other the subject became self-enclosed. From the "I think,'' world had to be derived by means of the geometrical method of inference. The self, in turn, while apodictic, became privatized and closed off from the world.

Now this model of interpretation also contains a notion of what must be the case for self-conceptions or self-interpretation. The Cartesian *cogito* with its own self-evidence *knows itself immediately;* it is self-transparent in some sense. But this immediacy contrasts with whatever is now "out there" and which must now be constructed. I may model this notion of self-understanding in the following way:

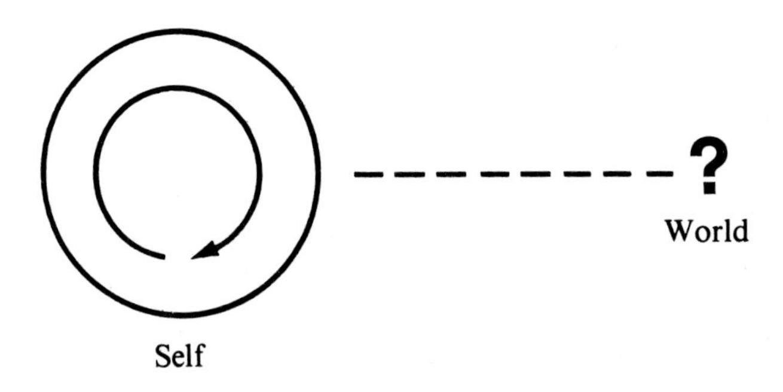


Here the self knows itself directly by being self-contained and presumably transparent. But the world becomes doubtful, unknown, and external, and the relation between the "subject" and the "object" becomes an enigma. Historically, of course, the solution to the enigma was the introduction of the "geometric method," which infers the world.

Phenomenology, at its inception through Edmund Husserl, explicitly rejects the Cartesian understanding of both self and world and formulates a radically alternative model of the relation between self and world. In his *Cartesian Meditations,* Husserl seemingly enacts a quasi-Cartesian reduction through a method of "doubt" now called a suspension—but the result is one which is directly counter to the Cartesian result. Without examining the steps along the way, I shall turn to that contrasting result: First, Husserl finds that the subject always already finds itself in a world. And although that world is always "there" in some sense, it no longer is "external" in the Cartesian sense. It is *present,* and its presence is what is to be interrogated first. Second, the subject now turns out never to be empty or self-contained but already correlated with a world. Indeed, what is primitive is that correlation itself, a correlation which in later phenomenology takes the form of being-in-a-world. Third, the task of phenomenology, then, becomes the examination of the correlation which Husserl termed *intentionality*. Thus the model for interpretation which contrasts with the Cartesian model may be diagramed as:

Here what appears or presents itself first is world and that to which it presents itself is "self." But the mode and form of presentation are the relation itself, which is intentionality. Now what this means is that for every *cogito,* or act of thought, there is a reference, a something thought, which is a something. To say that there are no empty thoughts ultimately means that there is no worldless subject. Phenomenologically world and self are *equally* certain. But in terms of analysis there is a certain primacy to the world. Phenomenologically it is *from* the world that I come to understand myself. Thus it is in interaction with the world that I come to any form of self-understanding, contrarily, without world I would understand nothing. In contrast to Cartesian "subjectivism" phenomenology emerges as a new kind of relational "objectivism."

For our purposes, however, what is needed is a notion of how this correlation functions with respect to a coming to self-conception. This may be formulated simply as: if I am always already *in* a world, and if it is by means of the world that I came to understand myself, then there is an essential sense in which self-understanding is always tied to an understanding of a world. Here a modification must be introduced to the first diagram of the intentional correlation:

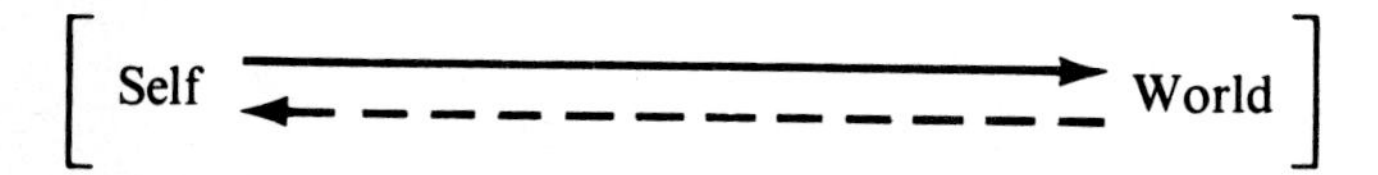


The intentional arrow turns out to be not single but interactional in form. It is both *projective,* a focused reference to world, and *reflective,* a movement from the world.

For interpreting self-understanding, the phenomenological self is seen to be neither self-contained nor separated from a context, a field, a world. In fact, to know oneself becomes a task, not a Cartesian given, since self-knowledge is no longer direct, autonomous, or self-contained.

My interpretation will follow this model of self-understanding, and I will examine certain specific features of what may be called a "technological world" and of self-understanding in that context. It will be suspected from the outset that, if self-knowledge is correlational or intentional, that something must happen at both ends of the relation with respect to human self-conception.

## TECHNOLOGY AND HUMAN SELF-CONCEPTION

If the intentional model of interpretation is the structure of phenomenology, the use of variations to isolate invariants is the dynamics of phenomenology. I shall use a set of variations in Husserlian style to point up what I believe are invariant features of human self-interpretation. The variants I shall use, however, are not strictly either the perceptual or the imaginative ones used in early phenomenological history. Rather I shall draw from both anthropological and history-of-religion disciplines to outline what may be called "world interpretations." I am seeking here an understanding both of what is variant and of what is invariant with respect to the ways human beings *project* and *reflect* a basic set of relations to a world.

What is not Husserlian in this analysis is that I am taking what I shall call *existential praxis* as a fundamental mode of human-world correlation. By this I mean a very broad, but concrete pattern of *actions* which always include relations with things of a material sort. Thus from the start I potentially include what come to be "technologies." The thesis is meant to illustrate both variations and invariants which relate to such human activities in the world.

To simplify the situation, I shall take only three somewhat idealized examples of a cultural human-world correlation with the specific "worlds" understood to be in scare quotes and seek to isolate what are the specific modes of understanding world and self in each example. My interpretation follows the phenomenological model of interpretation formerly outlined:

### Variation 1

Imagine a somewhat idealized partly nomadic Amerindian society. It is a hunting and gathering society whose form of existential praxis is one which gains its living from hunting animals and gathering plants and their products, but it is not primarily a sedentary society which practices agriculture or animal domestication. First, what is the "world" which is projected? Obviously it is a "world" which focuses upon animal and plant life. Of it there must be genuine knowledge, and this takes the shape of knowing what is good to eat, what is useful as a product for shelter, clothing, and so on. In short, this society adapts in a particular, focused or—phenomenologically—intentionally shaped way. Its project includes knowing migratory routes, animal behavior, plant distribution, and other patterns. Its "world," in short, is that lively world of animals and plants to which human beings focally relate for existence itself.

I have already suggested that the focal understanding of this "world" will take a particular shape. At its center will stand those items of the world which are of particular concern to the inhabitants. But if we now turn to the reflexive side of this project, we may note how this pattern of concern takes the specific form of human self-conception that it does. Here we may undertake a play upon the notion of reflection. A "reflective activity" in such a society takes the form of ritual and ritually transmitted knowledge. And here examination of societies of this sort reveal an interesting phenomenon. Often the reflection specifically projects an idealized form of existential praxis into some nonordinary time or place in the form of cultic behavior. In this case we find specific cults—of the dog, the buffalo, the snake, and so on. Young males, potential hunters, are initiated by secret ritual and dramatized form into such cults, and from this they learn the patterns of existential praxis which in fact form the basis of the society's life. But in the esoteric ritual one factor stands out—the initiate identifies or becomes "like" that to which he is primarily related. Thus in such a society the Indian thinks himself to be "brother" to the cult animal or, generalized, to be "brother" to the animal kingdom.

Similarly, the young women of the tribe undergo counterpart ritual initiations in which plant life and its patterns often are internalized and related to the existential praxis of gathering, making, and so on. In short, the actual form of life which projects a certain relation to world in the form of a "world" of the Amerindian also reflects into an idealized mode this form, and from it emerges a type of self-understanding in which the human becomes "like" the "world" which is projected.

### Variation 2

Now take an idealized agricultural society to note the form of life it takes. Here the existential praxis is changed. Both animal and plant life has been domesticated, and the basic patterns of life are those which revolve around a set of rhythms and relations which follow the cycles of grain and animal production. Now the familiar "world correlate" takes on an increasingly definite pattern in which constant repetition becomes basic and to which change is seen as a kind of threat.

The agricultural cycle is one of planting, growth, maturation, harvest, dormancy, and again planting. On the animal side there is a similar rhythm, more specifically related to sexual fertility in which there is birth, maturity, rutting and reproduction, slaughter, and again birth. This is the appearance of the ancient agricultural "world." The existential praxis is focused, and human beings here undertake and enhance those patterns which support life. But they also increasingly project this form upon a wider and wider set of phenomena—including, as we know, the projection of repeated cycles into the very heavens themselves as ancient agricultural society becomes urbanized and invents forms of astronomy and so on.

The ritual reflection of this form of life again takes the shape of humanity needing both to repeat its basic pattern of life and to be "like" that very "world" which is projected. The rites of agricultural society in ancient form are great cycle rites emphasizing fertility and identification with animal and plant life. The dying and rising gods, peasants copulating in fields to assure fertility, blood rituals of animal slaughter repeat and reinforce the very form of life which mirrors existential praxis.

At this point, however, both the variant and the invariant features begin to come clear, and variation 3 may be anticipated. Here it may be the case that objections begin to take shape precisely because we are quick to see where this leads. Both the Amerindian and the ancient agricultural variants, we might hold, are "primitive" and "prescientific" in that the apparent "anthropomorphization" of their "worlds" is too obvious to us. We have, we might hold, at least deanthropomorphized the world and thus are presumably nearer the truth than such "primitives." But if my thesis holds, we must apply the intentional method equally to all variations to see what presents itself.

### Variation 3

In this case the "world" is a technological one. What life-shape does it take, and how does "world" present itself and get understood? Of our dominant interpretation there can be little doubt. The contemporary "world" has been deanimated of both the nomadic and the agricultural projections. But, positively put, it has been mechanated. The "world" for us is interpreted as one of a vast system of impersonal relations, often explicitly conceived of in terms of mechanical metaphors. From Descartes on, the "world" has frequently been characterized as a "mechanism" which at one time was sometimes tinkered with by the Maker—but which today runs without tinkering. Even our bodies and those of animals were, and continue to be, interpreted along *technological* lines. We are contrivances of pumps (hearts), levers (arms), and electrical systems (nerves). What is this but a technological projection of "world"?

Furthermore, just as the Amerindian and the ancient agriculturalist became "like" their worlds through the projective-reflective process, we also internalize the external understanding. To understand itself is often to understand a "mechanism"; to follow rhythms of day and night must presuppose some form of "biological clock." Now although I could extend this virtually ad infinitum, the direction is clear enough. Do not our "world" and self-understanding in fact reflect the very form of existential praxis in which we have been engaged for several centuries?

To this point I have admittedly very sketchily made some suggestions, but a few preliminary observations may follow from these: First, I am suggesting that, far from clearly showing that twentieth-century life is superior to other forms of human life regarding self-understanding, I have suggested that the form of self-interpretation itself is *invariant* through all three examples. We project in terms of some focused form of existential praxis which influences, if not sets, the selected forms of knowledge which we will regard as central. Second, far from having deanthropomorphized our "world" compared with those of the other two variations, we have created only a new form of "anthropomorphization" now in the form of a technologically projected "world." Third, the emergent form of self-understanding is one which functionally is exactly equivalent to the other two variations as a kind of reflection and internalization of the very projected "world" we find ourselves in.

This, however, is still suggestive. What I must now do is turn more specifically to the theme of technology and self-understanding and demonstrate more intensely that our form of *existential praxis* is itself basic for our "world" and show how this reflects itself in contemporary "ritual" life.

## A TECHNOLOGICAL "WORLD"

Each of the variants sketched, I hold, contain genuine forms of knowledge. Amerindian nomadic societies knew the patterns of animal migration and behavior, often more thoroughly than even our contemporary territorialists do; ancient agricultural society not only came to know the cyclical patterns of plant and animal life but also invented a type of experimental genetics and breeding insights, just as we genuinely have come to know the vastly extended "mechanical" knowledge of the universe itself. Each projection aimed at the world carries with it a form of knowledge. And each projection reflects this back into the deeper form of life which revolves around the concrete praxis of the human "life-world." But the problem for the inhabitant of any given "world" is that it is so familiar to him or her that little distance is to be found. The familiar is simply taken as the "true."

To gain distance and insight one needs to see the variants within one's own "world" against a deeper, invariant structure, a structure which both takes account of the objective societal relations and can point up the forms of imaginative projection which depict the ideals and aims of the human world.

It is clearly not the case that we are lacking these insights altogether. For example, the social sciences have developed certain means of analysis which have pointed up some of the factors I have already alluded to. Focally, the ancient agricultural variant of a human world may be seen to be dependent upon the reiteration and repetition of a certain cyclical pattern which is repeated from the farmer, who repeats and practices the cycle of agriculture, all the way to the specific *interest* groups which symbolize the form of life, as for example a priesthood or royalty who ritually enact the fertility and renewal rites of the society.

Here the question naturally arises, how does this projection, repetition, and ritual renewal take place in our form of technological life? The recent lengthy discussions of alienation, depersonalization and anomie as resulting from or at least associated with technological society point to a certain lack of adjustment suffered by contemporary human beings in this "world." But where do we affirm and reenact our current existential praxis? What are our rituals of identification with the "world" we are projecting? Clearly we have nothing like either the esoteric initiation rituals or the annual repetition of the New Year's festivals which repeat the sacred time of agricultural society.

What is needed here, however, is to understand the functional nature of ritual as an imaginative projection-reflection. What happens functionally in affirming a self-understanding is that some form of existential praxis is projected into a not-now or other or sacred time and place and is then reflected back upon the inhabitants of a "world." And while it is the case that the first variant projected such an imaginative action into an esoteric realm and the agricultural society projected the same into an ancient pretime, as might be expected with technological time, we project our self-understanding into an *irreal future* time.

Permit me here a final and only partly cryptic speculation to illustrate what I take to be a certain fascination with technological self-understanding. Our "rituals," like all past ritual, is most appropriate if it is conveyed dramatically in clothing appropriate for our form of life. Just as the snake dancer or the eagle dancer creates the impression of a sacred and dramatized being, or as the cyclical birth-and-death rituals encapsule existential praxis in the New Year's festival, each embodying the imagination in a form appropriate to the respective "world" reflected, so we might expect our "ritual" to be embodied.

I therefore suggest that ours is embodied in various technological representations, the most notable candidates being film and television. And here I engage a bit of fancy to make a point: the current fascination of utopias, science fiction, UFOs, and such phenomena as *Star Wars,* "Battlestar Galactica," and the like illustrates a genre of projections placed in irreal future time. I propose that we take these momentarily as examples of rituals of self-interpretation and see what they say.

If they are instances of "ritual," they can at least be interpreted to show certain patterned notions which do relate to contemporary self-conceptions. I shall take account of only a few of these. First, note the role of what may be called "nature" compared with "culture" in the genre. Nature is at best a background, often spectacular but not itself a force to be reckoned with. Its limits have been conquered, including the absolute limit of our present science, the speed of light. What is foreground is totalized culture. Life takes shape within and often literally inside various forms of technological cocoons. Home is a spaceship. Outside one must wear a miniaturized spaceship in the form of a space suit. These rituals project a totalized technology which becomes the marvelous but then familiar home for future human beings. Never mind that future technology does not seem plagued with the constant repair, entropy, and breakdown of ours.

This idealized projection, however, is also a reflection of *present* existential praxis. The spaceship as "home" is a projection of the present ideal of the totally self-contained and hence totally controlled environment. This direction can be instanced in architecture. At Stony Brook the new Health Sciences Center, which looks like something out of an Antonioni movie, is a monstrous set of buildings many stories high—but it has virtually no windows. In the "megastructure" which is the main building, the windows it does have actually open either to hallways or to the floor where the heating and cooling machinery is installed (only elite physicians' offices have corner oval windows). The rationale, of course, is total control over the environment, and thus the self-enclosure. The hospital on the other hand does have windows. I asked someone why that was the case—were they being humane to patients? The answer was, "No, we would have preferred a windowless hospital, too, but state law requires windows in hospitals." My point is that the projection of the ideal future world, the spaceship inside of which humans live, move, and have their being, is an expression of current existential praxis. Moreover, this self-contained technological environment is not only instanced in buildings but approximated in such items as the mini-environments of such things as contemporary recreational vehicles, which take technological suburbia, complete with TV, flush toilets, gas cookers, and the rest right into the wilderness. Such technological cocoons exemplify the trajectory of our civilization to totality.

If "home" is the technological cocoon, what becomes of the "outside" which was nature? In the genre of projections mentioned, it is background, but in function it is something else. It is the realm of *resources,* not just there presumably for human use but there as stored energy. And nature, too, is to be reordered and transformed. Modernization includes the reordering of everything from a transportation system to fully developed agriculture with transformed seeds (hybridization) planting processes (fertilizers and pesticides) to products (white bread). Ultimately nature in its untransformed state is inverted and taken into the totality of technological culture in the form of natural museums. Wilderness areas are functionally large pictures in the museums, set aside for temporary enjoyment and relaxation or escape from daily life, but episodic with respect to the cultural and "normal" world of technologically transformed nature. Thus the ultimate victory of technological totalization would be this inversion in which nature is itself taken into culture. Then technological civilization would be complete, and we would be "one" with our "world."

## SUMMARY

The somewhat, but only somewhat, facetious examples I have been tracing are not meant to complete a picture of what human self-understanding would be in technological civilization, nor are they really meant to cast a negative tone on our form of life. They are instead intended to illustrate the one and only thesis I have been developing: that all self-interpretation takes its shape in a certain way with respect to some basic form of existential praxis which is projected upon the world and reflected back in ways which become dominant ways of understanding ourselves and our world.

I have tried to show that, far from being obviously and self-evidently superior in attaining self-knowledge—a worthy Socratic goal—we remain caught in some degree in precisely the dyanamism which characterized the seeming "anthropomorphizations" of the other variants. We end up modeling ourselves upon the very "world" we project and interpret ourselves in terms of technology.

Negatively, there is also a point not developed here. If human beings are, as the existentialists have claimed, always transcendent beings, then what is needed more than ever is a way of attaining a transcendence to any reduction which ties us too closely to a "world" in which we continually fall into some interpretation which forecloses us in a totality. What is needed is what I shall call a "loose" or maybe even "Zen" relation to technology.

# From *Phenomenology of Perception*

**Maurice Merleau-Ponty**

From the existential point of view, these two facts, which scientific induction contents itself with setting side by side, are linked internally and are understood in the light of one and the same idea. If man is not to be embedded in the matrix of that syncretic setting in which animals lead their lives in a sort of *ek-stase,* if he is to be aware of a world as the common reason for all settings and the theatre of all patterns of behaviour, then between himself and what elicits his action a distance must be set, and, as Malebranche put it, forms of stimulation from outside must henceforth impinge on him 'respectfully'; each momentary situation must cease to be, for him, the totality of being, each particular response must no longer fill his whole field of action. Furthermore, the elaboration of these responses, instead of occurring at the centre of his existence, must take place on the periphery and finally the responses themselves must no longer demand that on each occasion some special position be taken up, but they must be outlined once and for all in their generality. Thus it is by giving up part of his spontaneity, by becoming involved in the world through stable organs and pre-established circuits that man can acquire the mental and practical space which will theoretically free him from his environment and allow him to *see* it. And provided that even the realization of an objective world is set in the realm of existence, we shall no longer find any contradiction between it and bodily conditioning: it is an inner necessity for the most integrated existence to provide itself with an habitual body. What allows us to link to each other the 'physiological' and the 'psychic', is the fact that, when reintegrated into existence, they are no longer distinguishable respectively as the order of the *in-itself,* and that of the *for-itself,* and that they are both directed towards an intentional pole or towards a world. Doubtless the two histories never quite coincide: one is commonplace and cyclic, the other may be open and unusual, and it would be necessary to keep the term 'history' for the second order of phenomena if history were a succession of events which not only have a meaning, but furnish themselves with it. However, failing a true revolution which breaks up historical categories so far valid, the figure in history does not create his part completely: faced with typical situations he takes typical decisions and Nicholas II, repeating the very words of Louis XVI, plays the already written part of established power in face of a new power. His decisions translate the a priori of a threatened prince as our reflexes translate a specific a priori. These stereotypes, moreover, are not a destiny, and just as clothing, jewellery and love transfigure the biological needs from which they arise, in the same way within the cultural world the historical a priori is constant only for a given phase and provided that the balance of *forces* allows the same *forms* to remain. So history is neither a perpetual novelty, nor a perpetual repetition, but the *unique* movement which creates stable forms and breaks them up. The organism and its monotonous dialectical processes are therefore not alien to history and as it were inassimilable to it. Man taken as a concrete being is not a psyche joined to an organism, but the movement to and fro of existence which at one time allows itself to take corporeal form and at others moves towards personal acts. Psychological motives and bodily occasions may overlap because there is not a single impulse in a living body which is entirely fortuitous in relation to psychic intentions, not a single mental act which has not found at least its germ or its general outline in physiological tendencies. It is never a question of the incomprehensible meeting of two causalities, nor of a collision between the order of causes and that of ends. But by an imperceptible twist an organic process issues into human behaviour, an instinctive act changes direction and becomes a sentiment, or conversely a human act becomes torpid and is continued absentmindedly in the form of a reflex. Between the psychic and the physiological there may take place exchanges which almost always stand in the way of defining a mental disturbance as psychic *or* somatic. The disturbance described as somatic produces, on the theme of the organic accident, tentative psychic commentaries, and the 'psychic' trouble confines itself to elaborating the human significance of the bodily event. A patient feels a second person implanted in his body. He is a man in half his body, a woman in the other half. How are we to distinguish in this symptom the physiological causes and psychological motives? How are we to associate the two explanations and how imagine any point at which the two determinants meet? 'In symptoms of this kind, the psychic and the physical are so intimately linked that it is unthinkable to try to complete one of these functional domains by the other, and that both must be subsumed under a third...(We must)...move on from knowledge of psychological and physiological facts to a recognition of the animic event as a vital process inherent in our existence.'[1] Thus, to the question which we were asking, modern physiology gives a very clear reply: the psycho-physical event can no longer be conceived after the model of Cartesian physiology and as the juxtaposition of a process in itself and a *cogitatio*. The union of soul and body is not an amalgamation between two mutually external terms, subject and object, brought about by arbitrary decree. It is enacted at every instant in the movement of existence. We found existence in the body when we approached it by the first way of access, namely through physiology. We may therefore at this stage examine this first result and make it more explicit, by questioning existence this time on its own nature, which means, by having recourse to psychology.

The acquisition of habit as a rearrangement and renewal of the corporeal schema presents great difficulties to traditional philosophies, which are always inclined to conceive synthesis as intellectual synthesis. It is quite true that what brings together, in habit, component actions, reactions and 'stimuli' is not some external process of association.[2] Any mechanistic theory runs up against the fact that the learning process is systematic; the subject does not weld together individual movements and individual stimuli but acquires the power to respond with a certain type of solution to situations of a certain general form. The situations may differ widely from case to case, and the response movements may be entrusted sometimes to one operative organ, sometimes to another, both situations and responses in the various cases having in common not so much a partial identity of elements as a shared meaning. Must we then see the origin of habit in an act of understanding which organizes the elements only to withdraw subsequently?[3] For example, is it not the case that forming the habit of dancing is discovering, by analysis, the formula of the movement in question, and then reconstructing it on the basis of the ideal outline by the use of previously acquired movements, those of walking and running? But before the formula of the new dance can incorporate certain elements of general motility, it must first have had, as it were, the stamp of movement set upon it. As has often been said, it is the body which 'catches' (*kapiert*) and 'comprehends' movement. The acquisition of a habit is indeed the grasping of a significance, but it is the motor grasping of a motor significance. Now what precisely does this mean? A woman may, without any calculation, keep a safe distance between the feather in her hat and things which might break it off. She feels where the feather is just as we feel where our hand is.[4] If I am in the habit of driving a car, I enter a narrow opening and see that I can 'get through' without comparing the width of the opening with that of the wings, just as I go through a doorway without checking the width of the doorway against that of my body.[5] The hat and the car have ceased to be objects with a size and volume which is established by comparison with other objects. They have become potentialities of volume, the demand for a certain amount of free space. In the same way the iron gate to the Underground platform, and the road, have become restrictive potentialities and immediately appear passable or impassable for my body with its adjuncts. The blind man's stick has ceased to be an object for him, and is no longer perceived for itself; its point has become an area of sensitivity, extending the scope and active radius of touch, and providing a parallel to sight. In the exploration of things, the length of the stick does not enter expressly as a middle term: the blind man is rather aware of it through the position of objects than of the position of objects through it. The position of things is immediately given through the extent of the reach which carries him to it, which comprises besides the arm's own reach the stick's range of action. If I want to get used to a stick, I try it by touching a few things with it, and eventually I have it 'well in hand', I can see what things are 'within reach' or out of reach of my stick. There is no question here of any quick estimate or any comparison between the objective length of the stick and the objective distance away of the goal to be reached. The points in space do not stand out as objective positions in relation to the objective position occupied by our body; they mark, in our vicinity, the varying range of our aims and our gestures. To get used to a hat, a car or a stick is to be transplanted into them, or conversely, to incorporate them into the bulk of our own body. Habit expresses our power of dilating our being-in-the-world, or changing our existence by appropriating fresh instruments.[6] It is possible to know how to type without being able to say where the letters which make the words are to be found on the banks of keys. To know how to type is not, then, to know the place of each letter among the keys, nor even to have acquired a conditioned reflex for each one, which is set in motion by the letter as it comes before our eye. If habit is neither a form of knowledge nor an involuntary action, what then is it? It is knowledge in the hands, which is forthcoming only when bodily effort is made, and cannot be formulated in detachment from that effort. The subject knows where the letters are on the typewriter as we know where one of our limbs is, through a knowledge bred of familiarity which does not give us a position in objective space. The movement of her fingers is not presented to the typist as a path through space which can be described, but merely as a certain adjustment of motility, physiognomically distinguishable from any other. The question is often framed as if the perception of a letter written on paper aroused the representation of the same letter which in turn aroused the representation of the movement needed to strike it on the machine. But this is mythological language. When I run my eyes over the text set before me, there do not occur perceptions which stir up representations, but patterns are formed as I look, and these are endowed with a typical or familiar physiognomy. When I sit at my typewriter, a motor space opens up beneath my hands, in which I am about to 'play' what I have read. The reading of the word is a modulation of visible space, the performance of the movement is a modulation of manual space, and the whole question is how a certain physiognomy of 'visual' patterns can evoke a certain type of motor response, how each 'visual' structure eventually provides itself with its mobile essence without there being any need to spell the word or specify the movement in detail in order to translate one into the other. But this power of habit is no different from the general one which we exercise over our body: if I am ordered to touch my ear or my knee, I move my hand to my ear or my knee by the shortest route, without having to think of the initial position of my hand, or that of my ear, or the path between them. We said earlier that it is the body which 'understands' in the acquisition of habit. This way of putting it will appear absurd, if understanding is subsuming a sense-datum under an idea, and if the body is an object. But the phenomenon of habit is just what prompts us to revise our notion of 'understand' and our notion of the body. To understand is to experience the harmony between what we aim at and what is given, between the intention and the performance—and the body is our anchorage in a world. When I put my hand to my knee, I experience at every stage of the movement the fulfillment of an intention which was not directed at my knee as an idea or even as an object, but as a present and real part of my living body, that is, finally, as a stage in my perpetual movement towards a world. When the typist performs the necessary movements on the typewriter, these movements are governed by an intention, but the intention does not posit the keys as objective locations. It is literally true that the subject who learns to type incorporates the key-bank space into his bodily space.

The example of instrumentalists shows even better how habit has its abode neither in thought nor in the objective body, but in the body as mediator of a world. It is known[7] that an experienced organist is capable of playing an organ which he does not know, which has more or fewer manuals, and stops differently arranged, compared with those on the instrument he is used to playing. He needs only an hour's practice to be ready to perform his programme. Such a short preparation rules out the supposition that new conditioned reflexes have here been substituted for the existing sets, except where both form a system and the change is all-embracing, which takes us away from the mechanistic theory, since in that case the reactions are mediated by a comprehensive grasp of the instrument. Are we to maintain that the organist analyses the organ, that he conjures up and retains a representation of the stops, pedals and manuals and their relation to each other in space? But during the short rehearsal preceding the concert, he does not act like a person about to draw up a plan. He sits on the seat, works the pedals, pulls out the stops, gets the measure of the instrument with his body, incorporates within himself the relevant directions and dimensions, settles into the organ as one settles into a house. He does not learn objective spatial positions for each stop and pedal, nor does he commit them to 'memory'. During the rehearsal, as during the performance, the stops, pedals and manuals are given to him as nothing more than possibilities of achieving certain emotional or musical values, and their positions are simply the places through which this value appears in the world. Between the musical essence of the piece as it is shown in the score and the notes which actually sound round the organ, so direct a relation is established that the organist's body and his instrument are merely the medium of this relationship. Henceforth the music exists by itself and through it all the rest exists.[8] There is here no place for any 'memory' of the position of the stops, and it is not in objective space that the organist in fact is playing. In reality his movements during rehearsal are consecratory gestures: they draw affective vectors, discover emotional sources, and create a space of expressiveness as the movements of the augur delimit the *templum*.

The whole problem of habit here is one of knowing how the musical significance of an action can be concentrated in a certain place to the extent that, in giving himself entirely to the music, the organist reaches for precisely those stops and pedals which are to bring it into being. Now the body is essentially an expressive space. If I want to take hold of an object, already, at a point of space about which I have been quite unmindful, this power of grasping constituted by my hand moves upwards towards the thing. I move my legs not as things in space two and a half feet from my head, but as a power of locomotion which extends my motor intention downwards. The main areas of my body are devoted to actions, and participate in their value, and asking why common sense makes the head the seat of thought raises the same problem as asking how the organist distributes, through 'organ space', musical significances. But our body is not merely one expressive space among the rest, for that is simply the constituted body. It is the origin of the rest, expressive movement itself, that which causes them to begin to exist as things, under our hands and eyes. Although our body does not impose definite instincts upon us from birth, as it does upon animals, it does at least give to our life the form of generality, and develops our personal acts into stable dispositional tendencies. In this sense our nature is not long-established custom, since custom presupposes the form of passivity derived from nature. The body is our general medium for having a world. Sometimes it is restricted to the actions necessary for the conservation of life, and accordingly it posits around us a biological world; at other times, elaborating upon these primary actions and moving from their literal to a figurative meaning, it manifests through them a core of new significance: this is true of motor habits such as dancing. Sometimes, finally, the meaning aimed at cannot be achieved by the body's natural means; it must then build itself an instrument, and it projects thereby around itself a cultural world. At all levels it performs the same function which is to endow the instantaneous expressions of spontaneity with 'a little renewable action and independent existence.'[9] Habit is merely a form of this fundamental power. We say that the body has understood and habit has been cultivated when it has absorbed a new meaning, and assimilated a fresh core of significance.

To sum up, what we have discovered through the study of motility, is a new meaning of the word 'meaning'. The great strength of intellectualist psychology and idealist philosophy comes from their having no difficulty in showing that perception and thought have an intrinsic significance and cannot be explained in terms of the external association of fortuitously agglomerated contents. The *cogito* was the coming to self-awareness of this inner core. But all meaning was ipso facto conceived as an act of thought, as the work of a pure *I*, and although rationalism easily refuted empiricism, it was itself unable to account for the variety of experience, for the element of senselessness in it, for the contingency of contents. Bodily experience forces us to acknowledge an imposition of meaning which is not the work of a universal constituting consciousness, a meaning which clings to certain contents. My body is that meaningful core which behaves like a general function, and which nevertheless exists, and is susceptible to disease. In it we learn to know that union of essence and existence which we shall find again in perception generally, and which we shall then have to describe more fully.

## NOTES

1 E. Menninger-Lerchenthal, *Das Truggebilde der eigenen Gestalt,* pp. 174–75
2 See, on this point, *La Structure du Comportment,* pp. 125 and ff.
3 As Bergson, for example, thinks when he defines habit as 'the fossilized residue of a spiritual activity.'
4 Head, *Sensory disturbances from cerebral lesion,* p. 188.
5 Grünbaum, *Aphasie und Motorik,* p. 395.
6 It thus elucidates the nature of the body image. When we say that it presents us immediately with our bodily position, we do not mean, after the manner of empiricists, that it consists of a mosaic of 'extensive sensation'. It is a system which is open on to the world, and correlative with it.
7 Cf. Chevalier, *L'Habitude,* pp. 202 and ff.
8 'As though the musicians were not nearly so much playing the little phrase as performing the rites on which it insisted before it would consent to appear.' (Proust, *Swann's Way,* II, trans. C. K. Scott Moncrieff, Chatto & Windus, p. 180.)
'Its cries were so sudden that the violinist must snatch up his bow and race to catch them as they came.' (Ibid., p. 186.)
9 Valéry, *Introduction à la Méthode de Léonard de Vinci, Variété,* p. 177.

# From *In the Age of the Smart Machine*

**Shoshana Zuboff**

## THE AUTOMATIC DOORS

The bleach plant is one of the most complex and treacherous areas of a pulp mill. In Piney Wood, a large pulp plant built in the mid-1940s, railroad tank cars filled with chemicals used in the bleaching process pull up alongside the four-story structure in which dirty brown digested pulp is turned gleaming white. Each minute, 4,000 gallons of this brown mash flow through a labyrinth of pipes into a series of cylindrical vats, where they are washed, treated with chlorine-related chemicals, and bleached white. No natural light finds its way into this part of the mill. The fluorescent tubes overhead cast a greenish-yellow pall, and the air is laced with enough chemical flavor that as you breathe it, some involuntary wisdom built deep into the human body registers an assault. The floors are generally wet, particularly in the areas right around the base of one of the large vats that loom like raised craters on a moonscape. Sometimes a washer runs over, spilling soggy cellulose knee-deep across the floor. When this happens, the men put on their high rubber boots and shovel up the mess.

The five stages of the bleaching process include hundreds of operating variables. The bleach operator must monitor and control the flow of stock, chemicals, and water, judge color and viscosity, attend to time, temperature, tank levels, and surge rates—the list goes on. Before computer monitoring and control, an operator in this part of the mill would make continual rounds, checking dials and graph charts located on the equipment, opening and shutting valves, keeping an eye on vat levels, snatching a bit of pulp from a vat to check its color, sniff it, or squeeze it between his fingers ("Is it slick? Is it sticky?") to determine its density or to judge the chemical mix.

In 1981 a central control room was constructed in the bleach plant. A science fiction writer's fantasy, it is a gleaming glass bubble that seems to have erupted like a mushroom in the dark, moist, toxic atmosphere of the plant. The control room reflects a new technological era for continuous-process production, one in which microprocessor-based sensors linked to computers allow remote monitoring and control of the key process variables. In fact, the entire pulp mill was involved in this conversion from the pneumatic control technology of the 1940s to the microprocessor-based information and control technology of the 1980s.

Inside the control room, the air is filtered and hums with the sound of the air-conditioning unit built into the wall between the control room and a small snack area. Workers sit on orthopedically designed swivel chairs covered with a royal blue fabric, facing video display terminals. The terminals, which display process information for the purposes of monitoring and control, are built into polished oak cabinets. Their screens glow with numbers, letters, and graphics in vivid red, green, and blue. The floor here is covered with slate-gray carpeting; the angled countertops on which the terminals sit are rust brown and edged in black. The walls are covered with a wheat-colored fabric and the molding repeats the polished oak of the cabinetry. The dropped ceiling is of a bronzed metal, and from it is suspended a three dimensional structure into which lights have been recessed and angled to provide the right amount of illumination without creating glare on the screens. The color scheme is repeated on the ceiling—soft tones of beige, rust, brown, and gray in a geometric design.

The terminals each face toward the front of the room—a windowed wall that opens onto the bleach plant. The steel beams, metal tanks, and maze of thick pipes visible through those windows appear to be a world away in a perpetual twilight of steam and fumes, like a city street on a misty night, silent and dimly lit. What is most striking about the juxtaposition of these two worlds, is how a man (and there were only men working in this part of the mill) traverses the boundary between them.

The control room is entered through an automatic sliding-glass door. At the push of a button, the two panels of the door part, and when you step forward, they quickly close behind you. You then find yourself facing two more automatic doors at right angles to one another. The door on the right leads to a narrow snack area with booths, cabinets, a coffee machine, and a refrigerator. The door to the left leads into the control room. It will not open until the first door has shut. This ensures that the filtered air within the control room is protected from the fumes and heat of the bleach plant. The same routine holds in reverse. When a man leaves the control room, he presses a button next to the frame on the inner door, which opens electronically. He then steps through it into the tiny chamber where he must wait for the door to seal behind him so that he can push a second button on the outer door and finally exit into the plant.

This is not what most men do when they move from the control room out into the bleach plant. They step through the inner door, but they do not wait for that door to seal behind them before opening the second door. Instead, they force their fingertips through the rubber seal down the middle of the outer door and, with a mighty heft of their shoulders, pry open the seam and wrench the door apart. Hour after hour, shift after shift, week after week, too many men pit the strength in their arms and shoulders against the electronic mechanism that controls the doors. Three years after the construction of the sleek, glittering glass bubble, the outer door no longer closes tightly. A gap of several inches, running down the center between the two panels of glass, looks like a battle wound. The door is crippled.

"The door is broke now because the men pushed it too hard comin' in and out," says one operator. In talking to the men about this occurrence, so mundane as almost to defy reflection, I hear not only a simple impatience and frustration but also something deeper: a forward momentum of their bodies, whose physical power seems trivialized by the new circumstances of their work; a boyish energy that wants to break free; a subtle rebellion against the preprogrammed design that orders their environment and always knows best. Yet these are the men who also complained, "The fumes in the bleach plant will kill you. You can't take that chlorine no matter how big and bad you are. It will bleach your brains and no one (in management) gives a damn."

Technology represents intelligence systematically applied to the problem of the body. It functions to amplify and surpass the organic limits of the body; it compensates for the body's fragility and vulnerability. Industrial technology has substituted for the human body in many of the processes associated with production and so has redefined the limits of production formerly imposed by the body. As a result, society's capacity to produce things has been extended in a way that is unprecedented in human history. This achievement has not been without its costs, however. In diminishing the role of the worker's body in the labor process, industrial technology has also tended to diminish the importance of the worker. In creating jobs that require less human effort, industrial technology has also been used to create jobs that require less human talent. In creating jobs that demand less of the body, industrial production has also tended to create jobs that give less to the body, in terms of opportunities to accrue knowledge in the production process. These two-sided consequences have been fundamental for the growth and development of the industrial bureaucracy, which has depended upon the rationalization and centralization of knowledge as the basis of control.

These consequences also help explain the worker's historical ambivalence toward automation. It is an ambivalence that draws upon the loathing as well as the commitment that human beings can experience toward their work. Throughout most of human history, work has inescapably meant the exertion and often the depletion of the worker's body. Yet only in the context of such exertion was it possible to learn a trade and to master skills. Since the industrial revolution, the accelerated progress of automation has generally meant a reduction in the amount of effort required of the human body in the labor process. It has also tended to reduce the quality of skills that a worker must bring to the activity of making something. Industrial technology has been developed in a manner that increases its capacity to spare the human body, while at the same time it has usurped opportunities for the development and performance of skills that only the body can learn and remember. In their treatment of the automatic doors, the bleach plant workers have created a living metaphor that reflects this ambivalence toward automation. They want to be protected from toxic fumes, but they simultaneously feel a stubborn rebellion against a structure that no longer requires either the strength or the know-how lodged in their bodies.

The progress of automation has been associated with both a general decline in the degree of know-how required of the worker and a decline in the degree of physical punishment to which he or she must be subjected. Information technology, however, does have the potential to redirect the historical trajectory of automation. The intrinsic power of its informating capacity can change the basis upon which knowledge is developed and applied in the industrial production process by lifting knowledge entirely out of the body's domain. The new technology signals the transposition of work activities to the abstract domain of information. Toil no longer implies physical depletion. "Work" becomes the manipulation of symbols, and when this occurs, the nature of skill is redefined. The application of technology that preserves the body may no longer imply the destruction of knowledge; instead, it may imply the reconstruction of knowledge of a different sort.

## FROM ACTION-CENTERED TO INTELLECTIVE SKILL

The pulp and paper mills reveal the shift in the grounds of knowledge associated with a technology that informates. Men and women accustomed to an intimate physical association with the production process found themselves removed from the action. Now they had to know and to do based upon their ability to understand and manipulate electronic data. In Piney Wood, a $200 million investment in technology was radically altering every phase of mill life. Managers believed they were merely "upgrading" in order to modernize production and to improve productivity. Tiger Creek was undergoing a similar modernization process. In both cases, informating dynamics tended to unfold as an unintended and undermanaged consequence of these efforts. Cedar Bluff had been designed with a technological infrastructure based on integrated information and control systems. In that organization, managers were somewhat more self-conscious about using the informating capacity of the technology as the basis for developing new operating skills.

The experiences of the skilled workers in these mills provide a frame of reference for a general appraisal of the forms of knowledge that are required in an informated environment. My contention is that the skill demands that can be deciphered from their experiences have relevance for a wider range of organizational settings in both manufacturing and service sectors. Later chapters will compare the experiences of clerks and managers to those of the mill operators.

This joint appraisal will help to unravel the intrinsic and the contingent aspects of change and to gauge the generalizations that follow from the dilemmas of transformation described here.

A fundamental quality of this technological transformation, as it is experienced by workers and observed by their managers, involves a reorientation of the means by which one can have a palpable effect upon the world. Immediate physical responses must be replaced by an abstract thought process in which options are considered, and choices are made and then translated into the terms of the information system. For many, physical action is restricted to the play of fingers on the terminal keyboard. As one operator put it, "Your past physical mobility must be translated into a mental thought process." A Cedar Bluff manager with prior experience in pulping contemplates the distinct capacities that had become necessary in a highly computerized environment:

> In 1953 we put operation and control as close together as possible. We did a lot of localizing so that when you made a change you could watch the change, actually see the motor start up. With the evolution of computer technology, you centralize controls and move away from the actual physical process. If you don't have an understanding of what is happening and how all the pieces interact, it is more difficult. You need a new learning capability, because when you operate with the computer, you can't see what is happening. There is a difference in the mental and conceptual capabilities you need—you have to do things in your mind.

When operators in Piney Wood and Tiger Creek discuss their traditional skills, they speak of knowing things by habit and association. They talk about "cause-and-effect" knowledge and being able to see the things to which they must respond. They refer to "folk medicine" and knowledge that you don't even know you have until it is suddenly displayed in the ability to take a decisive action and make something work.

In plants like Piney Wood and Tiger Creek, where operators have relied upon action-centered skill, management must convince the operator to leave behind a world in which things were immediately known, comprehensively sensed, and able to be acted upon directly, in order to embrace a world that is dominated by objective data, is removed from the action context, and requires a qualitatively different kind of response. In this new world, personal interpretations of how to make things happen count for little. The worker who has relied upon an intimate knowledge of a piece of equipment—the operators talk about having "pet knobs" or knowing just where to kick a machine to make it hum—feels adrift. To be effective, he or she must now trade immediate knowledge for a more explicit understanding of the science that undergirds the operation. One Piney Wood manager described it this way:

> The workers have an intuitive feel of what the process needs to be. Someone in the process will listen to things, and that is their information. All of their senses are supplying data. But once they are in the control room, all they have to do is look at the screen. Things are concentrated right in front of you. You don't have sensory feedback. You have to draw inferences by watching the data, so you must understand the theory behind it. In the long run, you would like people who can take data and draw broad conclusions from it. They must be more scientific.

Many managers are not optimistic about the ability of experienced workers to trade their embodied knowledge for a more explicit, "scientific" inference.

> The operators today know if I do "x," then "y" will happen. But they don't understand the real logic of the system. Their cause-and-effect reasoning comes from their experience. Once we put things under automatic control and ask them to relate to the process using the computer, their personal judgments about how to relate to equipment go by the wayside. We are saying your intuition is no longer valuable. Now you must understand the whole process and the theory behind it.

Now a new kind of learning must begin. It is slow and scary, and many workers are timid, not wanting to appear foolish and incompetent. Hammers and wrenches have been replaced by numbers and buttons. An operator with thirty years of service in the Piney Wood Mill described his experience in the computer-mediated environment:

> Anytime you mash a button you should have in mind exactly what is going to happen. You need to have in your mind where it is at, what it is doing, and why it is doing it. Out there in the plant, you can know things just by habit. You can know them without knowing that you know them. In here you have to watch the numbers, whereas out there you have to watch the actual process.

"You need to have in your mind where it is at"—it is a simple phrase, but deceptive. What it takes to have things "in your mind" is far different from the knowledge associated with action-centered skill.

This does not imply that action-centered skills exist independent of cognitive activity. Rather, it means that the processes of learning, remembering, and displaying action-centered skills do not necessarily require that the knowledge they contain be made explicit. Physical cues do not require inference; learning in an action-centered context is more likely to be analogical than analytical. In contrast, the abstract cues available through the data interface do not require explicit inferential reasoning, particularly in the early phases of the learning process. It is necessary to reason out the meaning of those cues—what is their relation to each other and to the world "out there"?

It is also necessary to understand the procedures according to which these abstract cues can be manipulated to result in the desired effects. Procedural reasoning means having an understanding of the internal structure of the information system and its functional capacities. This makes it possible both to operate skillfully through the system and to use the system as a source of learning and feedback. For example, one operation might require sixteen control actions spread across four groups of variables. The operator must first think about what has to be done. Second, he or she must know how data elements (abstract cues) correspond to actual processes and their systemic relations. Third, the operator must have a conception of the information system itself, in order to know how actions taken at the information interface can result in appropriate outcomes. Fourth, having decided what to do and executed that command, he or she must scan new data and check for results. Each of these processes folds back upon a kind of thinking that can stand independent from the physical context. An operator summed it up this way:

> Before computers, we didn't have to think as much, just react. You just knew what to do because it was physically there. Now, the most important thing to learn is to think before you do something, to think about what you are planning to do. You have to know which variables are the most critical and therefore what to be most cautious about, what to spend time thinking about before you take action.

The vital element here is that these workers feel a stark difference in the forms of knowledge they must now use. Their experience of competence has been radically altered. "We never got paid to have ideas," said one Tiger Creek worker. "We got paid to work." Work was the exertion that could be known by its material results. The fact that a material world must be created required physical exertion. Most of the operators believed that some people in society are paid to "think," but they were not among them. They knew themselves to be the ones who gave their bodies in effort and skill, and through their bodies, they made things. Accustomed to gauging their integrity in intimate measures of strain and sweat, these workers find that information technology has challenged their assumptions and thrown them into turmoil. There was a gradual dawning that the rules of the game had changed. For some, this created panic; they did not believe in their ability to think in this new way and were afraid of being revealed as incompetent.

Such feelings are no mere accident of personality, but the sedimentation of long years of conditioned learning about who does the "thinking"—a boundary that is not meant to be crossed. As a Tiger Creek manager observed:

> Currently, managers make all the decisions. . . . Operators don't want to hear about alternatives. They have been trained to *do*, not to *think*. There is a fear of being punished if you think. This translates into a fear of the new technology.

In each control room, a tale is told about one or two old-timers who, though they knew more about the process than anyone else, "just up and quit" when they heard the new technology was coming. From one plant to another, reports of these cases were remarkably similar:

> He felt that because he had never graduated high school, he would never be able to keep up with this new stuff. We tried to tell him different, but he just wouldn't listen.

Despite the anxiety of change, those who left were not the majority. Most men and women need their jobs and will do whatever it takes to keep them. Beyond this, there were many who were honestly intrigued with the opportunity this change offered. They seemed to get pulled in gradually, observing their own experiences and savoring with secret surprise each new bit of evidence of their unexpected abilities. They discussed the newness and strangeness of having to act upon the world by exerting a more strictly intellectual effort. Under the gentle stimulus of a researcher's questions, they thought about this new kind of thinking. What does it feel like? Here are the observations of an operator who spent twenty years in one of the most manually intensive parts of the Tiger Creek Mill, which had recently been computerized:

> If something is happening, if something is going wrong, you don't go down and fix it. Instead, you stay up here and think about the sequence, and you think about how you

> want to affect the sequence. You get it done through your thinking. But dealing with information instead of things is very...well, very intriguing. I am very aware of the need for my mental involvement now. I am always wondering: Where am I at? What is happening? It all occurs in your mind now.

Another operator discussed the same experience but added a dimension. After describing the demand for thinking and mental involvement, he observed:

> Things occur to me now that never would have occurred to me before. With all of this information in front of me, I begin to think about how to do the job better. And, being freed from all that manual activity, you really have time to look at things, to think about them, and to anticipate.

As information technology restructures the work situation, it abstracts thought from action. Absorption, immediacy, and organic responsiveness are superseded by distance, coolness, and remoteness. Such distance brings an opportunity for reflection. There was little doubt in these workers' minds that the logic of their jobs had been fundamentally altered. As another worker from Tiger Creek summed it up, "Sitting in this room and just thinking has become part of my job. It's the technology that lets me do these things."

The thinking this operator refers to is of a different quality from the thinking that attended the display of action-centered skills. It combines abstraction, explicit inference, and procedural reasoning. Taken together, these elements make possible a new set of competencies that I call *intellective skills.* As long as the new technology signals only deskilling—the diminished importance of action-centered skills—there will be little probability of developing critical judgment at the data interface. To rekindle such judgment, though on a new, more abstract footing, a reskilling process is required. Mastery in a computer-mediated environment depends upon developing intellective skills.

### Trusting Symbols

To understand the significance of intellective skills and how they differ from action-centered skills in this new environment, some appreciation of the nature of symbols is required. The data interface is a symbolic medium through which one produces effects and on the basis of which one derives an interpretation of "what is happening." These symbols are abstractions; they are experienced as remote from the rich sensory reality to which people are accustomed. Because of this remoteness, the new medium is not spontaneously felt to be legitimate. People confront this new world of symbols and ask, what does it mean?

In a symbolic medium, meaning is not a given value; rather, it must be constructed. This is a problem that action-centered skills are not required to address. The medium of equipment and materials conveys its meaning in its own immediate context. An example can be found in any object. Consider a pot handle: the handle means that human hands grasp objects in particular ways, and the handle exists as an acknowledgment of how hands can best grasp pots. When the handle is made of a nonconductive material, it further means that heat is dangerous for the human hand. Hands can best accomplish the work of pot grasping when the handle is of a temperature congenial to the skin. Alternatively, one can see that the meaning of objects is contained within their situations when objects are lifted out of their ordinary contexts and placed in a different or "abnormal" environment. Photographers, painters, and sculptors frequently juxtapose familiar objects with unfamiliar contexts in order to evoke a deeper sense of reflection from the viewer. Once lifted from its meaningful context, the object can be regarded "in-itself." Its meaning becomes problematic and must be constructed by the observer.[1]

The civilizing process has increased the distance between behavior and the impulse life of the animal body. It has also produced symbolic media (for example, the alphabet, mathematic notation, printed text) that can both convey and absorb human meaning unfettered by the contextual limitations of embodied presence. With each new medium people have had to revisit the problem of meaning. Rich historical and anthropological materials testify to a profound initial disjuncture between symbol and experience; they reveal the mental effort with which human beings have had to construct the linkages that connect a new kind of symbolic medium to a meaningful context. With time, these linkages become so tightly wrought that it is difficult to recover the original problematic quality of the relationship.

Language itself is an excellent field with which to demonstrate this historical process. The spoken word represents one medium that human beings have developed to convey experience; it preserves a close relationship between the word and bodily presence. Spoken words emit from and are shaped by the body's immediate interior condition (for example, breathlessness, fright, grief, joy). Their communicative power was bounded by the presence of both speaker and listener. The historical progression to the written word plainly shows the crisis of meaning that emerged as language took on a life of its own at a distance from experience and independent of speakers and listeners.

The medieval historian M. T. Clanchy has illustrated the reluctant acceptance of written documentation in place of first-person witness as it occurred over more than three centuries of early English history. "Documents," he tell us, "did not immediately inspire trust." People had to be persuaded that written documentation was a reliable reflection of concrete, observable events. To the modern mind, the evanescence of the spoken word seems more plastic, quixotic, and undependable than the printed word. To members of a highly oral culture, however, the spoken word was connected to the incontrovertible realities of bodily experience, while the written word was a thin, substanceless scratching whose two-dimensionality seemed highly arbitrary. Clanchy quotes one twelfth-century scholar who struggled to define the relationship between the written word and the palpable reality associated with the spoken word: "Fundamentally letters are shapes indicating voices. Hence they represent things which they bring to mind through the windows of the eyes. Frequently they speak voicelessly the utterances of the absent."[2]

The skepticism and mistrust that greeted the advent of the written word were evident in the procedures used to legitimate legal transactions. For example, a land transaction required witnesses who heard the donor utter the words of the grant and saw him make the transfer with a symbolic object, such as a knife or a piece of turf from the land. If there was a dispute, the accounts of these witnesses provided the evidence with which to settle it.

> Before documents were used, the truth of an event or transaction had been established by personal statement, often made on oath, by the principals or witnesses. If the event were too far in the past for that, the oldest and wisest men were asked what they could remember about it...the establishment of what passed for truth was simple and personal, since it depended on the good word of one's fellows....Both to ignorant illiterates and to sophisticated Platonists, a written record was a dubious gift, because it seemed to kill living eloquence and trust and substitute for them a mummified semblance in the form of a piece of parchment. Henry I's partisans in the dispute with Anselm, who had called a papal bull a sheepskin "blackened with ink and weighted with a little lump of lead," were arguing for the priority of the personal testimony of the three bishops who exercised memory over the mere "external marks" of a writing.[3]

In a completely different context, J. R. Clammer's study of literacy in the Fiji islands reveals the same skepticism and bewilderment as people confronted the relationship between "external markings" and their wordly experience. He recounts the experience of a missionary who had written a note to his wife on a wood chip and asked a Fijian chief to deliver the message: "The Chief was scornful of the errand and asked, 'What must I say?'...You have nothing to say, I replied. The chip will say all I wish. With a look of astonishment and contempt, he held up the piece of wood and said, 'How can this speak? Has this a mouth?'"[4]

In the modern world, literate minds have long become accustomed to a comfortable unity of the written word and the world to which it refers. When we read, we barely confront the problem of meaning. As long as a word is understood, our minds are well adapted to experience the word in terms of what we know it "means." But when we come upon a word, *fardel,* for example, whose definition we may not know, the problem of meaning suddenly wells up. We look at the word, and it seems to be an arbitrary collection of letters—there is no sense to it. That combination of letters becomes a meaningful symbol only when we discover what it refers to. Then we can reenter the sense of connectedness between the word and the world—the same comfortable "sense" of a word that we have when we read *chair* or *book.* In order to achieve this comfort, it was first necessary to encounter the word as a problem and to make its meaning explicit. Once that explicit recognition is accomplished, it is possible to develop a more implicit sense of the word, knowing that, should it become necessary, it will be possible to reconstruct an explicit definition from our reservoir of implicit knowledge.

We have made our peace with the problem of meaning in other ways as well. The voice over the telephone, the image in a photograph, the scene on a television screen—each of these is tacitly treated as though they fully convey a reality. We no longer puzzle over their connectedness to the "real" thing, though it is still possible to find individuals who remember when people were frightened of telephones or looked behind the television set hoping to discover the source of its images.

## NOTES

1 Surrealist painters such as Magritte and Dali have developed this technique to an extreme. In another vein, David Smith's Voltri series uses hand-tool-like objects to similar effect.
2 M. T. Clanchy, *From Memory to Written Record: England, 1066–1307* (Cambridge: Harvard University Press, 1979), 202.
3 Ibid., 232–33.
4 J. R. Clammer, *Literacy and Social Change: A Case Study of Fiji* (Leiden, The Netherlands: E. J. Brill, 1976), 67.

# The Aesthetic Drama of the Ordinary

**John J. McDermott**

Traditionally, we think of ourselves as "in the world," as a button is in a box, a marble in a hole, a coin in a pocket, a spoon in a drawer; in, always in something or other. And yet, to the contrary, I seem to carry myself, to lead myself, to have myself hang around, furtive of nose, eye, and hand, all the while spending and wasting, eating and fouling, minding and drifting, engaging in activities more descriptive of a permeable membrane than of a box. To feel is to be felt. To be in the world is to "world" and to be "worlded." No doubt, the accepted language of expository prose severely limits us in this effort to describe our situation experientially. Were I to say, for example, my presence in the world or my being in the world, I would still fall prey to the container theory and once again be "in" as over against "out." Is this not why it is necessary to describe an unusual person, situation, or state of being as being "out of this world," or "spaced out" or simply "out of it." Why is it that ordinary language, or our language as used ordinarily, so often militates against the ways in which we actually have, that is, undergo, our experiencing? Why is it that we turn to the more specialized forms of discourse such as jokes, fiction, poetry, music, painting, sculpture, and dance, in order to say what we "really" mean? Does this situation entail the baleful judgment that the comparative bankruptcy of our ordinary language justly points to the comparable bankruptcy of our ordinary experience?

In gross and obvious empirical terms, it is difficult to say no to the necessity of this entailment. Surely it is true that we are surrounded by the banal, monumentalized in a miniature and trivial fashion by the American shopping center. And it is equally, yea, painfully true that the "things," of our everyday experience are increasingly de-aestheticized, not only by misuse and failure to maintain, but forebodingly in their very conception of design and choice of material, as witnessed by the recent national scandal in our urban bus fleet, when millions of dollars were spent on buses that were not built for city traffic, roads, or frequency of use. How striking, as well, is the contrast between those Americans at the turn of the century, who built the IRT subway in New York City, complete with a mosaic of inlaid tile, balustrades, and canopied entrances, over against their descendants, our peers, who seem not able to find a way to eradicate the stink and stain of human urine from those once proud and promising platforms and stairwells. So as not to contribute any further to the offensive and misleading assumption that our main aesthetic disasters are now found in the great urban centers of the Northwest, let us point to one closer to my home.

The city of Houston, in paying homage to a long outdated frontier myth of every "building" for itself, proceeds to construct an environment which buries an urban aesthetic in the wake of free enterprise. Houston gives rise to tall and imposing buildings whose eyes of window and light point to the surrounding plains, but whose feet are turned inward. These buildings do not open in a merry Maypole of neighborhood frolic and function. Houston buildings are truly sky-buildings, for they look up and out, leaving only the sneer of a curved lip to waft over the enervated neighborhoods below, most of them increasingly grimy and seedy. As an apparent favor to most of us, Houston provides a way for us to avoid these neighborhoods, allowing us to careen around the city, looking only at the bellies of the titans of glass and steel, astride the circular ribbon of concrete known appropriately as the beltway, marred only by the dead trees, broken car jacks, and the intrusive omnipresence of Texas-sized billboards. Perhaps it is just as well that we, too, rise above the madding crowd, for in that way we miss the awkwardness of wandering into one of those walled-off, sometimes covenanted and patrolled, fancy enclaves which make the city tolerable for the rich. And as we make our "beltway," we miss as well that strikingly sad experience of downtown Houston at 6 P.M. of a weekend evening, when the loneliness and shabbiness of the streets are cast into stark relief by the perimeter of empty skyscrapers and the hollow sounds of the feet of the occasional snow-belt emigré traveler, emerging from the Hyatt Regency in a futile search for action. What is startling and depressing about all of this is that the city of Houston is the nation's newest and allegedly most promising major city.

Actually, whether it is North, South, East, or West matters little, for in general the archons of aesthetic illiteracy have seen to it that on behalf of whatever other ideology they follow, the presence of aesthetic sensibility has been either ruled out or, where traditionally present, allowed to erode. Further, to the extent that we prehend ourselves as a thing among things or a functioning item in a box, then we get what we deserve. Supposing, however, we were to consider the major metaphorical versions of how we carry on our human experiencing and, in so doing, avoid using the imagery of the box. Instead, let us consider ourselves as being in a uterine situation, which binds us to nutrition in a distinctively organic way. James Marston Fitch, a premier architectural historian, writes about us as follows:

> Life is coexistent with the external natural environment in which the body is submerged. The body's dependence upon this external environment is absolute—in the fullest sense of the word—uterine.[1]

No box here. Rather we are floating, gestating organisms, transacting with our environment, eating all the while. The crucial ingredient in all uterine situations is the nutritional quality of the environment. If our immediate surroundings are foul, soiled, polluted harbors of disease and grime, ridden with alien organisms, then we falter and perish. The growth of the spirit is exactly analogous to the growth of the organism. It too must be fed and it must have the capacity to convert its experiences into a nutritious transaction. In short, the human organism has need of two livers. The one, traditional and omnipresent, transforms our blood among its 500 major functions and oversees the elimination from our body of ammonia, bacteria, and an assortment of debris, all of which would poison us. The second is more vague, having no physical analogue. But its function is similar and crucial. This second liver eats the sky and the earth, sorts out tones and colors, and provides a filter through which the experienced environment enters our consciousness. It is this spiritual liver which generates our feelings of queasiness, loneliness, surprise, and celebration. And it is this liver which monitors the tenuous relationship between expectations and anticipations on the one hand and realizations, disappointments, and failures on the other. We are not simply in the world so much as we are of and about the world. On behalf of this second type of livering, let us evoke the major metaphors of the fabric, of the uterus, through which we have our natal being. Our context for inquiry shall be the affairs of time and space, as well as the import of things, events, and relations. We shall avoid the heightened and intensified versions of these experiential filters and concentrate on the explosive and implosive drama of their ordinariness.

## TIME

Time passing is a death knell. With the license of a paraphrase, I ask, For whom does the bell toll? It tolls for thee and me and for ours. We complain about the studied repetition, which striates our lives, and yet, in honesty, we indulge this repetition as a way of hiding from the inexorability of time passing, as a sign equivalent to the imminence of our self-eulogy. Time is a shroud, often opaque, infrequently diaphanous. Yet, from time to time, we are able to bring time into our own self-awareness and to bring time to its knees. On those rare occasions when time is ours rather than we being creatures of time, we feel a burst of singularity, of independence, even perhaps of the eternal import of our being present to ourselves. How has it happened that we have become slaves to time? Surely as children of Kant and Einstein, we should know better. For them and for modern physics, time is a mock-up, an earth phenomenon, no more relevant cosmically than the watches which watch time, supposedly passing. Still, Kant not withstanding, time is the name given to the process of our inevitable dissolution. On the morrow, our kidney is less quick, our liver less conscientious, our lung less pulsatile, and our brain less alert. Is it possible, without indulging ourselves in a Walter Mittyesque self-deception, to turn this erosive quality of time passing to our own advantage?

I suggest that we can beat time at its own game. Having created time, let us obviate it. Time, after all, rushes headlong into the future, oblivious to its damages, its obsoleting, and its imperviousness to the pain it often leaves in its wake. A contrary view is that in its passing, time heals. But it is not time which heals us, it is we who heal ourselves by our retroactive reconstruction of history. It is here that time is vulnerable, for it has no history, no past. Time is ever lurching into the future. We, however, can scavenge its remains and make them part of ourselves. For us, the past is existentially present if we have the will and the attentiveness to so arrange. I offer here that we recover the detritus of time passing and clot its flow with our freighted self-consciousness. We can become like the giant balloons in the Macy's Thanksgiving Day parade, thick with history and nostalgia, forcing time passing to snake around us, assuring that it be incapable of enervating our deepest feelings of continuity. What, for example, could time do to us if every time we met a person, or thought a thought, or dreamt a dream, we involved every person ever met, every thought ever thought, and every dream ever dreamt? What would happen if every event, every place, every thing experienced, resonated all the events, places, and things of our lives? What would happen if we generated a personal environment in which the nostalgic fed into the leads of the present, a self-created and sustained environment with implications fore and aft? In so doing, we would reduce time passing to scratching on the externals of our Promethean presence. Time would revolve around us rather than passing through us. Time would provide the playground for our activities rather than the graveyard of our hopes. We would time the world rather than having the world time us. And we would reverse the old adage, to wit, if you have the place, I have the time, for time is mine to keep and to give. And, in addition to telling our children now is your time, we would tell ourselves, no matter how old, now is our time.

## SPACE

It is equally as difficult to extricate ourselves from the box of space as it is to escape from the penalties of time. Here too, we have failed to listen to Kant and Einstein, for space, just as time, has no existential reality other than our conception of it. Yet we allow the prepossessing character of space to dwarf us. Nowhere is this more apparent than in Texas, where the big sky of Montana is outdone by the scorching presence of a sun that seems never to set, frying our brains in the oven of its arrogance. In the spring of the year, the bluebonnets and Indian paintbrush state our position: fey, lovely, quiet, reserved, and delicate of manner. The Texas sun indulges this temporary human-scaled assertion while hovering in the background with vengeance on its mind. As the flowers fade, the horizon widens and the sun takes its place at the center of our lives, burning us with the downdraft of its rays. Listen to Larry King on the sun and sky in West Texas.

> The land is stark and flat and treeless, altogether as bleak and spare as mood scenes in Russian literature, a great dry-docked ocean with small swells of hummocky tan sand dunes or humpbacked rocky knolls that change colors with the hour and the shadows: reddish brown, slate gray, bruise colored. But it is the sky—God-high and pale, like a blue chenille bedspread bleached by seasons in the sun—that dominates. There is simply *too much* sky. Men grow small in its presence and—perhaps feeling diminished—they sometimes are compelled to proclaim themselves in wild or berserk ways. Alone in those remote voids, one may suddenly half believe he is the last man on earth and go

> in frantic search of the tribe. Desert fever, the natives call it.... The summer sun is as merciless as a loan shark: a blinding, angry orange explosion baking the land's sparse grasses and quickly aging the skin.[2]

Texans pride themselves as being larger than life. But this is just a form of railing against the sun. The centuries-long exodus from the Northeast and the coastal cities was in part an escape from urban claustrophobia. In that regard, the escape was short-lived and self-deceptive, for it soon became apparent that the West presented a claustrophobia of another kind—paradoxically, that of open space. The box was larger, the horizon deeper, but the human self became even more trivialized than it was among the skyscrapers and the crowded alleyways and alcoves of the teeming urban centers. No, to the extent that we are overshadowed by an external overhang, be it artifact or natural, we cower in the presence of an *other* which is larger, more diffuse, still threatening and depersonalizing. In response, just as we must seize the time, so too must we seize the space, and turn it into a place, our place.

The placing of space is the creating of interior space, of personal space, of your space and my space, of our space. I am convinced, painful though it be, that we as human beings have no natural place. We are recombinant organisms in a cosmic DNA chain. Wrapped in the mystery of our origins, we moved from natural places to artifactual ones, from caves to ziggurats to the Eiffel tower. We moved from dunes to pyramids and then to the World Trade Center. The history of our architecture, big and small, functional and grandiloquent, lovely and grotesque, is the history of the extension of the human body into the abyss. We dig and we perch. We level and we raise. We make our places round and square and angular. We make them hard and soft and brittle. We take centuries to make them and we throw them up overnight. In modern America, the new Bedouins repeat the nomadic taste of old and carry their places with them as they plod the highway vascular system of the nation, hooking up here and there.

Some of our idiomatic questions and phrases tell us of our concern for being in place. Do you have a place? Set a place for me. This is my place. Why do we always go to your place? Would you care to place a bet? I have been to that place. Wow, this is *some* place. Win, place, show. The trouble with him is that he never went any place and the trouble with her is that she never got any place. How are you doing? How is it going? Fine, I am getting someplace. Not so well, I seem to be no place.

Recall that poignant scene in *Death of a Salesman* when Willy Loman asks Howard for a place in the showroom rather than on the road. In two lines, Howard tells Willy three times that he has no "spot" for him. I knew your father, Howard, and I knew you when you were an infant. Sorry, Willy! No spot, no place, for you. Pack it in. You are out of time and have no place.

Listen lady, clear out. But this is my place. No lady, this place is to be replaced. The harrowing drama of eviction haunts all of us as we envision our future out of place and on the street.[3] Dorothy Day founded halfway houses, places somewhere between no place and my place, that is, at least, someplace. And, finally, they tell us that we are on the way to our resting place, a place from where there is no return.

These are only anecdotal bare bones, each of them selected from a myriad of other instances which point to our effort to overcome the ontological *angoisse* which accompanies our experience of *Unheimlichkeit*, a deep and pervasive sense of ultimate homelessness. We scratch out a place and we raise a wall. The windows look out but the doors open in. We hang a picture and stick a flower in a vase. We go from cradle and crib to a coffin, small boxes at the beginning and end of journeys through slightly larger boxes. Some of us find ourselves in boxes underneath and on top of other boxes in a form of apartmentalization. Some of our boxes are official boxes and we call them offices, slightly less prestigious than the advantage of a box seat. Everywhere in the nation, the majority of our houses are huddled together, sitting on stingy little pieces of ground, while we ogle the vast stretch of land held by absentees. One recalls here "Little Boxes," a folk song of the 1960s that excoriates the ticky-tacky boxes on the hillsides, as a preface to the yuppiedom of our own time. For the most part, our relation to external space is timid, even craven. From time to time, we send forth a camel, a schooner, a Conestoga wagon, or a space shuttle as probes into the outer reaches of our environ, on behalf of our collective body. Yet these geographical efforts to break out are more symbolic than real, for after our explorations we seem destined to repeat our limited variety of habitat.

The locus classicus for an explication of the mortal danger in a sheerly geographical response to space is found in a story by Franz Kafka, "The Burrow." In an effort to protect his food from an assumed intruder, the burrower walls off a series of mazes sure to confuse an opponent. This attempt is executed with such cunning and brilliance that his nonreflective anality is missed as a potential threat. The food is indeed walled off from the intruder—from the burrower as well. He dies of starvation, for he cannot find his own food.

The way out of the box is quite different, for it has to do not with the geography and physicality of space, but rather with our symbolic utilization of space for purposes of the human quest. We manage our ontological dwarfing and trivialization at the hands of infinite space, and the rush of time passing and obsoleting, by our construction, management, placing, and relating of *our* things. It is to our things, to creating our salvation in a world without guarantee of salvation, that we now turn.

## THINGS

Thing, orthographically and pronouncedly, is one of the ugly words in contemporary American usage. Yet it is also, inferentially and historically, one of the most subtle and beautiful of our words. It is lamentable that we do not speak the way Chaucer spoke. From the year 1400 and a work of Lydgate, *Troy-Book*, the text reads: "That thei with Paris to Greece schulde wende, To Brynge this thynge to an ende." The Trojan war was a thing? Of course it was a thing, for thing means concern, assembly, and, above all, an affair. Thing is a woman's menses and a dispute in the town. Thing is a male sex organ and a form of prayer. (The continuity is not intended, although desirable.) Thing is what is to be done or its doing. I can't give you any thing but love, baby. That is the only thing, I have plenty of, baby. When you come, bring your things. I forgot to bring my things. My things are packed away. Everything will be all right. And by the way, I hope that things will be better.

What and who are these things to which we cling? An old pari-mutuel ticket, a stub for game seven of the World Series, a class ring, a mug, a dead Havana cigar, loved but unsmoked. My snuff box, my jewelry drawer, an album, a diary, a yearbook, all tumbled into the box of memories, but transcendent and assertive of me and mine. Do not throw out his things, they will be missed. Put her things in the attic, for someday she will want them as a form of reconnoitering her experienced past. Do you remember those things? I know that we had them. Where are they? They are in my consciousness. Can we find them? We didn't throw them out, did we? How could we?

The making, placing, and fondling of our things is equivalent to the making, placing, and fondling of our world. We are our things. They are personal intrusions into the vast, impersonal reach of space. They are functional clots in the flow of time. They are living memories of experiences had but still viable. They are memorials to experiences undergone and symbolically still present. The renewed handling of a doll, a ticket, a toy soldier, a childhood book, a tea cup, a bubble-gum wrapper, evokes the flood of experiences past but not forgotten.[4] How we strive to say hello, to say here I am, in a cosmos impervious, unfeeling, and dead to our plaintive cry of self-assertion. To make is to be made and to have is to be had. My thing is not anything or something. Your thing is not my thing but it could be our thing. The ancients had it right, bury the things with the person. We should do that again. Bury me with a copy of the *New York Times,* a Willie Mays baseball card, a bottle of Jameson, my William James book, a pipe, some matches, and a package of Seven-Seas tobacco.

The twentieth-century artist Alexander Calder once said that no one is truly human who has not made his or her own fork and knife. Homemade or not, do you have your own fork, your own knife, your own cup, your own bed, desk, chair? You must have your own things! They are you. You are they. As the poet Rilke tells us, "Being here amounts to so much."[5]

Our things are our things. They do not belong to the cosmos or to the gods. They can be had by others only in vicarious terms. Commendable though it may be for those of us who are collectors of other people's things, nonetheless, those who burn their papers or destroy their things just before they die are a testament to both the radical self-presence and transiency of human life. Those of us, myself included, who collect other people's things, are Texas turkey vultures, seizing upon the sacred moments hammered out by transients and eating them in an effort to taste the elixir of memory for our own vapid personal life. Ironically, for the most part their experience of their things were similar efforts, sadly redeemed more by us than by them. Now to the crux of the matter before us.

It is not, I contend, humanly significant to have the primary meaning of one's life as posthumous. We and our things, I and my things, constitute our world. The nectar of living, losing, loving, maintaining, and caring for our things is for us, and for us alone. It is of time but not in time. It is of space but not in space. We and our things make, constitute, arrange, and determine space and time. The elixir garnered by the posthumous is for the survivors. It cannot be of any biological significance to us, although many of us have bartered our present for the ever absent lilt of being remembered. St. Francis of Assisi and John Dewey both taught us the same *thing:* time is sacred, live by the sacrament of the moment and listen to the animals. We may have a future. It is barely conceivable, although I doubt its existence. We do have, however, a present. It is the present, canopied by our hopefully storied past, that spells the only meaning of our lives. Still, the present would be empty without our things.

You, you out there, you have your things. Take note. Say hello, say hello, things. They are your things. Nay, they are you. No things, no you, or in correct grammar, you become no*thing*. So be it. Space and time are simply vehicles for things, our things, your things, my things. These things do not sit, however, in rows upon rows, like ducks in a shooting gallery. These things make love, hate, and tire. Like us, they are involved. We consider now this involvement of persons, things, things and persons, all struggling to time space and space time, namely, the emergence of events as relations.

## THINGS AS EVENTS AS AESTHETIC RELATIONS

We have been in a struggle to achieve nonderivative presence of ourselves and our things over against the dominating worlds of space and time. Fortunately, for us, space and time do not necessarily speak to each other. Our canniness can play them off, one against the other. The triumph is local, never ultimate, although it does give us staying power in our attempt to say I, me, you, we, us, and other asserted pronominal outrages against the abyss.

A happy phenomenon for human life is that things not only are; they also happen. I like to call these happenings events. The literal meaning of event is intended: a coming out, a party, a debutante dance, a bar mitzvah, a hooray for the time, given the circumstance. In my metaphysics, at least, things are bundles of relations, snipped at the edges to be sure. Usually, we give our things a name and this name takes the place of our experience of the thing. It does not take long to teach a child a list of nouns, each bent on obviating and blocking the rich way in which the child first comes upon and undergoes things. It is difficult to overcome this prejudice of language, especially since row upon row of nouns, standing for things, makes perfectly good sense, if you believe that space is a container and time is the measure of external motion. If, however, you believe as I do, that space and time are human instincts, subject to the drama of our inner lives, then things lose their inert form. Emerson says this best when he claims that every fact and event in our private history shall astonish us by "soaring from our body into the empyrean."[6]

The clue here is the presence of a person. Quite aside from the geographical and physical relationships characteristic of things and creatures, we further endow a whole other set of relations, the aesthetic. I refer to the rhythm of how we experience *what* we experience. The most distinctive human activity is the potentially affective dimension of our experiencing ourself, experiencing the world. I say potentially, for some of us all of the time and most of us most of the time are dead to the possible rhythms of our experiences. We are ghouls. We look alive but we are dead, dead to our things and dead even to ourselves. As John Cage warned us, we experience the names of sounds and not the sounds themselves. It is not the things as names, nouns, which are rich. It is how the things do and how they are done to. It is how they marry and divorce, sidle and reject. The aesthetic drama of the ordinary plays itself out as a result of allowing all things to become events, namely, by allowing all things the full run of their implications. This run may fulfill our anticipations and our expectations. This run may disappoint us. This run may surprise us, or blow us out. Implicitness is everywhere and everywhen. Were we to experience an apparently single thing in its full implicitness, as an event reaching out to all its potential relations, then, in fact, we would experience everything, for the leads and the hints would carry us into the nook and cranny of the implicitness of every experience.[7]

We are caught between a Scylla and Charybdis with regard to the drama of the ordinary. The scions of the bland and the anaesthetic convince us that nothing is happening, whereas the arbiters and self-announcers of high culture tell us that only a few can make it happen, so we are reduced to watching. My version is different. The world is already astir with happenings, had we the wit to let them enter our lives in their own way, so that we may press them backward and forward, gathering relations, novelties, all the while. Our affective presence converts the ordinary to the extraordinary. The world is made sacred by our *handling* of our things. We are the makers of our world. It is we who praise, lament, and celebrate. Out of the doom of obviousness and repetition shall come the light, a light lit by the fire of our eyes.

## NOTES

1 Cited in Serge Chermayeff and Christopher Alexander, *Community and Privacy*, New York, Anchor Books, 1965, p. 29.

2 Larry L. King, "The Last Frontier," *The Old Man and Lesser Mortals*, New York, Viking Press, 1975, p. 207.

3 Cf. the moving and poignant scene of "eviction" in Ralph Ellison, *Invisible Man*, New York, Vintage Books, 1972 [1952], pp. 261–77.

4 The master of "things" and "boxes" is, of course, Joseph Cornell. Indeed, he is the master of things in boxes, known forever as Cornell boxes. Only those who have experienced these "boxes" can appreciate Cornell's extraordinary ability to merge the surrealism of the imagination and the obviousness of things as a "memorial to experience." Cf. Diane Waldman, *Joseph Cornell* (New York: George Braziller, Inc., 1977), and Kynaston McShine, *Joseph Cornell* (New York: The Museum of Modern Art, 1981). As with Cornell, by "things" we mean, as does William James, bundles of relations. Things

are not construed here as Aristotelian essences, much less as conceptually rendered boxes.

**5** Rainer Maria Rilke, "The Ninth Elegy," *Duino Elegies,* New York, W. W. Norton, 1939, p. 73.

**6** Ralph Waldo Emerson, "The American Scholar," in *Works*, vol. I, Boston, Houghton Mifflin, 1903–1904, pp. 96–97.

**7** Cf. William Blake, "Auguries of Innocence," *The Poetry and Prose of William Blake,* ed. David V. Erdman, New York, Anchor Books, 1965, p. 481.

PART **FOUR**

# THE PHENOMENOLOGY OF EVERYDAY AFFAIRS

## INTRODUCTION

In the introduction to Part One, I suggested that because of their ubiquity, technological artifacts often become invisible. It is as if our alarm clocks, coffeepots, running shoes, automobiles, televisions, and computers constitute the materials from which we have spun a technological cocoon. We dwell in it. It isolates us from nature. It becomes for us another nature, a "second" nature. The essays in this section constitute an extensive exploration of this theme.

In his essay "Man the Technician" (included in Part Five), the late Spanish philosopher José Ortega y Gasset suggested that human beings have become like naive forest or jungle dwellers in the sense that we are so involved with our technological environment, so enmeshed in its richness, that we just take it for granted. Technology has become for us as nature is for them: a given, a bounty that is conspicuous only when for some reason or other it is no longer available to us.

Ortega suggested that when our environment, whether natural or artifactual, becomes invisible to us, then we have lost our way. It is in this context that he speaks of "autofabrication." Human beings are the only animals who significantly and consciously alter their environments. They construct machines and tools, as well as laws and institutions. But by doing these things they also construct or fabricate themselves. New artifacts transform human beings, and newly transformed human beings create new artifacts. A major part of this project of autofabrication is the development of critical skills, the assessment of implications and alternatives, the weighing of evidence—in short, the introduction of ordered and effective inquiry into a world that would become literally dehumanized in the absence of this uniquely human activity.

Consider, for example, the role that an adequate assessment of the implications of automobile technology could have had at the time of its introduction into American society. Melvin Kranzberg has reminded us that the automobile was uncritically hailed by early twentieth-century New Yorkers as a wonderful solution to their pollution problems. The transportation system then in use, horse-drawn trolleys and carriages, had created a prodigious health problem: thousands of gallons of horse urine and thousands of pounds of manure were being deposited daily on the streets of New York City.

Nevertheless, apart from the simplistic moralism catalogued by Glen Jeansonne in his essay "The Automobile and American Morality," there were apparently very few attempts to consider the change in the moral climate that was to follow the widespread use of the automobile. John Dos Passos, writing about America at midtwentieth century, remarked that a whole generation of Americans had been conceived—if indeed not born—in the backseats of Model T Fords. The mating rituals of American young people moved out of the closely monitored front porch swing into the private spaces within, or accessible only by means of, the automobile. As Jeansonne reports, a survey undertaken by sociologists at Southern Methodist University in 1936 produced the amazing results that 75 percent of the clientele of tourists courts in the Dallas area were not tourists at all but local couples.

The expectations and surprises registered by those who experienced the introduction of the automobile into American culture are to us today more than simply a source of amusement, although they are that. They give us clues about our own program of assessing the effects of technology, and they reveal significant aspects of what automobiles mean to us. Douglas Browning, for example, in his essay "Some Meanings of Automobiles," explores a distinctly human activity which he calls "personation"—that is, playing at being oneself (as opposed to "impersonation," or playing at being another). This is, of course, an activity that Ortega would include as a part of autofabrication. Browning suggests that the automobile provides one of our greatest opportunities to personate ourselves. The interior of the automobile is a world that is richly private and at the same time open to immediate and sudden intrusion. We become aware, as no individual in a natural setting could be, of the importance of the category of the sudden: not the suddenly *happened* but the suddenly *happening*. Being open to disclosure at any time, our automobiled lives become precarious; as when picking up a hitchhiker, when making love in an automobile, or when becoming aware that an accident is about to occur.

The essays discussed so far have examined the automobile from the viewpoint of its users. But automobiles, like other artifacts, are the result of concrete acts of production. In the selection taken from his book *The Assembly Line,* Robert Linhart catalogues the harsh conditions and the indignities associated with automobile production in a Citroën factory in 1968. Linhart left his position as a professor of economics at a French university in order to infiltrate an auto plant and to mobilize its workers against the inhuman conditions under which they labored. He has given us a vivid account of the dangers and the harsh repetitiveness of the assembly line. He encounters racism in hiring and promotion, and wages that are barely sufficient to enable a family to survive. He depicts managers who are at times insensitive to workers' needs, at times almost sadistic in their demands. His readers are encouraged to consider the phenomenon of extreme poverty and hardship in the midst of prosperous industrial societies and to ask themselves if such conditions are not indicative of a soft and rotting core at the very heart of those societies that are usually called "advanced."

Whereas at the present time no single academic discipline attempts to assess the impact of the automobile on the societies in which it has played an important role, the same cannot be said with respect to the impact of electronic technology, especially the electronic media. Taking their point of departure from the work of Norbert Wiener, the father of cybernetics, schools and departments of communications have now gone beyond the original sense of cybernetics as control (Wiener drew upon a Greek word for "helmsman" in coining the term) to explore media along lines that are historical, sociological, political, psychological, and aesthetic.

Perhaps one of the most interesting questions arising from the assessment of electronic technology is its relation to print. This is more than simply a question of how we get our information or what happens when we watch television as opposed to what happens when we read. The assessment of this relation provides the material for complex theories about the ways in which technology has developed and about its future. It is at this point that the essays by George Gerbner and Edmund Carpenter in this section relate intimately to the essays in Part Five, especially that of Paul Levinson, and to my essay and the one by David Edge and myself in Part Two.

Marshall McLuhan, who probably did more than anyone else to explore the transition from print to electronic media as the source of our dominant cultural metaphors, argued that books constitute a "hot" or "saturated" medium, a "linear" one with high definition. The print person, he suggested, is an interior person; we normally read silently and alone. Reading books emphasizes one of our senses, the visual, at the expense of the others. It habituates us to thought whose form is linear and unidimensional.

Television, on the other hand, is what McLuhan described as a "cool" or "unsaturated" medium. It engages us in ways that are social as opposed to individual, it develops not only the eye but also the ear, and it emphasizes the nonlinear and the multidimensional. It provides a sense of all-at-onceness with respect to events that are separated geographically by thousands of miles and historically by decades.

McLuhan's collaborator, Edmund Carpenter, develops these points in his essay "The New Languages," included in this section. He calls our attention to the way in which the book is digested line by line, one paragraph following upon another in a serial, structured fashion. "Print" thinking brings with it an "assembly line" perception of time; it involves an orderly flow from past to future.

Newspapers and television, on the other hand, present forms of time and space that are topsy-turvy in relation to those nourished by the book. Images and ideas are stacked on one another; they are juxtaposed without any apparent reason and even dissolve into one another. Before newspapers and television, it was expected that a proper novel should have a beginning, a middle, and an end. Since their development, however, the novel has undergone a radical transformation: with its flashbacks and stream-of-consciousness devices, it has come to imitate the "all-at-onceness" of television. In the absence of the well-defined narrative, the viewer is left to "saturate" the unsaturated material with his or her own contribution.

Not only does television convey different types of information than does the book, but we *see* television differently. A book is a sustained presentation of a series of related ideas. But it is not uncommon for a sixty-second television commercial to have as many as thirty or forty cuts—that is, for the camera to change its distance to its subject, its angle, or its locale about once per second. (Advertisements that appeal to younger audiences, such as Chevrolet commercials directed at young single adults, have a much quicker pace of cutting than do ads that market Cadillacs to viewers who are older and more established.)

Yet we do not see these changes unless we are made aware of them: we have been conditioned to view the total gestalt or form of the commercial without attending to its details. In order to "see" commercials, or for that matter any of the other ubiquitous technological features of our lifescapes, it is necessary to undertake a kind of figure-ground shift that yields new insights into our world of experience and generates new relations to it and within it.

Carpenter exhibits just this kind of phenomenological approach in his analysis of four versions of *The Caine Mutiny:* the book, the film, the television play, and the stage play. He finds that each of the versions had a different "hero." The hero of the book was the young naval officer whose development it chronicles. The movie, with its colorful shots of ships and the sea, turned into a kind of advertisement for another hero: the U.S. Navy. The constraints of the stage presentation confined most of the action to the courtroom scene: its hero consequently became the attorney for the defense. And the television show, designed for a mass audience, "emphasized patriotism, authority, allegiance."

For Carpenter and McLuhan, the latest technological revolution, the one brought about by the shift from print to electronic technology as primary medium, calls to mind an earlier technological revolution: the one occasioned by the invention in the late fifteenth century of the printing press. Carpenter argues that our "postliterate" era has much in common with the "preliterate" period—the era before the fifteenth century. Pre- and postprint men and women tend to be more socially oriented, more emphatic in their commitment to a deeply sensuous and sensual interaction with the world, and more open to nonstandard experience than were print men and women. At the same time, however, Carpenter suggests that our postliterate era will carry forward the critical attitudes and the commitment to quantification that were generated during the age of high literacy. In short, we are once again entering an aural-oral world, but this time with our eyes wide open.

These revolutions do not occur without casualties. Enormous problems plague lives lived in a period of transition between two overarching technological paradigms. How, for example, are print people to learn to assess (and access) electronic technology? The converse of this problem is one that is worrying many educators today. They want to know how to teach students who are primarily oriented toward electronic media to gain access to the subtleties of print. How can the children of television be taught to read critically?

Some have taken quite seriously the claim of McLuhan and Carpenter that electronic technology has produced a new "super system," a new cultural form that supplants the images, myths, and metaphors of the old print system with those of its own. The essays by Levinson in Part Five and by Edge in Part Two have this issue as their central theme.

In "Television: The New State Religion?" George Gerbner is worried that the elements of traditional religion that render the individual passive before a transcendent value-giver may be present in and on television. The myths and rituals associated with television, Gerbner claims, have "tremendous popular mobilizing power which holds the least informed and least educated most in its spell." In addition to assessing Gerbner's claim that television itself has become a new state religion, it is important that we take a critical look at the "electronic church," for televised religion as promulgated by the Jimmy Swaggarts and the Jerry Falwells has opened a new chapter in the history of mass ideology.

I shall mention two other approaches to television which, despite their importance, were not included in this collection for lack of space. Whereas Carpenter and McLuhan see television as one among many languages whose grammar requires further study, and whereas Gerbner sees television primarily as social indicator, Jerry Mander, in his widely read book *Four Arguments for the Elimination of Television,* is appalled that television even continues to exist. Mander, a former advertising executive, worries about the radiation that emanates from the television screen. He fears the de-education of the televiewing masses. And he compares television to sensory deprivation, suggesting that it robs its viewers of a healthy sense of reality.

Among those who share Mander's distaste of television, perhaps no one has built a more eloquent case against that medium than Frederick Exley. The main character of his vaguely autobiographical novel *A Fan's Notes* retreats to the fantasy world of television during bouts with mental illness. In a lucid period he writes:

> I watched—but there is no need to enumerate. Not once during those months did there emanate from the screen a genuine idea or emotion, and I came to understand the medium as subversive.... Television undermines strength of character, saps vigor, and irreparably perverts notions of reality. But it is a tender, loving medium, and when it has done its savage job completely and reduced one to a prattling salivating infant, like a buxom mother it stands always poised to take one back to the shelter of its brown-nippled bosom.[1]

The assessment of electronic media is also a task that has been undertaken by groups and individuals who are outside both the academic and literary communities, many of whom are working in fields such as video art, video documentary, film, and syntheses of these media. Especially significant in this area was the work in the 1970s of a group of artists known as TVTV (Top Value Television). Whereas scholarly articles on television such as those by Carpenter and Gerbner are usually addressed to a constituency deeply immersed in print, TVTV offered a critique of television from within that very medium. Persons working in such alternative ventures as was TVTV often differentiated between television and video, with the former term referring to the broadcast medium (or *narrow*cast, in the case of cable), and the latter to a kind of meta-television that quite often articulates a biting critique of the accepted assumptions and conventions of television. This use of the term "video" should be carefully differentiated from the same word as it is used by devotees of MTV, usually to refer to short visual segments accompanying rock music, almost exclusively shot on 16- or 35-millimeter film and then lip-synched. It may be that the only true "video" materials on MTV are the advertisements for MTV and certain special effects incorporated into the film sequences.

The coverage provided by TVTV of the 1972 Republican national convention in Miami, aptly and with prescient irony entitled "Four More Years," not only provided insights into the nature of television news reportage but also cleverly assessed the nature of television documentaries. Instead of covering the speeches on the platform of the convention, in the fashion of the major networks, the TVTV reporters interviewed the delegates, the protesters outside the hall, the students bused in to make posters for the delegates, and even the network reporters. This remarkable example of video documentary was broadcast on PBS and remains a classic in its field. Techniques invented by TVTV are now utilized by the major networks. The American people see their political conventions in new ways as a result of the innovative work of TVTV.

One of the most important messages of the TVTV phenomenon is that small-gauge (½-inch and 8-millimeter) videotape can have a prodigious effect upon the dissemination of information and cultural values. Although it is most frequently utilized for the production of home movies, portable videotape technology has been shown to have enormous potential as a truly democratic, decentralizing, and popular form of electronic technology.

An adequate assessment of television cannot be done entirely within the context of print. It is important that the works of video artists and documentarists be perceived as entering into dialogue with print-oriented critics. The works of Skip Blumberg, William Wegman, Peter Campus, Ed Emschwiller, TVTV, Ann Magnuson, and others would be a proper part of any serious investigation of television, as would the work of its print critics and the investigations of individual viewers. Each of these sources of insight must be taken into account if we are to become aware of the present effects and future possibilities of electronic media.

A third technological artifact, fully as important as the automobile or television to the structure of our everyday affairs, is the clock. Serious questions arise from a consideration of the roles of clocks in our lives. What do clocks measure? Are there varieties of duration for which the clock is not an adequate measure? What part did the clock play in the formation of western industrialized societies? Are there significant differences between analog clocks and ones that are digital?

Officially, a second is defined as 9,192,631,700 cycles of the frequency associated with the transition between the two energy levels of the isotope cesium 133. But the clock certainly does not measure this cycle: it is only correlated to it.

Neither does the clock measure human biological or psychological time. We now know that biological rhythms "speed up" and "slow down" and that uniformity must be imposed upon, not read out of, the living organism. If clock time measured psychological time, we could never take "time out" for recreation—to heal our disorders by, as it were, stepping outside time for a time. A more adequate view of clock time is that it is artificial time, a technological invention that is loosely tied to astronomical events and by means of which we can correlate numerous and varied alternative types of nonstandard duration.

This is not to say that there are no regularities within the universe that can be correlated to clock time. It is rather to say that the clock—especially the analog clock, with hour and minute hands—is a machine whose product is uniformity. The clock, then, is a machine for representing the passage of events within a spatial dimension.

But the clock is also a machine for socializing men and women and molding them to the constraints of mass production and mass consumption. In his essay "The Monastery and the Clock," Lewis Mumford argues that the medieval monk was the first "modern" person because he was the first whose day was organized by means of the clock. Mumford also argues that the clock, no less than the printing press, laid the conceptual foundations for the machine age, an era during which enormous value was placed upon uniform, repeatable products manufactured by assembly line methods. Jacques Le Goff, in *Time, Work, and Culture in the Middle Ages,* details the manner in which clock time began to change the nature of the workday, and the nature of the tasks it comprised, during the fourteenth century. As early as 1349, workers in the cloth trade were already calling strikes against the use of the clock-bells to regulate the workday.[2]

The history of timekeeping exhibits surprising twists and turns. Jean Gimpel, in his important book *The Medieval Machine,* argues that the first mechanical clocks "did not primarily tell time, but rather were built to forecast movements of the sun, the moon and the planets. Remarkably enough, the time telling clock which dominates our lives was only a by-product of the astronomical clock."[3] Earlier clocks, prior to the fourteenth century, he suggests, indicated unequal hours. "Egyptian, Greek, Roman, Byzantine and Islamic water clocks had indicated unequal, or temporal hours—in all these civilizations the day was divided into hours of light and hours of darkness, generally periods of twelve hours each."[4] Of course, the result of dividing day and night in this way was to produce "hours" that varied. At the latitude of London the daylight "hour" varied from thirty-eight minutes at the winter solstice to eighty-two minutes at the summer solstice! In "The Rise of the Equal Hour," Daniel Boorstin chronicles the artifactual and social changes that led to our own standard system. Along the way he speculates in an imaginative way on the origin of the twelve- and twenty-four-hour systems, and he provides an interesting account of the etymology of important timekeeping words.

Marshall McLuhan, following Mumford's lead, compared the clock to the blast furnace, which, he wrote, "speeded the melting of materials and the smooth conformity in the contours of social life."[5] Both Mumford and McLuhan have argued that perceptions of time vary radically from one technological stage to another.

Mumford points out that whereas the machine age, which he calls "paleotechnic," emphasized cheap labor, the preceding medieval wind, water, and agricultural or "eotechnic" period emphasized "cheap time." The medieval peasant worked fewer hours than did his or her counterpart in the English factory of the nineteenth century. Agricultural workers of the medieval period lived in an environment in which time was flexible and fluid. Their means of measuring time were closely associated with the rhythms of their own bodies and those of the rest of nature. During the time of paleotechnic technology, however, workers became a part of a structure modeled on the nonorganic features of the machine—a structure in which the only acceptable time was the machine-made time of the clock. Their workday expanded to the point that fourteen to sixteen hours a day, six days a week, year round, were spent in service to the machine.

Although some of his critics have accused Mumford of overromanticizing life during the Middle Ages, it seems clear that his main thesis is secure. The "first" industrial revolution (the technological revolution which began in England in the last half of the eighteenth century) did in fact produce the compression of organic human beings into mechanically uniform time structures, and this in turn led to a decline in the quality of life of most individuals. The average life span of workers during the depths of the paleotechnic or industrial age was some twenty years shorter than that of their middle-class counterparts during the same period.

Charlie Chaplin parodied this situation in his film *Modern Times,* in which he portrayed an assembly line worker whose work was accelerated to the point at which he suffered a nervous breakdown. The havoc he then wrought left the factory in shambles. In his tragicomic delirium, he saw the foreman's nose and the buttons on the dress of the plant secretary as just more nuts and bolts to be tightened.

What Mumford calls the "neotechnic" or postindustrial age has seen a move away from the tyranny of abstract clock time. For one thing, the workweek has been shortened, thus allowing more "unstructured" time for workers. In addition, some enlightened corporations are now using flextime as a part of their employee-relations programs. In situations where such practice is feasible, workers are given a time span of twelve or fourteen hours in which to accomplish eight hours of work. Flextime allows workers to align the workday to their own biological and psychological needs and preferences. The results of approaching problems of time and productivity in this nonlinear fashion have been impressive. Besides increasing productivity, flextime allows workers to adapt to social, familial, and personal needs that are outside the sphere of the workplace, and it may well have the added benefit of relieving overburdened transportation facilities.

The concept of the "electronic cottage," in which a worker does not physically commute to work on a daily basis but performs some work at home, linked to his or her office by means of a computer, is a special case of this "neotechnic" perception of time. It should be noted, however, that the electronic cottage concept is not without its critics. Some have argued that the lack of regulation of work done in the home may, and often does, lead to exploitation. They warn of the "electronic sweatshop."

In his essay "Urban Time," John J. McDermott speculates about the differences between urban or "thin" time, and rural or "fat" time, that is, "nature" time. In his view, it is possible to find in America's shift from baseball time to football time a rich metaphor of the changing consciousness of its individuals and institutions. Baseball is a game in which the time-out is relatively trivial and which could, in theory, go on to infinity. Football, on the other hand, is a game in which every second counts and in which the participants attempt to "steal" time by taking a time out or by making the game go into overtime.

Much could be said of the entirely new ways in which digital clocks orient us toward time. The sundial, the hourglass, the weighted clock, the pendulum clock, the clockwork or mainspring clock, even the quartz analog watch all represent time relative to space. But the digital clock represents time abstractly as the relation among digits, lines, and dots displayed on a surface. The comedian George Carlin expressed his marvelous insight into these different types of time when he noted that with the analog or mechanical clock it is much more difficult to work during the last half of an hour than during the first half, since during the last half of the hour, gravity itself appears to oppose the regular progression of time; its forces seem to retard the upward movement of the minute hand.

Novel social practices are demanded by the clock with integrated circuitry. If one is wearing an analog watch, the correct response to "What time is it?" is to round off to the nearest five-minute mark. To give 9:37 as an answer is to be thought overly fastidious. But when one's watch actually says 9:37:54, a new degree of precision has been introduced which is hard to overlook or to deny.

Even though we do not come into tangible contact with nuclear weapons on a daily basis, they have nevertheless become artifacts whose symbolic import casts a shadow over our quotidian lives. In "Nuclear Weapons and Everyday Life," Paul Thompson conducts us beyond the more common discussions of the moral problems associated with the construction of nuclear weapons and the strategies for their deployment. He invites us to analyze the ways in which nuclear weapons figure within the matrix of technologies that support and condition the tenor of everyday life. His account is richly furnished with examples from the essays of other authors in this collection, including those of Browning, Mumford, and Edge.

Taking his cue from McLuhan, one of Thompson's main assumptions is that we as human beings live immersed in technological media. This "mediation," as he calls it, subjects us to changes in our self-concepts and in the qualities of our experience that are difficult to perceive precisely because they subtly alter the very categories we use to interpret ourselves and our experiences. Entire societies become so immersed in technological media that they evolve along paths favored not by reigning ideology or trends in demography but rather by whatever technology is actually in use. Such mediation is richly symbolic: changes in technological symbols and metaphors function as both cause and effect of broader technological changes.

Thompson sees the symbolic fallout of nuclear technology expressed in phenomena as different as the current worldwide resurgence of religious and political fundamentalism and the plot devices of the horror movies of the 1950s. Indeed, the very concept of nuclear deterrence is symbolic: it says that it is better to send the message of intent than the message of the actual warheads. But because nuclear weapons are media rich in content, they send many unintended messages as well. Although the Soviet Union is a country which binds together a number of complex cultures, interests, and goals, its massive participation in the nuclear arms race sends a very simple message to the people of the United States: the Soviets are a people resolved to oppose us. Of course, the same could be said of the message sent by the government of the United States to the people of the Soviet Union. In each case, one component of a national agenda becomes the dominant message for information gatherers on the other side of the globe.

In the final essay in this section, Ruth Schwartz Cowan draws our attention to an area of technology that is frequently overlooked: housework and its tools. She reminds us that the century from the American Civil War to the 1960s was not only a period which saw the development of factories, railroads, and assembly lines but also one during which the technology of housework underwent profound change. If this matter has received scant attention, it may be because the writing of history, including the history of technology, has tended to exhibit certain biases, such as its definition of "industry" as only what happens outside the home. The undeniable fact, however, is that "kitchens are as much a locus for industrialized work as factories and coal mines are, and washing machines and microwave ovens are as much a product of industrialization as are automobiles and pocket calculators."

Cowan isolates three significant senses in which the industry of housework is different from the industrial work performed outside the home, which she calls "market" work: it is mostly unpaid, it is performed in isolated workplaces, and its practitioners are unspecialized workers. But there are also three important features which housework and market work have in common: both utilize nonhuman and nonanimal energy sources, both involve dependency on a network of social and economic institutions that make such work possible, and both are accompanied by the estrangement of the worker from the tools that make labor possible.

Cowan organizes her argument around two concepts. The first is what she calls the "work process," and the second is what she terms the "technological system." By calling housework a work process, she is suggesting that it is not so much a set of simply definable tasks as it is a whole network of interrelated ones. This becomes an important consideration when we try to ascertain whether household management has become more or less time-consuming during a given historical period. Her term "technological system" reminds us that each implement of household work must be linked to others in order to function properly. Further, many of the electrical power generators, water supply and drain pipes, and food processing plants on which household work depends are not only out of the sight of most householders but out of their control as well.

Each of the essays in this section takes a critical look at one or more of the most concrete artifacts within our technological cocoon. What binds these essays together is the underlying assumption of each of them that technology cannot be understood in abstract terms alone but only as we continually test our abstract theories against our most concrete experiences.

## NOTES

**1** Frederick Exley, *A Fan's Notes,* Simon and Schuster, New York, 1977, pp. 180–181.
**2** Jacques Le Goff, *Time, Work, and Culture in the Middle Ages,* Arthur Goldhammer (trans.), University of Chicago Press, Chicago, 1980, p. 45 ff.
**3** Jean Gimpel, *The Medieval Machine,* Penguin Books, New York, 1977, p. 155.
**4** Gimpel, op. cit., pp. 165–168.
**5** Marshall McLuhan, *Understanding Media,* McGraw-Hill, New York, 1965, p. 155.

## Some Meanings of Automobiles

**Douglas Browning**

Of course there are many meanings of automobiles. I am interested in just three interdependent meanings which I take to be philosophically instructive yet seldom if ever explicitly formulated. These three are the enlargement of the body, the introduction of the category of the sudden, and the deepening of the concept of anonymous privacy. To sum up these points, the distinctive meaning of automobiles with which I am concerned herein is self-enrichment through the anonymity of speed. With it is involved a new understanding of the human self.

Let me state emphatically that I am not concerned with the meanings of automobiles *to* the human being. If meanings come about only through human beings, still there is the living fact of the automobiled human being, just as there are the facts of the economic human being or the enfamilied human being. Meaning is a fact about the concrete fact of organism in environmental transaction and not a fact about a mind separate and distinct from environment.

The personal automobile, especially one with clutch and gearshift, serves to enlarge the body beyond the skin. This first point is the most obvious. Such enlargement begins with the simple fact of the instrumentality of the automobile as a means of conveyance. The automobile is a vehicle for satisfying demands, and is as such a tool like a hammer. But whereas hammers extend only our hands and arms, automobiles extend our legs, hands, fingers, arms, eyes, ears, and so on. The automobile is a thoroughgoing tool. Moreover, though all tools in a sense enlarge the body, some tools become more than simple means when they are chosen for the enjoyment of their functioning. Among those so chosen, some become especially personated. By the verb "to personate" I mean here the proper correlative to the verb "to impersonate." To impersonate is to play at being another; to personate is to play at being oneself. The automobile in the lives of many is a thoroughgoing tool within which the skinned body is absorbed and enjoyed for its functioning and in terms of which one plays at being a self.

Too, an automobile is like a suit of clothes. It is our traffic habit, our highway wear. Clothes serve the ends of modesty and decoration; the automobile gives us anonymity and status. Both protect us from the elements, natural and social. But as everyone is well aware, some garments come to be identified with one's character. They merge into the personality. They are stuck by some secret glue to one's privacies. They are personated.

The richness of this meaning of the automobile is not exhausted by the consideration of it as a personated, thoroughgoing tool and garment. In order for the automobile to enlarge the body, it must be demachined, for the automobile is a machine indeed while man is not. Many grave men once believed that all of nature was a vast machine and that the human physiological plant was one very marvelous machine within the larger one. Men seldom believe this today. But suppose it were true. It is obvious, is it not, that the physiological machine is amazingly personated, organic, pervaded with feeling, intimate. This could only be by some sleight-of-hand, some arch-trickery, whereby what is a machine is discovered to be functioning as though it were not a machine. This is what I mean by demachining. There is the science-fiction story of the humanoid robot that comes to serve as a personal friend of the hero. There is the companion story of the robot who asserts his rights to life, liberty, and property. There is more here than mere personification, for in each case the machine comes to function as what is no longer a machine. The automobile is not only personated, it is necessarily demachined in the process.

Contrast the power of automobiles as the grand machine of the turnpikes, still incompletely demachined, with the personation of the fine frame, which is so one with the protoplasm as to be uncommanded and unmastered. The swell machines are fun to drive, for we set ourselves up as bronco-busters to master their idiosyncratic powers. In fact any unfamiliar automobile is such a machine. It is the autonomous other for our mastery. However, the demachined automobile is not an automobile under control, as though it were another we have mastered; it functions as uncontrolled as our hands and feet. Who thinks of his hands as a thing to be mastered and controlled? And if one did, would it not thereby be meaningfully other than the body?

From what has been said so far it must be clear that not everyone who drives is an automobiled human being. Some of you may not know what I am talking about. The automobile remains a thing *for* you, merely a vehicle, an object of fear or astonishment, or perhaps a friend like a favorite dog, a child, or a mistress. The meanings of which I speak are, as yet, unrealized potentialities in your experience. You must not be disconcerted, for it is a simple enough truth that the meaning of things often eludes us. If you have never been a party to the kind of fact of which I speak, yet someday you may, and then perhaps you shall find this meaning within the fact.

I have said that the three meanings of automobiles with which I am concerned are interrelated. The initial point, the enlargement of the body, cannot be made adequately without anticipating the next two. Let me make two transitional comments.

First, the identification of automobile with self, its personations and demachining, seems to be at a peak at high speeds. This is, I suspect, analogous to the fact that one's identity with the flesh-and-bone body is experienced most profoundly in those periods of its peak yet effortless functioning. The demachining of the automobile requires, however, an added inducement, which is satisfied in the transformation of landscape at highway speeds into a uniquely alien environment characterized by anonymity and suddenness. Thus the dualism of self and the world is radical, and clearly the automobile itself exists at the still center which "I am." This experience of peak identification I call "the phenomenon of the Texas highway."

The second comment is closely related to the first. The automobile has an inside and an outside. This is a psychological as well as a spatial structure. What passes as ordinary visual and auditory sensations of externals in walking life becomes, in driving life, split between sensations of landscape and sensations of the automobile interior. The latter sensations take on an introspective aura, especially at high speeds, and the automobile interior becomes a nonpublic place of privacies. The character of one's internal life becomes enriched, and a larger self is realized. The dualism of self and world is compounded. This aspect of personation, whereby interiors of sensation become introspections, is the other side of the processes of certain psychoses in which the internals of phantasy are read wholeheartedly into an external world. I call this aspect introspectation. As an illustration consider the role of the automobile radio. Like our eyes and ears, it links us to the world, for through it we sense the happenings of the day; yet the sounds made are not themselves the link from dashboard to the outside, but the interiorized links from receptor to consciousness and as such fall within the body.

I now pass to the second meaning of automobiles, the introduction of the category of the sudden. I am referring to a new way of seeing things, a fresh principle that has arisen into public consciousness since the coming to be of the automobile and according to which the world takes on character and meaning of which human consciousness was not previously aware. I will not say that the basic structure of the universe has itself changed, for I know nothing of the universe, but I suggest that the world as the object of human consciousness has altered since the coming of the automobile. And the alteration that has come to be is simply this: the world presents itself as a place of the sudden, as an arena wherein the rapidly developing lurks to spring across the background of the slow, methodical passage of days. The basic category of processional growth, the natural way of the coming to be of man, trees, and love, is in no way replaced. But now the promise and fact of the sudden cuts across nature as over the surface of the waves of a patiently toiling sea. Birth, life, love, and death, spring, summer, autumn, and winter are there. The ripening of friendship, the rewards of long labor, digestion, pregnancy, creation are there. But also the scream of tires, the yellow light, the narrowly avoided accident.

I would like for you to make some efforts to understand me. I am not maintaining that the sudden is a fact of automobile experiencing alone. It is, no doubt, more apparent when one drives and clearest when one drives in traffic at high speeds. My suggestion is much more radical than this. I am suggesting that the sudden has become a category of all of our experience. I am also suggesting that the automobiled human being is the source and sanction of the publicity of this category and that this is one important meaning of automobiles. I suggest that contemporary man is the first man to live in a world of suddenness, that daily life, art, religion, philosophy, politics, conversation, education, and indeed all human concerns are infected with experience of and concern for crisis, and that the automobiled human being of traffic and highway is the breeder, infector, living reinforcement, and purest exemplification of the flash happening.

Now is the time to rectify one misunderstanding. Throughout this discussion you have suspected perhaps that the perspective of the sudden was neither new nor concerned with automobiles. What you have done is to confuse the sudden with the abrupt, the suddenly-happening with the suddenly-happened, rapidity with discontinuity. These are by no means the same. The category of the suddenly-happened has been native to thought and experience as long as men have lived in an environment in which important things suddenly-happened. And man always has. One suddenly dies, one suddenly wakes up screaming, tigers suddenly spring from trees. But the suddenly-happening is not the over and done with, the finished, the beginning abruptly or the instantly closed out. It is something going on, transpiring, but at the very limit of human capacities for adjustment or control. It directs our attention to a new facet of man, his reaction time. Man becomes, as does his age, crisis oriented. Perhaps this is not totally new. Maybe horse soldiers and gladiators participated in some such world as ours, but the fact is more pervasive now. It is public domain. It is common property. It is categorical.

A new phenomenon requires a new attitude. The attitude one takes up in the automobile is controlled by its adequacy to the landscape as a speedily shifting anonymous field of lurking suddenness, violence, and crisis. The attitude is an orientation to the appearance of the sudden, though the sudden does not constantly appear. When it does appear, the sudden moves a special object of concern across a background of speed and anonymity, as though it were a vividly red flash on a dull greyish ground. The attitude is directed to the occurrence of such flashes, but not in order to react to them, but in order to develop with and control them. This means that the attitude must be a constant tension which is both patient with long stretches of routines and capable of instant transformation into rapid and precise business. The attitude is not one of contemplation of the road, for the absorption in the object of attention precludes action. Nor is this attitude an ordinary tension or anxiety, for these allow neither long inactivity nor coolly perfect execution. In fact there is nothing quite like it. It is best understood by exemplification. Consider driving long distances at high speeds. What happens: There is an immense coolness, almost trancelike, in the performance of routine affairs. The times comes when the automobile is no longer handled; it handles itself. You, the driver, sit and await the exigencies. Now suppose you stop for a sandwich and a cup of coffee. Notice how difficult it is to make the transition. Your head rings: because your eyes are too wide to take in the slow subtle movements of human expressions, you have to give your order twice: you have to take a moment to get your bearings. You have to reorient yourself to a world of greater viscosity.

The category of the sudden is only understood by a glance backward at the meaning of the automobile as an extension of the body. You can see now that the new body serves to increase the scope of the manageable far beyond that in any previous age. At night on the highway the headlights pick out an anonymous landscape that extends approximately the length of control. The automobile moves into a world of secrets, anyone of which may emerge at the extreme of vision as a suddenly unfolding drama of collision. Such things may be managed. The mere punch of a foot serves to brake. The slight twitch of a hand swerves an enormous mass into another lane. The scope of the manageable is increased in distance, speed, mass, and momentum. The sudden as the advent of such management is precisely the initial impetus for the demachining of the automobile.

And now let me make a brief comment in transition to a discussion of the last meaning of automobiles to be considered. The new attitude is controlled by the scope of the manageable and posits a world of exigencies, of potential explosiveness, which is radically other than the automobiled human being himself. The human self, enlarged by the automobile, becomes a fact of mobile privacy. We are now in a position to understand this contemporary man.

Privacy, we sometimes feel, is a function of anonymity. The anonymity of the automobile, its faceless commonness, is mobile and hence it is a constant of human traffic, a mask for all occasions. It does not reveal us as do our face and speech, nor must the automobile assume the frightening fixity of assumed joviality, wittiness, or sternness. The automobiled person is from the outside the only typical specimen of the mass man. On the highway one always travels incognito.

In proportion to the perfection of anonymity in the automobile privacy is deepened. The automobile is spacious, like a spacious soul, and contains room for the greatest and the smallest of intimacies. We cannot sing loudly and with proper flourishes even in our homes, but the roar of engine and the passage through anonymous landscapes allow us such expressions in the automobile. The relaxation of the strains of holding steadily to our pedestrian masks effects release in the private car. These expressions no longer have the character of overt behavior but become as internal, demachined, personated, and introspectated as the automobile itself.

There is thus an inside and an outside to the automobiled human being. From the outside there is anonymity. From the inside there is privacy. The windshield is the frontier between the person and the world. The world is the place of exigencies, the coiled potentialities of violence, the possibilities of the sudden. The person is the autonomous master manipulator with the enriched intimacies of a mobile hideout.

But there is an ambiguity involved. Such utter facelessness on the outside and such complete privacy on the inside are only the typical facts of the flow of traffic; these facts are strained when an intrusion occurs. There are sometimes riders, sometimes love, sometimes death. These are intruders in the privacy and the anonymity, brutally contradicting both. If intrusion could destroy anonymity without destroying privacy, if intrusion could cancel privacy without canceling anonymity, then there would no ambiguity. But intrusion actually serves only for the destruction of the simple inside/outside structure of anonymity/privacy. The bringing of outsides into insides, the world into the person, gives the intruders a peculiar quality of intimacy not otherwise encountered. The peculiarity is due to the fact that the intruders are ambiguous—ultimate strangers in a private land yet ultimately personated interiors from a public landscape. Love, death, and riders are internal-externals, which if not contradictions are at least riddles. But the driver of the automobile is a party to the intrusion also. Without him intrusions would be only entries. The automobile man in his moment of intrusion is an ambiguity.

I will illustrate with three vignettes of intrusion.

**Vignette One: The Hitchiker.** I step into the automobile bidden and yet, as an outsider in his domain, an intruder. I am welcome; I should make myself at home. I am not welcome; I should take up as little of the privacy as possible. I am a question from the start. He does not know me. He does not trust me. I am bringing into his life the unpredictable, the unassimilated, the faceless, the strange. But now he speaks to me. He is running around on his wife; he hates his job; he fears death by cancer. How privileged I am to be the object of such confidence. I am in his deepest soul, a sharer in his hideout, yet surely no one has ever been so external to his person. Interloper and confidant, I am irremediably ambivalent to him. I become, even to myself, a pun.

**Vignette Two: Love-Play.** This is a mobile rendezvous. You enter myself as something stolen from the world. The secret is deep indeed. We may not linger, for the anonymity of our place is subject to immediate discovery. Wariness is mixed with the recklessness of intimacy. The outside peers into the inside. The ambiguity of sex is unique and exciting. By invitation the alien world admits to my trespass. A piece of the landscape transgresses my soul. I am myself the one who is most naked and vulnerable. I am at the same time the anonymous manipulator. Desperate passion is one with infinite reserve. The suddenness of the world wraps into our privacies, as though the automobile were turned inside out.

**Vignette Three: Death.** The traffic fatality is statistical. There is in every highway death the symbolic anonymity of a figure in a column of figures, an announcement on the radio. But the suddenly perceived inevitability of final intrusion unfolds the eternally personal moment. You will participate in your death. There in the stark aloneness of the quiet center of suddenness, you will be totally aware. The very enlargement of the body underlines the vulnerability, the contingency of your automobile. The death is public, for it is an object of traffic and the openly observable recovery of a swollen body by the anonymity of landscape. The automobile is still, no longer personated, but a simple inert piece of the environment. It is a personal death; it is a common end.

These vignettes of intrusion take us beyond the meanings of automobiles and introduce us to the meanings of contemporary men. Therefore they may be labeled metaphysical escapades.

## The Automobile and American Morality

**Glen Jeansonne**

The American automobile has traveled the whole circuit from hero to villain. Once enshrined as a liberating and democratizing agent, it is now condemned as a major cause of pollution and congestion. Ever since the early automobiles clanked along dusty roads Americans have both loved and hated their cars.

During the pioneering stages of automobiling, Tennessee law required that a motorist advertise his intention of going upon the road one week in advance.[1] Vermont enforced an ordinance compelling motor car drivers to hire a person to walk one-eighth of a mile ahead of the car, bearing a red flag.[2] A law introduced in the Illinois legislature stated:

On approaching a corner where he cannot command a view of the road ahead, the automobilist must stop not less than one hundred yards from the turn, toot his horn, fire a revolver, halloo, and send up three bombs at intervals of five minutes.[3]

The pioneering era of the automobile had ended by the twenties. There had been 8,000 cars registered by 1900. As the twenties opened, there were 8,000,000. At the close of the decade there were almost 23,000,000.[4]

Not a phase of American life was untouched by the automobile. Americans no longer measured distance in miles, but in minutes. Isolated wasteland became choice property when a major highway cut through it. The automobile made elopements easier, increased the income of parsons who specialized in quick matrimonials, and swelled the duties and the fees of village constables. It made bootlegging profitable and prohibition impractical. It enriched the American vocabulary with such words as "flivver," "skid," and "jaywalker." It captured the hearts, imaginations, and pocketbooks of Americans.[5]

Several factors contributed to the automobile's impact on society. It provided flexible, time-saving transportation. Engineering improvements and a great increase in service stations made it a reliable means of transportation. Highway expansion and improvements opened new areas to travelers. A great variety of both economical and luxury cars became available. Marketing techniques were improved and credit sales extended. Advertising benefited from the growth of movies and radio. General prosperity provided abundant purchasing power. Wartime standardization and pent-up demand helped spur production. Favorable government policies toward business and public adulation made the enterprise highly satisfying for those who could stand the competition. Finally, the spirit of the decade was characterized by a distaste for moral crusades and a readiness for adventure.

As automotive producers experienced prosperity or suffered decline, they pulled other industries along with them. In 1929 the automobile industry consumed 85% of the rubber imported, 19% of the iron and steel made in the United States, 67% of the plate glass, 18% of the hardwood lumber, 27% of the lead, and 80% of the gasoline. One person in every ten worked for the automotive industry or affiliated companies.[6]

Psychological effects were equally poignant, if less measurable. The manufacture of automobiles demanded streamlined mass production and dealt with a product which lent itself readily to standardization.[7] Little skill was needed to operate the automotive machines. The ability to maintain a methodical pace, eliminate wasted motion and follow instructions precisely constituted the requirements demanded of a machine operator. Social intercourse was limited by noise, fear of falling behind the cycle and the realization that failure to pay close attention could create safety hazards. Pride in achievement was lost in the endless drone of the assembly line. One worker complained:

There is nothing more discouraging than having a barrel beside you with 10,000 bolts in it and using them all up. Then you get a barrel with another 10,000 bolts, and you know every one of those bolts has to be picked up and put in exactly the same place as the last 10,000 bolts.[8]

High wages and abundant jobs partially compensated for the drabness of automotive work. Entire cities, such as Cork, Michigan, were revitalized by the automobile industry. When Henry Ford built his Fordson tractor works there in 1917, the community economy was languishing. By 1929 workers at the Fordson plant were receiving five million dollars yearly in wages. Salaries paid by the Ford organization, which operated on a five-day week, were higher than the best wages paid by plants operating on a six-day week—fifty percent higher in some cases.[9]

Henry Ford himself was considered, if not an outright deity, certainly an agent of God. Ford made money by mass-producing cars and selling them for less than his competitors. In 1914 the price of the Model T was about $450. By 1924 the price had been reduced to $290. A poll of college students that year voted Henry Ford the third greatest figure of all time—Jesus Christ winning first place and Napoleon running second. Upon being told this, Ford's only comment was that he personally would have voted Napoleon first, rather than Christ. "Napoleon," he said, "had real get-up-and-go."[10]

While stimulating the economy of many cities, the automobile helped to alleviate some of the isolation and loneliness which had made rural life bleak. The automobile provided the farmer with access to the city and threshers, tractors, reapers and combines afforded him the leisure to utilize his new-found opportunity.[11]

The genesis of the automobile also meant the death of the one-room school. In Montgomery County, Pennsylvania, for example, sixty one-room schools had been closed through the organization of eighteen consolidated schools by 1923. New York City employed 180 motor buses to transport students.[12]

Although the automobile centralized rural activity, it led to a decentralization of the city. The twenties saw the growth of a new urban phenomenon, the suburb. Previously, cities had by necessity been compact. Now they could sprawl beyond the shadows of factories. In New York City as early as 1920, 420,000 people traveled back and forth daily by automobile. A survey taken among Michigan industrial workers several years later showed that 650,000 of 850,000 polled depended entirely on their cars to get to and from work.[13]

In 1922 the National Department Store in St. Louis inaugurated a novel trend by locating a branch store three miles from the center of town. The day of the branch location had arrived, soon to be followed by the development of the shopping center.[14]

As dwellings mushroomed in the suburbs, office buildings rose to the sky in the city's core. With space in the city's business district at a premium, the automobile became a substantial liability, for it was a voracious space-consumer. The automobile, which had promised to provide workers with the luxury of the suburb, demanded severe recompense by accelerating the blight of the city.[15]

To escape city monotony or the strain of daily work, millions of Americans went vacationing. Everyone now had a chance to be as free as a tramp and to do it respectably. By 1923 motor touring was being termed "the great American summer sport." For the rich there were elegant and majestic hotels; for the moderate, boarding houses, modest lodges and private homes. For the poor and the adventuresome there was camping. Anyone who could afford a car could afford a motor camping trip. No longer were vacation resorts the monopoly of the rich. Frank Brimmer, auto camping editor of *Field and Stream,* termed motor camping "the finest melting pot of our American democracy."[16]

The lure of the road proved too great a temptation for many churchgoers. Motor trips began to rival church socials as leisure activities in small communities. The rural church experienced the same fate as the small rural school. Churches were forced to follow the highways. Those distant from good roads declined and disbanded. Those near well-traveled highways grew and prospered.[17]

Frustrated theologians and parents attributed much of the moral laxity as well as religious neglect to the advent of the auto. The motor car was amoral if not immoral. It was undeniably iconoclastic and irreverent. As one observer remarked: "A load of booze is no heavier to an auto than an equal weight of deacons."[18]

Some uses of the automobile especially shocked defenders of sexual purity. The car displaced the parlor in the scheme of courtship. Young people traveled to neighboring towns where they could cut loose from parental supervision and gossiping whispers. As a perceptive critic noted: "A bulwark of American morality had always been the difficulty of finding a suitable place for misconduct."[19]

The judge of a juvenile court called the automobile "a house of prostitution on wheels." He explained that of all the girls brought before him for sex crimes in the past year, one-third had committed their offense in cars.[20]

In 1927 a police officer in New York City arrested a pair of newlyweds for petting in the rear seat of their car. The case was dismissed when it came before court, but the arresting officer expressed dismay at the trend of public morals. He said: "...conditions have been getting worse, with all sorts of couples kissing and hugging in automobiles and the girls openly smoking cigarettes."[21]

American morality received a further loosening with the growth of roadside tourist courts. These made it easier for families to travel late and avoid having to dress formally. They also made it convenient for unmarried couples with other things in mind. F.B.I. Director J. Edgar Hoover called the tourist courts "little more than camouflaged brothels." A survey conducted by sociologists from Southern Methodist University in 1936 found that 75% of the patrons of tourist camps in the Dallas area were not tourists at all, but local couples. At one camp a succession of couples had rented a certain cabin sixteen times in one night.[22]

The automobile was also on trial in the academic world. It was banned at Penn State and the University of Illinois, severely restricted at Yale, and condemned at Princeton. In 1926 Princeton banned the use of automobiles by students holding scholarships. The Secretary of the University explained: "Boys who can afford to drive a car in Princeton are not justified in asking financial aid from the university." The decision was buttressed by the Secretary's opinion that ownership of an automobile was "detrimental to the academic career of a student."[23]

Women as well as young people felt themselves freed from ancient taboos by the motor car. In her automobile the lady could smoke in private or sample bootlegged liquor. She could vacation alone, view movies, shop in city fashion centers, and temporarily escape the frustrations of raising children.

Gasoline cars were at first considered an improper concern for the lady who wished to remain a lady. Ladies were not supposed to possess the muscle or technical ability to drive or even appreciate cars. Motoring entailed wearing rough clothes and getting dirty, which was neither very fashionable nor very comfortable.[24]

Both attitudes and automobiles were soon forced to change to accommodate the ladies. Women demanded that cars be handsome as well as sturdy. Advertisements began to appeal not only to the pocketbook but to the aesthetic sense of the buyer. Bright colors became stylish.[25]

The automobile helped emancipate women and young people, but it also offered unprecedented opportunities to criminals. It facilitated bank robberies, bootlegging and insurance fraud. Car theft itself became a major problem.

Thieves were able to strike any part of a city and flee quickly. State lines were obliterated as the auto enabled criminals to greatly expand their escape routes and increase their refuges.[26]

Henry Ford received many congratulatory letters after he introduced his V-8 model in 1932. One letter, handwritten, was postmarked Tulsa, Oklahoma, April 10th, 1934:

> Mr. Henry Ford, Detroit Michigan
>
> Dear Sir:
>
> While I still have got breath in my lungs I will tell you what a dandy car you make. I have drove Fords exclusively when I could get away with one. For sustained speed and freedom from trouble the Ford has got every other car skinned and even if my business hasn't been strictly legal it don't hurt anything to tell you what a fine car you got in the V-8.
>
> Yours truly,
>
> Clyde Barrow[27]

Although it aided criminals, the auto also helped law enforcement. Patrolmen were able to cover wider areas and render emergency aid faster. More equipment could be carried in patrol cars and they could be dispatched quickly from headquarters to the scene of a crime.[28]

The automobile wrought subtler changes in the attitudes of honest Americans. This had long been a nation of frugal workers, but the motor car helped alter the tradition which linked indebtedness with immorality. Consumption, not hoarding, was the new virtue. In 1926, 75% of all motor vehicles purchased were bought on the installment plan. Automobile purchases involved a hierarchical structure in which the buyer borrowed from the dealer, the dealer from a finance company, and the finance company from a commercial bank.[29]

The car became a leading status symbol. By 1925 there were 20% more cars than telephones. Only 12% of farm families had running water, but 60% had cars. When asked by a Department of Agriculture investigator why her family owned a car but not a bathtub, a farm wife replied: "Why, you can't go to town in a bathtub."[30]

Across the nation, the automobile helped to build a more fluid society. The frontier which historian Frederick Jackson Turner said had closed in 1890 now opened again. Dissatisfied wage earners could tramp the highways in search of better fortunes. Probably they did not often find them, but the mere opportunity made contemplation of the future more hopeful.

The automobile did not make Americans a mobile people. The pioneers who traversed the Atlantic and the "Forty-Niners" who crossed the continent to California were hardly sedentary. But the automobile did make a naturally mobile people more mobile. Like most technological advances, the automobile was not so much a social innovator as a catalyst to innovation. America would have become more democratic, materialistic, urban, industrial, and less straitlaced even without the automobile. But the auto stimulated change at an almost revolutionary rate. It accelerated the pace of American life. It enabled more people to do more things in less time at distant places. The things that they did were both admirable and mischievous. That depended on the person, not the car.

The most striking image generated by the auto was that of speed. The automobile was exciting and frightening, challenging and frustrating. It brought new solutions and multiplied old problems. It increased freedom but demanded responsibility. It freed workers to travel while it chained them to the assembly line. It ended the farmer's isolation but increased his sense of inferiority by revealing the affluence of his urban neighbors. It brought efficiency to industry, new opportunity to recreation, and anarchy to traditional morality.

When prosperity ended in the early thirties the frustrated demand for the automobile and the things it made possible increased tension. Those who could still afford to drive automobiles became objects of bitter jealousy. But the desire to own and operate one's personal car remained deeply entrenched. It had become a part of the American dream.

## NOTES


1 Dwight Lowell Dumond, *America in Our Time* (New York, 1947), 42.
2 William E. Leuchtenburg, *The Perils of Prosperity* (Chicago, 1958), 185.
3 "Motor Laws that Make Lawbreakers," *Literary Digest,* LXXVII (May 12, 1923), 58.
4 John Bell Rae, *The American Automobile: A Brief History* (Chicago, 1965), 87; 238.
5 "The Automobile: Its Province and Problems," *Annals of the American Academy of Political and Social Science,* CXVI (November, 1924), vii; *The New York Times,* November 25, 1923.
6 "Automobiles and Prosperity," *New Republic,* LXI (January 29, 1930), 263–64.
7 Charles Reitell, "Machinery and its effect upon the workers in the Automotive Industry," *Annals of the American Academy,* CXVI (November, 1924), 38–39.
8 Charles Rumford Walker and Robert H. Guest, *The Man on the Assembly Line* (Cambridge, 1952), 11; 69.
9 Detroit *News,* September 15, 1929.
10 Keith Sward, *The Legend of Henry Ford* (New York, 1948), 279.
11 John M. McKee, "The Automobile and American Agriculture," *Annals of the American Academy,* CXVI (November, 1924), 14; J. C. Long, "The Motor Car as the Missing Link between Country and Town," *Country Life,* XLIII (February, 1923), 112; Ralph Cecil Epstein, *The Automobile Industry: Its Economic and Commercial Development* (Chicago, 1928), 7; Lloyd R. Morris, *Not So Long Ago* (New York, 1949), 394–95.

**12** LeRoy A. King, "Consolidation of Schools and Public Transportation," *Annals of the American Academy,* CXVI (November, 1924), 74–77.
**13** New York *Times,* November 7, 1920.
**14** Morris, *Not So Long Ago,* 385.
**15** John Ihlden, "The Automobile and Community Planning," *Annals of the American Academy,* CXVI (November, 1924), 200–204.
**16** Frank E. Brimmer, "Vacationing on Wheels," *American Magazine,* CII (July, 1926), 176.
**17** Warren H. Wilson, "What the Automobile has done to and for the Country Church," *Annals of the American Academy,* CXVI (November, 1924), 84; "Where the Car has helped the Church," *Literary Digest,* LXX (July 16, 1921), 53.
**18** Joseph K. Hart, "The Automobile in the Middle Ages," *Survey,* LIV (August 1, 1926), 494.
**19** Frank Robert Donavan, *Wheels for a Nation* (New York, 1965), 164.
**20** Morris, *Not So Long Ago,* 382.
**21** New York *Times,* September 28, 1927.
**22** David Lewis Cohn, *Combustion on Wheels: An Informal History of the Automobile Age* (Boston, 1944), 225.
**23** New York *Times,* November 14, 1926; October 19, 1926.
**24** Morris, *Not So Long Ago,* 387.
**25** New York *Times,* January 6, 1928.
**26** Arch Mandel, "The Automobile and the Police," *Annals of the American Academy,* CXVI (November 24, 1924), 191–94; New York *Times,* January 7, 1923.
**27** *Time,* May 2, 1934, 45.
**28** Mandel, "The Automobile and the Police," 191–94; New York *Times,* January 7, 1923.
**29** Cohn, *Combustion on Wheels,* 259; Epstein, *The Automobile Industry,* 116–19.
**30** Morris, *Not So Long Ago,* 381.

## From *The Assembly Line*

**Robert Linhart**

### THE FIRST DAY: MOULOUD

"Show him, Mouloud."

The man in the white overalls (he's the foreman, called Gravier, as they'll tell me later) leaves me standing and goes off busily toward his glass-walled cage.

I look at the laborer who is working. I look at the shop floor. I look at the assembly line. No one speaks to me. Mouloud takes no notice of me. The foreman has gone. So I observe at random: Mouloud, the Citroën 2 CV car bodies passing in front of us, and the other laborers.

The assembly line isn't as I'd imagined it. I'd visualized a series of clear-cut stops and starts in front of each work position: with each car moving a few yards, stopping, the worker doing his job, the car starting again, another one stopping, the same operation being carried out again, etc. I saw the whole thing taking place rapidly—with those "diabolical rhythms" mentioned in the leaflets. *The assembly line:* the words themselves conjured up a jerky, rapid flow of movement.

The first impression, on the contrary, is one of a slow but continuous movement by all the cars. The operations themselves seem to be carried out with a kind of resigned monotony, but without the speed I expected. It's like a long, gray-green, gliding movement, and after a time it gives off a feeling of somnolence, interrupted by sounds, bumps, flashes of light, all repeated one after the other, but with regularity. The formless music of the line, the gliding movement of the unclad gray steel bodies, the routine movements: I can feel myself being gradually enveloped and anesthetized. Time stands still.

Three sensations form the boundaries of this new universe. The smell: an acrid odor of scorched metal and metallic dust. The sound: the drills, the roaring of the blowtorches, the hammer strokes on metal. And the grayness: everything's gray, the walls of the shop, the metallic bodies of the 2 CVs, the overalls and work clothes that the men wear. Even their faces look gray, as though the pale, greenish light from the cars passing in front of them were imprinted on their features.

The soldering shop, where I've just been allocated ("Put him to watch number 86," the sector manager had said) is fairly small. Thirty positions or so, arranged around a semicircular line. The cars arrive as nailed-up sections of coachwork, just pieces of metal joined together: here the steel sections have to be soldered, the joints eliminated and covered up; the object which leaves the workshop is still a gray skeleton, a car body, but a skeleton looking from now on as if it's all in one piece. The body is now ready for the chemical coatings, painting, and the rest of the assembly.

I note each stage of the work.

The position at the entrance to the shop is manned by a worker with special lifting gear. We're on the first floor, or rather a kind of mezzanine floor with only one wall. Each car body is attached to a rope, and the man drops it—roughly—onto a platform at the start of the assembly line. He secures the platform to one of the big hooks you can see moving slowly at ground level, a yard or two apart. These hooks make up the visible part of the perpetual motion mechanism: "the assembly line." Beside this worker stands a man in blue overalls supervising the start of the line, and he intervenes from time to time to speed things up: "O.K., that's it, fix it on now!" Several times during the day I'll see him at this spot, urging the man with the lifting gear to get more cars into the circuit. They'll tell me later that he's Antoine, the charge hand. He's a Corsican, small and excitable. "He makes a lot of noise, but he's not bad. The thing is, he's afraid of Gravier, the foreman."

The crash of a new car body arriving every three or four minutes marks out the rhythm of the work.

As soon as the car has been fitted into the assembly line it begins its half-circle, passing each successive position for soldering or another complementary operation, such as filing, grinding, hammering. As I said, it's a continuous movement and it looks slow: when you first see the line it almost seems to be standing still, and you've got to concentrate on one actual car in order to realize that car is moving, gliding progressively from one position to the next. Since nothing stops, the workers also have to move in order to stay with the car for the time it takes to carry out the work. In this way each man has a well-defined area for the operations he has to make, although the boundaries are invisible: as soon as a car enters a man's territory, he takes down his blowtorch, grabs his soldering iron, takes his hammer or his file, and gets to work. A few knocks, a few sparks, then the soldering's done and the car's already on its way out of the three or four yards of this position. And the next car's already coming into the work area. And the worker starts again. Sometimes, if he's been working fast, he has a few seconds' respite before a new car arrives: either he takes advantage of it to breathe for a moment, or else he intensifies his effort and "goes up the line" so that he can gain a little time, in other words, he works further ahead, outside his normal area, together with the worker at the preceding position. And after an hour or two he's amassed the incredible capital of two or three minutes in hand, that he'll use up smoking a cigarette, looking on like some comfortable man of means as his car moves past already soldered, keeping his hands in his pockets while the others are working. Short-lived happiness: the next car's already there: he'll have to work on it at his usual position this time, and the race begins again, in the hope of gaining one or two yards, "moving up" in the hope of another peaceful cigarette. If, on the other hand, the worker's too slow, he "slips back," that is, he finds himself carried progressively beyond his position, going on with his work when the next laborer has already begun his. Then he has to push on fast, trying to catch up. And the slow gliding of the cars, which seems to me so near to not moving at all, looks as relentless as a rushing torrent which you can't manage to dam up: eighteen inches, three feet, thirty seconds certainly behind time, this awkward join, the car followed too far, and the next one already appearing at the usual starting point of the station, coming forward with its mindless regularity and its inert mass. It's already halfway along before you're able to touch it, you're going to start on it when it's nearly passed through and reached the next station: all this loss of time mounts up. It's what they call "slipping" and sometimes it's as ghastly as drowning.

I'll learn this assembly line existence later, as the weeks go by. On the first day I must get the hang of it: through the tension on a face, some irritable gesture, the anxiety in a man's glance toward a car body that's appearing when the one before is not yet finished. As I look at the laborers one after another I'm beginning to see differences in what seemed at first glance to be a homogeneous human mechanism: one man is calm and precise, another sweats from overwork; I notice that some are ahead, some are behind; I see the minute, tactical details at each station, the men who put their tools down between cars and those who hold onto them, those who get out of step... And the perpetual, slow, implacable gliding of the 2 CVs under construction, minute by minute, movement by movement, operation by operation. The punch. The sparks. The drills. Scorched metal.

Once the car body has finished its circuit at the end of the curving line it's taken off its platform and pushed into a moving tunnel which takes it off to the paint shop. And there's the crash of a new car coming on the line to replace it.

Through the gaps in this gray, gliding line I can glimpse a war of attrition, death versus life and life versus death. Death: being caught up in the line, the imperturbable gliding of the cars, the repetition of identical gestures, the work that's never finished. If one car's done, the next one isn't, and it's already there, unsoldered at the precise spot that's just been polished. Has the soldering been done? No, it's waiting. Has it been done once and for all this time? No, it's got to be done again, it's always waiting to be done, it's never done—as though there were no more movement, no result from the movements, no change, only a ridiculous illusion of work which would be undone as soon as it's finished under the influence of some curse. And suppose you said to yourself that nothing matters, that you need only get used to making the same movements in the same way in the same period of time, aspiring to no more than the placid perfection of a machine? A temptation to death. But life kicks against it and resists. The organism resists. The muscles resist. The nerves resist. Something, in the body and the head, braces itself against repetition and nothingness. Life shows itself in more rapid movement, an arm lowered at the wrong time, a slower step, a second's irregularity, an awkward gesture, getting ahead, slipping back, tactics at the station; everything, in the wretched square of resistance against the empty eternity of the work station, indicates that there are still human incidents, even if they're minute; there's still time, even if it's dragged out to abnormal lengths. This clumsiness, this unnecessary movement away from routine, this sudden acceleration, this soldering that's gone wrong, this hand that has to do it all over again, the man who makes a face, the man who's out of step, this shows that life is hanging on. It is seen in everything that yells silently within every man on the line, "I'm not a machine!"

In fact, two stations beyond Mouloud, a worker—another Algerian, but more obviously so, he looks almost Asiatic—is in the process of "slipping." He's been gradually moving down toward the next station. He's getting nervous about his four bits of soldering. I can see him becoming more agitated, I can see the rapid movement of the blowtorch. All of a sudden he's had enough. He calls out to the charge hand, "Hey, not so fast, stop them a minute, it's no good like this!" And he unhooks the platform from the car he's working on, keeping it still as far as the next hook which will carry it forward again a few seconds later. The men working at the preceding stations unhook in their turns, to avoid a pile-up of cars. Everyone breathes for a moment. It makes a gap of a few yards in the line—a space a little bigger than the others—but the Algerian has caught up. This time Antoine, the section manager, says nothing: he's been pushing hard for an hour, and he's three or four cars in advance. But on other occasions he intervenes, gets after the man who's slipping back, won't let him unhook the car, or, if he's already done so, he rushes up to get the platform back to its original place.

This incident had to happen before I realized how tight the time schedule is. Yet the movement of the cars seems slow, and as a rule there's no sign of haste in the movements made by the workers.

So here I am at the factory. "Settled in." Being taken on was easier than I imagined. I'd thought out my story carefully: worked as a clerk in a grocery store owned by an imaginary uncle in Orléans, then storekeeper for a year (work certificate by special dispensation), military service with the engineers at Avignon (I gave the service record of a worker friend of mine of my own age and pretended I'd lost my record book). No diplomas, no, not even the BEPC. I could pass for a Parisian of provincial origin at loose ends in the capital, reduced to working in a factory because the family had lost all its money. I answered the questions briefly; I was silent and anxious. My wretched expression should not look out of place among the general appearance of the new intake. I hadn't put on the expression: the troubles of May 1968 and after—a summer of upset and quarrels—still showed on my worn face, just as others among my companions showed visibly the harsh conditions in which they lived. You feel right down when you come to beg for a little manual job—just enough money for food, please—and you timidly reply "none" to the questions about diplomas and qualifications, about anything special you can do. All my comrades in the job line were immigrants and I could read in their eyes the humiliations of this "none." As for me, I looked sufficiently wretched to seem like a would-be worker who was beyond suspicion. The man in charge of the new intake must have thought, "Yes, here's a fellow from the country, he's a bit dazed, really, that's good, he won't make any trouble." And he gave me my pass for the medical examination. Next please. And in any case, why should there be any complication about taking on a worker for the assembly line? That notion's typical of an intellectual who's used to complex appointments, a list of degrees, and "job analyses." That's what it's like when you're somebody. But if you're nobody... Everything moves very fast here: it's easy to assess two arms! A lightning medical inspection, with the little gang of immigrants. A few movements of the muscles. X-ray. You're weighed. The atmosphere's there already: "Stand there!" "Strip to the waist!" "Hurry up over there!" A doctor makes a few marks on a form. That's all, O.K. to work for Citroën. Next please.

It's a good moment: just now, in early September 1968, Citroën's devouring workers. Production's high and they're filling the gaps left among the immigrants by the month of August: some have not returned from their remote holidays, others will come back late and will learn to their despair that they've been fired ("We don't believe a damn word of it, that story about your old mother being ill, rubbish!") and already replaced. They replace people at once. In any case Citroën works in a state of flux: quickly in, quickly out. Average employment period at Citroën: a year. "A high turnover," say the sociologists. In fact, it's quick march. And for me there's no problem: I'm caught up in the flood of new entrants.

I left the recruitment office at Javel on Friday with a document: allocated to the plant at the Porte de Choisy. "Go to the section manager on Monday morning at seven o'clock." And now, this Monday morning, the Citroën 2 CVs moving past in the soldering shop.

Mouloud still doesn't say anything. I watch him work. It doesn't look too difficult. On each car shell arriving, the metal parts which form the curve over the windshield have been lined up and nailed into position but there's a gap between them. Mouloud's job is to get rid of this gap. In his left hand he takes the tin, which is shiny; in his right, a blowtorch. A flame bursts for a moment. Part of the tin melts in a little heap of soft stuff on the joint between the sheets of metal: Mouloud carefully spreads this stuff out, using a little stick which he picks up as soon as he puts the torch down. The crack disappears: the metal section above the windshield now looks as if it's all one piece. Mouloud has been alongside the car for two yards; when the job's done he leaves it and returns to his station, to wait for the next one. Mouloud works fast enough to have a few seconds free between cars, but he doesn't use the time to "move up." He prefers to wait. Here's a new car now. Shiny tin, blow torch, little stick, a few strokes to the left, to the right, up and down... Mouloud walks along as he works on the car. A final stroke with the stick and the soldering's smooth. Mouloud comes back toward me. A new car approaches. No, it doesn't look too difficult: why doesn't he let me have a go?

The line stops. The men take out their snacks. "Break," Mouloud tells me, "it's quarter past eight." Is that all? I felt that hours had gone by in this gray shop, divided between the monotonous gliding of the car bodies and the dim flares from the blowtorches. This interminable flow of sheet iron and metal outside time: only an hour and a quarter?

Mouloud offers me a share of the bread that he's carefully unwrapped from a piece of newspaper. "No, thanks. I'm not hungry."

"Where are you from?"

"Paris."

"Is this your first job with Citroën?"

"Yes, and my first time in a factory."

"I see. I'm from Kabylia. I've got a wife and children out there."

He takes out his wallet and shows me a faded family photograph. I tell him that I know Algeria. We talk about the winding roads in Greater Kabylia and the sheer cliffs which drop down to the sea near Collo. The ten minutes are up. The line starts again. Mouloud takes his blowtorch and goes toward the first approaching car.

We go on talking, intermittently, between cars.

"For the time being you only have to watch," Mouloud tells me. "You see, the soldering's done with tin. The 'stick' is made of tin. You have to get into the way of using it: if you put too much tin, it makes a lump on the coachwork and that's no good. If you don't put enough tin it doesn't fill up the hole and that's no good either. Watch what I do; you can try it this afternoon." And, after a silence: "You'll start soon enough..."

And we talk about Kabylia, Algeria, olive growing, the rich Mitidja plain, tractors and ploughing, variations in the harvest, and the little mountain village where Mouloud's family has remained. He sends three hundred francs a month, and he's careful not to spend too much on himself. This month it's difficult, an Algerian workmate has died, and the others have subscribed to pay for sending the body back home, and a little money for the family, too. It's made a hole in Mouloud's budget, but he's proud of the solidarity among the Algerians and especially among the Kabylians. "We support each other like brothers."

Mouloud must be about forty. A small moustache, hair graying at the temples, a slow, calm voice. He speaks as he works, with precision and regularity. No unnecessary gestures. No unnecessary words.

The car bodies go by, Mouloud does his soldering. Torch, tin, piece of wood. Torch, tin, piece of wood.

A quarter past twelve. The canteen. Three-quarters of an hour to eat. When I come back to my place, just before one o'clock, Mouloud's already there. I'm glad to see his face again, already familiar in the midst of this dirty gray workshop, this colorless metal.

It isn't one o'clock yet: they're waiting for work to start again. A little farther on a crowd has formed around the Algerian worker with the Asiatic features whom I saw "slipping back" this morning. "Hey, Sadok, let's have a look! Where did you get it?" I go closer. Sadok is cheerfully showing everyone a pornographic magazine, from Denmark or somewhere like that. The cover shows a girl sucking an erect penis. It's a close-up, in aggressive, realistic colors. I find it very ugly but Sadok seems delighted. He bought it from one of the truckdrivers who work for Citroën, transporting sheets of metal, engines, spare parts, containers, and finished cars, and who at the same time bring into the factory small supplies of cigars, cigarettes, and various other things.

Mouloud, who has taken in the cause of all this excitement with one glance, doesn't move. Someone calls out to him: "Hey, Mouloud, come and see a bit of rump, it'll do you good." He doesn't move, replies: "I'm not interested." And when I come back to join him he tells me in a lower voice: "It's not good. I've got a wife and kids out there in Kabylia. It's different for Sadok. He's a bachelor, he can amuse himself."

The porn magazine among the metallic dust and the filthy gray overalls: a painful impression. Prisoners' fantasies. I'm glad Mouloud stays away.

A metallic noise, everyone goes back to his place, the line starts to move again.

"Now it's your turn," Mouloud tells me. "You've seen what you've got to do." And he hands me the blowtorch and the tin.

"No, no! Not like that! And put the gloves on, you'll burn yourself. Whoa! mind the torch! Give it to me!..."

This is the tenth car I've fenced with in vain. Mouloud warns me, guides my hand, passes me the tin, holds the torch for me, all in vain; I can't do it.

On one car I flood the metal with tin because I've held the torch too near the tin and for too long: all Mouloud can do is scrape it all off and redo the operation rapidly while the car has already almost left our area. Then I don't put on enough tin and the first touch with the little piece of wood merely shows up the crack again that had to be covered. And when by some miracle I've put more-or-less the right amount of tin, I spread it so clumsily—damn that little piece of wood that my fingers obstinately refuse to control!—that the soldering's as bumpy as a fairground switchback and there's a horrible lump where Mouloud succeeded in producing a perfectly smooth curve.

I get mixed up about the order of operations: you have to put the gloves on to hold the torch, take them off to use the little piece of wood, not touch the burning tin with your bare hands, hold the "stick" with your left hand, the torch in your right hand, the little piece of wood in your right hand, the gloves you've just taken off in your left hand, with the tin. It looked obvious when Mouloud did it, with a succession of precise, coordinated movements. I can't manage it, I panic: I nearly burn myself ten times over and Mouloud moves rapidly to push the torch away.

Every one of my joints has to be redone. Mouloud takes the instruments from me and just manages to catch up, three yards farther on. I'm sweating and Mouloud's beginning to get tired: his rhythm has been interrupted. He shows no impatience and continues to do this double work—guiding mine, and then redoing it—but we're slipping back. We're moving inexorably down toward the next position, we're starting the next car one yard too late, then two yards; we finish it, or rather Mouloud finishes it, rapidly, three or four yards farther on, with the torch cable stretched almost to the limit, in the middle of the instruments belonging to the next position. The faster I try to go, the more I panic: I let the molten tin drip all over, I drop the little piece of wood, as I turn round the flame from my torch nearly gets Mouloud, he just manages to avoid it.

"No, not like that, now look!" There's nothing to be done. My fingers are awkward, I'm incurably clumsy. I'm wearing myself out. My arms are trembling. I press too hard with the wood, I can't control my hands, drops of sweat are beginning to get in my eyes. The car bodies seem to move at a frantic speed, there's no hope of getting ahead, Mouloud's finding it harder and harder to catch up.

"Now listen, getting in a state doesn't help. Stop a minute and watch what I do."

Mouloud takes the instruments from me and picks up the regular rhythm of his work again, a bit faster than before, in order gradually to make good the time we've lost: a few inches on each car; after ten or so he's almost back in his normal place. As for me, I get my breath back as I watch him work. His movements look so natural! What have his hands got that mine haven't? A car comes: tin, torch, piece of wood, and at the spot where the curving metal was split there's now a perfectly smooth surface. Why can he do the work and I can't?

The 3:15 break. Mouloud sacrifices it to me. The others stretch their legs, form groups, chat, come and go, sit on barrels, or lean against the cars which are standing still. Mouloud begins to explain again. The car in front of our position isn't moving, it's easier. That's the distance at which you hold the torch. And this is how you place your fingers on the stick. There. Press with your thumb to grasp the round part of the metal. In the middle you must press very lightly in order to stop the tin from escaping and you must press more and more firmly as you move away: that's how you get the graded effect. Take the wood to the left first, then to the right. Then a short stroke upward, and another downward. Mouloud repeats the movement slowly: four times, five times. My turn now: he guides my hand, arranges my fingers against the wood. Like that. There. Good, perhaps that'll be all right... My brain thinks it understands it all: will my hands obey?

End of the break, back to work. Din of the line. A new car comes forward, slow and menacing: I've got to carry out those movements again with the real thing. Quick, the torch, oh no! I forgot, the gloves first, where's the tin? Good God, how quickly it's coming, it's in the middle of our station already, use the torch, blast! too much tin, get rid of it with the piece of wood, it's gone all over... Mouloud removes it for me with his hands. One more try... No, it's no good. I'm dismayed, I must have looked at Mouloud in despair, he tells me: "Don't worry now, it's always a bit hard at first, have a rest, let me do it." Once more I stand on the edge, watching helplessly: the line has rejected me. And yet it seems to move so slowly...

Mouloud decides not to give me the tools again.

"It'll go better tomorrow, go on now, you mustn't worry about it." We talk about his own beginnings at this station, a long time ago: he got the hang of it fairly quickly, but at first it's not easy... By now he's had long experience with soldering and he does it mechanically.

In fact I've heard that soldering is a craft. What qualifications has Mouloud got? I ask him how Citroën classifies him. "M2," he replies in laconic fashion. Laborer.

I'm astonished. He's only a laborer? Yet soldering's not as easy as all that. I can't do anything, but I've been taken on as a "semiskilled worker" (OS2, says the contract): in the hierarchy of the not-very-important, semiskilled is still above laborer... Mouloud obviously doesn't want to go places. I don't say anything more. As soon as I can I'll find out on what principles Citroën makes their classifications. A few days later another worker told me. There are six categories of nonqualified workers. Starting at the bottom: three categories of laborers (M1, M2, M3); three categories of semiskilled workers (OS1, OS2, OS3). The distinction is made in a perfectly simple way: it's racist. The Blacks are M1, right at the bottom of the ladder. The Arabs are M2 or M3. The Spaniards, Portuguese, and other European immigrants are usually OS1. The French are automatically OS2. And you become OS3 just because of the way you look, depending on how the bosses want it. That's why I'm a skilled worker and Mouloud's a laborer, that's why I earn a few centimes an hour more, although I'm incapable of doing his work. And later they will draw up subtle statistics about the "classification grid," as the specialists say.

That's it. Mouloud has just finished his last car. The hundred and forty-eighth of the day. It's a quarter to six. The line stops. So does the noise. "So long," says Mouloud, "see you tomorrow... Don't worry now, it'll go better." He hurries off to the cloakroom. I remain for a moment in the shop, which is emptying, my head's throbbing, my legs are unsteady. When I get to the stairs, and I'm really the last, there's no one in sight any more. The lights are out and the car bodies are motionless dark masses, waiting for the dawn and a new day's work.

I come home, exhausted and anxious. Why are all my limbs so painful? Why does my shoulder ache, and my thighs? Yet the blowtorch and the stick weren't all that heavy to carry... It's no doubt due to the repetition of the same movement over and over again. And the tension needed to control my clumsiness. And it's because I've been standing all that time: ten hours. But the others do it as well. Are they as exhausted as I am?

I think: it's the intellectual's ineptitude for physical effort. That's naïve. It isn't just a question of physical effort. The first day in a factory terrifies everyone, many people will speak to me about it later, often with anguish. What mind, what body can accept this form of slavery, this destructive rhythm of the assembly line, without some show of resistance? It's against nature. The aggressive wear and tear of the assembly line is experienced violently by everyone, city workers and peasants, intellectual and manual workers, immigrants and Frenchmen. And it's not unusual to see a new recruit give up after his first day, driven mad by the noise, the sparks, the inhuman pressure of speed, the harshness of endlessly repetitive work, the authoritarianism of the bosses and the severity of the orders, the dreary prisonlike atmosphere which makes the shop so frigid. Months and years in there? How can one imagine such a thing? No: better escape, poverty, the insecurity of little odd jobs, anything!

And what about me, someone from the establishment, am I going to be able to cope? What will happen tomorrow if I still can't do that soldering? Will they throw me out? How ridiculous! A day and a half on the job... and then fired for being incapable! And what about the others, those who haven't any diplomas and who are neither strong nor good with their hands, how will they manage to earn their living?

Night. I can't sleep. As soon as I close my eyes I see piles of 2 CVs, a sinister procession of gray car bodies. I see again Sadok's porn magazine among the sandwiches and the oil drums and the metal. Everything's ugly. And those 2 CVs, that interminable string of 2 CVs... The alarm clock goes off. Six o'clock already? I ache all over, I'm just as worn out as I was last evening. What have I done with my night?

## Television: The New State Religion?

**George Gerbner**

Both classical electoral and classical Marxist theories of government are based on assumptions rooted in cultural developments of the eighteenth and nineteenth centuries. These developments also gave rise to mass communications and eventually to research on mass communications. In the past few decades, however, rapidly accumulating changes brought about a profound transformation of the cultural conditions on which modern theories of government and of mass communications rest. That change presents an historic challenge to these theories and to scientific workers concerned with these theories. I would like to sketch the nature of that challenge and to make a few tentative suggestions about the tasks ahead.

Human consciousness seems to differ from that of other animals chiefly in that humans experience reality in a symbolic context. Human consciousness is a fabric of images and messages drawn from those towering symbolic structures of a culture that express and regulate the relationships of a social system. When those relationships change, sooner or later the cultural patterns also change to express and maintain the new social order.

For most of humankind's existence, these systems of society and culture changed very slowly and usually under the impact of a collapse or invasion. The long-enduring, face-to-face, preindustrial, preliterate cultural patterns, relatively isolated from each other, encompassed most of the storytelling, and the rituals, art, science, statecraft, and celebrations of the tribe or larger community. They explained over and over again the nature of the universe and the meaning of life. Their repetitive patterns, memorized incantations, popular sayings, and stories demonstrated the values, roles, productive tasks, and power relationships of society. Children were born into them, old men and women died to their ministrations, and both rulers and the ruled acted out their respective roles according to their tenets. These organically integrated symbolic patterns permeated the life space of every member of the community. Nonselective participation of all in the same symbolic world generated mistrust of strangers, the quest for security through protection by the powerful, and a sense of apprehension of and resistance to change. Conflicts of interest were submerged and dissent suppressed in the interests of what to most people seemed to be the only possible design for life.

All of this changed when the industrial revolution altered the contours of power and the structure of society. The extension of mass production into symbol-making correspondingly altered the symbolic context of consciousness and created cultural conditions necessary for the rise of modern theories of government.

One of the first industrial products was the printed book. Printing made it possible to relieve memory of its formula-bound burdens and opened the way to the endless accumulation of information and innovation. "Packaged knowledge" (the Book) could be given directly to individuals, bypassing its previously all-powerful dispensers, and could cross the old boundaries of status and community. Images and messages could now be used *selectively*. They could be chosen to express and advance individual and group interests. Printed stories—broadsides, crime and news, mercantile intelligence, romantic novels—could now speak selectively to different groups in the population and explain the newly differentiated social relationships which emerged from the industrial revolution. Print made it possible for the newly differentiated consciousness to spread beyond the limiting confines of face-to-face communication. Selectivity of symbolic participation was the prerequisite to the differentiation of consciousness among class and other interest groups within large and heterogeneous societies.

Publics are created and maintained through *publication*. Electoral theories of government are predicated upon the assumption of cultural conditions in which each public can produce and select information suited to the advancement of its own interests. Representatives of those interests are then supposed to formulate laws and administer policies that orchestrate different group interests on behalf of society as a whole.

Marxist-Leninist theories of government similarly (albeit more implicitly) assume cultural conditions that permit selectivity of symbolic production and participation, and thus differentiation of consciousness along class lines. Lenin's characterization of the press as collective organizer and mobilizer assumes (not unlike advertisers do in capitalist countries) that the major mass media are the cultural organs of the groups that own and operate (or sponsor) them. Only in that way could working class organizations (or business corporations) produce ideologically coherent and autonomous symbol systems for their publics.

Before these theories of government came to full fruition, the cultural conditions upon which they were explicitly or implicitly based began to change. Private corporate organizations grew to the size and power of many governments. The increasingly massive mass media became their cultural arms and the First Amendment their shield. Commercial pressures made the service of many small, poor, or dissenting publics impractical. Public relations replaced the autonomous aggregation of many publics. Public opinion became the published opinions of cross-sections of atomized individuals rather than a differentiated mosaic reflecting the composite of organized publics, each conscious of its own interest.

In the young socialist countries and People's Democracies, mass media became centralized organs of revolutionary establishments. Their governing responsibilities made it difficult to cultivate a distinctly working class consciousness and to institutionalize the critical functions of the press.

These problems and difficulties arose under essentially print-based cultural conditions. But in the past few decades even those conditions began to change.

The harbinger of that change is television. The special characteristics of television set it apart from other mass media to such an extent that it is misleading to think of it in the same terms or to research it in the same terms. Furthermore, these special characteristics are only the forerunners of the prospect of an all-electronic organically composed and orchestrated total symbolic environment.

What are these special characteristics of television? My observations are based primarily on our research and experience in the United States. We do not yet know to what extent they are applicable to other countries. (That, I think, should be an early task for communications research to discover.)

**1** Television consumes more time and attention of more people than all other media and leisure time activities combined. The television set is on for six hours and fifteen minutes a day in the average American home, and its sounds and images now fill the living space and symbolic world of most Americans.

**2** Unlike the other media, you do not have to wait for, plan for, go out to, or seek out television. It comes to you directly at home and is there all the time. It has become a member of the family, telling its stories patiently, compellingly, untiringly. Few parents, teachers, or priests can compete with its vivid demonstrations of what people of all kinds are like and how society works.

**3** Just as television requires no mobility, it requires no literacy. In fact, it shows and tells about the world to the less educated and the nonreader—those who have never before shared the culture of the literate—with special authority and force. Television now informs most people in the United States—many of its viewers simply do not read—and much of its information comes from what is called entertainment. As in ancient times of great rituals, festivals, and circuses, the information-poor are again royally entertained by the organic symbolic patterns informing those who do not seek information.

**4** These organic patterns have to be seen—and analyzed—as total systems. The content differentiations of the print era, where there were sharp distinctions between information (news) and entertainment (drama, etc.) or fiction and documentary or other genres, no longer apply. Besides, viewers typically select not programs but hours of the day and watch whatever is on during those hours. Unlike books, newspapers, magazines, or movies, television's content and effects do not depend on individually crafted and selected works, stories, etc. Assembly line production fills total programming formulas whose structure encompasses all groups but serves one overall perspective. Storytelling (drama and legendary) is at the heart of this—as of any other—symbol system. "Real-life" demonstrations of the same value structure, as in television news, provide verisimilitude and "documentary" confirmation to the mythological world of television. All types of programming within the program structure complement and reinforce one another. It makes no sense to study the content or impact of one type of program in isolation from the others. The same viewers watch them all; the total system as a whole is absorbed into the mainstream of common consciousness.

**5** For the first time since the preindustrial age, or perhaps in all of history, there is little age-grading or separation of the symbolic materials that socialize members into the community. Television is truly a cradle-to-grave experience. Infants are born into a television home and learn from its sounds and images before they can speak, let alone read. By the time they reach school age they will have spent more hours with television than they would spend in a college classroom. At the far end of the life cycle, old people, and most institutionalized populations, are almost totally dependent on television for regular "human" contact and engagement in the larger world. Only a minority of children and older age groups watch the few programs (none in "prime time") especially designed for them. Unlike other media, television tells its stories to children, parents, and grandparents, all at the same time.

**6** Television is essentially in the business of assembling heterogeneous audiences and selling their time to advertisers or other institutional sponsors. The audiences include all age, sex, ethnic, racial, and other interest groups. They are all exposed to the same repetitive messages conveying the largest common denominator of values and conduct in society. Minority groups see their own image shaped by the dominant interests of the larger culture. This means the dissolution of the concept of autonomous publics and of any authentic group or class consciousness. Television provides an organically related synthetic symbolic structure which once again presents a total world of meanings for all. It is related to the State as only the church was in ancient times.

All this adds up to a non-selectively used cultural pattern which can no longer serve the tasks of cultivating selective and differentiated group, class, or other public consciousness. The pattern is formula-bound, ritualistic, repetitive. It thrives on novelty but is resistant to change, and it cultivates resistance to change. In that, too, television's social symbolic functions resemble preindustrial religions more than they do the media that preceded it. The process has tremendous popular mobilizing power which holds the least informed and least educated most in its spell. Results of our research (reported under the title "Living with Television: The Violence Profile" in the Spring 1976 issue of the *Journal of Communication*) indicate that television viewing tends to cultivate its own particular outlook on social reality even among the well-educated and traditionally "elite" groups.

Heavy viewers of television are more apprehensive, anxious, and mistrustful of others than light viewers in the same age, sex, and educational groups. The fear that viewing American television seems to generate, the consequent quest for security and protection by the authorities, the effective dissolution of autonomous publics, and the ease with which credible threats and scares can be used (or provoked) to justify almost any policy create a fundamentally new cultural situation. The new conditions of synthetic consciousness-making pose new problems, difficulties, and challenges for those who wish to realistically analyze or guide public understanding of society.

Researchers and scholars of communication and culture should now devote major attention to long-range cross-cultural comparative media studies that investigate the policies, processes, and consequences of the mass-production of major symbol systems in light of the respective structures and aims of different social systems. Do media really do what they are designed to do according to the theories governing (or used to explain) the societies in which they exist? What are the differences and similarities among them? What are the cultural and human consequences of the international exchange of media materials? What are the effects of changing cultural, technological, and institutional conditions upon the social functions of media, particularly television? What are the new organizational, professional, artistic, and educational requirements for the effective fulfillment of societal goals in different cultural and social systems? And, finally, how can liberation from the age-old bonds of humankind lead to cultural conditions that enrich rather than limit visions of further options and possibilities?

These are broad and difficult tasks but we can at least begin to tackle them. Much depends on the success of the effort.

## The New Languages

**Edmund Carpenter**

*Brain of the New World, What a task is thine, To formulate the modern ...to recast poems, churches, art*

WHITMAN

English is a mass medium. All languages are mass media. The new mass media—film, radio, TV—are new languages, their grammars as yet unknown. Each codifies reality differently; each conceals a unique metaphysics. Linguists tell us it's possible to say anything in any language if you use enough words or images, but there's rarely time; the natural course is for a culture to exploit its media biases.

Writing, for example, didn't record oral language; it was a new language, which the spoken word came to imitate. Writing encouraged an analytical mode of thinking with emphasis upon lineality. Oral languages tended to be polysynthetic, composed of great, tight conglomerates, like twisted knots, within which images were juxtaposed, inseparably fused; written communications consisted of little words chronologically ordered. Subject became distinct from verb, adjective from noun, thus separating actor from action, essence from form. Where preliterate man imposed form diffidently, temporarily—for such transitory forms lived but temporarily on the tip of his tongue, in the living situation—the printed word was inflexible, permanent, in touch with eternity: it embalmed truth for posterity.

This embalming process froze language, eliminated the art of ambiguity, made puns "the lowest form of wit," destroyed word linkages. The word became a static symbol, applicable to and separate from that which it symbolized. It now belonged to the objective world; it could be seen. Now came the distinction between being and meaning, the dispute as to whether the Eucharist *was* or only *signified* the body of the Sacrifice. The word became a neutral symbol, no longer an inextricable part of a creative process.

Gutenberg completed the process. The manuscript page with pictures, colors, correlation between symbol and space, gave way to uniform type, the black-and-white page, read silently, alone. The format of the book favored lineal expression, for the argument ran like a thread from cover to cover: subject to verb to object, sentence to sentence, paragraph to paragraph, chapter to chapter, carefully structured from beginning to end, with value embedded in the climax. This was not true of great poetry and drama, which retained multi-perspective, but it was true of most books, particularly texts, histories, autobiographies, novels. Events were arranged chronologically and hence, it was assumed, causally; relationship, not being, was valued. The author became an *authority;* his data were serious, that is, *serially* organized. Such data, if sequentially ordered and printed, conveyed value and truth; arranged any other way, they were suspect.

The newspaper format brought an end to book culture. It offers short, discrete articles that give important facts first and then taper off to incidental details, which may be, and often are, eliminated by the makeup man. The fact that reporters cannot control the length of their articles means that, in writing them, emphasis can't be placed on structure, at least in the traditional linear sense, with climax or conclusion at the end. Everything has to be captured in the headline; from there it goes down the pyramid to incidentals. In fact there is often more in the headline than in the article; occasionally, no article at all accompanies the banner headline.

The position and size of articles on the front page are determined by interest and importance, not content. Unrelated reports from Moscow, Sarawak, London, and Ittipik are juxtaposed; time and space, as separate concepts, are destroyed and the *here* and *now* presented as a single Gestalt. Subway readers consume everything on the front page, then turn to page 2 to read, in incidental order, continuations. A Toronto banner headline ran: TOWNSEND TO MARRY PRINCESS; directly beneath this was a second headline: *Fabian Says This May Not Be Sex Crime*. This went unnoticed by eyes and minds conditioned to consider each newspaper item in isolation.

Such a format lends itself to simultaneity, not chronology or lineality. Items abstracted from a total situation aren't arranged in casual sequence, but presented holistically, as raw experience. The front page is a cosmic *Finnegans Wake*.

The disorder of the newspaper throws the reader into a producer role. The reader has to process the news himself; he has to co-create, to cooperate in the creation of the work. The newspaper format calls for the direct participation of the consumer.

In magazines, where a writer more frequently controls the length of his article, he can, if he wishes, organize it in traditional style, but the majority don't. An increasingly popular presentation is the printed symposium, which is little more than collected opinions, pro and con. The magazine format as a whole opposes lineality; its pictures lack tenses. In *Life,* extremes are juxtaposed: spaceships and prehistoric monsters, Flemish monasteries and dope addicts. It creates a sense of urgency and uncertainty: the next page is unpredictable. One encounters rapidly a riot in Teheran, a Hollywood marriage, the wonders of the Eisenhower administration, a two-headed calf, a party on Jones beach, all sandwiched between ads. The eye takes in the page as a whole (readers may pretend this isn't so, but the success of advertising suggests it is), and the page—indeed, the whole magazine—becomes a single Gestalt where association, though not causal, is often lifelike.

The same is true of the other new languages. Both radio and TV offer short, unrelated programs, interrupted between and within by commercials. I say "interrupted," being myself an anachronism of book culture, but my children don't regard them as interruptions, as breaking continuity. Rather, they regard them as part of a whole, and their reaction is neither one of annoyance nor one of indifference. The ideal news broadcast has half a dozen speakers from as many parts of the world on as many subjects. The London correspondent doesn't comment on what the Washington correspondent has just said; he hasn't even heard him.

The child is right in not regarding commercials as interruptions. For the only time anyone smiles on TV is in commercials. The rest of life, in news broadcasts and soap operas, is presented as so horrible that the only way to get through life is to buy this product: then you'll smile. Aesop never wrote a clearer fable. It's heaven and hell brought up to date: Hell in the headline, Heaven in the ad. Without the other, neither has meaning.

There's pattern in these new media—not line, but knot; not lineality or causality or chronology, nothing that leads to a desired climax; but a Gordian knot without antecedents or results, containing within itself carefully selected elements, juxtaposed, inseparably fused; a knot that can't be untied to give the long, thin cord of lineality.

This is especially true of ads that never present an ordered, sequential, rational argument but simply present the product associated with desirable things or attitudes. Thus Coca-Cola is shown held by a beautiful blonde, who sits in a Cadillac, surrounded by bronze, muscular admirers, with the sun shining overhead. By repetition these elements become associated, in our minds, into a pattern of sufficient cohesion so that one element can magically evoke the others. If we think of ads as designed solely to sell products, we miss their main effect: to increase pleasure in the consumption of the product. Coca-Cola is far more than a cooling drink; the consumer participates, vicariously, in a much larger experience. In Africa, in Melanesia, to drink a Coke is to participate in the American way of life.

Of the new languages, TV comes closest to drama and ritual. It combines music and art, language and gesture, rhetoric and color. It favors simultaneity of visual and auditory images. Cameras focus not on speakers but on persons spoken to or about; the audience *hears* the accuser but *watches* the accused. In a single impression it hears the prosecutor, watches the trembling hands of the big-town crook, and sees the look of moral indignation on Senator Tobey's face. This is real drama, in process, with the outcome uncertain. Print can't do this; it has a different bias.

Books and movies only pretend uncertainty, but live TV retains this vital aspect of life. Seen on TV, the fire in the 1952 Democratic Convention threatened briefly to become a conflagration; seen on newsreel, it was history, without potentiality.

The absence of uncertainty is no handicap to other media, if they are properly used, for their biases are different. Thus it's clear from the beginning that Hamlet is a doomed man, but, far from detracting in interest, this heightens the sense of tragedy.

Now, one of the results of the time-space duality that developed in Western culture, principally from the Renaissance on, was a separation within the arts. Music, which created symbols in time, and graphic art, which created symbols in space, became separate pursuits, and men gifted in one rarely pursued the other. Dance and ritual, which inherently combined them, fell in popularity. Only in drama did they remain united.

It is significant that of the four new media, the three most recent are dramatic media, particularly TV, which combines language, music, art, dance. They don't, however, exercise the same freedom with time that the stage dares practice. An intricate plot, employing flash backs, multiple time perspectives and overlays, intelligible on the stage, would mystify on the screen. The audience has no time to think back, to establish relations between early hints and subsequent discoveries. The picture passes before the eyes too quickly; there are no intervals in which to take stock of what has happened and make conjectures of what is going to happen. The observer is in a more passive state, less interested in subtleties. Both TV and film are nearer to narrative and depend much more upon the episodic. An intricate time construction can be done in film, but in fact rarely is. The soliloquies of *Richard III* belong on the stage; the film audience was unprepared for them. On stage Ophelia's death was described by three separate groups: one hears the announcement and watches the reactions simultaneously. On film the camera flatly shows her drowned where "a willow lies aslant a brook."

Media differences such as these mean that it's not simply a question of communicating a single idea in different ways but that a given idea or insight belongs primarily, though not exclusively, to one medium, and can be gained or communicated best through that medium.

Thus the book was ideally suited for discussing evolution and progress. Both belonged, almost exclusively, to book culture. Like a book, the idea of progress was an abstracting, organizing principle for the interpretation and comprehension of the incredibly complicated record of human experience. The sequence of events was believed to have a direction, to follow a given course along an axis of time; it was held that civilization, like the reader's eye (in J. B. Bury's words), "has moved, is moving, and will move in a desirable direction. Knowledge will advance, and with that advance, reason and decency must increasingly prevail among men." Here we see the three main elements of book lineality: the line, the point moving along that line, and its movement toward a desirable goal.

The Western conception of a definite moment in the present, of the present as a definite moment or a definite point, so important in book-dominated languages, is absent, to my knowledge, in oral languages. Absent as well, in oral societies, are such animating and controlling ideas as Western individualism and three-dimensional perspective, both related to this conception of the definite moment, and both nourished, probably bred, by book culture.

Each medium selects its ideas. TV is a tiny box into which people are crowded and must live; film gives us the wide world. With its huge screen, film is perfectly suited for social drama, Civil War panoramas, the sea, land erosion, Cecil B. DeMille spectaculars. In contrast, the TV screen has room for two, at the most three, faces, comfortably. TV is closer to stage, yet different. Paddy Chayefsky writes:

> The theatre audience is far away from the actual action of the drama. They cannot see the silent reactions of the players. They must be told in a loud voice what is going on. The plot movement from one scene to another must be marked, rather than gently shaded as is required in television. In television, however, you can dig into the most humble, ordinary relationships; the relationship of bourgeois children to their mother, of middle-class husband to his wife, of white-collar father to his secretary—in short, the relationships of the people. We relate to each other in an incredibly complicated manner. There is far more exciting drama in the reasons why a man gets married than in why he murders someone. The man who is unhappy in his job, the wife who thinks of a lover, the girl who wants to get into television, your father, your mother, sister, brothers, cousins, friends—all these are better subjects for drama than Iago. What makes a man ambitious? Why does a girl always try to steal her kid sister's boy friends? Why does your uncle attend his annual class reunion faithfully every year? Why do you always find it depressing to visit your father? These are the substances of good television drama; and the deeper you probe into and examine the twisted, semi-formed complexes of emotional entanglements, the more exciting your writing becomes.[1]

This is the primary reason, I believe, why Greek drama is more readily adapted to TV than to film. The boxed-in quality of live TV lends itself to static literary tragedy with greater ease than does the elastic, energetic, expandable movie. Guthrie's recent movie of *Oedipus* favored the panoramic shot rather than the selective eye. It consisted of a succession of tableaux, a series of elaborate, unnatural poses. The effect was of congested groups of people moving in tight formation as though they had trained for it by living for days together in a self-service elevator. With the lines, "I grieve for the City, and for myself and you... and walk through endless ways of thought," the inexorable tragedy moved to its horrible "come to realize" climax as though everyone were stepping on everyone else's feet.

The tight, necessary conventions of live TV were more sympathetic to Sophocles in the Aluminum Hour's *Antigone*. Restrictions of space are imposed on TV as on the Greek stage by the size and inflexibility of the studio. Squeezed by physical limitations, the producer was forced to expand the viewer's imagination with ingenious devices.

When T. S. Eliot adapted *Murder in the Cathedral* for film, he noted a difference in realism between cinema and stage:

> Cinema, even where fantasy is introduced, is much more realistic than the stage. Especially in an historical picture, the setting, the costume, and the way of life represented have to be accurate. Even a minor anachronism is intolerable. On the stage much more can be overlooked or forgiven; and indeed, an excessive care for accuracy of historical detail can become burdensome and distracting. In watching a stage performance, the member of the audience is in direct contact with the actor playing a part. In looking at a film, we are much more passive; as audience, we contribute less. We are seized with the illusion that we are observing an actual event, or at least a series of photographs of the actual event; and nothing must be allowed to break this illusion. Hence the precise attention to detail.[2]

If two men are on a stage in a theater, the dramatist is obliged to motivate their presence; he has to account for their existing on the stage at all. Whereas if a camera is following a figure down a street or is turned to any object whatever, there is no need for a reason to be provided. Its grammar contains that power of statement of motivation, no matter what it looks at.

In the theater, the spectator sees the enacted scene as a whole in space, always seeing the whole of the space. The stage may present only one corner of a large hall, but that corner is always totally visible all through the scene. And the spectator always sees that scene from a fixed, unchanging distance and from an angle of vision that doesn't change. Perspective may change from scene to scene, but within one scene it remains constant. Distance never varies.

But in film and TV, distance and angle constantly shift. The same scene is shown in multiple perspective and focus. The viewer sees it from here, there, then over here; finally he is drawn inexorably into it, becomes part of it. He ceases to be a spectator. Balázs writes:

> Although we sit in our seats, we do not see Romeo and Juliet from there. We look up into Juliet's balcony with Romeo's eyes and look down on Romeo with Juliet's. Our eye and with it our consciousness is identified with the characters in the film, we look at the world out of their eyes and have no angle of vision of our own. We walk amid crowds, ride, fly or fall with the hero and if one character looks into the other's eyes, he looks into our eyes from the screen, for, our eyes are in the camera and become identical with the gaze of the characters. They see with our eyes. Herein lies the psychological act of identification. Nothing like this "identification" has ever occurred as the effect of any other system of art and it is here that the film manifests its absolute artistic novelty.
>
> ... Not only can we see, in the isolated "shots" of a scene, the very atoms of life and their innermost secrets revealed at close quarters, but we can do so without any of the intimate secrecy being lost, as always happens in the exposure of a stage performance or of a painting. The new theme which the new means of expression of film art revealed was not a hurricane at sea or the eruption of a volcano: it was perhaps a solitary tear slowly welling up in the corner of a human eye.
>
> ... Not to speak does not mean that one has nothing to say. Those who do not speak may be brimming over with emotions which can be expressed only in forms and pictures, in gesture and play of feature. The man of visual culture uses these not as substitutes for words, as a deaf-mute uses his fingers.[3]

The gestures of visual man are not intended to convey concepts that can be expressed in words, but inner experiences, nonrational emotions, which would still remain unexpressed when everything that can be told has been told. Such emotions lie in the deepest levels. They cannot be approached by words that are mere reflections of concepts, any more than musical experiences can be expressed in rational concepts. Facial expression is a human experience rendered immediately visible without the intermediary of word. It is Turgenev's "living truth of the human face."

Printing rendered illegible the faces of men. So much could be read from paper that the method of conveying meaning by facial expression fell into desuetude. The press grew to be the main bridge over which the more remote interhuman spiritual exchanges took place; the immediate, the personal, the inner, died. There was no longer need for the subtler means of expression provided by the body. The face became immobile; the inner life, still. Wells that dry up are wells from which no water is dipped.

Just as radio helped bring back inflection in speech, so film and TV are aiding us in the recovery of gesture and facial awareness—a rich, colorful language, conveying moods and emotions, happenings and characters, even thoughts, none of which could be properly packaged in words. If film had remained silent for another decade, how much faster this change might have been!

Feeding the product of one medium through another medium creates a new product. When Hollywood buys a novel, it buys a title and the publicity associated with it: nothing more. Nor should it.

Each of the four versions of the *Caine Mutiny*—book, play, movie, TV—had a different hero: Willie Keith, the lawyer Greenwald, the United States Navy, and Captain Queeg, respectively. Media and audience biases were clear. Thus the book told, in lengthy detail, of the growth and making of Ensign William Keith, American man, while the movie camera with its colorful shots of ships and sea, unconsciously favored the Navy as hero, a bias supported by the fact the Navy cooperated with the movie makers. Because of stage limitations, the play was confined, except for the last scene, to the courtroom, and favored the defense counsel as hero. The TV show, aimed at a mass audience, emphasized patriotism, authority, allegiance. More important, the cast was reduced to the principals and the plot to its principles; the real moral problem—the refusal of subordinates to assist an incompetent, unpopular superior—was clear, whereas in the book it was lost under detail, in the film under scenery. Finally, the New York play, with its audience slanted toward Expense Account patronage—Mr. Sampson, Western Sales Manager for the Cavity Drill Company—became a morality play with Willie Keith, innocent American youth, torn between two influences: Keefer, clever author but moral cripple, and Greenwald, equally brilliant but reliable, a businessman's intellectual. Greenwald saves Willie's soul.

The film *Moby Dick* was in many ways an improvement on the book, primarily because of its explicitness. For *Moby Dick* is one of those admittedly great classics, like *Robinson Crusoe* or Kafka's *Trial,* whose plot and situation, as distilled apart from the book by time and familiarity, are actually much more imposing than the written book itself. It's the drama of Ahab's defiance rather than

Melville's uncharted leviathan meanderings that is the greatness of *Moby Dick*. On film, instead of laborious tacks through leagues of discursive interruptions, the most vivid descriptions of whales and whaling become part of the action. On film, the viewer was constantly aboard ship: each scene an instantaneous shot of whaling life, an effect achieved in the book only by illusion, by constant, detailed reference. From start to finish, all the action of the film served to develop what was most central to the theme—a man's magnificent and blasphemous pride in attempting to destroy the brutal, unreasoning force that maims him and turns man-made order into chaos. Unlike the book, the film gave a spare, hard, compelling dramatization, free of self-conscious symbolism.

Current confusion over the respective roles of the new media comes largely from a misconception of their function. They are art forms, not substitutes for human contact. Insofar as they attempt to usurp speech and personal, living relations, they harm. This, of course, has long been one of the problems of book culture, at least during the time of its monopoly of Western middle-class thought. But this was never a legitimate function of books, nor of any other medium. Whenever a medium goes claim jumping, trying to work areas where it is ill-suited, conflicts occur with other media, or, more accurately, between the vested interests controlling each. But, when media simply exploit their own formats, they become complementary and cross-fertile.

Some people who have no one around talk to cats, and you can hear their voices in the next room, and they sound silly, because the cat won't answer, but that suffices to maintain the illusion that their world is made up of living people, while it is not. Mechanized mass media reverse this: now mechanical cats talk to humans. There's no genuine feedback.

This charge is often leveled by academicians at the new media, but it holds equally for print. The open-mouthed, glaze-eyed TV spectator is merely the successor of the passive, silent, lonely reader whose head moved back and forth like a shuttlecock.

When we read, another person thinks for us: we merely repeat his mental process. The greater part of the work of thought is done for us. This is why it relieves us to take up a book after being occupied by our own thoughts. In reading, the mind is only the playground for another's ideas. People who spend most of their lives in reading often lose the capacity for thinking, just as those who always ride forget how to walk. Some people read themselves stupid. Chaplin did a wonderful take-off of this in *City Lights,* when he stood up on a chair to eat the endless confetti that he mistook for spaghetti.

Eliot remarks: "It is often those writers whom we are lucky enough to know whose books we can ignore; and the better we know them personally, the less need we may feel to read what they write."

Frank O'Connor highlights a basic distinction between oral and written traditions: "'By the hokies, there was a man in this place one time by name of Ned Sullivan, and he had a queer thing happen to him late one night and he coming up the Valley Road from Durlas.' This is how a folk story begins, or should begin.... Yet that is how no printed short story should begin, because such a story seems tame when you remove it from its warm nest by the cottage fire, from the sense of an audience with its interjections, and the feeling of terror at what may lurk in the darkness outside."

Face-to-face discourse is not as selective, abstract, nor explicit as any mechanical medium; it probably comes closer to communicating an unabridged situation than any of them, and, insofar as it exploits the give-take of dynamic relationship, it's clearly the most indispensably human one.

Of course, there can be personal involvement in the other media. When Richardson's *Pamela* was serialized in 1741, it aroused such interest that in one English town, upon receipt of the last installment, the church bell announced that virtue had been rewarded. Radio stations have reported receiving quantities of baby clothes and bassinets when, in a soap opera, a heroine had a baby. One of the commonest phrases used by devoted listeners to daytime serials is that they "visited with" Aunt Jenny or Big Sister. BBC and *News Chronicle* report cases of women viewers who kneel before TV sets to kiss male announcers good night.

Each medium, if its bias is properly exploited, reveals and communicates a unique aspect of reality, of truth. Each offers a different perspective, a way of seeing an otherwise hidden dimension of reality. It's not a question of one reality being true, the others distortions. One allows us to see from here, another from there, a third from still another perspective; taken together they give us a more complete whole, a greater truth. New essentials are brought to the fore, including those made invisible by the "blinders" of old languages.

This is why the preservation of book culture is as important as the development of TV. This is why new languages, instead of destroying old ones, serve as a stimulant to them. Only monopoly is destroyed. When actor-collector Edward G. Robinson was battling actor-collector Vincent Price on art on TV's *$64,000 Challenge,* he was asked how the quiz had affected his life; he answered petulantly, "Instead of looking at the pictures in my art books, I now have to read them." Print, along with all old languages, including speech, has profited enormously from the development of the new media. "The more the arts develop," writes E. M. Forster, "the more they depend on each other for definition. We will borrow from painting first and call it pattern. Later we will borrow from music and call it rhythm."

The appearance of a new medium often frees older media for creative effort. They no longer have to serve the interests of power and profit. Elia Kazan, discussing the American theater, says:

> Take 1900–1920. The theatre flourished all over the country. It had no competition. The box office boomed. The top original fare it had to offer was *The Girl of the Golden West.* Its bow to culture was fusty productions of Shakespeare. . . . Came the moving pictures. The theatre had to be better or go under. It got better. It got so spectacularly better so fast that in 1920–1930 you wouldn't have recognized it. Perhaps it was an accident that Eugene O'Neill appeared at that moment—but it was no accident that in that moment of strange competition, the theatre had room for him. Because it was disrupted and hard pressed, it made room for his experiments, his unheard-of subjects, his passion, his power. There was room for him to grow to his full stature. And there was freedom for the talents that came after his.[4]

Yet a new language is rarely welcomed by the old. The oral tradition distrusted writing, manuscript culture was contemptuous of printing, book culture hated the press, that "slag-heap of hellish passions," as one nineteenth-century scholar called it. A father, protesting to a Boston newspaper about crime and scandal, said he would rather see his children "in their graves while pure in innocence, than dwelling with pleasure upon these reports, which have grown so bold."

What really disturbed book-oriented people wasn't the sensationalism of the newspaper, but its nonlineal format, its nonlineal codifications of experience. The motto of conservative academicians became: *Hold that line!*

A new language lets us see with the fresh, sharp eyes of the child; it offers the pure joy of discovery. I was recently told a story about a Polish couple who, though long resident in Toronto, retained many of the customs of their homeland. Their son despaired of ever getting his father to buy a suit cut in style or getting his mother to take an interest in Canadian life. Then he bought them a TV set, and in a matter of months a major change took place. One evening the mother remarked that "Edith Piaf is the latest thing on Broadway," and the father appeared in "the kind of suit executives wear on TV." For years the father had passed this same suit in store windows and seen it both in advertisements and on living men, but not until he saw it on TV did it become meaningful. This same statement goes for all media: each offers a unique presentation of reality, which when new has a freshness and clarity that is extraordinarily powerful.

This is especially true of TV. We say, "We have a radio" but "We have television"—as if something had happened to us. It's no longer "The skin you love to touch" but "The Nylon that loves to touch you." We don't watch TV; it watches us: it guides us. Magazines and newspapers no longer convey "information" but offer ways of seeing things. They have abandoned realism as too easy: they substitute themselves for realism. *Life* is totally advertisements: its articles package and sell emotions and ideas just as its paid ads sell commodities.

Several years ago, a group of us at the University of Toronto undertook the following experiment: 136 students were divided, on the basis of their overall academic standing of the previous year, into four equal groups who either (1) heard and saw a lecture delivered in a TV studio, (2) heard and saw this same lecture on a TV screen, (3) heard it over the radio, or (4) read it in manuscript. Thus there were, in the CBC studios, four controlled groups who simultaneously received a single lecture and then immediately wrote an identical examination to test both understanding and retention of content. Later the experiment was repeated, using three similar groups; this time the same lecture was (1) delivered in a classroom, (2) presented as a film (using the kinescope) in a small theater, and (3) again read in print. The actual mechanics of the experiment were relatively simple, but the problem of writing the script for the lecture led to a consideration of the resources and limitations of the dramatic forms involved.

It immediately became apparent that no matter how the script was written and the show produced, it would be slanted in various ways for and against each of the media involved; no show could be produced that did not contain these biases, and the only real common denominator was the simultaneity of presentation. For each communication channel codifies reality differently and thus influences, to a surprising degree, the content of the message communicated. A medium is not simply an envelope that carries any letter; it is itself a major part of that message. We therefore decided not to exploit the full resources of any one medium, but to try to chart a middle-of-the-road course between all of them.

The lecture that was finally produced dealt with linguistic codifications of reality and metaphysical concepts underlying grammatical systems. It was chosen because it concerned a field in which few students could be expected to have prior knowledge; moreover, it offered opportunities for the use of gesture. The cameras moved throughout the lecture, and took close-ups where relevant. No other visual aids were used, nor were shots taken of the audience while the lecture was in progress. Instead, the cameras simply focused on the speaker for twenty-seven minutes.

The first difference we found between a classroom and a TV lecture was the brevity of the latter. The classroom lecture, if not ideally, at least in practice, sets a slower pace. It's verbose, repetitive. It allows for greater elaboration and permits the lecturer to take up several *related* points. TV, however, is stripped right down; there's less time for qualifications or alternative interpretations and only time enough for *one* point. (Into twenty-seven minutes we put the meat of a two-hour classroom lecture.) The ideal TV speaker states his point and then brings out different facets of it by a variety of illustrations. But the classroom lecturer is less subtle and, to the agony of the better students, repeats and repeats his identical points in the hope, perhaps, that ultimately no student will miss them, or perhaps simply because he is dull. Teachers have had captive audiences for so long that few are equipped to compete for attention via the new media.

The next major difference noted was the abstracting role of each medium, beginning with print. Edmund M. Morgan, Harvard law professor, writes:

> One who forms his opinion from the reading of any record alone is prone to err, because the printed page fails to produce the impression or convey the idea which the spoken word produced or conveyed. The writer has read charges to the jury which he had previously heard delivered, and has been amazed to see an oral deliverance which indicated a strong bias appear on the printed page as an ideally impartial exposition. He has seen an appellate court solemnly declare the testimony of a witness to be especially clear and convincing which the trial judge had orally characterized as the most abject perjury.[5]

Selectivity of print and radio are perhaps obvious enough, but we are less conscious of it in TV, partly because we have already been conditioned to it by the shorthand of film. Balázs writes:

> A man hurries to a railway station to take leave of his beloved. We see him on the platform. We cannot see the train, but the questing eyes of the man show us that his beloved is already seated in the train. We see only a close-up of the man's face, we see it twitch as if startled and then strips of light and shadow, light and shadow flit across it in quickening rhythm. Then tears gather in the eyes and that ends the scene. We are expected to know what happened and today we do know, but when I first saw this film in Berlin, I did not at once understand the end of this scene. Soon, however, everyone knew what had happened: the train had started and it was the lamps in its compartment which had thrown their light on the man's face as they glided past ever faster and faster.[6]

As in a movie theater, only the screen is illuminated, and, on it, only points of immediate relevance are portrayed; everything else is eliminated. This explicitness makes TV not only personal but forceful. That's why stage hands in a TV studio watch the show over floor monitors, rather than watch the actual performance before their eyes.

The script of the lecture, timed for radio, proved too long for TV. Visual aids and gestures on TV not only allow the elimination of certain words, but require a unique script. The ideal radio delivery stresses pitch and intonation to make up for the absence of the visual. That flat, broken speech in "sidewalk interviews" is the speech of a person untrained in radio delivery.

The results of the examination showed that TV had won, followed by lecture, film, radio, and finally print. Eight months later the test was readministered to the bulk of the students who had taken it the first time. Again it was found that there were significant differences between the groups exposed to different media, and these differences were the same as those on the first test, save for the studio group, an uncertain group because of the chaos of the lecture conditions, which had moved from last to second place. Finally, two years later, the experiment was repeated, with major modifications, using students at Ryerson Institute. Marshall McLuhan reports:

> In this repeat performance, pains were taken to allow each medium full play of its possibilities with reference to the subject, just as in the earlier experiment each medium was neutralized as much as possible. Only the mimeograph form remained the same in each experiment. Here we added a printed form in which an imaginative typographical layout was followed. The lecturer used the blackboard and permitted discussion. Radio and TV employed dramatization, sound effects and graphics. In the examination, radio easily topped TV. Yet, as in the first experiment, both radio and TV manifested a decisive advantage over the lecture and written forms. As a conveyor both of ideas and information, TV was, in this second experiment, apparently enfeebled by the deployment of its dramatic resources, whereas radio benefited from such lavishness. "Technology is explicitness," writes Lyman Bryson. Are both radio and TV more explicit than writing or lecture? Would a greater explicitness, if inherent in these media, account for the ease with which they top other modes of performance?[7]

Announcement of the results of the first experiment evoked considerable interest. Advertising agencies circulated the results with the comment that here, at last, was scientific proof of the superiority of TV. This was unfortunate and missed the main point, for the results didn't indicate the superiority of one medium over others. They merely directed attention toward differences between them, differences so great as to be of kind rather than degree. Some CBC officials were furious, not because TV won, but because print lost.

The problem has been falsely seen as democracy vs. the mass media. But the mass media *are* democracy. The book itself was the first mechanical mass medium. What is really being asked, of course, is: can books' monopoly of knowledge survive the challenge of the new languages? The answer is: no. What should be asked is: what can print do better than any other medium and is that worth doing?

## NOTES

[1] *Television Plays,* New York, Simon and Schuster, 1955, pp. 176–78.
[2] George Hoellering and T. S. Eliot, *Film of Murder in the Cathedral,* New York, Harcourt, Brace & Co., 1952, p. vi; London, Faber & Faber, 1952.
[3] Béla Balázs, *Theory of Film,* New York, Roy Publishers, 1953, pp. 48, 31, 40; London, Denis Dobson, 1952.
[4] "Writers and Motion Pictures," *The Atlantic Monthly,* 199, 1957, p. 69.
[5] G. Louis Joughin and Edmund M. Morgan, *The Legacy of Sacco and Vanzetti,* New York, Harcourt, Brace & Co., 1948, p. 34.
[6] Béla Balázs, op. cit., pp. 35–36.
[7] From a personal communication to the author.

# The Monastery and the Clock

**Lewis Mumford**

Where did the machine first take form in modern civilization? There was plainly more than one point of origin. Our mechanical civilization represents the convergence of numerous habits, ideas, and modes of living, as well as technical instruments; and some of these were, in the beginning, directly opposed to the civilization they helped to create. But the first manifestation of the new order took place in the general picture of the world: during the first seven centuries of the machine's existence the categories of time and space underwent an extraordinary change, and no aspect of life was left untouched by this transformation. The application of quantitative methods of thought to the study of nature had its first manifestation in the regular measurement of time; and the new mechanical conception of time arose in part out of the routine of the monastery. Alfred Whitehead has emphasized the importance of the scholastic belief in a universe ordered by God as one of the foundations of modern physics: but behind that belief was the presence of order in the institutions of the Church itself.

The technics of the ancient world were still carried on from Constantinople and Baghdad to Sicily and Cordova: hence the early lead taken by Salerno in the scientific and medical advances of the Middle Age. It was, however, in the monasteries of the West that the desire for order and power, other than that expressed in the military domination of weaker men, first manifested itself after the long uncertainty and bloody confusion that attended the breakdown of the Roman Empire. Within the walls of the monastery was sanctuary: under the rule of the order surprise and doubt and caprice and irregularity were put at bay. Opposed to the erratic fluctuations and pulsations of the worldly life was the iron discipline of the rule. Benedict added a seventh period to the devotions of the day, and in the seventh century, by a bull of Pope Sabinianus, it was decreed that the bells of the monastery be rung seven times in the twenty-four hours. These punctuation marks in the day were known as the canonical hours, and some means of keeping count of them and ensuring their regular repetition became necessary.

According to a now discredited legend, the first modern mechanical clock, worked by falling weights, was invented by the monk named Gerbert who afterwards became Pope Sylvester II near the close of the tenth century. This clock was probably only a water clock, one of those bequests of the ancient world either left over directly from the days of the Romans, like the water-wheel itself, or coming back again into the West through the Arabs. But the legend, as so often happens, is accurate in its implications if not in its facts. The monastery was the seat of a regular life, and an instrument for striking the hours at intervals or for reminding the bell-ringer that it was time to strike the bells, was an almost inevitable product of this life. If the mechanical clock did not appear until the cities of the thirteenth century demanded an orderly routine, the habit of order itself and the earnest regulation of time-sequences had become almost second nature in the monastery. Coulton agrees with Sombart in looking upon the Benedictines, the great working order, as perhaps the original founders of modern capitalism: their rule certainly took the curse off work and their vigorous engineering enterprises may even have robbed warfare of some of its glamor. So one is not straining the facts when one suggests that the monasteries—at one time there were 40,000 under the Benedictine rule—helped to give human enterprise the regular collective beat and rhythm of the machine; for the clock is not merely a means of keeping track of the hours, but of synchronizing the actions of men.

Was it by reason of the collective Christian desire to provide for the welfare of souls in eternity by regular prayers and devotions that time-keeping and the habits of temporal order took hold of men's minds: habits that capitalist civilization presently turned to good account? One must perhaps accept the irony of this paradox. At all events, by the thirteenth century there are definite records of mechanical clocks, and by 1370 a well-designed "modern" clock had been built by Heinrich von Wyck at Paris. Meanwhile, bell towers had come into existence, and the new clocks, if they did not have, till the fourteenth century, a dial and a hand that translated the movement of time into a movement through space, at all events struck the hours. The clouds that could paralyze the sundial, the freezing that could stop the water clock on a winter night, were no longer obstacles to time-keeping: summer or winter, day or night, one was aware of the measured clank of the clock. The instrument presently spread outside the monastery; and the regular striking of the bells brought a new regularity into the life of the workman and the merchant. The bells of the clock tower almost defined urban existence. Time-keeping passed into time-serving and time-accounting and time-rationing. As this took place, Eternity ceased gradually to serve as the measure and focus of human actions.

The clock, not the steam-engine, is the key-machine of the modern industrial age. For every phase of its development the clock is both the outstanding fact and the typical symbol of the machine: even today no other machine is so ubiquitous. Here, at the very beginning of modern technics, appeared prophetically the accurate automatic machine which, only after centuries of further effort, was also to prove the final consummation of this technics in every department of industrial activity. There had been power-machines, such as the water-mill, before the clock; and there had also been various kinds of automata, to awaken the wonder of the populace in the temple, or to please the idle fancy of some Moslem caliph: machines one finds illustrated in Hero and Al-Jazari. But here was a new kind of power-machine, in which the source of power and the transmission were of such a nature as to ensure the even flow of energy throughout the works and to make possible regular production and a standardized product. In its relationship to determinable quantities of energy, to standardization, to automatic action, and finally to its own special product, accurate timing, the clock has been the foremost machine in modern technics: and at each period it has remained in the lead: it marks a perfection toward which other machines aspire. The clock, moreover, served as a model for many other kinds of mechanical works, and the analysis of motion that accompanied the perfection of the clock, with the various types of gearing and transmission that were elaborated, contributed to the success of quite different kinds of machine. Smiths could have hammered thousands of suits of armor or thousands of iron cannon, wheelwrights could have shaped thousands of great water-wheels or crude gears, without inventing any of the special types of movement developed in clockwork, and without any of the accuracy of measurement and fineness of articulation that finally produced the accurate eighteenth century chronometer.

The clock, moreover, is a piece of power-machinery whose "product" is seconds and minutes: by its essential nature it dissociated time from human events and helped create the belief in an independent world of mathematically measurable sequences: the special world of science. There is relatively little foundation for this belief in common human experience: throughout the year the days are of uneven duration, and not merely does the relation between day and night steadily change, but a slight journey from East to West alters astronomical time by a certain number of minutes. In terms of the human organism itself, mechanical time is even more foreign: while human life has regularities of its own, the beat of the pulse, the breathing of the lungs, these change from hour to hour with mood and action, and in the longer span of days, time is measured not by the calendar but by the events that occupy it. The shepherd measures from the time the ewes lambed; the farmer measures back to the day of sowing or forward to the harvest: if growth has its own duration and regularities, behind it are not simply matter and motion but the facts of development: in short, history. And while mechanical time is strung out in a succession of mathematically isolated instants, organic time—what Bergson calls duration—is cumulative in its effects. Though mechanical time can, in a sense, be speeded up or run backward, like the hands of a clock or the images of a moving picture, organic time moves in only one direction—through the cycle of birth, growth, development, decay, and death—and the past that is already dead remains present in the future that has still to be born.

Around 1345, according to Thorndike, the division of hours into sixty minutes and of minutes into sixty seconds became common: it was this abstract framework of divided time that became more and more the point of reference for both action and thought, and in the effort to arrive at accuracy in this department, the astronomical exploration of the sky focused attention further upon the regular, implacable movements of the heavenly bodies through space. Early in the sixteenth century a young Nuremberg mechanic, Peter Henlein, is supposed to have created "many-wheeled watches out of small bits of iron" and by the end of the century the small domestic clock had been introduced in England and Holland. As with the motor car and the airplane, the richer classes first took over the new mechanism and popularized it: partly because they alone could afford it, partly because the new bourgeoisie were the first to discover that, as Franklin later put it, "time is money." To become "as regular as clockwork" was the bourgeois ideal, and to own a watch was for long a definite symbol of success. The increasing tempo of civilization led to a demand for greater power: and in turn power quickened the tempo.

Now, the orderly punctual life that first took shape in the monasteries is not native to mankind, although by now Western peoples are so thoroughly regimented by the clock that it is "second nature" and they look upon its observance as a fact of nature. Many Eastern civilizations have flourished on a loose basis in time: the Hindus have in fact been so indifferent to time that they lack even an authentic chronology of the years. Only yesterday, in the midst of the industrializations of Soviet Russia, did a society come into existence to further the carrying of watches there and to propagandize the benefits of punctuality. The popularization of time-keeping, which followed the production of the cheap standardized watch, first in Geneva, then in America around the middle of the last century, was essential to a well-articulated system of transportation and production.

To keep time was once a peculiar attribute of music: it gave industrial value to the workshop song or the tattoo or the chantey of the sailors tugging at a rope. But the effect of the mechanical clock is more pervasive and strict: it presides over the day from the hour of rising to the hour of rest. When one thinks of the day as an abstract span of time, one does not go to bed with the chickens on a winter's night: one invents wicks, chimneys, lamps, gaslights, electric lamps, so as to use all the hours belonging to the day. When one thinks of time, not as a sequence of experiences, but as a collection of hours, minutes, and seconds, the habits of adding time and saving time come into existence. Time took on the character of an enclosed space: it could be divided, it could be filled up, it could even be expanded by the invention of labor-saving instruments.

Abstract time became the new medium of existence. Organic functions themselves were regulated by it: one ate, not upon feeling hungry, but when prompted by the clock: one slept, not when one was tired, but when the clock sanctioned it. A generalized time-consciousness accompanied the wider use of clocks: dissociating time from organic sequences, it became easier for the men of the Renascence to indulge the fantasy of reviving the classic past or of reliving the splendors of antique Roman civilization: the cult of history, appearing first in daily ritual, finally abstracted itself as a special discipline. In the seventeenth century journalism and periodic literature made their appearance: even in dress, following the lead of Venice as fashion-center, people altered styles every year rather than every generation.

The gain in mechanical efficiency through coordination and through the closer articulation of the day's events cannot be overestimated: while this increase cannot be measured in mere horsepower, one has only to imagine its absence today to foresee the speedy disruption and eventual collapse of our entire society. The modern industrial régime could do without coal and iron and steam easier than it could do without the clock.

## The Rise of the Equal Hour

**Daniel Boorstin**

While man allowed his time to be parsed by the changing cycles of daylight he remained a slave of the sun. To become the master of his time, to assimilate night into the day, to slice his life into neat, usable portions, he had to find a way to mark off precise small portions—not only equal hours, but even minutes and seconds and parts of seconds. He would have to make a machine. It is surprising that machines to measure time were so long in coming. Not until the fourteenth century did Europeans devise mechanical timepieces. Until then, as we have seen, the measuring of time was left to the shadow clock, the water clock, the sandglass, and the miscellaneous candle clocks and scent clocks. While there was remarkable progress five thousand years ago in measuring the year, and useful week clusters of days were long in use, the subdivided day was another matter. Only in modern times did we begin to live by the hour, much less by the minute.

The first steps toward the mechanical measurement of time, the beginnings of the modern clock in Europe, came not from farmers or shepherds, nor from merchants or craftsmen, but from religious persons anxious to perform promptly and regularly their duties to God. Monks needed to know the times for their appointed prayers. In Europe the first mechanical clocks were designed not to *show* the time but to *sound* it. The first true clocks were alarms. The first Western clockworks, which set us on the way to clockmaking, were weight-driven machines which struck a bell after a measured interval. Two kinds of clocks were made for this purpose. Probably the earlier were small monastic alarms, or chamber clocks—called *horologia excitatoria,* or awakening clocks—for the cell of the *custos horologii,* or guardian of the clock. These rang a small bell to alert a monk to summon the others to prayer. He would then go up to strike the large bell, usually set high in a tower, so that all could hear. About the same time much larger turret clocks began to be made and placed in the towers, where they would ring the large bell automatically.

These monastic clocks announced the canonical hours, the times of day prescribed by the Church's canons, or rules, for devotion. The number of these hours varied, of course, with the changing canons of the Church, with the varied customs from place to place, and with the rules of particular orders. In the sixth century, after Saint Benedict, the canonical hours were standardized at seven. Distinct prayers were specified to be said at the first light or dawn (Matins or Lauds), with the sunrise (*Hora Prima*), at midmorning (*Hora Tertia*), at noon (*Hora Sexta* or *Meridies*), at midafternoon (*Hora Nona*), at sunset (Vespers, or *Hora Vesperalis*), and at nightfall (Compline, or *Completorium*). The number of strokes of the bell varied from four at sunrise to one at noon and back to four again at nightfall. The precise hour, by our modern calculation, for each of these prayers depended in any given place on the latitude and the season. Despite the complexity of the problem, monastic clocks were adjusted to vary the time between bells according to the season.

Efforts to adapt earlier timekeeping devices to the making of sounds had never been quite successful. A clever Parisian fitted a lens into his sundial to act as a burning glass which precisely at noon focused on the touchhole of a small cannon, and so automatically saluted the sun at its apex. This elegant cannon clock, installed by the Duke of Orléans in the garden of the Palais Royal in 1786, is said to have fired the shot that started the French Revolution. Centuries before, complicated water clocks had been designed to mark passing time by tossing pebbles or blowing whistles. Some such devices were probably tried in monasteries.

But a new kind of timepiece, a mechanical timepiece that was a true clock, would be much better adapted to the new mechanical needs. The very word "clock" bears the mark of its monastic origins. The Middle English *clok* came from the Middle Dutch word for bell and is a cognate of the German *Glocke,* which means bell. Strictly speaking, in the beginning a timepiece was not considered to be a clock unless it rang a bell. It was only later that it came to mean any device that measured passing time.

These first mechanical clocks came into an age when sunlight circumscribed the times of life and movement, when artificial lights had not yet begun to confuse night with day. Medieval striking clocks remained silent during the dark hours. After the four strokes which announced Compline, the prayers at nightfall, the next bell was not sounded until the time for Matins, the prayers at sunrise the next morning. But in the long run the unintended consequence of the making of mechanical clocks, and a hidden imperative of the machine itself, was to incorporate both hours of darkness and hours of sunlight into a single equal-houred twenty-four-hour day. The monastic clock, specially designed for *sounding* the time, pointed the way to a new way of thinking about time.

The sundial, water clock, and hourglass were all designed primarily to *show* the passing time, by the visible gradual flow of a shadow across a dial, of water from a bowl, of sand through a glass. But the mechanical clock, in its monastic origins, was made for a decisive mechanical act, a stroke of a hammer on a bell. The needs of mechanical timekeeping, the logic of the machine itself, imposed a new feeling. Instead of being synonymous with repeated cycles of the sun, which varied as the cycles of the seasons commanded, or with the shorter cycles of other flowing media, time now was to be measured by the staccato of a machine. Making a machine to *sound* the canonical hours required, and achieved, mechanical novelties which would be the foundation of clockmaking for centuries to come.

The force that moved the arm that struck the bell was provided by falling weights. What made the machine truly novel was the device that prevented the free fall of the weights and interrupted their drop into regular intervals. The sundial had shown the uninterrupted movement of the sun's shadow, and the sandglass operated by the free-falling of water or sand. What gave this new machine a longer duration and measured off the units was a simple enough device, which has remained almost uncelebrated in history. It was called an escapement, since it was a way of regulating the "escape" of the motive power into the clock, and it held revolutionary import for human experience.

With the simplicity of the greatest inventions, the "escapement" was nothing more than an arrangement that would regularly interrupt the force of a falling weight. The interruptor was so designed that it would alternately check and then release the force of the weight on the moving machinery of the clock. This was the basic invention that made all modern clocks possible. Now a weight falling only a short distance could keep a clock going for hours as the regular downward pull of the falling weights was translated into the interrupted, staccato movement of the clock's machinery.

The earliest simple form was the "verge" escapement. An unknown mechanical genius first imagined a way of connecting the falling weight by intersecting cogged wheels to a vertical axle which carried a horizontal bar, or verge, with weights attached. These weights regulated the movement. When they were moved outward, the clock beat slower; when moved inward, the clock went faster. The back-and-forth movement of the bar (moved by the large falling weights) would alternately engage and disengage the cogs on the clock's machinery. These interrupted movements eventually measured off the minutes and, later, the seconds. When, in due course, clocks became common, people would think of time no longer as a flowing stream but as the accumulation of discrete measured moments. The sovereign time that governed daily lives would no longer be the sunlight's smooth-flowing elastic cycles. Mechanized time would no longer flow. The tick-tock of the clock's escapement would become the voice of time.

Such a machine plainly had nothing to do with the sun or the movements of the planets. Its own laws provided an endless series of uniform units. The "accuracy" of a clock—which meant the uniformity of its measured units—would depend on the precision and regularity of the escapement.

The canonical hours, which had measured out the daylight into the appropriate elastic units between divine services, were registered on clocks until about the fourteenth century. It was around 1330 that the hour became our modern hour, one of twenty-four equal parts of a day. This new "day" included the night. It was measured by the time between one noon and the next, or, more precisely, what modern astronomers call "mean solar time." For the first time in history, an "hour" took on a precise, year-round, everywhere meaning.

There are few greater revolutions in human experience than this movement from the seasonal or "temporary" hour to the equal hour. Here was man's declaration of independence from the sun, new proof of his mastery over himself and his surroundings. Only later would it be revealed that he had accomplished this mastery by putting himself under the dominion of a machine with imperious demands all its own.

The first clocks did not have dials or hands at all. They did not need them, since their use was simply to *sound* the hour. An illiterate populace that might have trouble reading a dial would not mistake the sound of bells. With the coming of the "equal" hour, replacing the "temporary" or "canonical" hour, the sounding of hours was ideally adapted for measurement by a simple machine. Sun time was translated into clock time.

By the fourteenth century in Europe large turret clocks in the belfries of churches and town halls were sounding the equal hours, heralding a new time-consciousness. Church towers, built to salute God and to mark man's heavenward aspirations, now became clock towers. The *torre* became the campanile. As early as 1335, the campanile of the Chapel of the Blessed Virgin in Milan was admired by the chronicler Galvano della Fiamma for its wonderful clock with many bells. "A very large hammer... strikes one bell twenty-four times according to the number of the twenty-four hours of the day and night; so that at the first hour of the night it gives one sound, at the second, two strokes, at the third, three, and at the fourth, four; and thus it distinguishes hour from hour, which is in the highest degree necessary for all conditions of men." Such equal-hour clocks became common in the towns of Europe. Now serving the whole community, they were a new kind of public utility, offering a service each citizen could not afford to provide himself.

People unwittingly recognized the new era when, noting the time of day or night, they said it was nine "o'clock"—a time "of the clock." When Shakespeare's characters mentioned the time, "of the clock," they recalled the hour they had heard last struck. Imogen, Cymbeline's daughter, explains that a faithful lover is accustomed "to weepe 'twixt clock and clock" for her beloved. While the populace now began to know the "hour," several centuries passed before they could talk of "minutes." During the whole fourteenth century, dials were seldom found on clocks, for the clocks' function was still to *sound* the hours. They are not found on the Italian campaniles, though there may have been one on St. Paul's Cathedral (1344) in London. Early dials were not like ours. Some showed hours only from I to VI with hands that moved around the dial four times in twenty-four hours. Others, like the famous work of Giovanni de' Dondi (1318–1389), enumerated the full twenty-four hours.

It was not too difficult to improve clocks that already struck the hour so that they could strike the quarter-hour. A dial, marked 1 to 4, was sometimes added to indicate the quarters. Later these were replaced by the figures 15, 30, 45, and 60 to indicate minutes. There was still no minute hand.

By 1500 the clock at Wells Cathedral in England was striking the quarter-hours, but had no way to mark the minutes. To measure minutes you still had to use a sandglass. A separate concentric minute hand, in addition to the hour hand, did not come into use until the pendulum was successfully applied to clocks. The pendulum also made it possible to indicate seconds. By 1670 it was not unusual for clocks to have a second hand whose movements were controlled by a 39-inch pendulum with a period of just one second.

More than any earlier invention the mechanical clock began to incorporate the dark hours of night into the day. In order to show the right time at daybreak this time machine had to be kept going continuously all night.

When does a "day" begin? Answers to this question have been almost as numerous as those to the question of how many days should be in a week. "The evening and the morning were the first day,'' we read in the first chapter of Genesis. The very first ''day'' then was really a night. Perhaps this was another way of describing the mystery of Creation, leaving God to perform his miraculous handiwork in the dark. The Babylonians and the early Hindus calculated their day from sunrise. The Athenians, like the Jews, began their ''day'' at sunset, and carried on the practice through the nineteenth century. Orthodox Muslims, literally following Holy Script, continue to begin their day at sunset, when they still set their clocks at twelve.

As we have seen, for most of history, mankind did not think of a day as a unit of twenty-four hours. Only with the invention and diffusion of the mechanical clock did this notion become common. The early Saxons divided their day into ''tides''—''morningtide,'' ''noontide,'' and ''eveningtide''—and some of the earliest English sundials are so marked.

Other widespread ways of dividing the day were much simpler than the system of ''temporary'' hours, which subdivided daylight and darkness. The seven canonical hours marked the passing time for Columbus and his crew.

Even after the arrival of the mechanical clock the sun left its mark on the measuring of the hours. The ''double-twelve'' system, by which Americans count the hours, is such a relic. When the *daylight* hours were measured off and subdivided, in contrast to the hours of night, the hours of each of the two parts were numbered separately. And so it remained, even after a machine required that time be measured continuously. The first twenty-four-hour clocks—while substituting equal mechanical hours for the elastic canonical or ''temporary'' hours—still remained curiously tied to the sun. They normally used sunset as the end of the twenty-fourth hour.

To ask how we came to our day, hour, minute, and second takes us deep into the archaeology of everyday life. Our English word ''day'' (no relation to the Latin *dies*) comes from an old Saxon word ''to burn,'' which also meant the hot season, or the warm time. Our ''hour'' comes from Latin and Greek words meaning season, or time of day. It meant one-twelfth part of the sunlight or the darkness—the ''temporary,'' or seasonal, hour—varying with season and latitude, long before it acquired its modern meaning of one twenty-fourth part of the equinoctial day.

Why the twenty-four? Historians do not help us much. The Egyptians did divide their day into twenty-four ''hours''—''temporary,'' of course. Apparently they chose this number because they used the sexagesimal system of numbers, based on multiples of six, which had been developed by the Babylonians. This pushes the mystery back into earlier centuries, for we have no clear explanation of why the Babylonians built their arithmetic as they did. But their use of the number sixty seems to have had nothing to do with astronomy or the movement of heavenly bodies. We have seen how the Egyptians fixed 360 days as the regular days of their year—12 months of 30 days each, supplemented by 5 additional days at the end of each year. They also marked off 360 degrees in a circle, perhaps by analogy to the yearly circuit of the sun. Sixty, being one-sixth of the 360 and so a natural subdivision in their sexagesimal system, became a convenient subdivision of the circle, and also of each ''degree'' or each hour. Perhaps the

Chaldean Babylonians, noticing five planets—Mercury, Venus, Mars, Jupiter, and Saturn—multiplied 12 (the number of the months, and a multiple of 6) by the planetary 5, and so arrived at the significant 60.

An everyday relic of the primitive identification of the circuit of the sun with the full circle is our sign for a "degree." The tiny circle we now use to designate a degree is probably a hieroglyph for the sun. If the degree sign ° was a picture of the sun, then 360°—a full circle—would also properly mean a cycle of 360 days, or a full year. The degree as a way of dividing the circle was first applied by ancient Babylonian and Egyptian astronomers to the circle of the zodiac to designate the stage or distance traveled by the sun each day, just as a *sign* described for them the astronomical space passed through in a month.

Our "minute," from the medieval Latin *pars minuta prima* (first minute or small part), originally described the one-sixtieth of a unit in the Babylonian system of sexagesimal fractions. And "second," from *partes minutae secundae,* was a further subdivision on the base of sixty. Since the Babylonian arithmetic was based on that unit, it was their version of a decimal and was easier to handle in their scientific calculations than other "vulgar fractions" (*minutae*) would have been. Ptolemy used this sixty-unit system for subdividing the circle, and he also used it to divide the day. Not until much later, perhaps in the thirteenth century with the arrival of the mechanical clock, did the minute become a division of the hour. The language, again, is a clue to the needs and capacities of timekeeping machinery. The "second" was at first an abbreviation for "second minute," and originally described the unit resulting from the second operation of sexagesimal subdivision. Long used for subdivisions of a circle, seconds were not applied to timekeeping until clockmaking was refined in the late sixteenth century.

The clock did not entirely liberate itself from the sun, from the dictates of light and darkness. In Western Europe the hours of the clock continued to be numbered from noon, when the sun was at the meridian, or from midnight midway between two noons. In most of Europe and in America a new day still begins at midnight by the clock.

The archaeology of our everyday life leads us all over the world. The 365 days of our year acknowledge our debt to ancient Egyptian priests, while the names of months—January, February, March—and of the days of our seven-day week—Saturday, Sunday, Monday—remain our tie to the early Hebrews and to Greek and Roman astrologers. When we mark each hour of our twenty-four-hour day, and designate the minutes after the hour, we are living, as a historian of ancient science reminds us, by "the results of a Hellenistic modification of an Egyptian practice combined with Babylonian numerical procedures."

The broadcasting medium of the medieval town was bells. Since the human voice could not reach all who needed to hear a civic announcement, bells told the hours, summoned help to extinguish a fire, warned of an approaching enemy, called men to arms, brought them to work, sent them to bed, knelled public mourning at the death of a king, sounded public rejoicing at the birth of a prince or a coronation, celebrated the election of a pope or a victory in war. "They may ring their bells now," Sir Robert Walpole observed in 1739 on hearing bells rung in London to announce the declaration of war against Spain, "before long they will be wringing their hands." Americans treasure a relic of that age of bells in the Liberty Bell, which announced Independence in Philadelphia.

There was supposed to be power and therapy in the sound of the bells that were rung to ward off an epidemic or to prevent a storm. Citizens of Lyons, France, in 1481 petitioned their town council that they "sorely felt the need for a great clock whose strokes could be heard by all citizens in all parts of the town. If such a clock were to be made, more merchants would come to the fairs, the citizens would be very consoled, cheerful and happy and would live a more orderly life, and the town would gain in decoration."

Community pride was a pride of bells. Churches, monasteries, and whole towns were judged by the reach and resonance of the peals from their towers. An inscription on an old bell boasted, "I mourn death, I disperse the lightning, I announce the Sabbath, I rouse the lazy, I scatter the winds, I appease the bloodthirsty" (*Funera plango, fulmina frango, Sabbath pango, Excito lentos, dissipo ventos, paco cruentos*). Paul Revere, the messenger of the American Revolution, made a reputation, and a fortune, as a caster of bells for proud New England towns. The art of bell-casting and experiments with bell-ringing devices advanced the art of the clockmaker and encouraged the elaboration of clocks.

Widespread illiteracy helps explain why dials were slow to appear on the exterior of public clocks. Not everybody could read even the simple numbers on a clockface. The very same factors that delayed the production of calibrated dials also encouraged experiments, ingenuity, and playfulness with clockwork performances. The great public clocks of the Middle Ages did not much advance the precision of clockworks, which, before the pendulum, lost or gained as much as an hour a day. It was technically difficult to improve the escapement hidden inside the machinery, and that regulated the accuracy of the movement. But it was easy to add wheels to wheels to improve the automated public display.

Nowadays the calendrical or astronomical indicators on antique clocks seem superfluous ornaments on a machine that should only show us hours, minutes, and seconds. For at least two centuries after the great mechanical clocks began to be built in Europe, it was quite otherwise. The magnificent clock made about 1350 for the Cathedral of Strasbourg served the public as both a calendar and an aid to astrology. Also an instructive and entertaining toy, it performed a variety show as it tolled the hours. In addition to a moving calendar and an astrolabe with pointers marking the movement of sun, moon, and planets, in its upper compartment the Three Magi bowed in procession before a statue of the Virgin Mary while a tune played on the carillon. At the end of the procession of the Magi, an enormous cock made of wrought iron with a copper comb and set on a gilded base opened its beak and stuck out its tongue, crowing as it flapped its wings. When rebuilt in 1574, the Strasbourg clock included a calendar showing movable feasts, a Copernican planetarium with revolutions of the planets, phases of the moon, eclipses, apparent and sidereal times, precession of the equinoxes, and equations for translating sun and moon indicators into local time. A special dial showed the saint's days. Each of the four quarters of each hour was struck by a figure showing one of the Four Ages of Man: Infancy, Adolescence, Manhood, and Old Age. Every day at noon the twelve Apostles passed before Christ to receive His blessing. The days of the week were indicated by chariots among clouds, each carrying the appropriate pagan god. The burghers of Strasbourg boasted that they had produced one of the Seven Wonders of Germany. In the late nineteenth century, German immigrant clockmakers in the Pennsylvania Dutch countryside produced Americanized versions of these "apostolic clocks" which added to the traditional procession of Magi and Apostles a patriotic parade of presidents of the United States.

The most popular dramas of the Middle Ages did not occur on a theater stage, nor even at the fairs or in the courtyards of churches, but were broadcast from clock towers. When the great turret clocks were in full display, they performed every hour on the hour, and every day, including Sundays and holidays. The Wells Cathedral clock, first built in 1392 and improved in the following centuries, offered a widely appealing show. Dials indicated the hour, the age and phases of the moon. Opposite the moon was a figure of Phoebus, for the sun, weighted to remain upright. Another dial showed a minute hand concentric with an hour hand that carried an image of the sun which made a full circle each twenty-four hours. In a niche above, two pairs of armored knights circled around in combat in opposite directions. As the bell struck the hour one of them was unhorsed and then, when out of view, regained his saddle. A conventional uniformed figure, "Jack Blandifet," struck each hour with a hammer but sounded the quarter-hours on two smaller bells with his heels.

Clockmakers lost no opportunity for drama. In place of a clapper hidden in the bell, they preferred vivid automata to strike the hours and the quarters. The striking figure became personified as "Jack," derived from Jacquemart, a shortened form of Jacques combined with *marteau* (hammer). This word later became generalized into "jack," meaning a tool that saved labor. A pair of such Jacks, two robust men of bronze, dating from 1499, still perform for us on the Piazza San Marco in Venice. Here was something for everyone. As the chronicler at Parma observed in 1431, to the whole populace (*"al popolo"*) the town clock told only the simple hours, while to the few who could understand (*"agli intelligenti"*) it showed the phases of the moon and all sorts of astronomical subtleties.

The clock dial, a convenience for the literate, and the first mechanical device for registering time visibly rather than aurally, is said to have been invented by Jacopo de' Dondi of Chioggia, Italy, in 1344. For this he was honored with the title of the Horologist (*Del Orologio*), which became his family name. "Gracious Reader," his epitaph boasted, "advised from afar from the top of a high tower how you tell the time and the hours, though their number changes, recognize my invention. . . ." His son, Giovanni de' Dondi, completed in 1364 one of the most complicated clockworks ever built, combining a planetarium and a timepiece. Although the clock itself has disappeared, Dondi left detailed descriptions and complete drawings from which this famous "astrarium" has been reconstructed, and can now be viewed in the Smithsonian Institution in Washington, D.C. An elegant heptagonal machine of brass activated by falling weights, it stands about five feet high. In many respects it was centuries ahead of its time, for it took account of such subtleties as the slightly elliptical orbit of the moon. On its numerous dials, it recorded the mean hour and minute, the times of the setting and rising of the sun, conversion of mean time to sidereal time, the "temporary" hours, the day of the month and month of the year, the fixed feasts of the Church, the length of daylight for each day, the dominical letter of the year, the solar and lunar cycles, the annual movement of the sun and moon in the ecliptic, and the annual movements of the five planets. In addition, Dondi provided the means to predict eclipses, indicated the movable feasts of the Church, and devised a perpetual calendar for Easter. People from everywhere came to Padua to see the clock and meet the genius who had spent sixteen years making it.

In that age the boundary was much less sharp than it would become later between the data of the heavens and the needs of everyday life. Night was more threatening, and darker, and the modern mechanical antidotes to darkness, heat, and cold had not yet been invented. For people on the seacoast or on a river the tide times were crucial. Over everybody and everything, the influence of the planets—the astral powers—governed. The Strasbourg clock of 1352 used the data of the heavens to provide the community also with medical advice. A conventional human figure was surrounded by the signs of the zodiac. Lines were drawn from each sign to the parts of the body over which it ruled and which should be treated only when that sign was dominant. The clock then offered information about the changing dominance of the signs, helping citizens and doctors to choose the best times for medical treatment. The astrological indications on the public clock in Mantua, Italy, impressed a visitor in 1473 with its display of "the proper time for phlebotomy, for surgery, for making dresses, for tilling the soil, for undertaking journeys, and for other things very useful in this world."

## Urban Time

**John J. McDermott**

Having considered in some detail the experiencing of urban space and artifact as a context for human life, we turn now to a beginning analysis of some characteristics of urban time. Once again we must forgo a full discussion of the contrasting experience of nature time, offering only some contentions gleaned from previous consideration. Nature time can be described as fat time, running in seasons, even in years and decades. To a young city boy the prognostications of the *Farmers' Almanac* were as strange and alien as if they were made for a millennium hence. Nature time gives room to regroup, to reassess, featuring a pace and a rhythm tuned to the longstanding, even ancient responsive habits of our bodies. Feeding off the confidence in the regenerative powers of nature, time is regarded as realizing, liberating, a source of growth.[1] The rhythm of nature time shares with nature space a sense of expansiveness such that in nature we believe that we "have time" and that "in time" we too shall be regenerated. As exemplified by the extraordinary journey of the nineteenth-century Mormons, their "trek," a distinctive nature phenomenon, points to salvation "in time." Space is the context for "in time," both taken as a long period of time, walking from Nauvoo, Illinois, to the Great Salt Lake, Utah, and being saved "in the nick of time." In nature time undergoing, doing, and reflection function simultaneously, thereby providing little need for "high culture," chunks of reflection taken out of the flow of experience. Perhaps the most obvious way to describe nature time is to call it baseball time, referring to a game in which the clock technically plays no role and which conceivably could last to infinity, tied to the end.

By contrast, urban time is thin time, tense, transparent, yielding no place to hide. Urban time is clock time, jagged, self-announcing time, bearing in on us from a variety of mediated sources, so often omnipresent and obtrusive that many people refuse to wear watches in an effort to ward off its domination. Why are clocks when worn on our bodies called watches? The first meaning of "watch" was to go without sleep, that is, to beat nature. Is the urban "watch" to "watch" time passing or is it to make sure that no one steals our time, as when we hoard time by saying that we have no time?

Clock time, like clock games, carries with it the threat of sudden death, an increasing urban phenomenon. But sudden death can be averted with but "seconds" to go. Urban time is "second" time, which may very well mean second chance or surprise time. We now have clocks which tell time in hundredths and even thousandths of seconds, reminding us how much faster we must go if we are not to be obsoleted, left behind, for cities have little patience with the past. Some people in the urban environment like to think that they live by nature time, but this assumption is self-deceptive, for such claims are relative only to the frantic urban pace. Clock time, after all, overrides nature time, as, for example, when at one second after midnight, in the pitch dark, your radio announcer says good morning and describes the events of your life that day as having happened yesterday.[2]

Beneath this somewhat anecdotal discussion of urban time there reside some significant implications. The rapid pace of urban time radically transforms the experience of our bodies, which often seem to lag behind. The network of communication media, which blankets a city like a giant octopus constantly tunes us in to sensorially multiple experience, even if vicariously undergone. In a broad sense, when the setting is urban we experience less identity in spatial terms than proximity to events; rather than having a place, we identify ourselves relative to events taking place. In a city, when giving directions, the question as to "how long it will take" is not answered by the spatial distance traversed; rather the allotted time is a function of potential interventions, for the time of city-space is activity measured. Our imagination, fed at all times by the messaging of electronic intrusions,[3] races far ahead of our body, which we often claim to drag around. Yet, despite the pace, the apparent garrulousness and noise, urban life is extraordinarily introspective, enabling us to carve out inner redoubts of personal space. The urban person must protect himself against the rampaging activity of time, which dismantles our environment with alarming speed. It is a cliché that you can't go home again, for the spiritual and psychological experiences of childhood are unrepeatable; but further, for one seeking his urban childhood there is added the almost inevitable burden of having his physical environment obliterated. The urban past is notoriously unstable, so that in urban experience we often outlive our environment.

Just as the prepossessing verticality of urban space encourages us to endow body-scale places as experiential landmarks, so too in the rush of urban time is it necessary to further endow such loci with the ability to act as functional clots in the flow of time, in short, to stop time for a time. In urban argot we call this "hanging out," and we might reflect on how it differs from a new version, oriented toward nature and called "dropping out."

The verticality of urban space turns vision inward and the speed of urban time revs up our capacity for multiple experiences, thereby intensifying the need for inner personal space to play out the experiences subsequently in our own "good" time. More often than is supposed, urban man does attain management of this inner personal space, and, contrary to the offensive cliché of anti-urban critics, anonymity is *not* a major urban problem. As a matter of fact, urban life is criss-crossed with rich interpersonal relations, brought about by the extraordinary short-cuts to interpersonal intimacy which flourish in the type of situation we have been describing, namely, a welter of experiences, rapidly undergone, yet transacted by the ability of the person to impact some of them in both space and time and convert them to sources of emotional nutrition. Having cut to a bare minimum the time-span required to forge urban interpersonal relationships, in general one does not expect that longevity be a significant quality of these relationships. The pace of urban time, coupled with the people density of urban space, churns up tremendous possibilities for interpersonal life, for the multiplicity of relations widens considerably the range and quality of the intersections and transactions operative in our daily lives. As we see it, then, the major problem in urban life centers not in the relations between persons but rather in the relations of persons to the urban environment and in the studied institutional insensitivity to the aesthetic qualities germane to the processes of urban space and urban time.[4]

If we come full circle and remind ourselves of the need for urban consciousness-raising, a warning is in order. The healing and amelioration of the contemporary American city will be stymied if our efforts betray an ignorance or insensitivity to the experiential demands of the original qualities of being human in urban space and in urban time. As in most human situations, the caution of William James is relevant here:

> Woe to him whose beliefs play fast and loose with the order which realities follow in his experience; They will lead him nowhere or else make false connexions.[5]

## NOTES

1 Nature time is not always kind and regenerative, as the Dakota sod-farmers of the nineteenth century and the Okies of the twentieth century discovered. A recent and extraordinarily powerful and original version of the systematic madness often found in nature time is the photographic essay by Michael Lesy, *Wisconsin Death Trip* (New York: Pantheon, 1973). The setting is rural Wisconsin from 1895 to 1900 and the common experience is laced with misery and affliction.

2 Cf. Robert Sommer, *Design Awareness* (San Francisco: Rinehart Press, 1972), p. 66. In his chapter on "Space-Time," Sommer tells us that "a San Francisco radio station announces the exact time 932 times a week."
3 Surveying big city newspapers, one finds that the screeching headlines of the first edition frequently do not merit even a paragraph in the last editions. Are these instances references to pseudo-events, or is it the pace?
4 We refer here not only to the erosion of aesthetic quality in the urban environment, symbolized by the faceless projects of the days of urban renewal, but to the more subtle and equally important fact that we fail to articulate, let alone sanction, the still existing aesthetically rich experiences of city life. The development of such an articulation is equivalent to an urban pedagogy. For an important step in this direction, cf. Jonathan Freedman, *Crowding and Behavior* (New York: Viking Press, 1975).
5 John J. McDermott (ed.), *The Writings of William James* (New York: Random House, 1967), p. 205.

## Nuclear Weapons and Everyday Life

**Paul B. Thompson**

In one sense, nuclear weapons are almost entirely absent from everyday life.[1] Most of us have never seen a nuclear weapon, much less experienced (even at a safe distance) a nuclear blast. Except for the relatively few Americans who work in or live near nuclear weapons plants or military installations where the weapons are based, we do not encounter nuclear weapons, even indirectly, as part of our daily routine. Compared with the standard run of "everyday" technologies (refrigerators, automobiles, and, now, even computers), a nuclear weapon is an exotic technology known primarily for its extraordinary character and its uncommon, unusual, and singular role in world affairs.

There is, nevertheless, a sense in which nuclear weapons pervade everyday life. They are always there; they are part of the background, part of the scenery, so to speak, in which the drama of mundane, day-to-day living is played. The mere fact that they are there, however, is not in itself reason to think that nuclear weapons have any important meaning for everyday life. The stars are there, too, even when obliterated by the bright blue sky of sunlight; their meaning is a nightly occurrence. Other weapons—tanks, rockets launchers, and bayonets—are there as well, but their meaning is obscured in peacetime. Their meaning is captured by how they might be used in another time, another place. Is the meaning of nuclear weapons also completed by knowing how they might be used? An answer to this question is the subject of the following essay.

### PHILOSOPHY AND THE PROBLEM OF DETERRENCE

Since their introduction into national arsenals at the end of World War II, nuclear weapons have been the subject of many studies in the humanities. While historical and literary treatments have traversed a wide range of real and imagined topics relevant to nuclear weapons, philosophical studies have been occupied with a much narrower set of concerns. There has always been some interest in the morality of nuclear weapons, as is evidenced in early essays by Reinhold Niebuhr and Bertrand Russell; but the issue did not really catch on until filtered through the conceptual apparatus of game theory, assurance problems, and the so-called paradoxes of deterrence.[2] An upsurge of interest in the morality of nuclear weapons was led by Jonathan Schell's 1982 book, *The Fate of the Earth*.[3] The recent spate of philosophical books and articles has led one commentator to conclude that "the term 'nuclear proliferation' is nowadays more readily applicable to the literature of nuclear strategy than to the technology of nuclear weapons."[4]

The most frequently discussed philosophical problems in this arena have to do with the moral justification of nuclear deterrence. Most authors accept the judgment that a defense policy reliant upon nuclear weapons is justified not by the actual combat capabilities of these weapons (since any actual use of the weapons would be wrong), but instead by their capacity to deter aggression by others. There have been philosophical discussions of the probability that deterrence will succeed, as well as studies of the analogies between nuclear deterrence and more conventional concepts of deterring aggression; but the most active philosophical discussion has involved the paradoxes of nuclear deterrence.[5]

The problem is that nuclear deterrence allegedly requires United States policy to involve an intention to do something that it would be wrong to do in fact (i.e., launch a nuclear strike). NATO agreements in force since the mid-1960s require United States armed forces not only to respond to attack or aggression on United States territory and citizens, but also to defend western European nations (the members of NATO) against aggression by countries in the Warsaw Pact. The United States has announced a policy of responding to such aggression not with conventional forces, but with nuclear weapons. It is the belief of NATO strategists that the Soviet Union and its allies will not undertake aggression as long as they know that the United States is capable of and sincere in carrying out this threat. The philosophical problem is the fact that it is the intention to strike, rather than the strike itself, that deters enemy aggression. Unlike conventional weapons, which are designed to achieve concrete military objectives when used in war, nuclear weapons will have failed to achieve their purpose if they are used in war. Their use is to deter, but they deter only if there is a conditional and credible intent to use them in fact.

This has seemed odd to a number of authors in that once there is an occasion to carry out this conditional intention, the goal sought in forming the intention can no longer be achieved. Indeed, acting on the intention to launch a strike not only fails to deter (since, by definition, if one is acting, the aggression one hoped to prevent has already occurred), but imposes grievous harm upon the victims of the nuclear attack.[6] It thus seems that in exercising the option of nuclear attack, one does great harm for no justifiable reason—so how can we, in good conscience, form and announce a serious intention to do just that?

Michael Novak has tried to explain how by suggesting that what is intended here is the deterrence of Soviet aggression. Nuclear strike is an option for us, but one that will only be exercised if conditions that are beyond the scope of our control transpire contrary to our sincere wishes. Hence, the use of weapons is not genuinely intended at all.[7]

J. M. Cameron has rebutted Novak's view, arguing that the conditional character of the intention does not keep it from being a genuine intention. In fact, he says, all intentions about future action have this conditional character. Even if we intend to pay for a friend's dinner tonight, or cheat on our income taxes, for example, our ability to act on these intentions depends upon circumstances beyond our control.[8]

The exchange of ideas between Novak and Cameron is typical of how philosophers have tended to analyze the deterrence problem as one in understanding intentions. Some have thought that it is possible to divide intention and action in a way that makes the intent to use nuclear weapons justified on the grounds that it has the beneficial effect of deterrence, even if forming the intention means that the weapons actually would be used, should deterrence fail.[9] The U.S. Catholic Bishop Conference, however, has argued that intending to do something wrong is itself wrong, so deterrence should be rejected.[10]

It is not my purpose here to enter this ongoing debate, for in turning so sharply to the philosophical analysis of intentions, many recent authors have lost sight of the weapons themselves. I have no wish to belittle the importance of this question; the tactical and strategic use of the nuclear tool is without question its core significance. Nevertheless, philosophical study of technology reveals that tools have meanings more subtle, more seductive, and perhaps even more far reaching than anything that can be identified with their principal use. How do nuclear weapons figure (if at all) in the matrix of technologies that support (but also condition) the tenor of everyday life? This alternative question is the subject matter for the following essay, though we shall find occasion to think a bit about deterrence, as well.

## TECHNOLOGY AS A MEDIUM FOR EVERYDAY LIFE

Although nuclear weapons are unarguably among the most significant technologies to emerge in the twentieth century, they are not thought of as "everyday" technologies. They do not pervade the average person's life in the way that the automobile, the clock, or the television does. Before considering the difficult subject of nuclear weapons, it will be helpful to warm up a bit by asking some more familiar questions. We shall take our cue from Marshall McLuhan's observation that technologies serve as media in shaping everyday life.[11] Technology serves as a medium in that, though not initiating events in the manner of an efficient cause, it affects the course of events by determining a grid of possibilities. Though central to McLuhan's thought, this idea can be traced back at least to Marx. It was systematically applied in Lewis Mumford's analysis of technology and history.

Mumford's insight was to see that social and cultural institutions are selected as much by the technology in use at a given time as technology is selected by people intending a use for it. Social history thus can be traced along a path of technological demands, as opposed to evolving along ideological, programmatic, or demographic lines. Classic examples of social mediation abound in Mumford's *Technics and Civilization*. In one passage he describes the transitions in social and cultural organization that followed steam power and the growth of mining operations in England:

> Since the steam engine requires constant care on the part of the stoker and engineer, steam power was more efficient in large units than in small ones: instead of a score of small units, working when required, one large engine was kept in constant motion. Thus steam power fostered the tendency toward large industrial plants already present in the subdivision of the manufacturing process.... Though the railroad increased travel distances and the amount of locomotion and transportation, it worked within relatively narrow regional limits: the poor performance of the railroad on grades over two percent caused the new lines to follow the watercourses and valley bottoms. This tended to drain the population out of the back country that had been served... by high roads and canals: with the integration of the railroad system and the growth of international markets, population tended to heap up in the great terminal cities, the junctions, the port towns.[12]

Although the creation of cities was neither an intended result nor a necessary physical consequence of steam power, the steam engine may arguably be said to have had more impact upon the everyday life of the newly urban populations through its influence on social structure than through the material consequences of its intended use.

Mumford's analysis leads to the notion of unintended consequences, but there is more to the concept of mediation than that. It was McLuhan who made a generation of students and scholars aware of a medium's influence, both on the substance of messages successfully disseminated throughout a society and on the character of the human beings who receive those messages. Since McLuhan, it has become commonplace to note the particular effectiveness of television for the communication of stories with strong visual content, such as fires and wars, or by the relative ineffectiveness of television when compared to print in covering abstract items such as economic reports. Why should the medium make any difference to the effectiveness or content of the transmitted information? A series of points must be grasped in coming to an answer to this question.

The first is a point which is somewhat analogous to Mumford's. Some stories are poorly served by the printed word, just as the hinterlands were poorly served by the steam locomotive. A filmed report of a football game tells more than any printed account about who went where and what everyone else was doing at the same time. But there are also messages, such as the economic story, that cannot be photographed precisely because they do not exist in a realm accessible to the representational powers of film and videotape technology. One might say, therefore, that the television's preference for sports over economics is simply an unintended consequence of what the technology does well, in contrast to what it does poorly. The point is valid, but it does not yet approach McLuhan's message.

Second, a technology's effect upon our ability to grasp and interpret information (like a sports story or an economic item) can become identified as a property of the information itself. The print medium has a serial method of presenting information in discrete bits, that is, letters. Combined with conventions of prose composition, these bits are hierarchically ordered into words, sentences, and paragraphs. Even the football game, when reported in print, will seem to take on properties of clarity and distinctness that it might have lacked when absorbed holistically by a spectator or television viewer.[13] Printed accounts of events may impose a narrative or argumentative structure upon stories that would be more faithfully rendered imagistically as slices of experience. Here the medium does not merely select for a certain kind of story, as when the steam engine selects for an urban-centered social structure. The medium affects the information it transmits; it imposes structures that become identified as part of the content of the information.

Even this more subtle point does not adequately express the intent of McLuhan's dictum "The medium is the message," for in saying that there *is* a message (the football game, the economic events) that is altered or affected by the medium, one implies that meanings or messages can be independent of media (and, subsequently, independent of media's subliminal effects). There is a deep analogy here to longstanding philosophical puzzles over primary and secondary qualities. Primary qualities have been thought to be properties of the objects themselves, while secondary qualities have been thought to arise only in our perception of them. Colors are a good example. Are colors properties of the things themselves, or is color a phenomenon that we associate with things only because that is how we are physiologically equipped to perceive them? The football game and the national economy become meaningful only insofar as they are apprehended either directly through sense perception (although it is not clear that anyone has ever sensed an economy) or through media. The philosophical problem of what is real and what is merely perceived can be as acute for mediated meanings as it is for sense perception. We begin to grasp McLuhan's point when we see that the ontological divide between medium and message is, at least, somewhat up for grabs.

In *Technics and Praxis* Don Ihde analyzes the way that tools inform human perception, and his work on this problem can take us a step further toward understanding media. The dentist's probe, for example, allows one to perceive the fissures and surface characteristics of the patient's tooth. The surface texture of the tooth can be felt dramatically through the probe (try rubbing a sharpened pencil across a variety of surfaces to get some sense of the kind of information the dentist obtains). At the same time, unwanted information such as the wetness or temperature of the tooth is filtered out by the probe. Ihde uses the terms "amplification" and "reduction" to describe the selection of information that is achieved by the use of such a tool. The topography of the tooth is amplified; the surface characteristics stand out and become perceptible in a much more dramatic fashion than if the dentist were simply to feel the tooth with his finger. Other characteristics—dampness, temperature, even color—recede into the background. They are not in the focus of the dentist's attention while the probe is used. The probe cooperates, so to speak, with the dentist's lack of interest in these "other" characteristics by reducing the extent to which the dentist is even aware of them. Perception through instruments is colored by the way that the instrument makes some features of the object readily available and easily noticed (amplified), while other features are obscured and have a tendency to be overlooked.[14]

We use the television to perceive a distant land just as the dentist uses a probe to perceive a tooth. The tool that augments perception changes perception, however, because any tool will tend to amplify some features of the object perceived while reducing others. Because the focus of our attention is directed toward the object and not the tool, there is a tendency (Ihde calls it "latent telos") to identify the object in terms of features amplified rather than in terms of features reduced. Thus a person who reads a report of African famine in the newspaper may tend to interpret it in terms of its political, economic, and ecological components, while a person who learns of the same famine through the medium of television may see it as a human tragedy. There is no prima facie reason why either of these interpretations should be regarded as more true or accurate than the other, though each will certainly serve ends that the other will not.[15]

Having moved through this succession of points, we are now in a position to ponder how technology in general (and not mere recognized media like print, radio, or television) acts as a medium in affecting everyday life. Here it will be helpful to think of a medium as something in which agents are immersed, something that enables them to act and react, as in chemistry. If the agents are human beings rather than chemicals, and if the medium is technology, rather than, say, saltwater, then it is possible to understand McLuhan's talk of automobiles and automation as media. The point is that, living immersed in technology as we do, we endure changes in self-conception and in the quality of human experience that are difficult to perceive, difficult precisely because the changes take place in the very categories we use to recognize change. An exquisite example of McLuhan's point can be found in Douglas Browning's account of how driving in fast traffic opens human experience to the category of the sudden:

> I am referring to a new way of seeing things, a fresh principle that has arisen into public consciousness since the coming to be of the automobile and according to which the world takes on character and meaning of which human consciousness was not previously aware.... And the alteration which has come to be is simply this: the world presents itself as a place of the sudden, as an arena wherein the rapidly developing lurks to spring across the background of the slow, methodical passage of days. The basic category of processional growth, the natural way of the coming to be of man, trees, and love, is no way replaced. But now the promise and fact of the sudden cuts across nature as the rapidly shifting shimmer of moonlight plays over the surface of a patiently toiling sea. Birth, life, love and death, spring, summer, autumn, and winter are there. The ripening of friendship, the rewards of long labor, digestion, pregnancy, creation are there. But also the scream of tires, the yellow light, the narrowly avoided accident.[16]

Browning points to a transformation of cognitive abilities caused by technology's expansion of the individual's universe of possibility. He goes on to speculate that once this category is initiated it may find other avenues of application, as in our watchful attitude toward foreign affairs, where war may break out at a moment's notice. This way of understanding the medium as the message can now be applied to nuclear weapons, but first it will be useful to note one final approach to the concept of mediation.

## TECHNOLOGY, MEDIATION, AND METAPHOR

Critics of McLuhan have said that in calling attention to the unintended and often unnoticed effects of media, he had a tendency to blur distinctions and to collapse disparate phenomena under an all-encompassing framework.[17] While defenders like to point out that this criticism pedantically insists upon adherence to the serial aesthetics of print communication that McLuhan has called into question,[18] it is certainly true that McLuhan himself was as likely to apply his dictum to the Mumford-like unintended consequences of technology as to the more subtle shifts in perceptual categories and ontological ambiguity that have been discussed above. He also used it in yet another sense in which mediation is entirely symbolic and metaphorical in a fairly conventional sense. Although experiences of nature provide the key source for metaphor as it is taught in high school English classes, it is clear that descriptions of technology also provide detachable linguistic or imagistic expressions which can stand for something altogether different from the technology to which they originally referred.

D. O. Edge has offered a penetrating outline of symbolic mediation in his essay "Technological Metaphor." He writes:

> The advent of the steam engine, and the development of railways, offers a striking example. We nowadays, quite naturally, say of someone that he is "letting off steam." This metaphorical description of human behavior was quite impossible before the spread of steam engines (the OED quotes 1811 as the first dated use), but now our familiarity with railway engines and their eccentricities readily allows us to conceive of human anger in these terms. We talk, too, of certain social phenomena as "safety valves" (dated 1797), with the implication that they "reduce the pressure of society to "safe level."... By use of this technological metaphor the danger of "repressing basic human energies" can be made more vivid, and human and social theorists who analyze civilization in terms of the (potentially dangerous) repression or restraint of primal urges can more easily gain an attentive audience.[19]

"Could this," Edge continues, "account, at least in part, for the popularity of Marxist and Freudian notions?" There is something quite like the expansion of experience going on in what Edge describes, but there is also an important difference. In Edge's work, technology expands the linguistic vocabulary by allowing for the construction of metaphors based on language that has technological artifacts as its literal referent. While this is an important source of impact upon the human psyche, it is not really different the way that linguistic names and descriptions of natural artifacts have always served to expand conceptual horizons.

It is clear in McLuhan's work that this symbolic or metaphorical form of mediation is a key to understanding media. He writes at one juncture, "It is...by seeing one set of relations through another set that we store and amplify experience in such forms as money. For money is also a metaphor. And all media as extensions of ourselves serve to provide new transforming vision and awareness."[20] This passage is typical of the way that McLuhan tends to equate "media" and "metaphor." Edge, on the other hand, is less willing to make such a strong statement with respect to the pervasiveness of metaphor. More specifically, he worries that the metaphorical character of expressions can be forgotten; when metaphorical expressions come to be understood as virtually (if not literally) accurate, they go uncriticized and become myths. Of Kepler's use of the clock as a metaphor for the universe, Edge writes,

> This new, scientific, metaphor was the vehicle for a great cognitive advance. It led inexorably to Newton, and to the basis for the whole powerful structure of modern physics. Clocks are things you can tinker with. You buy and sell clocks; wind and adjust them; redesign them; replace them; even, without too great a pang of guilt, smash them. If the world comes to be perceived as genuinely "clocklike," with man in the external role of "clockminder," these other attitudinal changes are likely to follow.[21]

It is probably true that McLuhan's work achieves much of its intellectual power by collapsing this series of philosophical observations under a single, unifying concept of metaphor. Edge's more sobering remarks, however, remind us that although media tell a large and essential part of the story for understanding human experience in the twentieth century, mediation and metaphor can only take us so far in understanding human events.

## NUCLEAR WEAPONS

Thankfully, few of us have direct contact with bombs, but it would be hasty to conclude that technological mediation is irrelevant to the question of nuclear weapons. It is beyond question that nuclear weapons have had and are currently having unintended consequences for world politics, for social structure, and finally for individuals. The nations of the world are divided into the nuclear haves and the have-nots, with the United States and the Soviet Union forming a subset within the haves because of their extensive support technologies for the weapons themselves. Ironically, this has left the superpowers in a surprisingly weak position. They are less able to influence events through direct military intervention. If the history of our period dwells upon wars waged for political purposes, it may well find more of significance in the conflicts of the nonnuclear nations or of those (such as Israel and South Africa) joining a new kind of nuclear club: assumed to possess nuclear weapons but unwilling to acknowledge the fact. A Lewis Mumford, looking back on civilization, can discern patterns of unintended social consequences, but it is very difficult for those of us in the thick of it to see them. Future historians, if there are any, will be in a far better position to tell that story.

We can, however, look to ourselves and ponder some of the more personal effects of life in the nuclear age. We need not look far to find metaphorical impacts like those noted by Edge. The very word "atomic," which, in colloquial uses, ought to mean "small" or, perhaps, "irreducible," means something altogether different following four decades of perennial coupling with the word "bomb." The Atomic Cafe in the mid-eighties pop documentary of the same name wasn't dubbed "atomic" because of its minuscule size. Those of us who grew up in the 1950s sucked hot candies called "Atomic Fireballs"; a company in Denver made "Atomic Pickles." Although these examples are trivial, the images are not. They worked as advertising techniques because they drew upon the idea that there were now human experiences that were unprecedented in their capacity to produce devastating and comprehensive effects.

If there is an idea that came from atomic imagery in the sense that Edge thinks the idea of repression came from metaphorical talk of the steam engine, it must be the notion that events and experiences of rather short duration can be powerful enough to disassemble an entire framework of values. Although this idea has precedent in religious thought, in the atomic age there was no anticipation that lost values would be replaced by anything new or better. The possibility that one could be left naked and without cultural support by a particularly explosive experience became a realistic (perhaps even attractive) scenario. One popular song of the sixties drew upon metaphors of nuclear attack to characterize the 1969 Woodstock festival as an attempt to "get back to the garden."

These quasi-religious meanings go beyond simple metaphors, analogic inferences supported by a similarity of form, relation, or properties. Robert J. Lifton and Richard Falk have suggested that living in the nuclear era ignites a yearning for fundamentalist experience, that is, for a life event that totally annihilates previous values and supplants the self in a new milieu.[22] Perhaps the large rock festivals and religious cults of the sixties substituted for nuclear annihilation in the symbolic matrix of expected life experiences. The nuclear metaphor itself, however, was more likely to appear in a more sinister form. Radiation and holocaust became a plot device in 1950s horror movies that played upon fear of the unknown. Strange and hitherto unanticipated forces stalked the countryside in these films, sometimes wreaking general havoc, but often working invisibly to replace the recognizable persons and values of traditional life with visually identical (but soulless) alien substitutes. One can see why Hollywood producers would want to make films like this (no expensive budget for special effects), but why would people want to see them? Though frequently artless, the alien horror film of our time provides a script for intangible fears that, in another time, might have been dismissed as hysterical.

A merely symbolic or metaphorical reading of the holocaust does not account for these deeper and more subtle effects. It is important to our way of understanding the world that these weapons are real and poised for use. The reality and impending use of nuclear weapons is a fact that "makes us doubt that anything we do or make will last."[23] Lifton describes a syndrome he calls "nuclear withdrawal," which is a radically revised concept of personal security: "One freezes in the manner of certain animals facing danger, becomes as if dead in order to prevent actual physical or psychic death. But all too frequently the inner death of numbing has dubious value to the organism. And it may itself become a grave source of danger."[24] Lifton concludes that the numbing of experience associated with life in a nuclearized world interrupts the symbolizing process. If he is right, one consequence of living with nuclear weapons may be an inability to think about nuclear weapons themselves.

Certainly, the existence of nuclear weapons—warheads and delivery systems—gives more poignancy to the category of the sudden, as defined by Browning. We may, as Browning suggests, possess the basic experience required to generate this category as a result of our dealings with automobiles, but the reason why the capacity for sudden danger stays with us so persistently must have something to do with nuclear weaponry. It is, after all, these weapons that give credibility to the idea of sudden and total war, of sudden and total destruction. It is interesting, though probably futile, to speculate on whether nuclear weapons would have had the ominous psychological and conceptual impacts without a prior conditioning to the category of the sudden through our use of automobiles. It cannot be doubted that other technologies of everyday life enter into the matrix of mediation along with nuclear weapons. Lifton himself cites mass media as a contribution to psychological numbing:

> [The mass media] deprive us of the channels of feeling that had existed around earlier rituals and symbols. We then grasp at an extraordinary array of images available to us, seeking to recover significant forms of feeling. But our successes are spasmodic, and we run the risk of diffusion and unconnectedness, potential sources of further numbing.[25]

The link between media and weapons demands some scrutiny.

## "THE MEDIUM IS THE MISSILE"

McLuhan's dictum, "The medium is the message," was intended to lead us on to see that "the medium is the message." The idea was to see that technology "roughs us up," that it is not merely as detachable symbols that technological experiences reshape everyday life. Television and automobiles mediate perception and experience in virtue of their material contact with the senses. Nuclear weapons provide a challenge to this component of McLuhan's thought, since few of us have had tactile experiences with nuclear weapons that are anything like the daily massage we get from our televisions or our automobiles.

Indeed, McLuhan himself wrote relatively little about nuclear weapons. His most suggestive comment is probably his comparison of hot and cold wars:

> The "hot" wars of the past used weapons that knocked off the enemy, one by one. Even ideological warfare in the eighteenth and nineteenth centuries proceeded by persuading individuals to adopt new points of view, one at a time. Electric persuasion by photo and movie and TV works, instead, by dunking entire populations in new imagery.[26]

Even in this passage, McLuhan is not referring to nuclear weapons per se, but to electronic communications media. Nevertheless, no weapon could be more suited to the cold war of electronic imagery than the atomic bomb (unless, of course, it is the hydrogen bomb, coupled with the ICBM). It is as an information technology that nuclear weapons "rough us up."

Throughout the history of their development and deployment, nuclear weapons generally have been associated with strategic goals, the sending of political messages to one's enemy, rather than with tactical concerns. As media, nuclear weapons are fairly imprecise in terms of the audience that can be selected. Although we might intend to send a message to the Russians with our weapons, we cannot avoid sending a message to the rest of the world as well. Similarly, we receive a message from every possessor of nuclear technology, but what message do we receive? Although the possession of any weapon can be an effective means for the communication of power and resolve, weapons are notoriously poor communicators of intention (a fact that has been exploited for years in the National Rifle Association's bumper stickers "Guns don't kill people; people kill people"). As such, it should hardly be surprising that problems of intention and paradox have occupied much of the nuclear weapons literature.

As information technologies, nuclear weapons have two types of mediating effect—one for the generals and experts who are the main senders and recipients of information, and another for the rest of us. The generals have long discussed counter-force nuclear strategies where a controlled exchange of nuclear weapons is a means of "sending a message." It is hoped that the message will lead to disengagement, but since Hiroshima world leaders have eschewed this means of "communication" even when the balance of nuclear forces might have allowed them to get away with it. There seems to be a deep-seated distrust of the information carrying capacity of a brief nuclear exchange.[27] But does the possibility of sending a message with nuclear weapons influence a general's worldview? George Kennan, for one, has suggested that it does. As military planning becomes the primary medium for interpretation and understanding of international relations, the media bias of the nuclear weapon has an unfortunate effect. It is the appropriate and justifiable bias of strategic planning that the opponent is seen as hostile and formidable. Kennan writes, "In tens of thousands of documents, this image of the opponent is re-created, and depicted in all its implacable formidability, until it becomes hopelessly identified with the real country in question. In this way, the planner's hypothesis becomes, imperceptibly, the politician's and the journalist's reality."[28]

Nuclear weapons figure heavily in this process, for an opponent armed with nuclear weapons, hostile to one's vital interests, makes for a vision that is formidable indeed. According to Kennan,

> ...what began as a limited political conflict of interests and aspirations has evolved into a perceived total military hostility; and what was in actuality a Soviet armed-forces established with many imperfections and many limitations on its capabilities has come gradually to be perceived as an overpowering paragon of military efficiency, standing at the beck and call of a political regime consumed with no other purpose than to do us maximum harm. This sort of distortion has magnified inordinately, in the public eye, the dimensions of what was initially a serious political problem and has created, and fed, the impression that the problem is not to be solved otherwise than by some sort of a military showdown.[29]

The ability to make this critical point depends upon recognizing the mediating aspects of weapons. When both parties possess nuclear weapons, the effect, in Kennan's words, is to "(obscure) the real political, and even military, conflicts of interest between the United States and the Soviet Union behind a fog of nuclear fears, suspicions, and fancied scenarios."[30] One does not apply the technology of nuclear weapons to fulfill one's own goals but rather has those goals shaped by the technology itself.

## NUCLEAR WEAPONS, DETERRENCE, AND EVERYDAY LIFE

The phenomenon Kennan notes in the thinking of defense experts is almost certainly present to some degree in all of us. While the generals must think through the media side effects of sending messages via missile, we, the general public, think of ourselves more readily on the receiving end. In the above discussion of media, we learned that the ambiguity between media and message runs deep. Close attention to media effects shows that it is not often easy to unambiguously tie amplified signals to the message, as opposed to its medium of transmission. Nuclear deterrence works because the will and seriousness of the respective superpowers is amplified, but nuclear weapons reduce many important components of national interest to nothing. Do nuclear weapons communicate our interest in trade, in cultural exchange, in environmental quality, in a stable world economy, in safe transit, or in peaceful coexistence? Do they even communicate mutual military interests such as the control of militant Islamic fundamentalism? They do not; they were never intended to do so. They communicate one nation's resolve in opposing the actions of another.

And isn't this the popular image of the Soviet Union? Don't we think of them as a nation, a people, resolved to oppose us? Actually, it is worse than that, for it is not as if we think of them as a nation with diverse and complicated national interests who resolutely refuse to tolerate our interference in their pursuit of these interests. It is as if opposition to Western democracy is their dominant goal, the interest that they pursue, even to the detriment of all others. The amplified message of resolute opposition becomes identified as the essence of the Soviet meaning for the average American. Is this the message they intended to send, or have we been paying too much attention to the medium? The question is rhetorical, for it is impossible *not* to pay attention to the medium. We know ourselves as, at heart, forced to divert our energies from our true interests to resist a Soviet threat, but we cannot see this theme in the face that the Soviet Union presents to us. The nuclear warhead blocks the field, and its properties become the properties of the Soviets' meaning for us. And, of course, we must expect that the Soviet people cannot see beyond *our* missiles, either.

If we return to the more typical philosophical discussion of deterrence, we may now evaluate nuclear deterrence strategies in a new light. What is it that we deter? Is it a realistically evaluated probability of a Soviet/Warsaw Pact strike at the United States or the NATO allies that we expect to deter? Or is it the amplified image of an evil empire, bent upon merciless and total destruction of our world? The contrast between these two views points to the way that nuclear weapons mediate the moral ambiguities of deterrence. Deterrence per se is, on the face of it, a reasonable and defensible military strategy. Even in the time of Napoleon, deterrence of invasion by building military power involved a conditional intention to perpetrate acts that are morally ambiguous (if not wrong) once the deterrent goal has failed; but nuclear weapons make an enemy appear monstrous and menacing beyond all proportion. The sheer intensity of the evil we must face (and, in facing, also project) distorts the logic of deterrence into a caricature of itself.

Does this mean that we should abandon our weapons, giving up the strategy of deterrence? McLuhan was reluctant to draw any moral implications from his work on technology, in part, one suspects, because he knew that moral messages are as laden with the latent telics of media as any other. Our image of other peoples is heavily influenced by more traditional media. Consider, for example, the way that the meaning of the United Kingdom is communicated by history books and public television, or the way that the meaning of African peoples is communicated by charity fund-raising appeals and primitive art. Other peoples must have a meaning for us that is heavily freighted with components of the medium on which the message arrives. There is no direct access to the "reality" of foreign peoples, only a plurality of imperfect messages. Granting that the freight conveyed by nuclear weapons is considerably less benign than that of *Masterpiece Theater,* isn't the resolve communicated by this medium a real and serious component of U.S. national character (as we must assume it is for the Soviets)? Wouldn't it be disastrous if the seriousness of this resolve were not communicated to anyone who thought of making us an enemy? The moral issue is not cut and dried.

Nevertheless, the coarse media effects of sending a message with nuclear weapons will, no doubt, count against them for many people. Kennan expresses frustration at the way that the weapons deadlock seems to preclude U.S.–Soviet relations that would, in his view, be more consistent with our own diverse interests. If the strategic use of nuclear weapons amplifies one component of national interest disproportionately, then this effect must certainly be regarded as a cost, at least. If we are to use nuclear weapons to send messages, we must accept as a trade-off that others will have an image of us that is quite at odds with the one we might like to project. Whether the cost is outweighed by the benefits of deterrence is another question, one that takes us back to the more frequently discussed issues of intentions and their consequences. Media effects of weapons *are* part of their consequences. If ignored, the cost-benefit comparison cannot be made accurately. How they are to be weighed is a subject for some other time, some other place.

At the end, we have returned to the question of deterrence that was abandoned at the outset. Nuclear weapons are a medium that sends the message of resolve and intent to destroy. The message is expected to be received and taken as a warning; as a warning, it is expected to deter aggressive acts by hostile governments. The point is that there are negatives, costs that we bear, even if the strategy succeeds. They reduce our capacity to both perceive and project the face of humanity in a troubled world. These costs are not "wrongful intentions"; it is not their intentional character that makes wrong. They are consequences, unwanted and unanticipated, in adopting the strategy of deterrence. As consequences, they deserve to be counted (and countered) whenever we evaluate our future with nuclear weapons.

## NOTES

1 This essay includes sections of a previous paper by Paul B. Thompson, "Technological Mediation and Nuclear Weapons," published in L. Hickman (ed.), *Philosophy, Technology and Human Affairs,* Ibis Press, College Station, Tex., 1985. Although sections of the previous paper are used here verbatim, the theses of the two papers are substantially different; the former is an attempt to examine the concept of technological mediation for its explanatory value.

2 The literature cannot be cited even in summary without violating space limitations. Special issues of *Ethics 95*(3) April 1985, *Social Philosophy and Policy 3*(1) Autumn

1985, and *The Monist 70*(3) July 1987 are devoted entirely to the subject. Schonsheck examines some of the main positions in his review paper, "Wrongful Threats, Wrongful Intentions, and Moral Judgments about Nuclear Weapons," *The Monist,* vol. 70, 1987, pp. 330–356.

**3** Jonathan Schell, *The Fate of the Earth,* Knopf, New York, 1982.

**4** Jonathan Schonsheck, "Philosophical Scrutiny of the Strategic 'Defense' Initiatives," *Journal of Applied Philosophy,* vol. 3, 1986, p. 151.

**5** See, for example, Douglas Lackey, "Missiles and Morals," *Philosophy and Public Affairs,* vol. 11, 1982, pp. 189–231; Steven Lee, "Does Deterrence Work?" *QQ,* vol. 8, no. 1, 1988, pp. 9–12; and Gregory Kavka, "Some Paradoxes of Deterrence," *Journal of Philosophy,* vol. 75, no. 6, 1978, pp. 285–302.

**6** Gregory Kavka, op. cit., 1978.

**7** Michael Novak, *Moral Clarity in the Nuclear Age,* Thomas Nelson, Nashville, 1983.

**8** J. M. Cameron, "Nuclear Catholicism," *New York Review of Books,* Dec. 22, 1983, pp. 38–42.

**9** Robert W. Tucker, "Morality and Deterrence," *Ethics,* vol. 95, 1985, pp. 461–478.

**10** U.S. Catholic Bishop Conference: The Challenge of Peace: "God's Promice and Our Response," *Origins,* 12 (Oct., 1982). This issue contains the entire text of the pastoral letter. Richard Wasserstrom, "War, Nuclear War, and Nuclear Deterrence: Some Conceptual and Moral Issues," *Ethics,* vol. 95, 1985, pp. 424–444.

**11** Marshall McLuhan, *Understanding Media: The Extensions of Man,* McGraw-Hill, New York, 1964.

**12** Lewis Mumford, *Technics and Civilization,* Harbinger Books, New York, 1934, p. 162.

**13** John Culkin, "A Handful of Postulates," in G. E. Stearn (ed.), *McLuhan: Hot & Cool,* New American Library, New York, 1967, p. 55.

**14** Don Ihde, *Technics and Praxis,* Reidel, Boston, 1979, pp. 20–21.

**15** This is most certainly not to say that all messages are equal as far as truth or accuracy is concerned. A printed account of African famine that described it as "a Caribbean holiday" would be as surely false as a video account that used stock footage from *Lawrence of Arabia.* McLuhan's point makes truth and falsity relative to certain features of the representational capacities inherent in media, but it does not entail the more radical, deconstructionist move of making all messages into interpretively equivalent events.

**16** Douglas Browning, "Some Meanings of Automobiles," in L. Hickman and A. Al-Hibri (eds.), *Technology and Human Affairs,* C. V. Mosby, St. Louis, 1981, p. 15.

**17** Dwight MacDonald, "Running It Up the Totem Pole," in R. Rosenthal (ed.), *McLuhan: Pro & Con,* Penguin, New York, 1968, pp. 29–37.

**18** Irving J. Weiss, "Sensual Reality in the Mass Media," in R. Rosenthal (ed.), *McLuhan: Pro & Con,* Penguin, New York, 1968, pp. 38–57.

**19** D. O. Edge, "Technological Metaphor," in D. O. Edge and J. N. Wolfe (eds.), *Meaning and Control: Essays in the Social Aspects of Science and Technology,* Tavistock Publications, London, 1973, p. 34.

**20** McLuhan, op. cit., 1964, p. 60.

**21** Edge, op. cit., 1973, p. 37.

**22** Robert J. Lifton and Richard Falk, *Indefensible Weapons,* Basic Book, New York, 1982, pp. 85–90.

**23** Ibid., p. 71.

**24** Ibid., p. 104.

**25** Ibid., pp. 105–106.

**26** McLuhan, op. cit., p. 339.

**27** David Kaplan, *The Wizards of Armageddon,* Simon and Schuster, New York, 1983.

**28** George Kennan, *The Nuclear Delusion,* Pantheon, New York, 1983, p. 220.
**29** Ibid., pp. 220–221.
**30** Ibid., p. 221.

## Housework and Its Tools

**Ruth Schwartz Cowan**

Industrialization transformed every American household sometime between 1860 and 1960. For some families, this transition occurred very slowly: each generation lived in homes that were just a bit "more modern" than the generation immediately before it, and the working lives of the members of each adjacent generation were not so profoundly different as to leave unbridgeable communication gaps between them. For other families, the transition was more rapid; in these families, as the result of immigration or urbanization or sudden affluence, one generation of people may have been living and working in conditions that would have been familiar in the Middle Ages, and the very next generation may have been completely modernized—inhabitants, as it were, of a totally different world. Yet despite these differences in pacing, if we consider the broad spectrum of American households, from rich to poor, from the most urban to the most rural, a simple generalization can describe what happened in the century that was ushered in by the Civil War: before 1860 almost all families did their household work in a manner that their forebears could have imitated—to wit, in a preindustrial mode; after 1960 there were just a few families (and those either because they were very poor or very isolated or ideologically committed to agrarianism) who were not living in industrialized homes and pursuing industrialized forms of labor within them.

Now usually, when we think of the word *industrialization,* we think in terms not of homes but of factories and assembly lines and railroads and smokestacks. In our textbooks of history and economics and sociology, the terms *industrialization* and *home* are usually connected by the word *impact,* and we are usually asked to consider what happened when one term (industrialization) caused some significant economic process (productive work or the manufacture of goods for sale in the marketplace) to be removed from the domain of the other term (home). Implicitly (and sometimes explicitly) we are given the impression that industrialization occurred *outside* the four walls of home. The popular imagination goes one step farther; industrialization is conceived as being not just *outside* the home but virtually in *opposition* to it. Homes are idealized as the places to which we would like to retreat when the world of industrialization becomes too grim to bear; home is where the "heart" is; industry is where "dogs are eating dogs" and "only money counts."

Under the sway of such ideas, we have had some difficulty in acknowledging that industrialization has occurred just as rapidly within our homes as outside them. We resolutely polish the Early American cabinets that hide the advanced electronic machines in our kitchens and resolutely believe that we will escape the horrors of modernity as soon as we step under the lintels of our front doors. We are thus victims of a form of cultural obfuscation, for in reality kitchens are as much a locus for industrialized work as factories and coal mines are, and washing machines and microwave ovens are as much a product of industrialization as are automobiles and pocket calculators. A woman who is placing a frozen prepared dinner into a microwave oven is involved in a work process that is as different from her grandmother's methods of cooking as building a carriage from scratch differs from turning bolts on an automobile assembly line; an electric range is as different from a hearth as a pneumatic drill is from a pick and shovel. As industrialization took some forms of productive work out of our homes, it left other forms of work behind. That work, which we now call "housework," has been transformed in the preceding hundred years, and so have the implements with which it is done; this is the process that I have chosen to call the "industrialization of the home."

Households did not become industrialized in the same way that other workplaces did; there are striking differences between housework and other forms of industrialized labor. Most of the people who do housework do not get paid for it, despite the fact that it is, for many of them, a full-time job. They do not have job descriptions or time clocks or contractual arrangements; indeed, they cannot fairly be said even to have employers. Most of their work is performed in isolation, whereas most of their contemporaries work in the company of hundreds, perhaps even thousands of other adults. Over the years, market labor has become increasingly specialized, and the division of labor has become increasingly more minute; but housework has not been affected by this process. The housewife is the last jane-of-all trades in a world from which the jack-of-all trades have more or less disappeared; she is expected to perform work that ranges from the most menial physical labor to the most abstract of mental manipulations and to do it all without any specialized training. These various characteristics of household work have led some analysts to suggest that housework (or the household economy) is the last dying gasp of feudalism, a remnant of precapitalist conditions somehow (miraculously) vaulting the centuries unimpaired, the last surviving indicator of what the Western world was like before the market economy reared its ugly head.[1]

Perhaps this is true, but there are other sides to the coin; industrialized housework resembles industrialized market labor in significant ways. Modern housework depends upon nonhuman energy sources, just as advanced industrialized manufacturing systems do. Those of us who regularly perform household chores may regard this as an erroneous, or at least an ironic, statement, but it is nonetheless true. The computer programmer turns an electric switch in order to power the tool that makes his or her labor possible—and so does the houseworker; we are all equally dependent upon the supply lines that keep these energy sources flowing to us. We may be thoroughly exhausted by our labors at the end of a day of housework, but without electricity or the combustion of certain organic compounds (like natural gas or liquid petroleum or gasoline), our work could not be performed at all. None of us relies any longer solely on animal or human energy to do our work.

Thus, even if the household is an isolated work environment, it is also part of a larger economic and social system; and if it did not constantly interact with this system, it could not function at all—making it no different from the manufacturing plant outside the city or the supermarket down the street. The preindustrial household could, if necessary, function without a supportive community—as is demonstrated, most clearly, in the settlement pattern of our frontiers. Individual families were capable, when need arose, of supplying themselves with their own subsistence and protective needs, year in and year out. Very few families are capable of doing that any longer. Very few of us, for example, would know how to make our own bread, even if our lives (quite literally) depended upon it; if we could find and follow a recipe for making the bread, it is highly unlikely that we could (1) grow the wheat, (2) prepare it properly for use in bread, (3) obtain and keep the yeast alive, or (4) build and maintain a suitable fixture for baking it. We live in isolated households and do our marketing for the tiniest of consumption units; but, to get our bread to the table, we still need bakers, agribusiness, utility companies, and stove manufacturers. This is the second significant sense in which household work and market work have come to resemble one another.

Finally, both household labor and market labor are today performed with tools that can be neither manufactured nor understood by the workers who use them. Industrialized households contain vastly more implements than preindustrial ones did, and those implements are much more likely to have been made by persons and in locales that are totally foreign to their eventual users. Preindustrial households purchased some of their tools (especially those made of pottery, glass, or metal), but today we buy almost everything we use—from forks to microwave ovens. As a result, despite the diversity of what is available for purchase, almost nothing that we buy has been made "for us," to fit special needs that we may have. In addition, the implements that we have today are more complicated than the implements with which our foreparents worked—so much more complicated that most of us either cannot or will not repair them ourselves. If a brick fell out of an eighteenth-century fireplace, someone in the household would probably have known how to make and apply the mortar with which to replace the brick. If, on the other hand, a resistance coil comes loose on a twentieth-century electric oven, no one in the household is likely either to know what to do or to have the appropriate tools at hand. In these senses houseworkers are as alienated from the tools with which they labor as assembly line people and blast furnace operators.

In sum, we can say that there are three significant senses in which housework differs from market work (in being—most commonly—unpaid labor, performed in isolated workplaces, by unspecialized workers) and three significant senses in which the two forms of work resemble each other (in utilizing nonhuman—or nonanimal—energy sources, which create dependency on a network of social and economic institutions and are accompanied by alienation[2] from the tools that make the labor possible). If we take all six of these criteria and group them together, we will have a good definition of industrialization. Then we might be able to see that, in the West over the last two hundred years, women's work has been differentiated from men's by being incompletely industrialized or by being industrialized in a somewhat different manner.

How—and why—this situation came to pass is one of the great unresolved puzzles of Western history. Although the social arrangements to which we have become accustomed seem sometimes to have a rationale and a life of their own, there really is no a priori reason why things should have worked out in quite the way they did. Even if we assume, as the anthropologists tell us we should, that every society will construct some sexual division of labor for itself, there is no apparent reason why, for example, men's work could not have been incompletely industrialized instead of women's. We might then have had communal kitchens, to which we would repair for all of our food needs, but household metal goods that we forged in smithies in our own backyards; or perhaps electronic looms in every kitchen and communal nurseries in which children of our female physicians could be cared for and reared. Clearly we have the technological and the economic capacity to have constructed our society this way, but for some complex of reasons we did not do so.

... We all know that work is one of the activities through which we define ourselves as we mature; by analogy we might say that a society does the same thing, defining itself through the work that it does as it matures. Social scientists know that the industrialization of work has been one of the most traumatic processes of recent Western history, and yet work has not been a particularly popular focus for historical attention—and housework even less so. I regard this omission as unfortunate, even tragic. In the last decade or two, some historians have attempted to repair the damage and to write the history of work as it has altered for different classes of people in the last few centuries; but, as admirable as these studies have been, they have focused almost exclusively on market labor—work that is done in order to produce products or services to be sold.[3] Yet in many ways housework is more characteristic of our society than market work is. It is the first form of work that we experience as infants, the form of work that the largest proportion of us (to wit, almost all women) identify as the work that will be the principal definition of our adulthood. It is also the form of work that each of us—male and female, adult and child—pursues for at least some part of every week; and it is the occupational category that encompasses the single largest fraction of our population—to wit, full-time housewives. The absolute number of full-time housewives may be decreasing with every passing year, but more people spend their days in this "peculiar" form of labor than in either of the two more "standard" forms—blue-collar or white-collar work. If work shapes individual lives and social forms, and if industrialization has reshaped work in the past two centuries, then to fail to understand the history of housework is to fail to understand ourselves. If housework is a dominant social activity, and if it has been only incompletely industrialized, then, as a society, we may not be as industrialized as we think we are, or as "modern" as our pundits would have us believe.

... Human beings are tool-using animals; indeed, some anthropologists believe that, along with speech, the ability to use and to refine our tools is precisely what sets us apart from other species of primates. One of the few generalizations that can be made about people living under vastly different social conditions is that they all use tools to do their work. Because of our peculiar set of cultural blinders, we do not ordinarily associate "tools" with "women's work"—but household tools there nonetheless are and always have been.

Tools are not passive instruments, confined to doing our bidding, but have a life of their own. Tools set limits on our work; we can use them in many different ways, but not in an infinite number of ways. We try to obtain the tools that will do the jobs that we want done; but, once obtained, the tools organize our work for us in ways that we may not have anticipated. People use tools to do work, but tools also define and constrain the ways in which it is possible and likely that people will behave. Here is a simple example. In my house, we recently installed standard wall cabinets with doors above the counters in our kitchen; these cabinets are tools that we intended to use as containers for other tools (most notably our dishes), with the specific intention that they would make those other tools easy to locate when needed and would keep them clean between washings. Before we had the cabinets we kept our dishes on a remodeled floor-to-ceiling bookcase that did not have cabinet doors; we thought our new cabinets would make our housework easier to perform. Before we installed our new cabinets, the process of having our table set for dinner involved: (1) an adult's decision that it was time to have the table set; (2) the communication of that decision to children—which communication needed to be repeated more than once and in increasingly insistent tones; (3) the removal of the dishes by the children and their placement, in appropriate order, on the table. The adults in the family functioned as managers and decision makers; the children, as workers—often workers under duress. Our new cabinets have changed all of our behavior patterns. Since the children are too small to reach the shelves on which the dishes are now placed, the adults must become involved in the work process. Not only must my husband and I make the decision that it is time to set the table, but we must also do part of the physical labor; we have ceased to be the managers of the work and have been forced to become unwilling participants in it. In addition, if we have erred in our labor ("But, Mommy, you didn't *give* me the water pitcher!"), then we must be responsible for correcting our errors. The acquisition of this one new tool has temporarily (at least until the children grow taller) altered our domestic work process as well as the set of emotional entanglements that that work process entailed. At the very least, the acquisition of that new tool will now require us to acquire yet another tool (a stool) in order to return to the *status quo ante*—a behavioral alteration that was also unintended.

Multiply this small example millions of times, and you will have some sense of what it means to say that tools are not entirely passive instruments. This is precisely the lesson that the sorcerer was trying to teach his apprentice in the famous fable. Our tools are not always at our beck and call. The less we know about them, the more likely it is that they will command us, rather than the other way around.

Thus the history of housework cannot properly be understood without the history (which is separate) of the implements with which it is done—and vice versa. The relation is reciprocal, perhaps even dialectic. Tools have set limits on what could be done in households, but inventors have repeatedly broken through those limits by fashioning new tools. The tools have reorganized the work process, creating new needs, for which some people have attempted to provide new tools—and so on. What makes the history of household technology separate and distinguishable from the history of housework is the existence of social institutions that mediate the availability of tools to households. In times past, these mediators were institutions such as blacksmith shops and blacksmith guilds, peddlers, and international trade arrangements. As industrialization has progressed, the nature of the institutions has changed—we now have manufacturing firms and advertising agencies and market researchers; but the impact of the institutions remains structurally the same. They mediate the availability of tools by keeping some tools off the market and promoting others, or by organizing the pricing and distribution of tools. Just as the history of industrialization cannot properly be written without the history of housework, so the history of household technology cannot be written without the history of the social and economic institutions that have affected the character and the availability of the tools with which housework is done.

In order to make the complex task of writing this multifocused history less daunting, I have made use of two organizing concepts: *work process* and *technological system*. Both awkward phrases need to be explained before I proceed farther. The phrase *work process* is used instead of the simpler term *work* in order to highlight the fact that no single part of housework is a simple, homogenous activity. One might be tempted to say that housework can be divided into a series of separable tasks—cooking, cleaning, laundering, child care, et cetera. This analysis does not go far enough, however, because each of these tasks is linked to others that it does not resemble. Cooking, for example, involves the treatment of raw or semi-raw foodstuffs so that they can, or will, be consumed; that much is obvious. Perhaps not so obviously, cooking also involves the procurement of those foodstuffs (by buying them or raising them), and their storage and prior preparation (by canning, salting, freezing, refrigerating, et cetera), the maintenance of the energy source (stoking the hearth, damping the stove, adding the coal) that is used to do the cooking, the maintenance and cleaning of the tools that are used to do the cooking, and the disposal of the waste that results from the process. Similarly, laundering is a matter not just of washing clothes but of moving them from place to place, of drying them, perhaps ironing them and putting them away, as well as acquiring the chemical agents—most notably soap and water—that will assist in the process. The concept of work process reminds us that housework (indeed, all work) is a series not simply of definable tasks but of definable tasks that are necessarily linked to one another: you cannot cook without an energy source, and you cannot launder without water. This concept also becomes important when we try to discover whether industrialization has made housework easier. We must ask not only whether one activity has been altered, but also whether the chain in which that activity is a link has been transformed. If, for example, we view cleaning rugs as work, then we might reasonably argue that this work can be done faster and with less expenditure of human energy with a vacuum cleaner instead of a broom. If, on the other hand, we view cleaning rugs as a work process, then we might see that it is composed of several activities (moving the instrument, moving the rugs, removing the accumulated dust, and so on), and that at least one of these (moving the instrument) is much harder to do with the vacuum cleaner than with the broom. In addition, if it is more likely that the presence of the vacuum cleaner will increase the frequency with which the work is done (once a week instead of once a season or once a year), or will involve fewer people in the work (for example, by releasing the stronger members of the household from the obligation to move the rugs outside, or the younger members from the obligation of beating them), then the question of whether cleaning a rug has been made easier or faster by the advent of vacuum cleaners becomes considerably more difficult to answer. Easier for whom? Faster for whom? Under what conditions? The history of housework studied in the light of the concept of work process, turns up some surprises, and some of these surprises will be central to my analysis.

Just as the activities of which housework is composed are complex, linked, and heterogeneous, so are the implements with which it is done—a situation that justifies my using the second of those awkward phrases, *technological system*. Each implement used in the home is part of a sequence of implements—a system—in which each must be linked to others in order to function appropriately. To put it bluntly: an electric range will not be much good if electric current is missing, and a washing machine cannot function in the absence of running water and grated soap. I have often thought that if the concept of a technological system were more generally understood, no one would have poked fun at the inhabitants of Appalachia who were reported (perhaps apocryphally) to have put coal in the bathtubs that were given them through federal largesse during the Depression. If you were obliged to haul your bathwater from stream or pump to stove and tub, what would you want with a four-footed, enamel-over-cast-iron bathtub on your porch? A stream, a pond, a lake, or a lightweight zinc tub would be infinitely preferable. Heavy bathtubs (indeed, recessed unmovable bathtubs) are part of a technological system that contains (among other things) municipal reservoirs, underground pipelines, hot water heaters, not to speak of soap-manufacturing plants and textile mills (would you bathe very often if you had no towels?). Some of those items could be dropped from the system without entirely altering it (one could make one's own soap if it came to that), but others (the drainpipes, for example) are absolutely essential.

The concept of a technological system becomes important in understanding the processes by which the American home became industrialized. On a superficial level, the industrialization of the home appears to have been composed of millions of individual decisions freely made by householders: the Jones's down the block decided to junk their washtub and buy a washing machine, and the Smiths around the corner fired the maid and bought a vacuum cleaner. On this level, industrialization of the home seems to have been the product of the perpetually rising expectations of American consumers—expectations that had been rising from at least the 1830s, when de Tocqueville toured the country, if not before. But the matter is not as simple as that. The Jones's washing machine would not have done them a bit of good if the town fathers had not decided to create a municipal water system several years earlier, and if the local gas and electric company had not gotten around to running wires and pipes into the neighborhood. Similarly, the Smiths' new vacuum cleaner might have cost a good deal more than it did (and might have thus forestalled the Smiths' decision to replace a maid with it) if the managers of the company that made it had not earlier decided to shift to assembly line modes of production. To put the case more generally: the industrialization of the home was determined partly by the decisions of individual householders but also partly by social processes over which the householders can be said to have had no control at all, or certainly very little control. Householders did their share in determining that their homes would be transformed (indeed, we have very few records of any who actively *resisted* the process), but so did politicians, landlords, industrialists, and managers of utilities.

These two concepts, the work process and the technological system, are the warp and the woof with which I hope to weave a description of the changes that occurred in the work that was done in American homes in the last one hundred years. The phrases are awkward, but the concepts that they denote are important. On one level, they seem to introduce complications that may be annoying; but on another level, they simplify descriptions and analyses so that certain essential features can emerge more clearly—to put it another way, they help us to see the forest through the trees. Housework is as difficult to study as it is to do. The student, like the houseworker, is hard pressed to decide where the activity begins and where it ends, what is essential and what is unessential, what is necessary and what is compulsive. If you are doing a time study of housewives, are you supposed to define the time they spend watching their children play in the park as leisure or as work? If you are trying to keep house yourself, is it really necessary to remove the chocolate stains from the front of a toddler's playsuit? The two problems have many conceptual similarities. I have found that the dual notions of a work process and a technological system have helped me to deal effectively with the scholarly problem of thinking and writing about the history of housework.

## NOTES

1 This is one of the many interesting insights about housework which can be derived from reading the Marxist debate about the relations between household and market labor. For a brief introduction to this literature, see Eli Zaretsky *Capitalism, the Family and Personal Life* (New York, 1974); Margaret Benston, "The Political Economy of Women's Liberation," *Monthly Review* 21 (September, 1969): 13–27; Lise Vogel, "The Earthly Family," *Radical America* 7 (July, 1973): 9–50; Wally Seccombe, "The Housewife and Her Labor under Capitalism," *New Left Review* 83 (January, 1974): 3–24.

2 I am using the term *alienation* here in the psychosocial sense of "strangeness."

3 Recent historical studies of market labor include Harry Braverman, *Labor and Monopoly Capital: The Degradation of Work in the Twentieth Century* (New York, 1975); Herbert G. Gutman, *Work, Culture, and Society in Industrializing America* (New York, 1976); Susan E. Hirsch, *Roots of the American Working Class* (Philadelphia, 1978); David Montegomery, *Workers' Control in America* (New York, 1979). For a sound introduction to the history of housework, see Susan Strasser, *Never Done: A History of American Housework* (New York, 1982).

PART **FIVE**

# PATTERNS OF TECHNOLOGICAL HISTORY

## INTRODUCTION

What is the relation of contemporary technology to the technologies of the past? Do changes in technological paradigms, including our attitudes toward technology and the metaphors and myths we use to characterize it, emerge from hardware innovations, or do such innovations arise from our changing metaphors, myths, and paradigms? Does the historical development of technology exhibit recognizable stages?

For the late Spanish philosopher José Ortega y Gasset, an understanding of the patterns of technological history is based not on a chronicle of technological inventions but on an examination of humanity's changing consciousness of its own making and doing. He argues that there have been three such stages, which he calls the technology of chance, the technology of the craftsman, and the technology of the technician.

During early human history, he argues, technological innovation was largely perceived either as chance occurrence or as a gift of the gods. Most major mythic systems contain such accounts of the provenance of tools and methods. A good example is the account in Greek mythology of the theft of fire from the gods and its gift to humankind by Prometheus, discussed by Autumn Stanley in her essay in this section. Human beings during this stage were only minimally aware of themselves as creators and inventors.

But at some point, when a sufficient store of technological skills had been accumulated, technology came to be thought of as a body of knowledge, or "know-how," to be preserved. Human beings at this stage were still not aware of technology in the abstract but identified all technology with the work of this or that artisan. Ortega thinks that this second stage conflates two essentially different aspects of technology. The first of these is what he calls "method or procedure," and the second is the execution of a plan. The reason for confusion of these two aspects of technology was that they were consolidated in the work of the craftsman, who, during this stage, was both technician and worker.

In its current phase, the technology of the technician, methodology and execution have become the proper tasks of the engineer and the worker, respectively. Moreover, there is a distinct awareness on the part of the technician of his or her own inventive power within an abstract system of invention called "technology." Ortega mentions Socrates in this context and characterizes his task as an attempt to get the people of his time to understand that beyond the concrete work of the craftsman there is an abstract or reified sense of technology that must be acknowledged.

Ortega warns of certain dangers that he sees emerging as a part of our current technological phase. For one thing, technology has become a "second nature"—we have become too comfortable within our technological cocoon. He suggests that we are in danger of becoming "denaturalized." We are losing our roots in nature even before we have secured our place within the sphere of technology. He is also concerned that we will forget that human life is a form of production and that cognition, science, and theory have their place in human activities only as a part of our ongoing production of ourselves and our institutions, that is, in the context of our "autofabrication."

In an essay that stands in sharp contrast to Ortega's, Lynn White, Jr., invites us to consider the concrete technological inventions which he thinks have conditioned our more abstract conceptions of technology. He rejects as "rubbish" much of the literature of technological history, pointing to its lack of concrete data to buttress its theories.

White is interested in tracing the development of concrete artifacts, among them the alphabet, coinage, the pennon (the medieval precursor of the flag), the stirrup, the crank, the trebuchet (a medieval fighting machine), and even the button. He links the very rise of philosophy in Ionia in the sixth century B.C. to the prosperity and leisure that resulted from the development of coinage. He even suggests that the metaphors which influenced the first attempts at physics may ultimately be technological, that when Heraclitus claimed that all things may be reduced to fire and fire to all things, just as gold may be turned into all things and all things into gold, he was drawing on a metaphor that had its roots in the technology of coinage.

White proposes that if historians and philosophers of technology wish to understand the influence of concrete artifacts on human values, they should examine even such humble objects as the button. He argues that besides furnishing an excellent example of a solution for a complex technological problem that is simple only in retrospect, the importance of the invention of the button to human welfare would be attested to by the many generations of mothers in boreal climes who have attempted to prepare their children for the out-of-doors. He wonders if the importance for "human values" of the invention of the button is not fully the equal of the importance of the invention of the sonnet.

Another influential voice among historians of technology has been that of Lewis Mumford. In the material reprinted here, taken from his pioneering 1934 book *Technics and Civilization,* Mumford contends that technological change has tended to be the result not, as he puts it, "of some mystical inner drive of the Faustian soul" but of cultural "syncretism"—the collection, reorganization, and implementation of bits and pieces from diverse cultures and times. He warns us that "the notion that a handful of British inventors suddenly made the wheels hum in the eighteenth century is too crude even to dish up as a fairy tale to children."

Like Ortega, Mumford divides the history of technology into three overlapping and interpenetrating stages: the *eotechnic,* the *paleotechnic,* and the *neotechnic.* Each of these phases is identifiable in terms of its employment of a unique mix of material resources, sources of energy, and attitudes toward production.

Mumford places the eotechnic, or "dawn-age" of modern technology, between the years 1000 and 1750. Its material resources were primarily wood and stone; its sources of energy were wind, water, and animals; and its dominant attitude toward production was that of the craftsman. This was a period of enormous invention but of very little integration of those inventions with one another. Most of the major developments of the second technological stage, the paleotechnic, germinated during the eotechnic. Major eotechnic inventions include the iron horseshoe (which increased the horse's pulling power by increasing its grip), the modern form of the harness (which allowed horses and oxen to pull greater loads without choking themselves), glass, mechanical clocks, the telescope, cheap paper, the printing press, the magnetic compass, and perhaps most important of all, the scientific method.

During the late eotechnic period, however, from the fifteenth century to the eighteenth, the momentum for change increased dramatically. By 1750, industry passed into a new phase with new material resources, new sources of energy, and new attitudes toward production. Wood and stone gave way to coal and iron. Wind, water, and animal power gave way to steam. The independent craftsman all but vanished as more and more people worked in factories. Some historians have suggested that with the new machinery of the paleotechnic, a worker could perform nearly 1,000 times as much work as his or her eotechnic predecessor.

Although this was a period that saw the accumulation of great fortunes, the lives of factory workers were nasty, brutish, and short. As Mumford puts it, "People lived and died within sight of the coal pit or the cotton mill in which they spent from fourteen to sixteen hours of their daily life, lived and died without either memory or hope, happy for the crusts that kept them alive or the sleep that brought them the brief uneasy solace of dreams." This was an age of unregulated trade, unregulated enterprise, and unregulated exploitation of working men, women, and children. The later portion of this period was what American economist Milton Friedman has called "the golden age of capitalism."

Mumford finds high irony in the paleotechnic outlook. The watchword of the Victorians was "progress," and they thought their culture was its pinnacle. Seen in retrospect, however, their culture seems more like what Mumford calls "an upthrust into barbarism, aided by the very forces and interests which originally had been directed toward the conquest of the environment and the perfection of human culture.''

When did the paleotechnic end? No single date can be given. It did not even get under way in the United States until the 1850s or in the Soviet Union until the 1920s. For most of western Europe its end was probably marked by the close of the First World War in 1918, but there are some third world countries that have never experienced it to any significant extent and whose technology is still best described as eotechnic.

Mumford views the paleotechnic period as a kind of spur track onto which the engine of technological progress was shunted and from which it was able to be extricated only with great difficulty. In time, however, the paleotechnic materials, coal and iron, were replaced with synthetics of all sorts. Steam power was replaced with electricity. Automation began to effect a shift from factory economies to service economies. But Mumford warned in 1934 that the neotechnic period was just beginning and that the ideals of the paleotechnic still dominated the industry and politics of the western industrial democracies.

When Mumford wrote, great technological inventions such as the gasoline engine, the phonograph, the motion picture, and the airplane had already altered the cultural landscape. Since that time, the neotechnic period has seen the invention and development of television, computers, space-age alloys, and rockets to the moon. What lies ahead? We can already begin to see the outlines of enormous advances in medicine and biotechnology, in materials research, and in microelectronics.

If the essence of the neotechnic period turns on its application of the scientific method, that is, systematic invention conducted by means of objective, repeatable experimentation, then it is clear that we still have a long way to go. For there are vast areas of human experience—in politics, in religion, and in other social arrangements—where men and women have scarcely begun to think and act scientifically. What form might such applications take? The application of scientific thinking to the management of the world population explosion could have enormous political consequences. Scientific thinking in religion could vitiate the growing creationist attack on evolutionary theory and undercut the worldwide fundamentalist assault on technology. And scientific thinking applied to the problems of global pollution and economic disparities could produce new and stronger social alliances between the societies of the developed northern hemisphere and those of the developing south.

There are some who fear that the application of scientific thinking to human problems will lead to vast technocracies—societies run by men and women who are committed to narrow forms of technology assessment and who view themselves as elite decision makers. They recall the decade of the 1960s, when American technocrats waged destructive wars abroad and constructed unworkable, paternalistic projects at home. But science is nothing if not democratic: its hypotheses must be objective, testable, and public, and all interested and affected parties must be heard from before its data collection is complete. It is therefore possible to agree with those who fear technocratic social arrangements and at the same time to argue for more scientific technology as the only legitimate response to political, religious, and other social problems. It is not just that science offers the *best* method of solving problems: it is the *only* method—short of luck.

Besides the three-stage accounts of the history of technology offered by Ortega and Mumford, there are many others. In order to cast some light on the context in which Ortega and Mumford wrote and in which they must now be understood, I shall give a brief account of two other theories.

One of the first coherent three-stage views was advanced by Karl Marx. He argued that the history of technology is generally the history of economic relationships and, more specifically, that it is the history of the means of production and the social relations which form around those means. In order to survive, men and women must transform what they find in nature into objects of use. Since they cannot do this as individuals, they enter into social relations. History thus becomes an account of those relations, which he calls relations of production, and of the means of production which give rise to them and condition them.

In *The German Ideology,* Marx presents his three stages of production, as follows.[1] The first stage is the remote precapitalist past, which he in turn analyzes into three substages. First, there is the primitive communism of the tribe. Production consists of hunting, fishing, and simple agriculture. What little property there is is owned in common, and the tribal organization is generally patriarchal. During the second precapitalist stage, tribes form larger groups and towns are established. Private ownership of property begins, and there is a division of labor between citizens and slaves. The third and final precapitalist stage is the feudal system of the medieval Latin West. The producing classes are no longer slaves but serfs. They are bound to the land. But as the serfs are gradually freed from the land, they move to the towns and cities and become artisans and tradespeople. For the most part, their capital consists of a house and the tools of their trade. As their capital accumulates, merchant classes begin to form, and production sites grow into factories.

The rise of large-scale manufacture ushers in the second of Marx's major economic epochs, capitalism. Competition among various peoples for domination of markets for manufactured goods gives rise to nation-states and to the invention of new machines of production. Legislators and major capitalists in these states enact laws that have the effect of undercutting the livelihood of the artisans and small tradespeople: those who were once free tradespeople and artisans are forced into bondage to the new machines. A sharply defined class structure develops: capitalist owners (the bourgeoisie) and former aristocrats dominate a poverty-ridden and overworked laboring class (the proletariat).

Lewis Mumford has succinctly described this, Marx's second great stage of technology: "The first requirement for the factory system, then, was the castration of skill. The second was the discipline of starvation. The third was the closing up of alternative occupations by means of land-monopoly and dis-education."[2] One of the foremost apologists for this unfortunate situation was Andrew Ure. "By the infirmity of human nature," he wrote, "it happens that the more skillful the workman, the more self-willed and intractable he is apt to become, and of course the less fit and component of the mechanical system in which...he may do great damage to the whole."[3]

Marx thought that there were contradictions within the capitalist stage of technology that would eventually lead to its dissolution and to the development of a third historical stage.[4] For one thing, the very technological change which is necessary to the ever greater production on which the capitalist system rests renders that system unstable. New machines (means of production) engender new forms of association (relations of production) that demand further changes in the means of production. For another, overproduction demands wider, international markets, and this undercuts the nationalistic underpinnings of the bourgeoisie.

In socialism, Marx's third technological stage, private property, education, religion, and occupation would not determine social status. Ownership of the means of production would pass from those whose goal is the accumulation of capital into the hands of those who do the actual work of production. Means of production would be more humane, and relations of production would be more egalitarian.

It does not require a profound knowledge of the history of western industrial societies to see that many of the predictions made by Marx have come true. Since he wrote, the workweek has been shortened, women have received the vote, graduated taxation has helped enlarge the middle class, and religion plays an ever smaller role in public affairs. But it must also be noted that the same inhuman working conditions he described with such force still exist in many industrializing countries, that there are still societies in which women are chattel, that the economic disparity between rich and poor nations is still extreme, and that religious fundamentalism still constitutes a threat to democratic governments in many parts of the world.

Another important three-stage view of the history of technology was advanced in the 1960s by media theorist Marshall McLuhan. Whereas Marx had been interested in the means of production, McLuhan was interested in the means of communication. The remote past is for McLuhan the period before the invention of printing in the late fifteenth century. The immediate past is what he calls "the Gutenberg" era, named for the inventor of the printing press. The present and future are the result of a second communications revolution in which print is supplanted by electronics as dominant medium. For McLuhan, as for Marx, the remote past breaks naturally into three sections: the strictly oral-aural period before the invention of the alphabet and writing, the ideographic east, and the manuscript culture of medieval Europe.

The key to McLuhan's understanding of the differences manifested by these stages of the history of communications is what he, following Walter Ong, called "sense-ratios."[5] In what are perhaps his most important works, *The Gutenberg Galaxy*[6] and *Understanding Media,*[7] McLuhan put forth the thesis that western culture during the fifteenth century began to move from a nonlinear culture, in which information was transmitted by mouth and received by ear, to a linear one, in which visual metaphors were dominant. He argued that printing laid the conceptual groundwork for the main features of the paleotechnic period, including the railway and the assembly line. For McLuhan, visual metaphors focus and differentiate in ways that oral-aural metaphors do not. Nationalism, individualism, scientific objectivity, and industrial capitalism all owe their rise to the use of visual metaphors. In the post-Gutenberg era, however, visual metaphors are giving way to ones that are once oral-aural and electronic: these include instantaneity, simultaneity, all-information-all-the-time.

McLuhan's work has been viewed by many of his academic critics as difficult and provocative and by some as outrageous and contradictory. But there are also those, such as journalist Tom Wolfe and communications theorist Paul Levinson, who see in McLuhan's work a source of brave and daring insights that offer a wealth of opportunities for understanding and reinterpreting our technological milieu.

Levinson's essay "Toy, Mirror and Art: The Metamorphosis of Technological Culture" draws upon McLuhan's insights in order to construct a life history of the technological artifact. Not all artifacts go through all three developmental stages, since some atrophy from lack of interest or even from an inherent lack of versatility. The history of the automobile serves as a good example of Levinson's thesis. In its early days, the automobile was a toy for the rich, a tinkerer's delight, a technological artifact that only very daring and playful people bought and used. With Henry Ford's Model T, however, the automobile became a mirror of reality: it came in any color as long as it was black, it was easily adaptable to farm and commercial use, and it was designed for strength and durability—not for glamor or speed. By the late 1950s automobiles sported tail fins, massive arrays of chrome, and other nonutilitarian features. In our own time, the automobile functions in all three ways: one can buy a fiberglass kit (a toy) to assemble and install over a Volkswagen chassis; one can select a practical, fuel-efficient sedan; or one can customize an automobile to express his or her aesthetic sensibilities. The freeways of Los Angeles and Houston sport vans with elaborate scenes of deserts or tropical paradises painted on their sides, and both cities host "low-rider" festivals.

Levinson richly illustrates his thesis with many other examples of this three-stage developmental cycle of the technological artifact. He includes the cinema, the telephone, and the phonograph. He also ties his own work to that of other three-stage theorists, including not only McLuhan, but also Jean Piaget, Sigmund Freud, Arthur Koestler, and Walter Ong.

In the final essay in this section, Autumn Stanley invites us to reconsider Voltaire's claim (and the common perception) that "there have never been women inventors." She thinks that once we rid ourselves of certain biases, this judgment will no longer seem accurate. One of these biases is nested in our use of language. We tend to use active verbs to describe the technological activities of men ("the men *choose* the wood for their bows with care") but passive verbs to describe equally significant activities undertaken by women ("cooking is *done* in watertight baskets"). Her perceptive question is "How did the baskets get to be watertight?"

Another bias regards what we accept as *significant* technology. Among peoples with primitive technologies, for example, the work of the male hunter probably provided only 20 to 40 percent of the food of a group. The remaining 60 to 80 percent was provided by gathering and by primitive agriculture, which was a more reliable source of food. At a more technologically advanced level, Stanley finds the origins of the lever, the crank, and the flywheel within the domain of domestic instruments such as the digging stick, the hand mill, and the potter's wheel. These domestic instruments were most likely invented and developed by women; but because they were "domestic" inventions, they do not seem to count as "technological" inventions.

Stanley develops her thesis by means of a careful analysis of three technological innovations: fire, machines, and medicine. First, mythological evidence from the most diverse of human societies strongly connects women with the taming of fire. In one of the deviations from this pattern, the Greek myth of Prometheus's theft of fire from Hephaestus, Prometheus also steals mechanical skill from the goddess Athena, who shared a workshop with Hephaestus. Second, the origins of rotary motion, the conceptual basis for all modern machinery, has evident roots in women's work: their use of the spindle whorl (for spinning fibers into thread), the rotary quern (for milling grain), and the potter's wheel. Third, most healers in antiquity were women. They had extensive knowledge of herbs and they probably even developed the first surgical instruments. Long before Sir Alexander Fleming discovered the antibiotic properties of penicillin, a type of mold, women healers were applying moldy bread to treat wounds.

Special attention should be paid to the historical struggle between men and women for the domination of medical practice. The wise women whom the medieval Christian church called "witches" and sought to exterminate were often practitioners of effective herbal medicine that did not rely on current theological fashions, which were dominated almost exclusively by males. The intensity of their persecution may well be an indication of their threat to the "establishment," and, as Stanley points out, "innumerable precious medical secrets no doubt burned with these women at the stake."

Stanley concludes her essay with a long list of the contributions made to technology by women. Her conclusion is an undeniable one: the history of technology needs to be rethought and rewritten to give credit to some of its most innovative, and most neglected, figures. But it also needs to be rewritten so as to provide an understanding of the role of women as a group in the development, the maturation, and the transformation of technology.

If we are to make informed decisions about the future of technology, we must seek to understand its past. Some of the essays in this section offer interpretations of the history of technology that are speculative, and each exhibits its own unique emphasis. Taken together, they offer important clues for helping us locate ourselves within our technological milieu. The unspoken assumption of each is that the history of humankind will remain opaque to us as long as we neglect the patterns of technological history.

## NOTES

1 An accessible source of Marx's treatment of this subject is the excerpt from *The German Ideology* found in *The Marx-Engels Reader,* 2d ed., Robert C. Tucker (ed.), W. W. Norton & Co., New York, 1978, pp. 146–200.

**2** Lewis Mumford, *Technics and Civilization,* Harcourt, Brace and World, New York, 1934, p. 173.
**3** Quoted in Mumford, op. cit., p. 173.
**4** A good account of these contradictions can be found in *The Manifesto of the Communist Party,* by Karl Marx and Friedrich Engels, reprinted in *The Marx-Engels Reader,* edited by Robert C. Tucker. See note 1 above.
**5** See Walter J. Ong, S.J., *The Presence of the Word,* Simon and Schuster, New York, 1967.
**6** Marshall McLuhan, *The Gutenberg Galaxy,* New American Library, New York, 1962.
**7** Marshall McLuhan, *Understanding Media,* New American Library, New York, 1964.

## From "Man the Technician"

**José Ortega y Gasset**

The answers which have been given to the question, what is technology, are appallingly superficial; and what is worse, this cannot be blamed on chance. For the same happens to all questions dealing with what is truly human in human beings. There is no way of throwing light upon them until they are tackled in those profound strata from which everything properly human evolves. As long as we continue to speak of the problems that concern man as though we knew what man really is, we shall only succeed in invariably leaving the true issue behind. That is what happens with technology. We must realize into what fundamental depths our argument will lead us. How does it come to pass that there exists in the universe this strange thing called technology, the absolute cosmic fact of man the technician? If we seriously intend to find an answer, we must be ready to plunge into certain unavoidable profundities.

We shall then come upon the fact that an entity in the universe, man, has no other way of existing than by being in another entity, nature or the world. This relation of being one in the other, man in nature, might take on one of three possible aspects. Nature might offer man nothing but facilities for his existence in it. That would mean that the being of man coincides fully with that of nature or, what is the same, that man is a natural being. That is the case of the stone, the plant, and, probably, the animal. If it were that of man, too, he would be without necessities, he would lack nothing, he would not be needy. His desires and their satisfaction would be one and the same. He would wish for nothing that did not exist in the world and, conversely, whatever he wished for would be there of itself, as in the fairy tale of the magic wand. Such an entity could not experience the world as something alien to himself; for the world would offer him no resistance. He would be in the world as though he were in himself.

Or the opposite might happen. The world might offer to man nothing but difficulties, i.e., the being of the world and the being of man might be completely antagonistic. In this case the world would be no abode for man; he could not exist in it, not even for the fraction of a second. There would be no human life and, consequently, no technology.

The third possibility is the one that prevails in reality. Living in the world, man finds that the world surrounds him as an intricate net woven of both facilities and difficulties. Indeed, there are not many things in it which, potentially, are not both. The earth supports him, enabling him to lie down when he is tired and to run when he has to flee. A shipwreck will bring home to him the advantage of the firm earth—a thing grown humble from habitude. But the earth also means distance. Much earth may separate him from the spring when he is thirsty. Or the earth may tower above him as a steep slope that is hard to climb. This fundamental phenomenon—perhaps the most fundamental of all—that we are surrounded by both facilities and difficulties gives to the reality called human life its peculiar ontological character.

For if man encountered no facilities it would be impossible for him to be in the world, he would not exist, and there would be no problem. Since he finds facilities to rely on, his existence is possible. But this possibility, since he also finds difficulties, is continually challenged, disturbed, imperiled. Hence, man's existence is no passive being in the world; it is an unending struggle to accommodate himself in it. The stone is given its existence; it need not fight for being what it is—a stone in the field. Man has to be himself in spite of unfavorable circumstances; that means he has to make his own existence at every single moment. He is given the abstract possibility of existing, but not the reality. This he has to conquer hour after hour. Man must earn his life, not only economically but metaphysically.

And all this for what reason? Obviously—but this is repeating the same thing in other words—because man's being and nature's being do not fully coincide. Because man's being is made of such strange stuff as to be partly akin to nature and partly not, at once natural and extranatural, a kind of ontological centaur, half immersed in nature, half transcending it. Dante would have likened him to a boat drawn up on the beach with one end of its keel in the water and the other in the sand. What is natural in him is realized by itself; it presents no problem. That is precisely why man does not consider it his true being. His extranatural part, on the other hand, is not there from the outset and of itself; it is but an aspiration, a project of life. And this we feel to be our true being; we call it our personality, our self. Our extra- and antinatural portion, however, must not be interpreted in terms of any of the older spiritual philosophies. I am not interested now in the so-called spirit (*Geist*), a pretty confused idea laden with speculative wizardry.

If the reader reflects a little upon the meaning of the entity he calls his life, he will find that it is the attempt to carry out a definite program or project of existence. And his self—each man's self—is nothing but this devised program. All we do we do in the service of this program. Thus man begins by being something that has no reality, neither corporeal nor spiritual; he is a project as such, something which is not yet but aspires to be. One may object that there can be no program without somebody having it, without an idea, a mind, a soul, or whatever it is called. I cannot discuss this thoroughly because it would mean embarking on a course of philosophy. But I will say this: although the project of being a great financier has to be conceived of in an idea, "being" the project is different from holding the idea. In fact, I find no difficulty in thinking this idea but I am very far from being this project.

Here we come upon the formidable and unparalleled character which makes man unique in the universe. We are dealing—and let the disquieting strangeness of the case be well noted—with an entity whose being consists not in what it is already, but in what it is not yet, a being that consists in not-yet-being. Everything else in the world is what it is. An entity whose mode of being consists in what it is already, whose potentiality coincides at once with his reality, we call a "thing." Things are given their being ready-made.

In this sense man is not a thing but an aspiration, the aspiration to be this or that. Each epoch, each nation, each individual varies in its own way the general human aspiration.

Now, I hope, all terms of the absolute phenomenon called "my life" will be clearly understood. Existence means, for each of us, the process of realizing, under given conditions, the aspiration we are. We cannot choose the world in which to live. We find ourselves, without our previous consent, embedded in an environment, a here and now. And my environment is made up not only by heaven and earth around me, but by my own body and my own soul. I am not my body; I find myself with it, and with it I must live, be it handsome or ugly, weak or sturdy. Neither am I my soul; I find myself with it and must use it for the purpose of living although it may lack will power or memory and not be of much good. Body and soul are things; but I am a drama, if anything, an unending struggle to be what I have to be. The aspiration or program I am, impresses its peculiar profile on the world about me, and that world reacts to this impress, accepting or resisting it. My aspiration meets with hindrance or with furtherance in my environment.

At this point one remark must be made which would have been misunderstood before. What we call nature, circumstance, or the world is essentially nothing but a conjunction of favorable and adverse conditions encountered by man in the pursuit of this program. The three names are interpretations of ours; what we first come upon is the experience of being hampered or favored in living. We are wont to conceive of nature and world as existing by themselves, independent of man. The concept "thing" likewise refers to something that has a hard and fast being and has it by itself and apart from man. But I repeat, this is the result of an interpretative reaction of our intellect upon what first confronts us. What first confronts us has no being apart from and independent of us; it consists exclusively in presenting facilities and difficulties, that is to say, in what it is in respect to our aspiration. Only in relation to our vital program is something an obstacle or an aid. And according to the aspiration animating us the facilities and difficulties, making up our pure and fundamental environment, will be such or such, greater or smaller.

This explains why to each epoch and even to each individual the world looks different. To the particular profile of our personal project, circumstance answers with another definite profile of facilities and difficulties. The world of the businessman obviously is different from the world of the poet. Where one comes to grief, the other thrives; where one rejoices, the other frets. The two worlds, no doubt, have many elements in common, viz., those which correspond to the generic aspiration of man as a species. But the human species is incomparably less stable and more mutable than any animal species. Men have an intractable way of being enormously unequal in spite of all assurances to the contrary.

### LIFE AS AUTOFABRICATION—TECHNOLOGY AND DESIRES

From this point of view, human life, the existence of man, appears essentially problematic. To all other entities of the universe existence presents no problem. For existence means actual realization of an essence. It means, for instance, that "being a bull" actually occurs. A bull, if he exists, exists as a bull. For a man, on the contrary, to exist does not mean to exist at once as the man he is, but merely that there exists a possibility of, and an effort towards, accomplishing this. Who of us is all he should be and all he longs to be? In contrast to the rest of creation, man, in existing, has to make his existence. He has to solve the practical problem of transferring into reality the program that is himself. For this reason "my life" is pure task, a thing inexorably to be made. It is not given to me as a present; I have to make it. Life gives me much to do; nay, it is nothing save the "to do" it has in store for me. And this "to do" is not a thing, but action in the most active sense of the word.

## THE STAGES OF TECHNOLOGY—TECHNOLOGY OF CHANCE

The subject is difficult. It took me some time to decide upon the principle best suited to distinguish periods of technology. I do not hesitate to reject the one readiest to hand, viz., that we should divide the evolution according to the appearance of certain momentous and characteristic inventions. All I have said in this essay aims to correct the current error of regarding such or such a definite invention as the thing which matters in technology. What really matters and what can bring about a fundamental advance is a change in the general character of technology. No single invention is of such caliber as to bear comparison with the tremendous mass of the integral evolution. We have seen that magnificent advances have been achieved only to be lost again, whether they disappeared completely or whether they had to be rediscovered.

Nay more, an invention may be made sometime and somewhere and still fail to take on its true technical significance. Gunpowder and the printing press, unquestionably two discoveries of great pith and moment, were known in China for centuries without being of much use. It is not before the fifteenth century in Europe that gunpowder and the printing press, the former probably in Lombardy, the latter in Germany, became historical powers. With this in view, when shall we say they were invented? No doubt, they grew effective in history only when they appeared incorporated in the general body of late medieval technology, serving the purposes of the program of life operative in that age. Firearms and the printing press are contemporaries of the compass. They all bear the same marks, so characteristic, as we shall shortly see, of that hour between Gothic and Renaissance, the scientific endeavors of which culminated in Copernicus. The reader will observe that, each in its own manner, they establish contact between man and things at a distance from him. They belong to the instruments of the *actio in distans,* which is at the root of modern technology. The cannon brings distant armies into immediate touch with each other. The compass throws a bridge between man and the cardinal points. The printing press brings the solitary writer into the presence of the infinite orbit of possible readers.

The best principle of delimiting periods in technical evolution is, to my judgment, furnished by the relation between man and technology, in other words by the conception which man in the course of history held, not of this or that particular technology but of the technical function as such. In applying this principle we shall see that it not only clarifies the past, but also throws light on the question we have asked before: how could modern technology give birth to such radical changes, and why is the part it plays in human life unparalleled in any previous age?

Taking this principle as our point of departure we come to discern three main periods in the evolution of technology: technology of chance; technology of the craftsman; technology of the technician.

What I call technology of chance, because in it chance is the engineer responsible for the invention, is the primitive technology of pre- and protohistoric man and of the contemporary savage, viz., of the least-advanced groups of mankind—as the Vedas in Ceylon, the Semang in Borneo, the pigmies in New Guinea and Central Africa, the Australian Negroes, etc.

How does primitive man conceive technology? The answer is easy. He is not aware of his technology as such; he is unconscious of the fact that there is among his faculties one which enables him to refashion nature after his desires.

The repertory of technical acts at the command of primitive man is very small and does not form a body of sufficient volume to stand out against, and be distinguished from, that of his natural acts, which is incomparably more important. That is to say, primitive man is very little man and almost all animal. His technical acts are scattered over and merged into the totality of his natural acts and appear to him as part of his natural life. He finds himself with the ability to light a fire as he finds himself with the ability to walk, swim, use his arms... His natural acts are a given stock fixed once and for all; and so are his technical. It does not occur to him that technology is a means of virtually unlimited changes and advances.

The simplicity and scantiness of these pristine technical acts account for their being executed indiscriminately by all members of the community, who all light fires, carve bows and arrows, and so forth. The one differentiation noticeable very early is that women perform certain technical functions and men certain others. But that does not help primitive man to recognize technology as an isolated phenomenon. For the repertory of natural acts is also somewhat different in men and women. That the woman should plow the field—it was she who invented agriculture—appears as natural as that she should bear the children.

Nor does technology at this stage reveal its most characteristic aspect, that of invention. Primitive man is unaware that he has the power of invention; his inventions are not the result of a premeditated and deliberate search. He does not look for them; they seem rather to look for him. In the course of his constant and fortuitous manipulation of objects he may suddenly and by mere chance come upon a new useful device. While for fun or out of sheer restlessness he rubs two sticks together a spark springs up, and a vision of new connections between things will dawn upon him. The stick, which hitherto has served as weapon or support, acquires the new aspect of a thing producing fire. Our savage will be awed, feeling that nature has inadvertently loosed one of its secrets before him. Since fire had always seemed a godlike power, arousing religious emotions, the new fact is prone to take on a magic tinge. All primitive technology smacks of magic. In fact, magic, as we shall shortly see, is nothing but a kind of technology, albeit a frustrated and illusory one.

Primitive man does not look upon himself as the inventor of his inventions. Invention appears to him as another dimension of nature, as part of nature's power to furnish him—nature furnishing man, not man nature—with certain novel devices. He feels no more responsible for the production of his implements than for that of his hands and feet. He does not conceive of himself as *homo faber*. He is therefore very much in the same situation as Mr. Koehler's monkey when it suddenly notices that the stick in his hands may serve an unforeseen purpose. Mr. Koehler calls this the "aha-impression" after the exclamation of surprise a man utters when coming upon a startling new relation between things. It is obviously a case of the biological law of trial and error applied to the mental sphere. The infusoria "try" various movements and eventually find one with favorable effects on them which they consequently adopt as a function.

The inventions of primitive man, being, as we have seen, products of pure chance, will obey the laws of probability. Given the number of possible independent combinations of things, a certain possibility exists of their presenting themselves some day in such an arrangement as to enable man to see performed in them a future implement.

## TECHNOLOGY AS CRAFTSMANSHIP—TECHNOLOGY OF THE TECHNICIAN

We come to the second stage, the technology of the artisan. This is the technology of Greece, of preimperial Rome, and of the Middle Ages. Here are in swift enumeration some of its essential features.

The repertory of technical acts has grown considerably. But—and this is important—a crisis and setback, or even the sudden disappearence of the principal industrial arts, would not yet be a fatal blow to material life in these societies. The life people lead with all these technical comforts and the life they would have to lead without them are not so radically different as to bar, in case of failures or checks, retreat to a primitive or almost primitive existence. The proportion between the technical and the nontechnical is not yet such as to make the former indispensable for the supporting of life. Man is still relying mainly on nature. At least, and that is what matters, so he himself feels. When technical crises arise he does therefore not realize that they will hamper his life, and consequently fails to meet them in time and with sufficient energy.

Having made this reservation we may now state that technical acts have by this time enormously increased both in number and in complexity. It has become necessary for a definite group of people to take them up systematically and make a full-time job of them. These people are the artisans. Their existence is bound to help man become conscious of technology as an independent entity. He sees the craftsman at work—the cobbler, the blacksmith, the mason, the saddler—and therefore comes to think of technology in terms and in the guise of the technician, the artisan. That is to say, he does not yet know that there is technology, but he knows that there are technicians who perform a peculiar set of activities which are not natural and common to all men.

Socrates in his struggle, which is so appallingly modern, with the people of his time began by trying to convince them that technology is not the same as the technician, that it is an abstract entity of its own not to be mixed up with this or that concrete man who possesses it.

At the second stage of technology everybody knows shoemaking to be a skill peculiar to certain men. It can be greater or smaller and suffer slight variations as do natural skills, running for instance, or swimming or, better still, the flying of a bird, the charging of a bull. That means shoemaking is now recognized as exclusively human and not natural, i.e., animal; but it is still looked upon as a gift granted and fixed once and for all. Since it is something exclusively human it is extranatural, but since it is something fixed and limited, a definite fund not admitting of substantial amplification, it partakes of nature; and thus technology belongs to the nature of man. As man finds himself equipped with the unexchangeable system of his bodily movements, so he finds himself equipped with the fixed system of the "arts." For this is the name technology bears in nations and epochs living on the technical level in question; and this also is the original meaning of the Greek word *techne*.

The way technology progresses might disclose that it is an independent and, in principle, unlimited function. But, oddly enough, this fact becomes even less apparent in this than in the primitive period. After all, the few primitive inventions, being so fundamental, must have stood out melodramatically against the workaday routine of animal habits. But in craftsmanship there is no room whatever for a sense of invention. The artisan must learn thoroughly in long apprenticeship—it is the time of masters and apprentices—elaborate usages handed down by long tradition. He is governed by the norm that man must bow to tradition as such. His mind is turned towards the past and closed to novel possibilities. He follows the established routine. Even such modifications and improvements as may be brought about in his craft through continuous and therefore imperceptible shifts present themselves not as fundamental novelties, but rather as differences of personal style and skill. And these styles of certain masters again will spread in the forms of schools and thus retain the outward character of tradition.

We must mention another decisive reason why the idea of technology is not at this time separated from the idea of the person who practices it. Invention has as yet produced only tools and not machines. The first machine in the strict sense of the word—and with it I anticipate the third period—was the weaving machine set up by Robert in 1825. It is the first machine because it is the first tool that works by itself, and by itself produces the object. Herewith technology ceases to be what it was before, handiwork, and becomes mechanical production. In the crafts the tool works as a complement of man; man with his natural actions continues to be the principal agent. In the machine the tool comes to the fore, and now it is no longer the machine that serves man but man who waits on the machine. Working by itself, emancipated from man, the machine, at this stage, finally reveals that technology is a function apart and highly independent of natural man, a function which reaches far beyond the bounds set for him. What a man can do with his fixed animal activities we know beforehand; his scope is limited. But what the machine man is capable of inventing may do, is in principle unlimited.

One more feature of craftsmanship remains to be mentioned which helps to conceal the true character of technology. I mean this: technology implies two things. First, the invention of a plan of activity, of a method or procedure—*mechane,* said the Greeks—and, secondly, the execution of this plan. The former is technology strictly speaking, the latter consists merely in handling the raw material. In short, we have the technician and the worker who between them, performing very different functions, discharge the technical job. The craftsman is both technician and worker; and what appears first is a man at work with his hands, and what appears last, if at all, is the technology behind him. The dissociation of the artisan into his two ingredients, the worker and the technician, is one of the principal symptoms of the technology of the third period.

We have anticipated some of the traits of this technology. We have called it the technology of the technician. Man becomes clearly aware that there is a capacity in him which is totally different from the immutable activities of his natural or animal part. He realizes that technology is not a haphazard discovery, as in the primitive period; that it is not a given and limited skill of some people, the artisans, as in the second period; that it is not this or that definite and therefore fixed "art"; but that it is a source of practically unlimited human activity.

This new insight into technology as such puts man in a situation radically new in his whole history and in a way contrary to all he has experienced before. Hitherto he has been conscious mainly of all the things he is unable to do, i.e., of his deficiencies and limitations. But the conception our time holds of technology—let the reader reflect a moment on his own—places us in a really tragicomic situation. Whenever we imagine some utterly extravagant feat, we catch ourselves in a feeling almost of apprehension lest our reckless dream—say a voyage to the stars—should come true. Who knows but that tomorrow morning's paper will spring upon us the news that it has been possible to send a projectile to the moon by imparting to it a speed great enough to overcome the gravitational attraction. That is to say, present-day man is secretly frightened by his own omnipotence. And this may be another reason why he does not know what he is. For finding himself in principle capable of being almost anything makes it all the harder for him to know what he actually is.

In this connection I want to draw attention to a point which does not properly belong here, that technology for all its being a practically unlimited capacity will irretrievably empty the lives of those who are resolved to stake everything on their faith in it and it alone. To be an engineer and nothing but an engineer means to be potentially everything and actually nothing. Just because of its promise of unlimited possibilities technology is an empty form like the most formalistic logic and is unable to determine the content of life. That is why our time, being the most intensely technical, is also the emptiest in all human history.

### RELATION BETWEEN MAN AND TECHNOLOGY IN OUR TIME—THE ENGINEER IN ANTIQUITY

This third stage of technical evolution, which is our own, is characterized by the following features:

Technical acts and achievements have increased enormously. Whereas in the Middle Ages—the era of the artisan—technology and the nature of man counterbalanced each other and the conditions of life made it possible to benefit from the human gift of adapting nature to man without denaturalizing man, in our time the technical devices outweigh the natural ones so gravely that material life would be flatly impossible without them. This is no manner of speaking, it is the literal truth. In *The Revolt of the Masses* I drew attention to the most noteworthy fact that the population of Europe between 500 and 1800 A.D., i.e., for thirteen centuries, never exceeded 180 million; whereas by now, in little over a century, it has reached 500 million, not counting those who have emigrated to America. In one century it has grown nearly three and a half times its size. If today 500 million people can live well in a space where 180 lived badly before, it is evident that, whatever the minor causes, the immediate cause and most necessary condition is the perfection of technology. Were technology to suffer a setback, millions of people would perish.

Such fecundity of the human animal could occur only after man had succeeded in interposing between himself and nature a zone of exclusively technical provenance, solid and thick enough to form something like a supernature. Present-day man—I refer not to the individual but to the totality of men—has no choice of whether to live in nature or to take advantage of this supernature. He is as irremediably dependent on, and lodged in, the latter as primitive man is in his natural environment. And that entails certain dangers. Since present-day man, as soon as he opens his eyes to life, finds himself surrounded by a superabundance of technical objects and procedures forming an artificial environment of such compactness that primordial nature is hidden behind it, he will tend to believe that all these things are there in the same way as nature itself is there without further effort on his part: that aspirin and automobiles grow on trees like apples. That is to say, he may easily lose sight of technology and of the conditions—the moral conditions, for example—under which it is produced and return to the primitive attitude of taking it for the gift of nature which is simply there. We thus have the curious fact that, at first, the prodigious expansion of technology made it stand out against the sober background of man's natural activities and allowed him to gain full sight of it, whereas by now its fantastic progress threatens to obscure it again.

Another feature helping man to discover the true character of his own technology we found to be the transition from mere tools to machines, i.e., mechanically working apparatus. A modern factory is a self-sufficient establishment waited on occasionally by a few persons of very modest standing. In consequence, the technician and the worker, who were united in the artisan, have been separated and the technician has grown to be the live expression of technology as such—in a word, the engineer.

Today technology stands before our mind's eye for what it is, apart, unmistakable, isolated, and unobscured by elements other than itself. And this enables certain persons, called engineers, to devote their lives to it. In the paleolithic age or in the Middle Ages technology, that is invention, could not have been a profession because man was ignorant of his own inventive power. Today the engineer embraces as one of the most normal and firmly established forms of activity the occupation of inventor. In contrast to the savage, he knows before he begins to invent that he is capable of doing so, which means that he has "technology" before he has "a technology." To this degree and in this concrete sense our previous assertion holds that technologies are nothing but concrete realizations of the general technical function of man. The engineer need not wait for chances and favorable odds; he is sure to make discoveries. How can he be?

The question obliges us to say a word about the technique of technology. To some people technique and nothing else is technology. They are right in so far as without technique—the intellectual method operative in technical creation—there is no technology. But with technique alone there is none either. As we have seen before, the existence of a capacity is not enough to put that capacity into action.

I should have liked to talk at leisure and in detail about both present and past techniques of technology. It is perhaps the subject in which I myself am most interested. But it would have been a mistake to let our investigations gravitate entirely around it. Now that this essay is breathing its last I must be content to give the matter brief consideration—brief, yet, I hope, sufficiently clear.

No doubt, technology could not have expanded so gloriously in these last centuries, nor the machine have replaced the tool, nor the artisan have been split up into his components, the worker and the engineer, had not the method of technology undergone a profound transformation.

Our technical methods are radically different from those of all earlier technologies. How can we best explain the diversity? Perhaps through the following question: How would an engineer of the past, supposing he was a real engineer and his invention was not due to chance but deliberately searched for, go about his task? I will give a schematic and therefore exaggerated example which is, however, historical and not fictitious. The Egyptian architect who built the pyramid of Cheops was confronted with the problem of lifting stone blocks to the highest parts of the monument. Starting as he needs must from the desired end, namely to lift the stones, he looked around for devices to achieve this. "This," I have said, meaning he is concerned with the result as a whole. His mind is absorbed by the final aim in its integrity. He will therefore consider as possible means only such procedures as will bring about the total result at once, in one operation that may take more or less time but which is homogeneous in itself. The unbroken unity of the end prompts him to look for a similarly uniform and undifferentiated means. This accounts for the fact that in the early days of technology the instrument through which an aim is achieved tends to resemble the aim itself. Thus in the construction of the pyramid the stones are raised to the top over another pyramid, an earthen pyramid with a wider base and a more gradual slope, which abuts against the first. Since a solution found through this principle of similitude—*similia similibus*—is not likely to be applicable in many cases, the engineer has no general rule and method to lead him from the intended aim to the adequate means. All he can do is to try out empirically such possibilities as offer more or less hope of serving his purpose. Within the circle defined by his special problem he thus falls back into the attitude of the primitive inventor.

# The Act of Invention

**Lynn White, Jr.**

The rapidly growing literature on the nature of technological innovation and its relation to other activities is largely rubbish because so few of the relevant concrete facts have thus far been ascertained. It is an inverted pyramid of generalities, the apex of which is very nearly a void. The five plump volumes of *A History of Technology,*[1] edited under the direction of Charles Singer, give the layman a quite false impression of the state of knowledge. They are very useful as a starting point, but they are almost as much a codification of error as of sound information.[2] It is to be feared that the physical weight of these books will be widely interpreted as the weight of authority and that philosophers, sociologists, and others whose personal researches do not lead them into the details of specific technological items may continue to be deceived as to what is known.

Since man is a hypothesizing animal, there is no point in calling for a moratorium on speculation in this area of thought until more firm facts can be accumulated. Indeed, such a moratorium—even if it were possible—would slow down the growth of factual knowledge because hypothesis normally provokes counter-hypotheses, and then all factions adduce facts in evidence, often new facts. The best that we can do at present is to work hard to find the facts and then to think cautiously about the facts which have been found.

In view of our ignorance, then, it would seem wise to discuss the problems of the nature, the motivations, the conditioning circumstances, and the effects of the act of invention far less in terms of generality than in terms of specific instances about which something seems to be known.

1. The beginning of wisdom may be to admit that even when we know some facts in the history of technology, these facts are not always fully intelligible, i.e., capable of "explanation," simply because we lack adequate contextual information. The Chumash Indians of the coast of Santa Barbara County built plank boats which were unique in the pre-Columbian New World: their activity was such that the Spanish explorers of California named a Chumash village "La Carpintería."[3] A map will show that this tribe had a particular inducement to venture upon the sea: they were enticed by the largest group of off-shore islands along the Pacific Coast south of Canada. But why did the tribes of South Alaska and British Columbia, of Araucanian Chile, or of the highly accidented eastern coast of the United States never respond to their geography by building plank boats? Geography would seem to be only one element in explanation.

Can a plank-built East Asian boat have drifted on the great arc of currents in the North Pacific to the Santa Barbara region? It is entirely possible; but such boats would have been held together by pegs, whereas the Chumash boats were lashed, like the dhows of the Arabian Sea or like the early Norse ships. Diffusion seems improbable.

Since a group can conceive of nothing which is not first conceived by a person, we are left with the hypothesis of a genius: a Chumash Indian who at some unknown date achieved a break-away from log dugout and reed balsa to the plank boat. But the idea of "genius" is itself an ideological artifact of the age of the Renaissance when painters, sculptors, and architects were trying to raise their social status above that of craftsmen.[4] Does the notion of genius "explain" Chumash plank boats? On the contrary, it would seem to be no more than a traditionally acceptable way of labeling the great Chumash innovation as unintelligible. All we can do is to observe the fact of it and hope that eventually we may grasp the meaning of it.

2. A symbol of the rudimentary nature of our thinking about technology, its development, and its human implications, is the fact that while the *Encyclopaedia Britannica* has an elaborate article on "Alphabet," it contains no discussion of its own organizational presupposition, alphabetization. Alphabetization is the basic invention for the classification and recovery of information: it is fully comparable in significance to the Dewey decimal system and to the new electronic devices for these purposes. Modern big business, big government, big scholarship are inconceivable without alphabetization. One hears that the chief reason why the Chinese Communist regime has decided to Romanize Chinese writing is the inefficiency of trying to classify everything from telephone books to tax registers in terms of 214 radicals of ideographs. Yet we are so blind to the nature of our technical equipment that the world of Western scholars, which uses alphabetization constantly, has produced not even the beginning of a history of it.

Fortunately, Dr. Sterling Dow of Harvard University is now engaged in the task. He tells me that the earliest evidence of alphabetization is found in Greek materials of the third century B.C. In other words, there was a thousand-year gap between the invention of the alphabet as a set of phonetic symbols and the realization that these symbols, and their sequence in individual written words, could be divorced from their phonetic function and used for an entirely different purpose: an arbitrary but very useful convention for storage and retrieval of verbal materials. That we have neglected thus completely the effort to understand so fundamental an invention should give us humility whenever we try to think about the larger aspects of technology.

3. Coinage was one of the most significant and rapidly diffused innovations of Late Antiquity. The dating of it has recently become more conservative than formerly: the earliest extant coins were sealed into the foundation of the temple of Artemis at Ephesus c. 600 B.C., and the invention of coins, i.e., lumps of metal the value of which is officially certified, was presumably made in Lydia not more than a decade earlier.[5]

Here we seem to know something, at least until the next archaeological spades turn up new testimony. But what do we know with any certainty about the impact of coinage? We are compelled to tread the slippery path of *post hoc ergo propter hoc*. There was a great acceleration of commerce in the Aegean, and it is hard to escape the conviction that this movement, which is the economic presupposition of the Periclean Age, was lubricated by the invention of coinage.

If we dare to go this far, we may venture further. Why did the atomic theory of the nature of matter appear so suddenly among the philosophers of the Ionian cities? Their notion that all things are composed of different arrangements of identical atoms of some "element," whether water, fire, ether, or something else, was an intellectual novelty of the first order, yet its sources have not been obvious. The psychological roots of atomism would seem to be found in the saying of Heraclitus of Ephesus that "all things may be reduced to fire, and fire to all things, just as all goods may be turned into gold and gold into all goods."[6] He thought that he was just using a metaphor, but the metaphor had been possible for only a century before he used it.

Here we are faced with a problem of critical method. Apples had been dropping from trees for a considerable period before Newton discovered gravity:[7] we must distinguish cause from occasion. But the appearance of coinage is a phenomenon of a different order from the fall of an apple. The unprecedented element in the general life of sixth-century Ionia, the chief stimulus to the prosperity which provided leisure for the atomistic philosophers, was the invention of coinage: the age of barter was ended. Probably no Ionian was conscious of any connection between this unique new technical instrument and the brainstorms of the local intellectuals. But that a causal relationship did exist can scarcely be doubted, even though it cannot be "proved" but only perceived.

4. Fortunately, however, there are instances of technological devices of which the origins, development, and effects outside the area of technology are quite clear. A case in point is the pennon.[8]

The stirrup is first found in India in the second century B.C. as the big-toe stirrup. For climatic reasons its diffusion to the north was blocked, but it spread wherever India had contact with barefoot aristocracies, from the Philippines and Timor on the east to Ethiopia on the west. The nuclear idea of the stirrup was carried to China on the great Indic culture wave which also spread Buddhism to East Asia, and by the fifth century the shod Chinese were using a foot stirrup.

The stirrup made possible, although it did not require, a new method of fighting with the lance. The unstirrupped rider delivered the blow with the strength of his arm. But stirrups, combined with a saddle equipped with pommel and cantle, welded rider to horse. Now the warrior could lay his lance at rest between his upper arm and body: the blow was delivered not by the arm but by the force of a charging stallion. The stirrup thus substituted horse-power for man-power in battle.

The increase in violence was tremendous. So long as the blow was given by the arm, it was almost impossible to impale one's foe. But in the new style of mounted shock combat, a good hit might put the lance entirely through his body and thus disarm the attacker. This would be dangerous if the victim had friends about. Clearly, a baffle must be provided behind the blade to prevent penetration by the shaft of the lance and thus permit retraction.

Some of the Central Asian peoples attached horse tails behind the blades of lances—this was probably being done by the Bulgars before they invaded Europe. Others nailed a piece of cloth, or pennon, to the shaft behind the blade. When the stirrup reached Western Europe c. 730 A.D., an effort was made to meet the problem by adapting to military purposes the old Roman boar-spear which had a metal crosspiece behind the blade precisely because boars, bears, and leopards had been found to be so ferocious that they would charge up a spear not so equipped.

This was not, however, a satisfactory solution. The new violence of warfare demanded heavier armor. The metal crosspiece of the lance would sometimes get caught in the victim's armor and prevent recovery of the lance. By the early tenth century Europe was using the Central Asian cloth pennon, since even if it got entangled in armor it would rip and enable the victor to retract his weapon.

Until our dismal age of camouflage, fighting men have always decorated their equipment. The pennons on lances quickly took on color and design. A lance was too long to be taken into a tent conveniently, so a knight usually set it upright outside his tent, and if one were looking for him, one looked first for the flutter of his familiar pennon. Knights riding held their lances erect, and since their increasingly massive armor made recognition difficult, each came to be identified by his pennon. It would seem that it was from the pennon that distinctive "connoissances" were transferred to shield and surcoat. And with the crystallization of the feudal structure, these heraldic devices became hereditary, the symbols of status in European society.

In battle, vassals rallied to the pennon of their liege lord. Since the king was, in theory if not always in practice, the culmination of the feudal hierarchy, his pennon took on a particular aura of emotion: it was the focus of secular loyalty. Gradually a distinction was made between the king's two bodies,[9] his person and his "body politic," the state. But a colored cloth on the shaft of a spear remained the primary symbol of allegiance to either body, and so remains even in polities which have abandoned monarchy. The grimly functional rags first nailed to lance shafts by Asian nomads have had a great destiny. But it is no more remarkable than that of the cross, a hideous implement in the Greco-Roman technology of torture, which was to become the chief symbol of the world's most widespread religion.

In tracing the history of the pennon, and of many other technological items, there is a temptation to convey a sense of inevitability. However, a novel technique merely offers opportunity; it does not command. As has been mentioned, the big-toe stirrup reached Ethiopia. It was still in common use there in the nineteenth century, but at the present time Muslim and European influences have replaced it with the foot stirrup. However, travellers tell me that the Ethiopian gentleman, whose horse is equipped with foot stirrups, rides with only his big toes resting in the stirrups.

5. Indeed, in contemplating the history of technology, and its implications for our understanding of ourselves, one is as frequently astonished by blindness to innovation as by the insights of invention. The Hellenistic discovery of the helix was one of the greatest of technological inspirations. Very quickly it was applied not only to gearing but also to the pumping of water by the so-called Archimedes screw.[10] Somewhat later the holding screw appears in both Roman and Germanic metal work.[11] The helix was taken for granted thenceforth in western technology. Yet Joseph Needham of Cambridge University assures me that, despite the great sophistication of the Chinese in most technical matters, no form of helix was known in East Asia before modern times: it reached India but did not pass the Himalayas. Indeed, I have not been able to locate any such device in the Far East before the early seventeenth century when Archimedes screws, presumably introduced by the Portuguese, were used in Japanese mines.[12]

6. Next to the wheel, the crank is probably the most important single element in machine design, yet until the fifteenth century the history of the crank is a dis-

mal record of inadequate vision of its potentialities.[13] It first appears in China under the Han dynasty, applied to rotary fans for winnowing hulled rice, but its later applications in the Far East were not conspicuous. In the West the crank seems to have developed independently and to have emerged from the hand quern. The earliest querns were fairly heavy, with a handle, or handles, inserted laterally in the upper stone, and the motion was reciprocating. Gradually the stones grew lighter and thinner, so that it was harder to insert the peg-handle horizontally: its angle creeps upward until eventually it stands vertically on top. All the querns found at the Saalburg had horizontal handles, and it is increasingly clear that the vertical peg is post-Roman.

Seated before a quern with a single vertical handle, a person of the twentieth century would give it a continuous rotary motion. It is far from clear that one of the very early Middle Ages would have done so. Crank motion was a kinetic invention more difficult than we can easily conceive. Yet at some point before the time of Louis the Pious the sense of the appropriate motion changed; for out of the rotary quern came a new machine, the rotary grindstone, which (as the Latin term for it, *mola fabri,* shows) is the upper stone of a quern turned on edge and adapted to sharpening. Thus, in Europe at least, crank motion was invented before the crank, and the crank does not appear before the early ninth century. As for the Near East, I find not even the simplest application of the crank until al-Jazarī's book on automata of 1206 A.D.

Once the simple crank was available, its development into the compound crank and connecting rod might have been expected quite quickly. Yet there is no sign of a compound crank until 1335, when the Italian physician of the Queen of France, Guido da Vigevano, in a set of astonishing technological sketches, which Rupert Hall has promised to edit,[14] illustrates three of them.[15] By the fourteenth century Europe was using crankshafts with two simple cranks, one at each end; indeed, this device was known in Cambodia in the thirteenth century. Guido was interested in the problem of self-moving vehicles: paddlewheel boats and fighting towers propelled by windmills or from the inside. For such constricted situations as the inside of a boat or a tower it apparently occurred to him to consolidate the two cranks at the ends of the crankshaft into a compound crank in its middle. It was an inspiration of the first order, yet nothing came of it. Evidently the Queen's physician, despite his technological interests, was socially too far removed from workmen to influence the actual technology of his time. The compound crank's effective appearance was delayed for another three generations. In the 1420's some Flemish carpenter or shipwright invented the bit-and-brace with its compound crank. By c. 1430 a German engineer was applying double compound cranks and connecting rods to machine design: a technological event as significant as the Hellenistic invention of gearing. The idea spread like wildfire, and European applied mechanics was revolutionized.

How can we understand the lateness of the discovery, whether in China or Europe, of even the simple crank, and then the long delay in its wide application and elaboration? Continuous rotary motion is typical of inorganic matter, whereas reciprocating motion is the sole movement found in living things. The crank connects these two kinds of motion; therefore we who are organic find that crank motion does not come easily to us. The great physicist and philosopher Ernst Mach noticed that infants find crank motion hard to learn.[16] Despite the rotary grindstone, even today razors are whetted rather than ground: we find rotary motion a bar to the greatest sensitivity. Perhaps as early as the tenth century the hurdy-gurdy was played with a cranked resined wheel vibrating the strings. But by the thirteenth century the hurdy-gurdy was ceasing to be an instrument for serious music. It yielded to the reciprocating fiddle bow, an introduction of the tenth century which became the foundation of modern European musical development. To use a crank, our tendons and muscles must relate themselves to the motion of galaxies and electrons. From this inhuman adventure our race long recoiled.

7. A sequence originally connected with the crank may serve to illustrate another type of problem in the act of technological innovation: the fact that a simple idea transferred out of its first context may have a vast expansion. The earliest appearance of the crank, as has been mentioned, is found on a Han-dynasty rotary fan to winnow husked rice.[17] The identical apparatus appears in the eighteenth century in the Palatinate,[18] in upper Austria and the Siebenbürgen,[19] and in Sweden.[20] I have not seen the exact channel of this diffusion traced, but it is clearly part of the general Jesuit-inspired *Chinoiserie* of Europe in that age. Similarly, I strongly suspect, but cannot demonstrate, that all subsequent rotary blowers, whether in furnaces, dehydrators, wind tunnels, air conditioning systems, or the simple electric fan, are descended from this Han machine which seems, in China itself, to have produced no progeny.

8. Doubtless when scholarship in the history of technology becomes firmer, another curious device will illustrate the same point. To judge by its wide distribution,[21] the fire piston is an old invention in Malaya. Dr. Thomas Kuhn of the University of California at Berkeley, who has made careful studies of the history of our knowledge of adiabatic heat, assures me that when the fire piston appeared in late eighteenth-century Europe not only for laboratory demonstrations but as a commercial product to light fires, there is no hint in the purely scientific publications that its inspiration was Malayan. But the scientists, curiously, also make no mention of the commercial fire pistons then available. So many Europeans, especially Portuguese and Netherlanders, had been trading, fighting, ruling, and evangelizing in the East Indies for so long a time before the fire piston is found in Europe, that it is hard to believe that the Malayan fire piston was not observed and reported. The realization of its potential in Europe was considerable, culminating in the diesel engine.

9. Why are such nuclear ideas sometimes not exploited in new and wider applications? What sorts of barriers prevent their diffusion? Why, at times, does what appeared to be a successful technological item fall into disuse? The history of the faggoted forging method of producing sword blades[22] may assist our thinking about such questions.

In late Roman times, north of the Alps, Celtic, Slavic, and Germanic metallurgists began to produce swords with laminations produced by welding together bundles of rods of different qualities of iron and steel, hammering the resulting strip thin, folding it over, welding it all together again, and so on. In this way a fairly long blade was produced which had the cutting qualities of steel but the toughness of iron. Although such swords were used at times by barbarian auxiliaries in the Roman army, the Roman legions never adopted them. Yet as soon as the Western Empire crumbled, the short Roman stabbing sword vanished and the laminated slashing blade alone held the field of battle. Can this conservatism in military equipment have been one reason for the failure of the Empire to stop the Germanic invasions? The Germans had adopted the new type of blade with enthusiasm, and by Carolingian times were manufacturing it in quantities in the Rhineland for export to Scandinavia and to Islam where it was much prized. Yet, although such blades were produced marginally as late as the twelfth century, for practical purposes they ceased to be used in Europe in the tenth century. Does the disappearance of such sophisticated swords indicate a decline in medieval metallurgical methods?

We should be cautious in crediting the failure of the Romans to adopt the laminated blade to pure stupidity. The legions seem normally to have fought in very close formation, shield to shield. In such a situation, only a stabbing sword could be effective. The Germans at times used a "shield wall" formation, but it was probably a bit more open than the Roman and permitted use of a slashing sword. If the Romans had accepted the new weapon, their entire drill and discipline would have been subject to revision. Unfortunately, we lack studies of the development of Byzantine weapons sufficiently detailed to let us judge whether, or to what extent, the vigorously surviving Eastern Roman Empire adapted itself to the new military technology.

The famous named swords of Germanic myth, early medieval epic and Wagnerian opera were laminated blades. They were produced by the vast patience and skill of smiths who themselves became legendary. Why did they cease to be made in any number after the tenth century? The answer is found in the rapid increase in the weight of European armor as a result of the consistent Frankish elaboration of the type of mounted shock combat made possible by the stirrup. After the turn of the millennium a sword in Europe had to be very nearly a club with sharp edges: the best of the earlier blades was ineffective against such defenses. The faggoted method of forging blades survived and reached its technical culmination in Japan[23] where, thanks possibly to the fact that archery remained socially appropriate to an aristocrat, mounted shock combat was less emphasized than in Europe and armor remained lighter.

10. Let us now turn to a different problem connected with the act of invention. How do methods develop by the transfer of ideas from one device to another? The origins of the cannon ball and the cannon may prove instructive.[24]

Hellenistic and Roman artillery was activated by the torsion of cords. This was reasonably satisfactory for summer campaigns in the Mediterranean basin, but north of the Alps and in other damper climates the cords tended to lose their resilience. In 1004 A.D. a radically different type of artillery appeared in China with the name *huo p'ao*. It consisted of a large sling-beam pivoted on a frame and actuated by men pulling in unison on ropes attached to the short end of the beam away from the sling. It first appears outside China in a Spanish Christian illumination of the early twelfth century, and from this one might assume diffusion through Islam. But its second appearance is in the northern Crusader army attacking Lisbon in 1147 where a battery of them were operated by shifts of one hundred men for each. It would seem that the Muslim defenders were quite unfamiliar with the new engine of destruction and soon capitulated. This invention, therefore, appears to have reached the West from China not through Islam but directly across Central Asia. Such a path of diffusion is the more credible because by the end of the same century the magnetic needle likewise arrived in the West by the northern route, not as an instrument of navigation but as a means of ascertaining the meridian, and Western Islam got the compass from Italy.[25] When the new artillery arrived in the West it had lost its name. Because of structural analogy, it took on a new name borrowed from a medieval instrument of torture, the ducking stool or *trebuchetum*.

Whatever its merits, the disadvantages of the *huo p'ao* were the amount of man-power required to operate it and the fact that since the gang pulling the ropes would never pull with exactly the same speed and force, missiles could not be aimed with great accuracy. The problem was solved by substituting a huge counterweight at the short end of the sling-beam for the ropes pulled by men. With this device a change in the weight of the caisson of stones or earth, or else a shift of the weight's position in relation to the pivot, would modify the range of the projectile and then keep it uniform, permitting concentration of fire on one spot in the fortifications to be breeched. Between 1187 and 1192 an Arabic treatise written in Syria for Saladin mentions not only Arab, Turkish, and Frankish forms of the primitive trebuchet, but also credits to Iran the invention of the trebuchet with swinging caisson. This ascription, however, must be in error; for from c. 1220 onward oriental sources frequently call this engine *magribī,* i.e., "Western." Moreover, while the counterweight artillery has not yet been documented for Europe before 1199, it quickly displaced the older forms of artillery in the West, whereas this new and more effective type of siege machinery became dominant in the Mameluke army only in the second half of the thirteenth century. Thus the trebuchet with counterweights would appear to be a European improvement on the *huo p'ao*. Europe's debt to China was repaid in 1272 when, if we may believe Marco Polo, he and a German technician, helped by a Nestorian Christian, delighted the Great Khan by building trebuchets which speedily reduced a besieged city.

But the very fact that the power of a trebuchet could be so nicely regulated impelled Western military engineers to seek even greater exactitude in artillery attack. They quickly saw that until the weight of projectiles and their friction with the air could be kept uniform, artillery aim would still be variable. As a result, as early as 1244 stones for trebuchets were being cut in the royal arsenals of England calibrated to exact specifications established by an engineer: in other words, the cannon ball before the cannon.

The germinal idea of the cannon is found in the metal tubes from which, at least by the late ninth century, the Byzantines had been shooting Greek fire. It may be that even that early they were also shooting rockets of Greek fire, propelled by the expansion of gases, from bazooka-like metal tubes. When, shortly before 673, the Greek-speaking Syrian refugee engineer Callinicus invented

Greek fire, he started the technicians not only of Byzantium but also of Islam, China, and eventually the West in search of ever more combustible mixtures. As chemical methods improved, the saltpeter often used in these compounds became purer, and combustion tended toward explosion. In the thirteenth century one finds, from the Yellow Sea to the Atlantic, incendiary bombs, rockets, firecrackers, and fireballs shot from tubes like Roman candles. The flame and roar of all this has made it marvelously difficult to ascertain just when gunpowder artillery, shooting hard missiles from metal tubes, appeared. The first secure evidence is a famous English illumination of 1327 showing a vase-shaped cannon discharging a giant arrow. Moreover, our next certain reference to a gun, a "pot de fer à traire garros de feu" at Rouen in 1338, shows how long it took for technicians to realize that the metal tube, gunpowder, and the calibrated trebuchet missile could be combined. However, iron shot appear at Lucca in 1341; in 1346 in England there were two calibres of lead shot; and balls appear at Toulouse in 1347.

The earliest evidence of cannon in China is extant examples of 1356, 1357, and 1377. It is not necessary to assume the miracle of an almost simultaneous independent Chinese invention of the cannon: enough Europeans were wandering the Yuan realm to have carried it eastward. And it is very strange that the Chinese did not develop the cannon further, or develop hand guns on its analogy. Neither India nor Japan knew cannon until the sixteenth century when they arrived from Europe. As for Islam, despite several claims to the contrary, the first certain use of gunpowder artillery by Muslims comes from Cairo in 1366 and Alexandria in 1376; by 1389 it was common in both Egypt and Syria. Thus there was roughly a forty-year lag in Islam's adoption of the European cannon.

Gunpowder artillery, then, was a complex invention which synthesized and elaborated elements drawn from diverse and sometimes distant sources. Its impact upon Europe was equally complex. Its influences upon other areas of technology such as fortification, metallurgy, and the chemical industries are axiomatic, although they demand much more exact analysis than they have received. The increased expense of war affected tax structures and governmental methods; the new mode of fighting helped to modify social and political relationships. All this has been self-evident for so long a time that perhaps we should begin to ask ourselves whether the obvious is also the true.

For example, it has often been maintained that a large part of the new physics of the seventeenth century sprang from concern with military ballistics. Yet there was continuity between the thought of Galileo or Newton and the fundamental challenge to the Aristotelian theory of impetus which appeared in Franciscus de Marchia's lectures at the University of Paris in the winter of 1319–20,[26] seven years before our first evidence of gunpowder artillery. Moreover, the physicists both of the fourteenth and of the seventeenth centuries were to some extent building upon the criticisms of Aristotle's theory of motion propounded by Philoponus of Alexandria in the age of Justinian, a time when I can detect no new technological stimulus to physical speculation. While most scientists have been aware of current technological problems, and have often talked in terms of them, both science and technology seem to have enjoyed a certain autonomy in their development.

It may well be that continued examination will show that many of the political, economic, and social as well as intellectual developments in Europe which have traditionally been credited to gunpowder artillery were in fact taking place for quite different reasons. But we know of one instance in which the introduction of firearms revolutionized an entire society: Japan.[27]

Metallurgical skills were remarkably high in Japan when, in 1543, the Portuguese brought both small arms and cannon to Kyushu. Japanese craftsmen quickly learned from the gunsmiths of European ships how to produce such weapons, and within two or three years were turning them out in great quantity. Military tactics and castle construction were rapidly revised. Nobunaga and his successor, Hideyoshi, seized the new technology of warfare and utilized it to unify all Japan under the shogunate. In Japan, in contrast to Europe, there is no ambiguity about the consequences of the arrival of firearms. But from this fact we must be careful not to argue that the European situation is equally clear if only we would see it so.

11. In examining the origins of gunpowder artillery, we have seen that its roots are multiple, but that all of them (save the European name *trebuchet*) lie in the soil of military technology. It would appear that each area of technology has a certain self-contained quality: borrowings across craft lines are not as frequent as might be expected. Yet they do occur, if exceptionally. A case in point is the fusee.

In the early fifteenth century clock makers tried to develop a portable mechanical timepiece by substituting a spring drive for the weight which powered stationary clocks. But this involved entirely new problems of power control. The weight on a clock exerted equal force at all times, whereas a spring exerts less force in proportion as it uncoils. A new escapement was therefore needed which would exactly compensate for this gradual diminution of power in the drive.

Two solutions were found, the stackfreed and the fusee, the latter being the more satisfactory. Indeed, a leading historian of horology has said of the fusee: "Perhaps no problem in mechanics has ever been solved so simply and so perfectly."[28] The date of its first appearance is much in debate, but we have a diagram of it from 1477.[29] The fusee equalizes the changing force of the mainspring by means of a brake of gut or fine chain which is gradually wound spirally around a conical axle, the force of the brake being dependent upon the leverage of the radius of the cone at any given point and moment. It is a device of great mechanical elegance. Yet the idea did not originate with the clock makers: they borrowed it from the military engineers. In Konrad Keyser's monumental, but still unpublished, treatise on the technology of warfare, *Bellifortis,* completed c. 1405, we find such a conical axle in an apparatus for spanning a heavy crossbow.[30] With very medieval humor, this machine was called "the virgin," presumably because it offered least resistance when the bow was slack and most when it was taut.

In terms of eleven specific technological acts, or sequences of acts, we have been pondering an abstraction, the act of technological innovation. It is quite possible that there is no such thing to ponder. The analysis of the nature of creativity is one of the chief intellectual commitments of our age. Just as the old unitary concept of "intelligence" is giving way to the notion that the individual's mental capacity consists of a large cluster of various and varying factors mutually affecting each other, so "creativity" may well be a lot of things and not one thing.

Thirteenth century Europe invented the sonnet as a poetic form and the functional button[31] as a means of making civilized life more nearly possible in boreal climes. Since most of us are educated in terms of traditional humanistic presuppositions, we value the sonnet but think that a button is just a button. It is doubtful whether the chilly northerner who invented the button could have invented the sonnet then being produced by his contemporaries in Sicily. It is equally doubtful whether the type of talent required to invent the rhythmic and phonic relationships of the sonnet-pattern is the type of talent needed to perceive the spatial relationships of button and buttonhole. For the button is not obvious until one has seen it, and perhaps not even then. The Chinese never adopted it: they got no further than to adapt the tie-cords of their costumes into elaborate loops to fit over cord-twisted knobs. When the Portuguese brought the button to Japan, the Japanese were delighted with it and took over not only the object itself but also its Portuguese name. Humanistic values, which have been cultivated historically by very specialized groups in quite exceptional circumstances, do not encompass sufficiently the observable human values. The billion or more mothers who, since the thirteenth century, have buttoned their children snugly against winter weather might perceive as much of spirituality in the button as in the sonnet and feel more personal gratitude to the inventor of the former than of the latter. And the historian, concerned not only with art forms but with population, public health, and what S. C. Gilfillan long ago identified as "the coldward course" of culture,[32] must not slight either of these very different manifestations of what would seem to be very different types of creativity.

There is, indeed, no reason to believe that technological creativity is unitary. The unknown Syrian who, in the first century B.C., first blew glass was doing something vastly different from his contemporary who was building the first water-powered mill. For all we now know, the kinds of ability required for these two great innovations are as different as those of Picasso and Einstein would seem to be.

The new school of physical anthropologists who maintain that *Homo* is *sapiens* because he is *faber,* that his biological differentiation from the other primates is best understood in relation to tool making, are doubtless exaggerating a provocative thesis. *Homo* is also *ludens, orans,* and much else. But if technology is defined as the systematic modification of the physical environment for human ends, it follows that a more exact understanding of technological innovation is essential to our self-knowledge.

## REFERENCES


1 (Oxford, 1954–58).

2 Cf. the symposium in *Technology and Culture,* I (1960), 299–414.

3 E. G. Gudde, *California Place Names,* 2nd ed. (Berkeley and Los Angeles, 1960), 52; A. L. Kroeber, "Elements of Culture in Native California," in *The California Indians,* ed. R. F. Heizer and M. A. Whipple (Berkeley and Los Angeles, 1951), 12–13.

**4** E. Zilsel, *Die Entstehung des Geniebegriffes* (Tübingen, 1926).

**5** E. S. G. Robinson, "The Date of the Earliest Coins," *Numismatic Chronicle,* 6th ser., XVI (1956), 4, 8, arbitrarily dates the first coinage c. 640–630 B.C. allowing "the Herodotean interval of a generation" for its diffusion from Lydia to the Ionian cities. But, considering the speed with which coinage appears even in India and China, such an interval is improbable.

D. Kagan, "Pheidon's Aeginetan Coinage," *Transactions and Proceedings of the American Philological Association,* XCI (1960), 121–136, tries to date the first coinage at Aegina before c. 625 B.C. when, he believes, Pheidon died; but the argument is tenuous. The tradition that Pheidon issued a coinage is late, and may well be no more than another example of the Greek tendency to invent culture-heroes. The date of Pheidon's death is uncertain: the belief that he died c. 625 rests solely on the fact that he is not mentioned by Strabo in connection with the war of c. 625–600 B.C.; but if Pheidon, then a very old man, was killed in a revolt of 620 (cf. Kagan's note 21) his participation in this long war would have been so brief and ineffective that Strabo's silence is intelligible.

**6** H. Diels, *Fragmente der Vorsokratiker,* 6th ed. (Berlin, 1951), 171 (B. 90).

**7** The story of the apple is authentic: Newton himself told William Stukeley that when "the notion of gravitation came into his mind [it] was occasion'd by the fall of an apple, as he sat in a contemplative mood"; cf. I. B. Cohen, "Newton in the Light of Recent Scholarship," *Isis,* LI (1960), 490.

**8** The materials on pennons, and other baffles behind the blade of a lance, are found in L. White, Jr., *Medieval Technology and Social Change* (Oxford, 1962), 8, 33, 147, 157.

**9** See the classic work of Ernst Kantorowicz, *The King's Two Bodies* (Princeton, 1957).

**10** W. Treue, *Kulturgeschichte der Schraube* (Munich, 1955), 39–43, 57, 109.

**11** F. M. Feldhaus, *Die Technik der Vorzeit, der Geschichtlichen Zeit und der Naturvölker* (Leipzig, 1914), 984–987.

**12** E. Treptow, "Der älteste Bergbau und seiner Hilfsmittel," *Beiträge zur Geschichte der Technik und Industrie,* VIII (1918), 181, fig. 48; C. N. Bromehead, "Ancient Mining Processes as Illustrated by a Japanese Scroll," *Antiquity,* XVI (1942), 194, 196, 207.

**13** For a detailed history of the crank, cf. White, op. cit., 103–115.

**14** A. R. Hall, "The Military Inventions of Guido da Vigevano," *Actes du VIII$^{e}$ Congrès International d'Histoire des Sciences* (Florence, 1958), 966–969.

**15** Bibliothèque Nationale, MS latin 11015, fols. 49$^{r}$, 51$^{v}$, 52$^{v}$. Singer, op. cit., II, figs. 594 and 659, illustrates the first and third of these, but with wrong indications of folio numbers.

**16** H. T. Horwitz, "Uber die Entwicklung der Fahigkeit zum Antreib des Kurbelmechanismus," *Geschichtsblätter fur Technik und Industrie,* XI (1927), 30–31.

**17** White, op. cit., 104 and fig. 4. For what may be a slightly earlier specimen, now in the Seattle Art Museum, see the catalogue of the exhibition *Arts of the Han Dynasty* (New York, 1961), No. 11, of the Chinese Art Society of America.

**18** I am so informed by Dr. Paul Leser of the Hartford Theological Foundation.

**19** L. Makkai, in *Agrártörténeti Szemle,* I (1957), 42.

**20** P. Leser, "Plow Complex; Culture Change and Cultural Stability," in *Man and Cultures: Selected Papers of the Fifth International Congress of Anthropological and Ethnological Sciences,* ed. A. F. C. Wallace (Philadelphia, 1960), 295.

**21** H. Balfour, "The Fire Piston," in *Anthropological Essays Presented to E. B. Tylor* (Oxford, 1907), 17–49.

**22** E. Salin, *La Civilisation Mérovingienne,* III (Paris, 1957), 6, 55–115.

**23** C. S. Smith, "A Metallographic Examination of Some Japanese Sword Blades," *Quaderno II del Centro per la Storia della Metallurgia* (1957), 42–68.

**24** White, op. cit., 96–103, 165.
**25** Ibid., 132.
**26** A. Maier, *Zwei Grundprobleme der scholastischen Naturphilosophie,* 2nd ed. (Rome, 1951), 165, n. 11.
**27** D. M. Brown, "The Impact of Firearms on Japanese Warfare, 1543–98," *Far Eastern Quarterly,* VII (1948), 236–253.
**28** G. Baillie, *Watches* (London, 1929), 85.
**29** Singer, op. cit., III, fig. 392.
**30** Göttingen University Library, Cod. phil. 63, fol. 76[v]; cf. F. M. Feldhaus, "Uber den Ursprung vom Federzug und Schnecke," *Deutsche Uhrmacher-Zeitung,* LIV (1930), 720–723.
**31** Some buttons were used in antiquity for ornament, but apparently not for warmth. The first functional buttons are found c. 1235 on the "Adamspforte" of Bamberg Cathedral, and in 1239 on a closely related relief at Bassenheim; cf. E. Panofsky, *Deutsche Plastik des 11. bis 13. Jahrhundert* (Munich 1924), pl. 74; H. Schnitsler, "Ein unbekanntes Reiterrelief aus dem Kreise des Naumburger Meisters," *Zeitschrift des Deutschen Vereins fur Kunstwissenschaft,* I (1935), 413, fig. 13.
**32** In *The Political Science Quarterly,* XXXV (1920), 393–410.

# From *Technics and Civilization*

**Lewis Mumford**

## TECHNICAL SYNCRETISM

Civilizations are not self-contained organisms. Modern man could not have found his own particular modes of thought or invented his present technical equipment without drawing freely on the cultures that had preceded him or that continued to develop about him.

Each great differentiation in culture seems to be the outcome, in fact, of a process of syncretism. Flinders Petrie, in his discussion of Egyptian civilization, has shown that the admixture which was necessary for its development and fulfillment even had a racial basis; and in the development of Christianity it is plain that the most diverse foreign elements—a Dionysian earth myth, Greek philosophy, Jewish Messianism, Mithraism, Zoroastrianism—all played a part in giving the specific content and even the form to the ultimate collection of myths and offices that became Christianity.

Before this syncretism can take place, the cultures from which the elements are drawn must either be in a state of dissolution, or sufficiently remote in time or space so that single elements can be extracted from the tangled mass of real institutions. Unless this condition existed the elements themselves would not be free, as it were, to move over toward the new pole. Warfare acts as such an agent of dissociation, and in point of time the mechanical renascence of Western Europe was associated with the shock and stir of the Crusades. For what the new civilization picks up is not the complete forms and institutions of a solid culture, but just those fragments that can be transported and transplanted: it uses inventions, patterns, ideas, in the way that the Gothic builders in England used the occasional stones or tiles of the Roman villa in combination with the native flint and in the entirely different forms of a later architecture. If the villa had still been standing and occupied, it could not have been conveniently quarried. It is the death of the original form, or rather, the remaining life in the ruins, that permits the free working over and integration of the elements of other cultures.

One further fact about syncretism must be noted. In the first stages of integration, before a culture has set its own definite mark upon the materials, before invention has crystallized into satisfactory habits and routine, it is free to draw upon the widest sources. The beginning and the end, the first absorption and the final spread and conquest, after the cultural integration has taken place, are over a worldwide realm.

These generalizations apply to the origin of the present-day machine civilization: a creative syncretism of inventions, gathered from the technical debris of other civilizations, made possible the new mechanical body. The waterwheel, in the form of the Noria, had been used by the Egyptians to raise water, and perhaps by the Sumerians for other purposes; certainly in the early part of the Christian era watermills had become fairly common in Rome. The windmill perhaps came from Persia in the eighth century. Paper, the magnetic needle, gunpowder, came from China, the first two by way of the Arabs: algebra came from India through the Arabs, and chemistry and physiology came via the Arabs, too, while geometry and mechanics had their origins in pre-Christian Greece. The steam engine owed its conception to the great inventor and scientist, Hero of Alexandria: it was the translations of his works in the sixteenth century that turned attention to the possibilities of this instrument of power.

In short, most of the important inventions and discoveries that served as the nucleus for further mechanical development, did not arise, as Spengler would have it, out of some mystical inner drive of the Faustian soul: they were windblown seeds from other cultures. After the tenth century in Western Europe the ground was, as I have shown, well plowed and harrowed and dragged, ready to receive these seeds; and while the plants themselves were growing, the cultivators of art and science were busy keeping the soil friable. Taking root in medieval culture, in a different climate and soil, these seeds of the machine sported and took on new forms: perhaps, precisely because they had *not* originated in Western Europe and had no natural enemies there, they grew as rapidly and gigantically as the Canada thistle when it made its way onto the South American pampas. But at no point—and this is the important thing to remember—did the machine represent a complete break. So far from being unprepared for in human history, the modern machine age cannot be understood except in terms of a very long and diverse preparation. The notion that a handful of British inventors suddenly made the wheels hum in the eighteenth century is too crude even to dish up as a fairy tale to children.

## THE TECHNOLOGICAL COMPLEX

Looking back over the last thousand years, one can divide the development of the machine and the machine civilization into three successive but *overlapping and interpenetrating phases:* eotechnic, paleotechnic, neotechnic. The demonstration that industrial civilization was not a single whole, but showed two marked, contrasting phases, was first made by Professor Patrick Geddes and published a generation ago. In defining the paleotechnic and neotechnic phases, he however neglected the important period of preparation, when all the key inventions were either invented or foreshadowed. So, following the archeological parallel he called attention to, I shall call the first period the eotechnic phase: the dawn age of modern technics.

While each of these phases roughly represents a period of human history, it is characterized even more significantly by the fact that it forms a technological complex. Each phase, that is, has its origin in certain definite regions and tends to employ certain special resources and raw materials. Each phase has its specific means of utilizing and generating energy, and its special forms of production. Finally, each phase brings into existence particular types of workers, trains them in particular ways, develops certain aptitudes and discourages others, and draws upon and further develops certain aspects of the social heritage.

Almost any part of a technical complex will point to and symbolize a whole series of relationships within that complex. Take the various types of writing pen. The goose-quill pen, sharpened by the user, is a typical eotechnic product: it indicates the handicraft basis of industry and the close connection with agriculture. Economically it is cheap; technically it is crude, but easily adapted to the style of the user. The steel pen stands equally for the paleotechnic phase: cheap and uniform, if not durable, it is a typical product of the mine, the steel mill and of mass production. Technically, it is an improvement upon the quill pen; but to approximate the same adaptability it must be made in half a dozen different standard points and shapes. And finally the fountain pen—though invented as early as the seventeenth century—is a typical neotechnic product. With its barrel of rubber or synthetic resin, with its gold pen, with its automatic action, it points to the finer neotechnic economy: and in its use of the durable iridium tip the fountain pen characteristically lengthens the service of the point and reduces the need for replacement. These respective characteristics are reflected at a hundred points in the typical environment of each phase; for though the various parts of a complex may be invented at various times, the complex itself will not be *in working order* until its major parts are all assembled. Even today the neotechnic complex still awaits a number of inventions necessary to its perfection: in particular an accumulator with six times the voltage and at least the present amperage of the existing types of cell.

Speaking in terms of power and characteristic materials, the eotechnic phase is a water-and-wood complex: the paleotechnic phase is a coal-and-iron complex, and the neotechnic phase is an electricity-and-alloy complex. It was Marx's great contribution as a sociological economist to see and partly to demonstrate that each period of invention and production had its own specific value for civilization, or, as he would have put it, its own historic mission. The machine cannot be divorced from its larger social pattern; for it is this pattern that gives it meaning and purpose. Every period of civilization carries within it the insignificant refuse of past technologies and the important germs of new ones: but the center of growth lies within its own complex.

The dawn-age of our modern technics stretches roughly from the year 1000 to 1750. During this period the dispersed technical advances and suggestions of other civilizations were brought together, and the process of invention and experimental adaptation went on at a slowly accelerating pace. Most of the key inventions necessary to universalize the machine were promoted during this period; there is scarcely an element in the second phase that did not exist as a germ, often as an embryo, frequently as an independent being, in the first phase. This complex reached its climax, technologically speaking, in the seventeenth century, with the foundation of experimental science, laid on a basis of mathematics, fine manipulation, accurate timing, and exact measurement.

The eotechnic phase did not of course come suddenly to an end in the middle of the eighteenth century: just as it reached its climax first of all in Italy in the sixteenth century, in the work of Leonardo and his talented contemporaries, so it came to a delayed fruition in the America of 1850. Two of its finest products, the clipper ship and the Thonet process of making bentwood furniture, date from the 1830s. There were parts of the world, like Holland and Denmark, which in many districts slipped directly from an eotechnic into the neotechnic economy, without feeling more than the cold shadow of the paleotechnic cloud.

With respect to human culture as a whole, the eotechnic period, though politically a chequered one, and in its later moments characterized by a deepening degradation of the industrial worker, was one of the most brilliant periods in history. For alongside its great mechanical achievements it built cities, cultivated landscapes, constructed buildings, and painted pictures, which fulfilled, in the realm of human thought and enjoyment, the advances that were being decisively made in the practical life. And if this period failed to establish a just and equitable polity in society at large, there were at least moments in the life of the monastery and the commune that were close to its dream: the afterglow of this life was recorded in More's Utopia and Andreae's Christianopolis.

Noting the underlying unity of eotechnic civilization, through all its superficial changes in costume and creed, one must look upon its successive portions as expressions of a single culture. This point is now being reenforced by scholars who have come to disbelieve in the notion of the gigantic break supposed to have been made during the Renaissance: a contemporary illusion, unduly emphasized by later historians. But one must add a qualification: namely, that with the increasing technical advances of this society there was, for reasons partly independent of the machine itself, a corresponding cultural dissolution and decay. In short, the Renaissance was not, socially speaking, the dawn of a new day, but its twilight. The mechanical arts advanced as the humane arts weakened and receded, and it was at the moment when form and civilization had most completely broken up that the tempo of invention became more rapid, and the multiplication of machines and the increase of power took place.

## NEW SOURCES OF POWER

At the bottom of the eotechnic economy stands one important fact: the diminished use of human beings as prime movers, and the separation of the production of energy from its application and immediate control. While the tool still dominated production energy and human skill were united within the craftsman himself: with the separation of these two elements the productive process itself tended toward a greater impersonality, and the machine-tool and the machine developed along with the new engines of power. If power machinery be a criterion, the modern industrial revolution began in the twelfth century and was in full swing by the fifteenth.

The eotechnic period was marked first of all by a steady increase in actual horsepower. This came directly from two pieces of apparatus: first, the introduction of the iron horseshoe, probably in the ninth century, a device that increased the range of the horse, by adapting him to other regions besides the grasslands, and added to his effective pulling power by giving his hoofs a grip. Second: by the tenth century the modern form of harness, in which the pull is met at the shoulder instead of at the neck, was reinvented in Western Europe—it had existed in China as early as 200 B.C.—and by the twelfth century, it had supplanted the inefficient harness the Romans had known. The gain was a considerable one, for the horse was now not merely a useful aid in agriculture or a means of transport: he became likewise an improved agent of mechanical production: mills utilizing horsepower directly for grinding corn or for pumping water came into existence all over Europe, sometimes supplementing other forms of non-human power, sometimes serving as the principal source. The increase in the number of horses was made possible, again, by improvements in agriculture and by the opening up of the hitherto sparsely cultivated or primeval forest areas in northern Europe. This created a condition somewhat similar to that which was repeated in America during the pioneering period: the new colonists, with plenty of land at their disposal, were lacking above all in labor power, and were compelled to resort to ingenious labor-saving devices that the better settled regions in the south with their surplus of labor and their easier conditions of living were never forced to invent. This fact perhaps was partly responsible for the high degree of technical initiative that marks the period.

But while horse power ensured the utilization of mechanical methods in regions not otherwise favored by nature, the greatest technical progress came about in regions that had abundant supplies of wind and water. It was along the fast-flowing streams, the Rhône and the Danube and the small rapid rivers of Italy, and in the North Sea and Baltic areas, with their strong winds, that this new civilization had its firmest foundations and some of its most splendid cultural expressions.

## ENGLAND'S BELATED LEADERSHIP

By the middle of the eighteenth century the fundamental industrial revolution, that which transformed our mode of thinking, our means of production, our manner of living, had been accomplished: the external forces of nature were harnessed and the mills and looms and spindles were working busily through Western Europe. The time had come to consolidate and systematize the great advances that had been made.

At this moment the eotechnic regime was shaken to its foundations. A new movement appeared in industrial society which had been gathering headway almost unnoticed from the fifteenth century on: after 1750 industry passed into a new phase, with a different source of power, different materials, different social objectives. This second revolution multiplied, vulgarized, and spread the methods and goods produced by the first: above all, it was directed toward the quantification of life, and its success could be gauged only in terms of the multiplication table.

For a whole century the second industrial revolution, which Geddes called the paleotechnic age, has received credit for many of the advances that were made during the centuries that preceded it. In contrast to the supposedly sudden and inexplicable outburst of inventions after 1760 the previous seven hundred years have often been treated as a stagnant period of small-scale petty handicraft production, feeble in power resources and barren of any significant accomplishments. How did this notion become popular? One reason, I think, is that the critical change that actually did take place during the eighteenth century threw into shadow the older technical methods: but perhaps the main reason is that this change took place first and most swiftly in England, and the observations of the new industrial methods, after Adam Smith—who was too early to appraise the transformation—were made by economists who were ignorant of the technical history of Western Europe, or who were inclined to belittle its significance. The historians failed to appreciate the debt of England's navy under Henry VIII to Italian shipbuilders, of her mining industry to imported German miners, of her waterworks and land-clearance schemes to Dutch engineers, and her silk spinning mills to the Italian models which were copied by Thomas Lombe.

The fact is that England, throughout the Middle Ages, was one of the backward countries of Europe: it was on the outskirts of the great continental civilization and it shared in only a limited way in the great industrial and civic development that took place in the South from the tenth century onward. As a wool-raising center, in the time of Henry VIII, England was a source of raw materials, rather than a well-rounded agricultural and manufacturing country; and with the destruction of the monasteries by the same monarch, England's backwardness was only accentuated. It was not until the sixteenth century that various traders and enterprises began to develop mines and mills and glassworks on any considerable scale. Few of the decisive inventions or improvements of the eotechnic phase—one excepts knitting—had their home in England. England's first great contribution to the new processes of thought and work came through the marvelous galaxy of distinguished scientists it produced in the seventeenth century: Gilbert, Napier, Boyle, Harvey, Newton, and Hooke. Not until the eighteenth century did England participate in any large degree in the eotechnic advances: the horticulture, the landscape gardening, the canal building, even the factory organization of that period, correspond to developments that had taken place from one to three centuries earlier in other parts of Europe.

Since the eotechnic regime had scarcely taken root in England, there was less resistance there to new methods and new processes: the break with the past came more easily, perhaps, because there was less to break with. England's original backwardness helped to establish her leadership in the paleotechnic phase.

## THE NEW BARBARISM

As we have seen, the earlier technical development had not involved a complete breach with the past. On the contrary, it had seized and appropriated and assimilated the technical innovations of other cultures, some very ancient, and the pattern of industry was wrought into the dominant pattern of life itself. Despite all the diligent mining for gold, silver, lead and tin in the sixteenth century, one could not call the civilization itself a mining civilization; and the handicraftsman's world did not change completely when he walked from the workshop to the church, or left the garden behind his house to wander out into the open fields beyond the city's walls.

Paleotechnic industry, on the other hand, arose out of the breakdown of European society and carried the process of disruption to a finish. There was a sharp shift in interest from life values to pecuniary values: the system of interests which only had been latent and which had been restricted in great measure to the merchant and leisure classes now pervaded every walk of life. It was no longer sufficient for industry to provide a livelihood: it must create an independent fortune: work was no longer a necessary part of living: it became an all-important end. Industry shifted to new regional centers in England: it tended to slip away from the established cities and to escape to decayed boroughs or to rural districts which were outside the field of regulation. Bleak valleys in Yorkshire that supplied water power, dirtier bleaker valleys in other parts of the land which disclosed seams of coal, became the environment of the new industrialism. A landless, traditionless proletariat, which had been steadily gathering since the sixteenth century, was drawn into these new areas and put to work in these new industries: if peasants were not handy, paupers were supplied by willing municipal authorities: if male adults could be dispensed with, women and children were used. These new mill villages and milltowns, barren of even the dead memorials of an older humaner culture, knew no other round and suggested no other outlet, than steady unremitting toil. The operations themselves were repetitive and monotonous; the environment was sordid; the life that was lived in these new centers was empty and barbarous to the last degree. Here the break with the past was complete. People lived and died within sight of the coal pit or the cotton mill in which they spent from fourteen to sixteen hours of their daily life, lived and died without either memory or hope, happy for the crusts that kept them alive or the sleep that brought them the brief uneasy solace of dreams.

Wages, never far above the level of subsistence, were driven down in the new industries by the competition of the machine. So low were they in the early part of the nineteenth century that in the textile trades they even for a while retarded the introduction of the power loom. As if the surplus of workers, ensured by the disfranchisement and pauperization of the agricultural workers, were not enough to reenforce the Iron Law of Wages, there was an extraordinary rise in the birthrate. The causes of this initial rise are still obscure; no present theory fully accounts for it. But one of the tangible motives was the fact that unemployed parents were forced to live upon the wages of the young they had begotten. From the chains of poverty and perpetual destitution there was no escape for the new mine worker or factory worker: the servility of the mine, deeply engrained in that occupation, spread to all the accessory employments. It needed both luck and cunning to escape those shackles.

Here was something almost without parallel in the history of civilization: not a lapse into barbarism through the enfeeblement of a higher civilization, but an upthrust into barbarism, aided by the very forces and interests which originally had been directed toward the conquest of the environment and the perfection of human culture. Where and under what conditions did this change take place? And how, when it represented in fact the lowest point in social development Europe had known since the Dark Ages did it come to be looked upon as a humane and beneficial advance? We must answer those questions.

The phase one here defines as paleotechnic reached its highest point, in terms of its own concepts and ends, in England in the middle of the nineteenth century: its cock crow of triumph was the great industrial exhibition in the new Crystal Palace at Hyde Park in 1851: the first World Exposition, an apparent victory for free trade, free enterprise, free invention, and free access to all the world's markets by the country that boasted already that it was the workshop of the world. From around 1870 onwards the typical interests and preoccupations of the paleotechnic phase have been challenged by later developments in technics itself, and modified by various counterpoises in society. But like the eotechnic phase, it is still with us: indeed, in certain parts of the world, like Japan and China, it even passes for the new, the progressive, the modern, while in Russia an unfortunate residue of paleotechnic concepts and methods has helped misdirect, even partly cripple, the otherwise advanced economy projected by the disciples of Lenin. In the United States the paleotechnic regime did not get under way until the 1850s, almost a century after England; and it reached its highest point at the beginning of the present century, whereas in Germany it dominated the years between 1870 and 1914, and, being carried to perhaps fuller and completer expression, has collapsed with greater rapidity there than in any other part of the world. France, except for its special coal and iron centers, escaped some of the worst defects of the period; while Holland, like Denmark and in part Switzerland, skipped almost directly from an eotechnic into a neotechnic economy, and except in ports like Rotterdam and in the mining districts, vigorously resisted the paleotechnic blight.

In short, one is dealing with a technical complex that cannot be strictly placed within a time belt; but if one takes 1700 as a beginning, 1870 as the high point of the upward curve, and 1900 as the start of an accelerating downward movement, one will have a sufficiently close approximation to fact. Without accepting any of the implications of Henry Adams's attempt to apply the phase rule of physics to the facts of history, one may grant an increasing rate of change to the processes of invention and technical improvement, at least up to the present; and if eight hundred years almost defines the eotechnic phase, one should expect a much shorter term for the paleotechnic one.

## CARBONIFEROUS CAPITALISM

The great shift in population and industry that took place in the eighteenth century was due to the introduction of coal as a source of mechanical power, to the use of new means of making that power effective—the steam engine—and to new methods of smelting and working up iron. Out of this coal and iron complex, a new civilization developed.

Like so many other elements in the new technical world, the use of coal goes back a considerable distance in history. There is a reference to it in Theophrastus: in 320 B.C. it was used by smiths; while the Chinese not merely used coal for baking porcelain but even employed natural gas for illumination. Coal itself is a unique mineral: apart from the precious metals, it is one of the few unoxidized substances found in nature; at the same time it is one of the most easy to oxidize: weight for weight it is of course much more compact to store and transport than wood.

As early as 1234 the freemen of Newcastle were given a charter to dig for coal, and an ordinance attempting to regulate the coal nuisance in London dates from the fourteenth century. Five hundred years later coal was in general use as a fuel among glassmakers, brewers, distillers, sugar bakers, soap boilers, smiths, dyers, brick makers, lime burners, founders, and calico printers. But in the meanwhile a more significant use had been found for coal: Dud Dudley at the beginning of the seventeenth century sought to substitute coal for charcoal in the production of iron: this aim was successfully accomplished by a Quaker, Abraham Darby, in 1709. By that invention the high-powered blast furnace became possible; but the method itself did not make its way to Coalbrookdale in Shropshire to Scotland and the North of England until the 1760s. The next development in the making of cast-iron awaited the introduction of a pump which should deliver to the furnace a more effective blast of air: this came with the invention of Watt's steam pump, and the demand for more iron, which followed, in turn increased the demand for coal.

Meanwhile, coal as a fuel for both domestic heating and power was started on a new career. By the end of the eighteenth century coal began to take the place of current sources of energy as an illuminant through Murdock's devices for producing illuminating gas. Wood, wind, water, beeswax, tallow, sperm oil—all these were displaced steadily by coal and derivatives of coal, albeit an efficient type of burner, that produced by Welsbach, did not appear until electricity was ready to supplant gas for illumination. Coal, which could be mined long in advance of use, and which could be stored up, placed industry almost out of reach of seasonal influences and the caprices of the weather.

In the economy of the earth, the large-scale opening up of coal seams meant that industry was beginning to live for the first time on an accumulation of potential energy, derived from the ferns of the carboniferous period, instead of upon current income. In the abstract, mankind entered into the possession of a capital inheritance more splendid than all the wealth of the Indies; for even at the present rate of use it has been calculated that the present known supplies would last three thousand years. In the concrete, however, the prospects were more limited, and the exploitation of coal carried with it penalties not attached to the extraction of energy from growing plants or from wind and water. As long as the coal seams of England, Wales, the Ruhr, and the Alleghenies were deep and rich the limited terms of this new economy could be overlooked: but as soon as the first easy gains were realized the difficulties of keeping up the process became plain. For mining is a robber industry: the mine owner, as Messrs. Tryon and Eckel point out, is constantly consuming his capital, and as the surface measures are depleted the cost per unit of extracting minerals and ores becomes greater. The mine is the worst possible local base for a permanent civilization: for when the seams are exhausted, the individual mine must be closed down, leaving behind its debris and its deserted sheds and houses. The by-products are a befouled and disorderly environment; the end product is an exhausted one.

Now, the sudden accession of capital in the form of these vast coal fields put mankind in a fever of exploitation: coal and iron were the pivots upon which the other functions of society revolved. The activities of the nineteenth century were consumed by a series of rushes—the gold rushes, the iron rushes, the copper rushes, the petroleum rushes, the diamond rushes. The animus of mining affected the entire economic and social organism: this dominant mode of exploitation became the pattern for subordinate forms of industry. The reckless, get-rich-quick, devil-take-the-hindmost attitude of the mining rushes spread everywhere: the bonanza farms of the Middle West in the United States were exploited as if they were mines, and the forests were gutted out and mined in the same fashion as the minerals that lay in their hills. Mankind behaved like a drunken heir on a spree. And the damage to form and civilization through the prevalence of these new habits of disorderly exploitation and wasteful expenditure remained, whether or not the source of energy itself disappeared. The psychological results of carboniferous capitalism—the lowered morale, the expectation of getting something for nothing, the disregard for a balanced mode of production and consumption, the habituation to wreckage and debris as part of the normal human environment—all these results were plainly mischievous.

## THE DEGRADATION OF THE WORKER

Kant's doctrine, that every human being should be treated as an end, not as a means, was formulated precisely at the moment when mechanical industry had begun to treat the worker solely as a means—a means to cheaper mechanical production. Human beings were dealt with in the same spirit of brutality as the landscape: labor was a resource to be exploited, to be mined, to be exhausted, and finally to be discarded. Responsibility for the worker's life and health ended with the cash payment for the day's labor.

The poor propagated like flies, reached industrial maturity—ten or twelve years of age—promptly, served their term in the new textile mills or the mines, and died inexpensively. During the early paleotechnic period their expectation of life was twenty years less than that of the middle classes. For a number of centuries the degradation of labor had been going on steadily in Europe; at the end of the eighteenth century, thanks to the shrewdness and nearsighted rapacity of the English industrialists, it reached its nadir in England. In other countries, where the paleotechnic system entered later, the same brutality emerged: the English merely set the pace. What were the causes at work?

By the middle of the eighteenth century the handicraft worker had been reduced, in the new industries, into a competitor with the machine. But there was one weak spot in the system: the nature of human beings themselves: for at first they rebelled at the feverish pace, the rigid discipline, the dismal monotony of their tasks. The main difficulty, as Ure pointed out, did not lie so much in the invention of an effective self-acting mechanism as in the "distribution of the different members of the apparatus into one cooperative body, in impelling each organ with its appropriate delicacy and speed, and above all, in training human beings to renounce their desultory habits of work and to identify themselves with the unvarying regularity of the complex automaton." "By the infirmity of human nature," wrote Ure again, "it happens that the more skillful the workman, the more self-willed and intractable he is apt to become, and of course the less fit and component of the mechanical system in which...he may do great damage to the whole."

The first requirement for the factory system, then, was the castration of skill. The second was the discipline of starvation. The third was the closing up of alternative occupations by means of land monopoly and dis-education.

In actual operation, these three requirements were met in reverse order. Poverty and land monopoly kept the workers in the locality that needed them and removed the possibility of their improving their position by migration: while exclusion from craft apprenticeship, together with specialization in subdivided and partitioned mechanical functions, unfitted the machine-worker for the career of pioneer or farmer, even though he might have the opportunity to move into the free lands in the newer parts of the world. Reduced to the function of a cog, the new worker could not operate without being joined to a machine. Since the workers lacked the capitalists' incentives of gain and social opportunity, the only things that kept them bound to the machine were starvation, ignorance, and fear. These three conditions were the foundations of industrial discipline, and they were retained by the directing classes even though the poverty of the worker undermined and periodically ruined the system of mass production which the new factory discipline promoted. Therein lay one of the inherent "contradictions" of the capitalist scheme of production.

It remained for Richard Arkwright, at the beginning of the paleotechnic development, to put the finishing touches upon the factory system itself: perhaps the most remarkable piece of regimentation, all things considered, that the last thousand years have seen.

Arkwright, indeed, was a sort of archetypal figure of the new order: while he is often credited, like so many other successful capitalists, with being a great inventor, the fact is that he was never guilty of a single original invention: he appropriated the work of less astute men. His factories were located in different parts of England, and in order to supervise them he had to travel with Napoleonic diligence, in a post-chaise, driven at top speed: he worked far into the night, on wheels as well as at his desk. Arkwright's great contribution to his personal success and to the factory system at large was the elaboration of a code of factory discipline: three hundred years after Prince Maurice had transformed the military arts, Arkwright perfected the industrial army. He put an end to the easy, happy-go-lucky habits that had held over from the past: he forced the one-time independent handicraftsman to "renounce his old prerogative of stopping when he pleases, because," as Ure remarks, "he would thereby throw the whole establishment into disorder."

Following upon the earlier improvements of Wyatt and Kay, the enterpriser in the textile industries had a new weapon of discipline in his hands. The machines were becoming so automatic that the worker himself, instead of performing the work, became a machine-tender, who merely corrected failures in automatic operation, like a breaking of the threads. This could be done by a woman as easily as by a man, and by an eight-year-old child as well as by an adult, provided discipline were harsh enough. As if the competition of children were not enough to enforce low wages and general submission, there was still another police-agent: the threat of a new invention which would eliminate the worker altogether.

From the beginning, technological improvement was the manufacturer's answer to labor insubordination, or, as the invaluable Ure reminded his readers, new inventions "confirmed the great doctrine already propounded that when capital enlists science into its service the refractory hand of labor will be taught docility." Nasmyth put this fact in its most benign light when he held, according to Smiles, that strikes were more productive of good than of evil, since they served to stimulate invention. "In the case of many of our most potent self-acting tools and machines, manufacturers could not be induced to adopt them until compelled to do so by strikes. This was the case of the self-acting mule, the wool-combing machine, the planning-machine, the slotting-machine, Nasmyth's steam-arm, and many others."

At the opening of the period, in 1770, a writer had projected a new scheme for providing for paupers. He called it a House of Terror: it was to be a place where paupers would be confined at work for fourteen hours a day and kept in hand by a starvation diet. Within a generation, this House of Terror had become the typical paleotechnic factory: in fact the ideal, as Marx well says, paled before the reality.

Industrial diseases naturally flourished in this environment: the use of lead glaze in the potteries, phosphorus in the match-making industry, the failure to use protective masks in the numerous grinding operations, particularly in the cutlery industry, increased to enormous proportions the fatal forms of industrial poisoning or injury: mass consumption of china, matches, and cutlery resulted in a steady destruction of life. As the pace of production increased in certain trades, the dangers to health and safety in the industrial process itself increased: in glass-making, for example, the lungs were overtaxed, in other industries the increased fatigue resulted in careless motions and the maceration of a hand or the amputation of a leg might follow.

With the sudden increase of population that marked the opening years of the paleotechnic period, labor appeared as a new natural resource: a lucky find for the labor-prospector and labor-miner. Small wonder that the ruling classes flushed with moral indignation when they found that Francis Place and his followers had endeavored to propagate a knowledge of contraceptives among the Manchester operatives in the eighteen-twenties: these philanthropic radicals were threatening an otherwise inexhaustible supply of raw material. And in so far as the workers were diseased, crippled, stupefied, and reduced to apathy and dejection by the paleotechnic environment they were only, up to a certain point, so much the better adapted to the new routine of factory and mill. For the highest standards of factory efficiency were achieved with the aid of only partly used human organisms—in short, of defectives.

With the large scale organization of the factory it became necessary that the operatives should at least be able to read notices, and from 1832 onwards measures for providing education for the child laborers were introduced in England. But in order to unify the whole system, the characteristic limitations of the House of Terror were introduced as far as possible into the school: silence, absence of motion, complete passivity, response only upon the application of an outer stimulus, rote learning, verbal parroting, piecework acquisition of knowledge—these gave the school the happy attributes of jail and factory combined. Only a rare spirit could escape this discipline, or battle successfully against this sordid environment. As the habituation became more complete, the possibility of escaping to other occupations and other environments became more limited.

One final element in the degradation of the worker must be noted: the maniacal intensity of work. Marx attributed the lengthening of the working day in the paleotechnic period to the capitalist's desire to extract extra surplus value from the laborer: as long as values in use predominated, he pointed out, there was no incentive to industrial slavery and overwork: but as soon as labor became a commodity, the capitalist sought to obtain as large a share of it as possible for himself at the smallest expense. But while the desire for gain was perhaps the impulse uppermost in lengthening the worker's day—as it happened, a mistaken method even from the most limited point of view—one must still explain the sudden intensity of the desire itself. This was not a result of capitalist production's unfolding itself according to an inner dialectic of development: the desire for gain was a causal factor in that development. What lay behind its sudden impetus and fierce intensity was the new contempt for any other mode of life or form of expression except that associated with the machine. The esoteric natural philosophy of the seventeenth century had finally become the popular doctrine of the nineteenth. The gospel of work was the positive side of the incapacity for art, play, amusement, or pure craftsmanship which attended the shriveling up of the cultural and religious values of the past. In the pursuit of gain, the ironmasters and textile masters drove themselves almost as hard as they drove their workers: they scrimped and stinted and starved themselves at the beginning, out of avarice and the will-to-power, as the workers themselves did out of sheer necessity. The lust for power made the Bounderbys despise a humane life: but they despised it for themselves almost as heartily as they despised it for their wage slaves. If the laborers were crippled by the doctrine, so were the masters.

For a new type of personality had emerged, a walking abstraction: the Economic Man. Living men imitated this penny-in-the-slot automaton, this creature of bare rationalism. These new economic men sacrificed their digestion, the interests of parenthood, their sexual life, their health, most of the normal pleasures and delights of civilized existence to the untrammeled pursuit of power and money. Nothing retarded them; nothing diverted them...except finally the realization that they had more money than they could use, and more power than they could intelligently exercise. Then came belated repentance: Robert Owen founds a utopian cooperative colony, Nobel, the explosives manufacturer, a peace foundation, Carnegie free libraries, Rockefeller medical institutes. Those whose repentance took a more private form became the victims of their mistresses, their tailors, their art dealers. Outside the industrial system, the Economic Man was in a state of neurotic maladjustment. These successful neurotics looked upon the arts as unmanly forms of escape from work and business enterprise: but what was their one-sided, maniacal concentration upon work but a much more disastrous escape from life itself? In only the most limited sense were the great industrialists better off than the workers they degraded: jailer and prisoner were both, so to say, inmates of the same House of Terror.

Yet though the actual results of the new industrialism were to increase the burdens of the ordinary worker, the ideology that fostered it was directed toward his release. The central elements in that ideology were two principles that had operated like dynamite upon the solid rock of feudalism and special privilege: the principle of utility and the principle of democracy. Instead of justifying their existence by reason of tradition and custom, the institutions of society were forced to justify themselves by their actual use. It was in the name of social improvement that many obsolete arrangements that had lingered on from the past were swept away, and it was likewise by reason of their putative utility to mankind at large that the most humane and enlightened minds of the early nineteenth century welcomed machines and sanctioned their introduction. Meanwhile, the eighteenth century had turned the Christian notion of the equality of all men in Heaven into an equality of all men on earth: they were not to achieve it by conversion and death and immortality, but were supposed to be "born free and equal." While the bourgeoisie interpreted these terms to their own advantage, the notion of democracy nevertheless served as a psychological rationalization for machine industry: for the mass production of cheap goods merely carried the principle of democracy on to the material plane, and the machine could be justified because it favored the process of vulgarization. This notion took hold very slowly in Europe; but in America, where class barriers were not so solid, it worked out into a leveling upward of the standard of expenditure. Had this leveling meant a genuine equalization of the standard of life, it would have been a beneficent one: but in reality it worked out spottily, following the lines most favorable for profits, and thus often leveling downward, undermining taste and judgment, lowering quality, multiplying inferior goods.

## THE BEGINNINGS OF NEOTECHNICS

The neotechnic phase represents a third definite development in the machine during the last thousand years. It is a true mutation: it differs from the paleotechnic phase almost as white differs from black. But on the other hand, it bears the same relation to the eotechnic phase as the adult form does to the baby.

During the neotechnic phase, the conceptions, the anticipations, the imperious visions of Roger Bacon, Leonardo, Lord Verulam, Porta, Glanvill, and the other philosophers and technicians of that day at last found a local habitation. The first hasty sketches of the fifteenth century were now turned into working drawings: the first guesses were now reenforced with a technique of verification: the first crude machines were at last carried to perfection in the exquisite mechanical technology of the new age, which gave to motors and turbines properties that had but a century earlier belonged almost exclusively to the clock. The superb animal audacity of Cellini, about to cast his difficult Perseus, or the scarcely less daring work of Michelangelo, constructing the dome of St. Peter's, was replaced by a patient cooperative experimentalism: a whole society was now prepared to do what had heretofore been the burden of solitary individuals.

Now, while the neotechnic phase is a definite physical and social complex, one cannot define it as a period, partly because it has not yet developed its own form and organization, partly because we are still in the midst of it and cannot see its details in their ultimate relationships, and partly because it has not displaced the older regime with anything like the speed and decisiveness that characterized the transformation of the eotechnic order in the late eighteenth century. Emerging from the paleotechnic order, the neotechnic institutions have nevertheless in many cases compromised with it, given way before it, lost their identity by reason of the weight of vested interests that continued to support the obsolete instruments and the anti-social aims of the middle industrial era. *Paleotechnic ideals still largely dominate the industry and the politics of the Western world:* the class struggles and the national struggles are still pushed with relentless vigor. While eotechnic practices linger on as civilizing influences, in gardens and parks and painting and music and the theater, the paleotechnic remains a barbarizing influence. To deny this would be to cling to a fool's paradise. In the seventies Melville framed a question in fumbling verse whose significance has deepened with the intervening years:

... Arts are tools;
But tools, they say, are to the strong:
Is Satan weak? Weak is the wrong?
No blessed augury overrules:
Your arts advanced in faith's decay:
You are but drilling the new Hun
Whose growl even now can some dismay.

To the extent that neotechnic industry has failed to transform the coal-and-iron complex, to the extent that it has failed to secure an adequate foundation for its humaner technology in the community as a whole, to the extent that it has lent its heightened powers to the miner, the financier, the militarist, the possibilities of disruption and chaos have increased.

But the beginnings of the neotechnic phase can nevertheless be approximately fixed. The first definite change, which increased the efficiency of prime movers enormously, multiplying it from three to nine times, was the perfection of the water-turbine by Fourneyron in 1832. This came at the end of a long series of studies, begun empirically in the development of the spoon-wheel in the sixteenth century and carried on scientifically by a series of investigators, notably Euler in the middle of the eighteenth century. Burdin, Fourneyron's master, had made a series of improvements in the turbine type of waterwheel—a development for which one may perhaps thank France's relative backwardness in paleotechnic industry—and Fourneyron built a single turbine of 50 hp as early as 1832. With this, one must associate a series of important scientific discoveries made by Faraday during the same decade. One of these was his isolation of benzine: a liquid that made possible the commercial utilization of rubber. The other was his work on electromagnetic currents, beginning with his discovery in 1831 that a conductor cutting the lines of force of a magnet created a difference in potential: shortly after he made this purely scientific discovery, he received an anonymous letter suggesting that the principle might be applied to the creation of great machines. Coming on top of the important work done by Volta, Galvani, Oersted, Ohm, and Ampére, Faraday's work on electricity, coupled with Joseph Henry's exactly contemporary research on the electro-magnet, erected a new basis for the conversion and distribution of energy and for most of the decisive neotechnic inventions.

By 1850 a good part of the fundamental scientific discoveries and inventions of the new phase had been made: the electric cell, the storage cell, the dynamo, the motor, the electric lamp, the spectroscope, the doctrine of the conservation of energy. Between 1875 and 1900 the detailed application of these inventions to industrial processes was carried out in the electric power station and the telephone and the radio telegraph. Finally, a series of complementary inventions, the phonograph, the moving picture, the gasoline engine, the steam turbine, the airplane, were all sketched in, if not perfected, by 1900: these in turn effected a radical transformation of the power plant and the factory, and they had further effects in suggesting new principles for the design of cities and for the utilization of the environment as a whole. By 1910 a definite countermarch against paleotechnic methods began in industry itself.

The outlines of the process were blurred by the explosion of the World War and by the sordid disorders and reversions and compensations that followed it. Though the instruments of a neotechnic civilization are now at hand, and though many definite signs of an integration are not lacking, one cannot say confidently that a single region, much less our Western Civilization as a whole, has entirely embraced the neotechnic complex: for the necessary social institutions and the explicit social purposes requisite even for complete technological fulfillment are lacking. The gains in technics are never registered automatically in society: they require equally adroit inventions and adaptations in politics; and the careless habit of attributing to mechanical improvements a direct role as instruments of culture and civilization puts a demand upon the machine to which it cannot respond. Lacking a cooperative social intelligence and good-will, our most refined technics promises no more for society's improvement than an electric bulb would promise to a monkey in the midst of a jungle.

True: the industrial world produced during the nineteenth century is either technologically obsolete or socially dead. But unfortunately, its maggoty corpse has produced organisms which in turn may debilitate or possibly kill the new order that should take its place: perhaps leave it a hopeless cripple. One of the first steps, however, toward combating such disastrous results is to realize that even technically the Machine Age does not form a continuous and harmonious unit, that there is a deep gap between the paleotechnic and neotechnic phases, and that the habits of mind and the tactics we have carried over from the old order are obstacles in the way of our developing the new.

## THE IMPORTANCE OF SCIENCE

The detailed history of the steam engine, the railroad, the textile mill, the iron ship, could be written without more than passing reference to the scientific work of the period. For these devices were made possible largely by the method of empirical practice, by trial and selection: many lives were lost by the explosion of steam boilers before the safety valve was generally adopted. And though all these inventions would have been the better for science, they came into existence, for the most part, without its direct aid. It was the practical men in the mines, the factories, the machine shops and the clockmakers' shops and the locksmiths' shops or the curious amateurs with a turn for manipulating materials and imagining new processes, who made them possible. Perhaps the only scientific work that steadily and systematically affected the paleotechnic design was the analysis of the elements of mechanical motion itself.

With the neotechnic phase, two facts of critical importance become plain. First, the scientific method, whose chief advances had been in mathematics and the physical sciences, took possession of other domains of experience: the living organism and human society also became the objects of systematic investigation, and though the work done in these departments was handicapped by the temptation to take over the categories of thought, the modes of investigation, and the special apparatus of quantitative abstraction developed for the isolated physical world, the extension of science here was to have a particularly important effect upon technics. Physiology became for the nineteenth century what mechanics had been for the seventeenth: instead of mechanism forming a pattern for life, living organisms began to form a pattern for mechanism. Whereas the mine dominated the paleotechnic period, it was the vineyard and the farm and the physiological laboratory that directed many of the most fruitful investigations and contributed to some of the most radical inventions and discoveries of the neotechnic phase.

Similarly, the study of human life and society profited by the same impulses toward order and clarity. Here the paleotechnic phase had succeeded only in giving rise to the abstract series of rationalizations and apologies which bore the name of political economy: a body of doctrine that had almost no relation to the actual organization of production and consumption or to the real needs and interests and habits of human society. Even Karl Marx, in criticizing these doctrines, succumbed to their misleading verbalisms: so that whereas *Das Kapital* is full of great historic intuitions, its description of price and value remains as prescientific as Ricardo's. The abstractions of economics, instead of being isolates and derivatives of reality, were in fact mythological constructions whose only justification would be in the impulses they excited and the actions they prompted. Following Vico, Condorcet, Herder and G. F. Hegel, who were philosophers of history, Comte, Quetelet, and Le Play laid down the new science of sociology; while on the heels of the abstract psychologists from Locke and Hume onward, the new observers of human nature, Bain, Herbart, Darwin, Spencer, and Fechner integrated psychology with biology and studied the mental processes as a function of all animal behavior.

In short, the concepts of science, hitherto associated largely with the cosmic, the inorganic, the "mechanical" were now applied to every phase of human experience and every manifestation of life. The analysis of matter and motion, which had greatly simplified the original tasks of science, now ceased to exhaust the circle of scientific interests: men sought for an underlying order and logic of events which would embrace more complex manifestations. The Ionian philosophers had long ago had a clue to the importance of order itself in the constitution of the universe. But in the visible chaos of Victorian society Newlands' original formulation of the periodic table as the Law of Octaves was rejected, not because it was insufficient, but because Nature was deemed unlikely to arrange the elements in such a regular horizontal and vertical pattern.

During the neotechnic phase, the sense of order became much more pervasive and fundamental. The blind whirl of atoms no longer seemed adequate even as a metaphorical description of the universe. During this phase, the hard and fast nature of matter itself underwent a change: it became penetrable to newly discovered electric impulses, and even the alchemist's original guess about the transmutation of the elements was turned, through the discovery of radium, into a reality. The image changed from "solid matter" to "flowing energy."

Second only to the more comprehensive attack of the scientific method upon aspects of existence hitherto only feebly touched by it, was the direct application of scientific knowledge to technics and the conduct of life. In the neotechnic phase, the main initiative comes, not from the ingenious inventor, but from the scientist who establishes the general law: the invention is a derivative product. It was Henry who in essentials invented the telegraph, not Morse; it was Faraday who invented the dynamo, not Siemens; it was Oersted who invented the electric motor, not Jacobi; it was Clerk-Maxwell and Hertz who invented the radio telegraph, not Marconi and De Forest. The translation of the scientific knowledge into practical instruments was a mere incident in the process of invention. While distinguished individual inventors like Edison, Baekeland and Sperry remained, the new inventive genius worked on the materials provided by science.

Out of this habit grew a new phenomenon: deliberate and systematic invention. Here was a new material: problem—find a new use for it. Or here was a necessary utility: problem—find the theoretic formula which would permit it to be produced. The ocean cable was finally laid only when Lord Kelvin had contributed the necessary scientific analysis of the problem it presented: the thrust of the propeller shaft on the steamer was finally taken up without clumsy and expensive mechanical devices, only when Michell worked out the behavior of viscous fluids: long-distance telephony was made possible only by systematic research by Pupin and others in the Bell Laboratories on the several elements in the problem. Isolated inspiration and empirical fumbling came to count less and less in invention. In a whole series of characteristic neotechnic inventions the thought was father to the wish. And typically, this thought is a collective product.

## Toy, Mirror and Art: The Metamorphosis of Technological Culture

**Paul Levinson**

The varied produce of our technological media—the amalgam of television shows, movies, books, recordings, etc., known collectively as "mass," "popular," or "technological" culture—has been the subject of considerable recent study and controversy. Arguing primarily from aesthetic and sociological perspectives, champions and critics of popular culture have alternately praised and condemned it as aesthetically democratizing and degrading, socially stabilizing and stultifying, and so forth.[1] Curiously missing from such discussions, however, is any serious analysis of the technological basis of popular culture. While theorists usually acknowledge that it is technology that makes most popular culture possible, they have apparently been content to view the connection as axiomatic and undeserving of further research.[2] Thus, an otherwise comprehensive summary of "Theories and Methodologies of Popular Culture" in a recent *Journal of Popular Culture* issue discussed everything from structuralism to cultural geography and popular culture, with barely a mention of technological underpinnings.[3] The omission is even more remarkable when one considers that Marshall McLuhan, one of the first to write seriously of popular culture, was also one of the first to point out that technological media are much more than passive conveyors of information and content.[4]

It is perhaps understandable that subtleties in the technological shaping of popular culture have gone unexplored, since the broad outlines of the relationship are so obvious. There seems little profit in pursuing, for example, the fact that without the invention of the motion picture camera there would be no film industry, and without the technological achievement of the phonograph, no popular recording culture. Yet upon closer examination, such simple connections begin to display an increasing number of complications. Why, to stay with the same example, did film attain a cultural prominence forty years before music recording, when the motion picture camera was in fact perfected shortly *after* the phonograph? While differences in societal receptivity, economics, and the like were no doubt in part responsible, it seems plausible that certain elements in the very mechanics of film and record production may have stimulated the first and inhibited the second as they arose in cultural impact and esteem.

Film, as the first product of the nineteenth-century electrochemical revolution to achieve artistic notice in the twentieth, might be a good place to begin an inquiry into the technological determination of technological culture. In tracing the changing usages and perceptions of film from its first appearance in society, it may be possible to discern a relationship between the level of technological sophistication in film and the type of popular culture each technological level engendered. To the extent that such observations are generalizable, they may suggest a series of principles that describe a step-by-step development, common to all technological culture, from new medium to widespread influence. Such principles may also have some bearing on the aesthetic controversies about culture, helping to explain how and why some technologies facilitate more "artistic" creations than others. The inquiry may also have some implications for the evolution and appliance of more "practical" technologies, and elucidate the distinctions and similarities between technologies used primarily for work, and those used for entertainment.

## STAGE ONE: TECHNOLOGY AS TOY

Writing of the inception of film technology in *A Short History of the Movies,* Gerald Mast describes an interesting pattern:

> The first film makers were not artists but tinkerers.... Their goal in making a movie was not to create beauty but to display a scientific curiosity. The invention of the first cameras and projectors set a trend that was to repeat itself with the introduction of every new movie invention: the invention was first exploited as a novelty in itself....[5]

A survey of early "talkies" like *The Jazz Singer,* first efforts in animation such as Disney's "Laugh-O-Gram" cartoons, and indeed the supposed debut of the motion picture itself in *Fred Ott's Sneeze,* supports Mast's observation of technology's supremacy in the beginning stages of technological culture. In each instance elements of plot, characterization, and what little content there is, play a subservient role to the exposition of the new gimmick, and perform in effect as low-key vehicles for a highly visible technique. The enjoyment in these primal forms lies in a fascination with the process—not the product of the process, but the process itself—in seeing and hearing a man sing on film, for example, rather than caring *what* the man sings on film. Thus, in the medium of film, at least, new technologies have made their entrances like the brash new kid on the block, in a flexing of muscle and raw technique that transcends and for all purposes *becomes* the content. In a sense, then, the most important content—or popular culture—of a new medium is the medium itself.

McLuhan has explored the concept of technology-as-the-content-of-technology, suggesting that *outmoded* or postfunctional media often serve as the content for newer media. (Plays and books, for example, become the content of the new medium, film, and film becomes, in turn, the content of the newer medium, television.[6]) McLuhan goes as far as to say that usually invisible technologies only become visible when no longer in use—a proposition that at first seems to contradict Mast's contention, but may in fact serve to complement it. For incipient media are as out of the mainstream as obsolescent forms—it's the familiar equation of childhood and old age—and as such occupy equivalent if opposing positions in the medium development cycle. Thus, McLuhan's model may be reduced to a basic expectation that the discernibility or observable impact of any medium will vary *inversely* with the usage or functioning of that medium in the overall society; that is, the workings of technology, like the blades of whirring fans, are most visible both before and after they reach the peak of their function, and the triviality and trickery of prepubescent media are as nonfunctional and hence ostentatious as the ritual pomp and funeral exhibition of media in demise.

The role of neonate medium as societal plaything is perhaps best documented in the history of film. But it is also readily apparent in the trajectories of most other communications media, and indeed in the invention and implementation of many technologies, used for entertainment and otherwise. William Orton's celebrated refusal to inexpensively buy up Bell's early telephone patents on grounds that the new device would never be more than an "electrical toy";[7] the corporate decision to initially promote the phonograph as a "novelty" music box *in spite of* Edison's early assertions that his new invention could perform more practical tasks;[8] the amateur crystal set radio fad of the 1920s and the gawking at televisions in department store windows in the 1940s; the continuing popularity, in our own time, of computer "games,"[9] as well as the propensity of programmers to couch computer terminology and printouts in cute phrases and configurations; and, most recently, the Citizen's Band or "CB" radio "craze"[10]—all testify to the tenacity with which the novel medium is perceived and tends to be employed as a toy. Moreover, examples of technologies not specifically concerned with communication, but nonetheless at first utilized for peripheral amusement, are even more varied and abundant. The Chinese discovery of gunpowder and the principles of rocketry, and their use solely as holiday and children's entertainment; the initial application of Newtonian mechanics to devise intricate dolls or "automatons" in the eighteenth century;[11] the debut of ether as a giddy party drug well before its medical properties were exploited[12]—all suggest that the "toy principle" may far exceed the province of popular culture and communications media. It is even tempting to suggest that *all* new technologies may gain first admittance into society as court jesters and Trojan horses, with their physical presence clearly visible, but their potentialities poorly understood.[13]

But if practical technologies indeed begin as playthings, what forces are needed to transform the playthings into practical technologies? That this transformation is by no means inevitable is documented by various curiosities of history, such as the failure of the Aztec civilization to use the wheel other than in children's toys, and the Chinese confinement of their gunpowder, rocket, and printing inventions to use on only ceremonial occasions. Abbott Payson Usher, the technological historian, sees this problem as central to an understanding of technology. "The history of invention," he writes, "is a study of the circumstances that have converted the simple but relatively inefficient mechanisms of early periods into the complex and more effective mechanisms of today."[14] Usher views economic needs and perception of technological potentials as the most important of these circumstances, but numerous other factors have been linked by theorists to the development, and nondevelopment, of specific technologies. Victor von Hagen, for example, points out that any Aztec attempt to use the wheel in more practical ways would have been foiled by Mexico's steep-walled landscape,[15] or the lack of a physical environment conducive to the technology of the wheel. McLuhan emphasizes the importance of compatible *media* environments, proposing that the Chinese ideograph was the main impediment to Chinese use of the printing press for mass communication, on grounds that ideographic writing is not well suited for reproduction on mass, movable type.[16] And Friedrich Hayek contends that the massive application of invention to industrial tasks in the nineteenth century—aptly characterized by Alfred North Whitehead as "the invention of the method of invention"[17]—was made possible by the uniquely invigorating climate of free capitalism.[18] It is thus fairly plain that the ignition of technological growth has often come from outside the specific technology itself, in a combination of supportive social, economic, media, and even physical conditions. It also follows that while the toy phase may be prerequisite to subsequent technological development, its existence by no means *guarantees* that development: lacking the proper environment, the technological toy may long endure in a case of "arrested" development. (In this sense, the confinement of ESP phenomena to largely show business and "magic" roles in our own society may constitute a failure to exploit a potentially useful "mental" technology due to lack of a proper attitude on our part. Future historians may well regard this failure in the same way we regard the Aztec "failure" with the wheel.[19])

In the case of the popular culture technologies of the past hundred years, however, the emergence of new communication toys has almost always led to their more extensive use as practical media and/or mass art. The lesson of the toy principle for popular culture, then, is that mass art forms don't spring full-blown from the head of new technologies, but rather pass through a series of developmental stages beginning with a naive, raw, almost "contentless" flexing of hardware. That there is little "mass" about these incipient mass media is obvious in the bygone kinetoscope parlors of primitive film and the ear horns of early victrolas—devices that doled out entertainment on a purely personal, one-to-one, fragmentary basis, characteristic more of the individual experience of toys than the mass experience of popular media. And yet with surprising regularity, these primordial media evolved into technologies and cultures of universal impact.

## STAGE TWO: TECHNOLOGY AS MIRROR OF REALITY

In the history of film, it wasn't long before Edison's kinetoscopic oddities faced stiff and ultimately overwhelming competition from the Lumieres' presentation of "actualities" on the screen. Rather than photographing sneezes and what amounted to other filmic gag lines, the Lumieres pointed their cameras at real-life events—workers leaving a factory, a baby's meal, and the famous train entering the station. Although the novelty of movie technology undoubtedly played a major role in the appreciation of these early "documentaries," the cries and jolts of audiences upon viewing *L'Arrive d'un train en gare* in 1895 (approximately six years after Ott's nasal acrobatics) clearly indicated a new focusing on content—in this case, a real train chugging into a real station, at an angle such that the audience could almost believe the train was chugging right in at *them*. The superficial amusement and curiosity characteristic of the earlier gimmick films were replaced with the deeper emotions of fright, sorrow, and so on—emotions that one would expect in a replication of a real-world interaction. In effect, the adoption of reality as film content distracted from the technology and artificiality of the film experience, directing attention to the nontechnological content—the events depicted on the screen—and in turn enhancing the believability of the content, i.e., belief that the events on the film were "really" happening. The co-option of reality in media thus becomes a self-fulfilling loop, in which the very mirroring of actuality tends to disguise the mirroring process and promote the actuality. It is perhaps the spiraling power of this media loop that accounts for the riveting impact of media technology once it left the infant toy stage and evolved into the succeeding reality/mirror phase.

If, as Whitehead said, the most important invention of the nineteenth century was invention itself, the most important development of the twentieth century was, as Bertrand Russell saw, the suspension of disbelief.[20] The public's willingness to respond to an electronic transcription of a voice as if it were a *live* voice, and to a photochemical likeness of a face as if it were a *real* face, soon enabled communications technology to effectively recapture or substitute for the real world on a massive scale. And it was this mirroring of the real world, with the attendant prominence of content and invisibility of technique (qualities which media theorists from McLuhan to Jacques Ellul have long seen as the defining traits of mature technologies),[21] that became the modus operandi of communications media. The telephone, of course, shortly confounded William Orton's assessment and became a major artery for both business and personal conversation. And while the phonograph's transition from toy to reality-transcriber may have been less obvious than that of the telephone, it was nonetheless profound. By 1893, the attempt to sell the phonograph as a "novelty" had run its course, and Edison, regaining control of his invention, planned for the introduction of popular music records.[22] This signaled a shift in phonographic emphasis from technique to content, the content in this case being the reality of a past musical performance retrieved and captured on the record. Television and radio, after even briefer tenures in the toy stage, began functioning as rather mature transcribers of reality—where they for the most part continue today. For in the broadcasting of film, video tapes, or recordings, television and radio perform as much of a reality transmission as when actual events and live performances are broadcast. The nonreality component, if it exists in the film or recording broadcast, is in the original film or recording, *not* in its transmission on television or radio. In fact, film and recordings on television and radio satisfy all the requirements of the stage two "reality" mirror: the film is the reality/content and the film/content is paramount in the audience's awareness, even as the underlying transmitting technology of television goes unnoticed. The audience suspends its disbelief and pretends it's seeing *a film* rather than a televised *broadcast* of a film.

Conveyance or interaction with reality is apparently the second and terminal phase of development, not only for most communications technology, but for most other applications of technology as well. When such curios as rocketry, electricity, and phosphorescence were finally harnessed for practical purposes—when the little toys finally grew up—they extended our physical control of the real world and, as Fuller, Hall, McLuhan, and many others have pointed out, in effect acted and continue to act as surrogates for our arms, legs, hands, and bodies.[23] There thus appears to be some merit to the proposition that communications media differ from other technologies only in specific application and content, sharing essentially the same developmental patterns and dynamics.

There does appear, however, to be at least one interesting difference in the development of film and most other technologies. Whereas, as indicated before, the transformation of most techniques from side show toys to mainstream appliances seems to have been sparked primarily by forces and attitudes *outside* of technology, the growth of film from gimmick to replicator was apparently in large part dependent upon a new technological component. As described earlier, the "toy" film played to individuals who peeked into individual kinetoscopes; but the "reality" film reached out to mass audiences, who viewed the reality-surrogate in group theaters. The connection between mass audience and reality simulation, moreover, was no accident. Unlike the perception of novelties, which is inherently subjective and individualized, reality perception is a fundamentally objective, group process—tested in the social consensus, as George Herbert Mead and social psychologists have long stressed—and as such is strengthened and even predicated upon mass experience. The creation of a group audience for simultaneous and reinforcing perception of the reality-surrogate film required that a new technology of film *projection* had to be devised and hooked into the existing communication chain. This suggests that technological determinism may have played a greater role in the development of film culture, and perhaps of all popular culture by extension, than it did in the appliance of more practical technologies.

In addition, film (and recording) seem to be distinguishable from most other media and technologies in one other significant respect: as implied earlier, it is film and recording that now provide the wellspring of imagination and originality for television and radio broadcasting. This nonreality is not the pre-reality of technological toys, but the postreality of media that have mastered the straight transcription of the real world and have gone on to something beyond—the rearrangement of the real world to create fantasy, eloquence, and art. To accomplish this feat, a technology must evolve to yet a third phase—a phase that can copy reality, dissect it, and put it back together again in new and intriguing ways.[24]

## STAGE THREE: TECHNOLOGY AS MIDWIFE TO ART

The present discussion began with an inquiry into the relationship between technology and popular art, and a specific question as to why film and recording, which were first introduced into the culture at approximately the same time, developed into popular art forms at such different rates of speed. Thus far, however, the discussion has had little to do with popular art—talking on the phone certainly doesn't constitute a popular art (at least, not for the average speaker), and, as indicated, what aesthetic content there is in radio and television derives from the content of the film, play, record, script, and not from its broadcast. The connection between technological process and technological art, then, remains yet to be defined.

Once again, the history of film might prove instructive. An oft-told story has it that George Melies, another French film pioneer, was shooting his camera at pedestrians and vehicles on the Place de l'Opera in Paris one fine spring day in 1898 (in the "actuality" fashion of the Lumieres), when his camera jammed. Thinking his film ruined, Melies nevertheless cleared the aperture gate, reset the film, and started shooting again—taking the film home for development just for the amusement of it (apparently one of the essential ingredients of many great discoveries). When the print came out, and Melies projected it, he received a little surprise: there, at the spot in the action where the camera had stopped and then started, was a magical transformation—men turning into women, children into adults, and a passing bus instantly materializing into a hearse! The film, in other words, gave no discernible indication that the camera had stopped and started with several seconds elapsing; all that was apparent was the continuous "reality" of a bus suddenly changing into a hearse. Melies had inadvertently hit upon the potent intervention of editing.[25]

The edit of course proved to be the key in the transformation of film from reality transcription to popular art: the discovery that disparate pieces of film, shot at different times and places and reflective of different realities, could be spliced together so as to project what would be accepted as one continuous *new* reality, freed film from dependence upon literal reality.[26] Film no longer need be wedded to the natural rhythms of time and space to create a natural, flowing experience; the editing room could create its own rhythms, which were equally "natural" and palatable to the perceiver. Within less than twenty years, Griffith and others exploited this opportunity to mold, bend, shape, fracture, and reconstruct realities to the dictates only of the writer/director/editor's imagination. Film now had a life of its own.

From Altamiran cave paintings to Victorian literature, the ability not only to retell but refashion reality in the retelling has been a hallmark of "art."[27] It is not surprising, then, that film's transcendence of reality touched off its explosion as a popular art form, which by the 1920s had become both global and golden. Conventional film history, of course, traces the employment of editing and its artistic vistas to the chance tinkering of Melies and his followers; but the serendipity of film editing can perhaps be better understood as not so much the personal fortune of Melies, as the serendipity of film's original *technology*. The mechanics of film were never *intended* to allow for an alteration of reality—celluloid was used for its flexibility in projection, not for its amenability to splicing—and the "chance" discovery of editing was thus a chance uncovering of a hidden capacity already present in the technology of the medium. In this specific sense, then, the popular art of film can be seen as a direct outgrowth of its peculiar technology.

The development of technological art thus appears dependent upon the special capacity of a technology, first designed as a toy and second used as a reality-substitute, to transcend reality and make new ones. This toy/mirror/art or prereality/reality/postreality dialectic of technological development bears some interesting resemblances to several well-known models of human development, including Piaget's sensorimotor, concrete, and formal (abstract) stages of intellectual growth;[28] McLuhan's oral, written, and electronic eras of communication;[29] Freud's oral, anal, and genital stages of psychosexual development (which Walter Ong has intriguingly compared to McLuhan's stages of communication, e.g., written and anal are retentive and reality-oriented);[30] and Arthur Koestler's Jester, Sage, and Artist as the three unfolding expressions of human creativity.[31] Note how technology as toy displays the subjectivity of the oral, the nonseriousness of the joke, the flexing of muscle for its own sake characteristic of sensorimotor activity, and the emphasis upon technique or delivery common to humor, oral communication, and sensorimotor behavior. Technology as mirror stresses accuracy, objectivity, and prominence of content or "knowledge" as befits both the sage and the scribe, as well as the literal transaction with reality basic to concrete operations. And technology as art, combining elements of the previous two stages, is both serious *and* subjective, capable of the emotional intensity of the genital stage (the multidimensionality of electronic communication) and the abstraction and restructuring of reality—the triumph of form over content—of the formal stage of intellectual functioning.

Most technologies, however, perhaps too well-suited to the second stage mirror task, simply lack the ability to make the artistic jump. Thus the telephone is purely a medium of reality communication, and still life photography, for all its aesthetic aspirations, remains essentially a medium of literal replication.[32] Other technologies, such as radio and television possess the ability to restructure reality and create art, but are limited by convention and economic pressures to simple reality-mirroring of previously created filmic, theatrical, or musical art, as discussed earlier. (The quick switching of video cameras and perspectives on talk shows like "The Tonight Show" often create ambiences that don't exist on the set, and thus may constitute a bona fide "art form" *produced*—rather than merely transmitted—by television. Experimentation with video editing, computer character generation, and so forth may also be a source of potential art.)

The case of phonograph/recording technology is even more unusual. Initially, the hardware used to record was hopelessly reality-bound—sound was stored and reproduced first on electric wires and then on discs, neither of which allowed for splicing, alteration, or rearrangement once the recording was made. Thus lacking the hidden potential of celluloid film, rubber records continued for better than forty years as a glorified Xerox operation for musical performances. It wasn't until the addition of a completely new tape technology in the 1940s—a spliceable medium which was initially introduced for remedial purposes, so as to make more accurate replications—that recording attained a faculty for reality alteration or art. Magnetic sound tape not only allowed for easy editing, but for overdubbing, multitrack sel-syncing, and a general reshuffling of recorded sounds to make for imaginative new combinations. Thus, within twenty years of the introduction of tape recording, artist/producer/songwriters like Phil Spector, the Beatles, and others turned recording into the popular art form of the generation—a two-billion-dollar industry that has at times surpassed even film and television in combined sales and cultural impact. (Note that the twenty years from the inception of tape technology to the Beatles parallels the twenty years from Melies to Griffith in film.)

The differential in the rise of film and recording as popular art forms can thus now be explained as follows: both were initially conceived as toys, and both were quickly adopted for reality transcription; the same technology that enabled film to adequately replicate the real world enabled film to reconstruct the real world, so film soon evolved into a popular art; but the technology that enabled recording to adequately replicate its real world contained no such double advantage, so recording remained a simple transcription device until the addition of a new mechanism capable of reality alteration. It is thus apparent that, as suggested at the outset of the present discussion, the relationship between technology and technological art is no simple cause-and-effect matter.

Indeed, the addition of a new component to an already-productive technology cannot even always be depended upon to enhance the medium's capacity for art: as suggested in the earlier reference to the first "talkie" films, the introduction of sound technology to the silent film in effect *reduced* the medium to the state of a toy—setting the whole technology back to stage one by creating a new medium, as it were. In this regard, some critics insist that to the present day, film has never recovered from the introduction of sound—that speech and dialogue have been used at worst as a toy and at best as an unimaginative, literal exposition of plot (some of the work of Orson Welles and perhaps Robert Altman might be an exception), to the detriment and even destruction of lofty artistic styles developed during the silent era.[33]

Moreover, the connection between technology and popular culture is further complicated by the tendency of various technologies to operate, not singly or in isolation, but in conjunction, often cross-influencing one another. Edison, it is said, invented the phonograph to perfect the telephone, and a motion picture process to enhance the phonograph.[34] The role of phonetic writing as a prerequisite to the mass usage of the printing press has already been alluded to (see note 16). And the rise of music recording culture, though clearly a product of its own technology, was augmented by, of all media, the television: when television co-opted radio as a medium of live entertainment in the early 1950s, radio was forced to rely much more extensively on recorded music to attract its listeners—and thus provided a sustaining forum for a recording technology already ripe for popular art. (In a similar fashion, the FCC's decision in 1965 that all FM radio stations must broadcast programs different from their AM radio affiliates hastened the development of the LP record as an art form—for many FM stations turned to what was previously considered "noncommercial" album music, thus giving the LP a much needed public forum.) This type of technological interaction of course demands an eventual analysis of the economic and social factors that mediate the technologies.

A complete discussion of technology and art inevitably invites some consideration of aesthetics—yet the complexity of the technology-to-culture equation makes an aesthetic of technological art and culture rather difficult. Criticisms of technological art have often been insensitive to gradations in technological process, and have been frequently directed at immature media that are physically incapable of, and make no pretense to, any type of technological "art." Thus, José Ortega y Gasset, for example, condemns "modern art" as, among other things, "play and nothing else," and "of no transcending consequence"[35]—qualities that, in the perspective of the present discussion, can be seen as more properly belonging to the technological toy than to technological art. This confusion of technological stages—an error of premature judgment and "mistaken identity" born of an inability to see technology as an evolving, developmental process—was recognized by Susanne Langer, whose assessment of filmic evolution aptly complements Mast's observations on novelties that served as the springboard for the present analysis. Langer writes:

> With every new invention—montage, the sound track, Technicolor—its [film's] devotees have raised a cry of fear that now its 'art' must be lost. Since every such novelty is, of course, promptly exploited before it is even technically perfected, and flaunted in its rawest state, as a popular sensation,...there is usually a tidal wave of particularly bad rubbish in association with every important advance. But the art goes on.[36]

The problem, of course, is that while stages of technological culture are indeed distinctive and successive, they by no means are mutually exclusive—the current popularity of "Sensurround" gimmickry and "wildlife saga" reality-mimicry in the movies suggests that, having attained the *capacity* for technological art, film need not necessarily always *produce art*. Instead, technologic evolution, like its biological model, allows for the coexistence of earlier and later designs, with an assortment of aesthetic ramifications. But if an awareness of media evolution cannot provide a definitive aesthetic for popular culture, it can at least offer a useful yardstick for making such judgments. Moreover, it perhaps at last reveals a common ground between the critics and champions of popular culture—the first looking at the caterpillar, the second at the butterfly, of the same technological process.

## SUMMARY AND CONCLUSIONS

The ways in which technologies engender and encourage mass culture and art are complex and multifaceted, yet have often been oversimplified or taken for granted. Although a physical invention must lie at the root of every technological art, very rarely if ever do inventions have immediate mass cultural impact or flowering. Instead, new technologies usually make their first appearance in the culture as novelties, gadgets, gimmicks, and toys. The content here is dominated by, and an exposition for, the new technique; the perceptual experience is personal, subjective, and highly individual rather than "mass"; and the toys usually perform on the sidelines of the overall society. Due more often than not to shifts in societal attitudes rather than developments in technology, the novelty item eventually (though not always) becomes a more practical device, used for various types of literal transactions with reality. The content in this phase attains a high prominence while the visible technology recedes; the perceptual experience is markedly social, objective, and "mass," as the entity of "audience" comes into play for the first time; and such transcribers of reality usually occupy significant and often central positions in the society. At this point, the evolutions of practical and artistic technologies are virtually indistinguishable—the difference being that practical technologies remain at the reality level, while artistic media must evolve to yet a third stage. To achieve this artistic jump, a medium must have the capacity not only to replicate reality, but to rearrange it in imaginative ways. Performance at this level entails a blending of features from the previous two stages: technological art is nonfunctional and subjective like the toy, yet nontrivial and (in most cases) group oriented and content dominated like the reality-surrogate. In the case of film, the supra-realism ability was inherent in the original realism technology, so the development of a popular film art was relatively swift. In the case of music recording, the capacity for reality alteration came only with the nonpurposeful addition of a new technological component, so the rise of a popular recording culture was correspondingly delayed. In yet another case, radio and especially television have the capacities for technological art, yet function as aesthetic parasites in relying upon other media for creativity and art. It is thus apparent that the technology for transcending reality, and its two antecedent stages, are necessary but not sufficient conditions for the fostering of mass culture and art; the remaining conditions probably lie in the interaction of various technologies both among themselves and with more abstract, nontechnological elements of society.

## NOTES

1 Herbert J. Gans, *Popular Culture and High Culture* (New York: Basic, 1975), summarizes many of the extant criticisms and defenses of popular culture.

2 A few theorists have argued that technology is *not* the necessary basis of popular culture, citing such non-technological cultures as the oral folk music tradition; see Ray B. Brown, "Popular Culture: Notes Towards a Definition," in *Side Saddle on the Golden Calf,* George H. Lewis, ed. (Pacific Palisades, Cal.: Goodyear, 1972), pp. 5–11; and Bruce A. Lohof, "Popular Culture: The *Journal* and the State of the Study," *J. of Popular Culture,* 6, No. 3 (1972): 438–455. Since even folk music, however, has become increasingly dependent upon the technology of electronic instruments and recording, it seems fair to say that such cases represent a diminishing series of exceptions to the technological rule.

3 Vol. 9, No. 2 (1975): 353–508.

4 Principally in *The Gutenberg Galaxy* (Toronto: University of Toronto, 1962), and *Understanding Media,* 2nd ed. (New York: Mentor, 1964).

5 (New York: Pegasus, 1971), p. 15. Unless otherwise indicated, examples of film history to be cited in the ensuing discussion come from Mast's account.

6 In the McLuhan schema, technology once liberated from function often becomes not only "content" but "art." "The machine turned Nature into an art form," McLuhan writes, by making humans nondependent upon natural technologies for survival. (See *Understanding Media,* p. ix.) Note that the art here is not the direct product of a technology at work—as is film montage, for example, from the technology of editing—but rather the peculiar result of a technology *not* at work. A fine recent example of this unusual type of art genesis appeared in a *New York Times* travel piece that seriously described New York City's increasingly nonfunctional subway system as a "delight-

fully elevated *tour de force,*'' and ''a scenic delight,'' featuring ''track-wheel music.'' (See Stan Fischler and Richard Friedman, ''Subways,'' 23 May 1976, Section 10, pp. 1, 22.) With ridership diminishing and service reduced, the mechanics of the subway system are now apparently capable of being appreciated not for what they do (or don't do), but for what they ''are.''

**7** Matthew Josephson, *Edison* (New York: McGraw-Hill, 1959), p. 141. Moreover, Orton, as president of the Western Union Telegraph Company, was apparently steadfast in his low regard for the ''talking telegraph.'' According to an amusing little article aptly entitled ''Three Great Mistakes'' by S. H. Hogarth in the November 1926 issue of *Blue Bell,* Orton counseled his hapless friend Chauncey M. Depew to pass up a chance to purchase one-sixth of the new Bell telephone enterprise for a mere $10,000, because, in Orton's view, ''the invention was a toy'' with no ''commercial possibilities.'' Meanwhile, John Brooks relates that use of the telephone in England was delayed for at least a decade due to the conviction that it was only a ''scientific toy.'' (See *Telephone: The First Hundred Years* (New York: Harper & Row, 1976), p. 92.)

**8** Josephson, p. 172. Among the more practical applications of the phonograph envisioned by Edison but long unimplemented were recording of letters and books for the blind, preservation of lectures and public addresses, and permanent transcription of telephone conversations.

**9** A recent *New York Post* article reports that sales of home computers are ''spreading like wildfire,'' and mostly to ''techno-fetishists'' who play a variety of visual and intellectual games with computers. David Ahl, editor of *Creative Computing* who was interviewed in the *Post* story, sees the current computer phase as consistent with a more general pattern of media development: ''When the principle of radio was first discovered,'' Ahl explains, ''it was the amateurs who developed the first sets.... it's the same here [with computers].'' Peter Keepnews, ''The Latest Do-It-Yourself Fetish: Computers,'' *New York Post,* 9 June 1976, p. 47.

**10** Although C.B. was first introduced by the F.C.C. in 1958, it was relatively unknown by the general public until the recent fanfare. Predictably, C.B.'s first burst into public awareness has been accompanied by a jargon of code names and passwords, both accoutrements of gimmick usage. And perhaps most significant is C.B. enthusiast and writer Michael Harwood's assertion that the messages relayed on Citizen's Band ''are often inconsequential''—clear evidence, again, of new technology overpowering content, or being operated just for the fun of it. See Michael Harwood, ''America With Its Ears On,'' *The New York Times Magazine,* 25 April 1976, pp. 28, 60, ff.

**11** Siegfried Giedeon provides a colorful account of ''invention in the service of the miracle'' from Alexandrian religious plays to mechanical ducks that defecated in the eighteenth- and nineteenth-century courts of Europe, in *Mechanization Takes Command* (New York: Norton, 1948), pp. 32–35. See also Robert S. Brumbaugh, *Ancient Greek Gadgets and Machines* (New York: Thomas Crowell, 1966), who points out that ''the Greeks invented the steam engine, but to them it was only a toy'' (p. 4). Brumbaugh then documents the Greek invention of numerous other mechanical devices used primarily to amuse and amaze.

**12** Rene Fulop-Miller, *Triumph Over Pain,* trans. by Eden Paul and Cedar Paul (New York: Literary Guild of America, 1938), pp. 95–97.

**13** Cyril Stanley Smith argues along similar lines in ''On Arts, Invention, and Technology,'' *Technology Review* 78, No. 7 (June 1976): 36–41, pointing out that practical metallurgy began with the making of ornamental necklaces, wheels first appeared on toys, lathes were used to carve snuff boxes a century before their use in heavy industry, and metal casting was first perfected for making bells rather than cannon—all of which suggests to Smith that technology may originate more from playful and aesthetic impulses than practical need.

**14** *The History of Mechanical Inventions,* 2nd ed. (Cambridge, Mass.: Harvard University, 1954), p. 117.

**15** "The wheel, had the Mexicans had it," von Hagen writes, "would have done them no good as all is up and down, and high valley is walled from high valley almost throughout the length and breadth of the land; its heights are only passable to foot traffic." *The Aztec: Man and Tribe* (New York: Mentor, 1961), p. 18.

**16** *The Gutenberg Galaxy,* p. 185. In a similar way, Roger Burlingame explains the failure to actualize most of Leonardo's inventions as due to a lack of "collateral" technology. (See "The Hardware of Culture," *Technology and Culture* 1, No. 1 (1959), p. 16.)

**17** *Science and the Modern World* (New York: Macmillan, 1925), p. 136.

**18** "That the inventive faculty of man had been no less in earlier periods," Hayek explains, "is shown by the many highly ingenious automatic toys and mechanical devices constructed....But the few attempts towards a more extended industrial use of mechanical inventions, some extraordinarily advanced, were promptly suppressed...the beliefs of the great majority of what was right and proper were allowed to bar the way of the individual innovator. Only since industrial freedom opened the path to the free use of new knowledge...has science made the great strides which in the last hundred and fifty years have changed the face of the world." *The Road to Serfdom* (Chicago: University of Chicago, 1944), pp. 15–16.

**19** Of course, not everyone agrees that application of technology to practical tasks is socially desirable. Lewis Mumford, for example, sees preindustrial, "aesthetic" inventions as more fundamentally human, and more beneficial to society, than "utilitarian" appliances; he thus views pre-nineteenth-century incipient technologies not as "arrested" development at all, but as the finest expressions of the human inventive impulse, and laments their absence in our modern culture. See *The Myth of the Machine,* vol. 1: *Technics and Human Development* (New York: Harcourt Brace Jovanovich, 1966), pp. 252–253.

**20** As cited by McLuhan in *Understanding Media,* p. 68. It was Samuel Taylor Coleridge who first identified "that willing suspension of disbelief for the moment, which constitutes poetic faith," *Biographia Literaria,* ed. J. Shawcross (1817; reprint ed.: London: Oxford University, 1907), vol. 2, p. 6.

**21** See William Kuhns, *The Post-Industrial Prophets* (New York: Harper Collophon, 1971), for a comparison of the work of McLuhan, Ellul, and other media theorists. For more on the narcotic capacity of media mirrors to disguise their operation, see McLuhan, *Understanding Media,* "The Gadget-Lover," pp. 51–56.

**22** Josephson, pp. 330–333. Edison's interest in the popular record was apparently due not only to foresight, but to the pressure of rival inventors—most notably Emile Berliner, whose "flat-disc" record in the 1890s took a sizeable bite out of Edison's "novelty" phonograph market.

**23** See, for example, R. Buckminster Fuller, *Nine Chains to the Moon* (Carbondale, Ill.: Southern Illinois University, 1938), pp. 38–39; Edward Hall, *The Silent Language* (New York: Fawcett, 1959), p. 60; and McLuhan, *The Gutenberg Galaxy* and *Understanding Media: The Extensions of Man.*

**24** This distinction between prereality and postreality technology will perhaps run contrary to conceptual frameworks that distinguish primarily between reality and nonreality, and are thus prone to view both toys and art as a same technological expression belonging to a single nonreality or nonpractical class. Thus Mumford, as suggested earlier, views pre-industrial technologies as the source of both games *and* the most genuine art (see note 19 above), and indeed later argues that art attempted by postindustrial technologies is in effect a contradiction of terms, or an "anti-art." (See *The Myth of the Machine,* vol. 2: *The Pentagon of Power* (New York: Harcourt Brace Jovanovich, 1970), pp. 361–368, et passim.) From a rather different perspective, Freud has equated the artist, child, primitive, and psychotic, in his "Relation of the Poet to Day-Dreaming," reprinted in *On Creativity and the Unconscious* (New York: Harper & Row, 1958), pp. 44–54, and *Totem and Taboo,* trans. A. A. Brill (New York: Vintage, 1918). And McLuhan's description of postoperative technologies as art, and their similarity to the prefunctional technological toy (see note 6 above), suggests at least one type of art which may be analogous to the toy. For the most part, however, the evidence of popular culture as well as intuition points to deep divergences between technological toys and art. While both *Fred Ott's Sneeze* and the movie *Chinatown,* for example, are indeed nonpractical, the first merely distracts from the real world through razzle-dazzle, whereas the second subtly restructures reality through imperceptible technique; the sneezing nose in the first is diversion, the bandaged nose in the second is commentary and symbol. Thus, the discussion which follows will accept Susanne Langer's observation that while both games and art are nonutilitarian, art is serious and games are not. *Philosophy in a New Key,* 2nd ed. (New York: Mentor, 1951), p. 42.

**25** There is apparently a sliver of suspicion that the Melies anecdote may be apocryphal. Eric Rhode, for example, in his recent *History of the Cinema* (New York: Hill and Wang, 1976), p. 34, prefaces his recounting of the episode with a weighty "it is alleged that. . . ." On the other hand, Maurice Bardeche and Robert Brasillach, writing much closer to the source in *The History of Motion Pictures* (New York: Norton, 1938), p. 11—as well as Lewis Jacobs' "George Melies: Artificially Arranged Scenes," first published in 1939 and reprinted in *The Emergence of Film Art,* ed. Lewis Jacobs (New York: Hopkinson and Blake, 1969), p. 11—present the Melies story without qualification. In any event, the specific manner in which the editing principle was discovered, whether accidental or other, is not as important as the fact of discovery itself—which, as the discussion will shortly emphasize, was an all but inevitable if unintended consequence of the particular technology of film.

**26** It is Edwin S. Porter, not Melies, who is usually credited with being the first to *physically* splice the film for story construction, as in Porter's *The Great Train Robbery* made in 1903. (See Jacobs, pp. 20–21.) Melies' technique of stopping the camera, rearranging the scene, and starting to shoot again was a cruder method of rearranging or "editing" reality.

**27** Langer refers to this supra-reality quality as "semblance," and defines the "artist's task" as follows: "to produce and sustain the essential illusion, set it off clearly from the surrounding world of actuality, and articulate its form to the point where it coincides with forms of feeling and living." *Feeling and Form* (New York: Scribner's, 1953), p. 68. See Sergei Eisenstein, *Film Form,* trans. and ed. Jay Leyda (New York: Harcourt, Brace and World, 1949), for a discussion of film art as montage or creation of new realities through editing.

**28** See Howard Gardner, *The Quest for Mind* (New York: Knopf, 1973), pp. 51–110, for a summary of Piaget's model.

**29** See both *The Gutenberg Galaxy* and *Understanding Media.*

**30** Walter Ong, *The Presence of the Word* (New Haven: Yale University, 1967), pp. 92–110.

**31** *The Act of Creation* (New York: Dell, 1964).

**32** As Stanley Milgram has recently pointed out, "The English language is blunt about the nature of photography. A photographer *takes* a picture. He does not *create* it." "The Image-Freezing Machine," *Psychology Today,* 10, No. 8 (January 1977), p. 52.

33 See, for example, Rudolf Arnheim, *Film as Art* (Berkeley, Cal.: University of California, 1968), foreword and pp. 229–230; and Francois Truffaut, *Hitchcock* (New York: Simon and Schuster, 1966).

34 Josephson, pp. 161, 385; also Mast, pp. 25–26.

35 *The Dehumanization of Art* (1925 reprint ed.: Princeton: Princeton University, 1968), p. 14.

36 *Feeling and Form,* p. 412.

## Women Hold Up Two-Thirds of the Sky: Notes for a Revised History of Technology

**Autumn Stanley**

Over two centuries ago, Voltaire declared, "There have been very learned women as there have been women lawyers, but there have never been women inventors" (1764, *s.v.* "Femmes"). Just three decades ago, Edmund Fuller wrote, "For whatever reason, there are few women inventors, even in the realm of household arts. . . . I cannot find a really conspicuous exception to cite" (1955, 301). Although Voltaire and Fuller were both mistaken, their view permeates most available accounts of human technological development. A revised account of that development, fairly and fully evaluating women's contributions through the ages, is long overdue.

What would such a revised history of technology look like? In the first place, the very *definition of technology would change,* from what men do to what *people* do. We would no longer find anthropological reports using the active voice to describe male activities (the men *choose* the wood for their bows with care) and the passive voice to describe women's activities (cooking is *done* in watertight baskets: But by whom? And how did the baskets get to be watertight?) Nor would any anthropologist say, as George Murdock did in 1973, "The statistics reveal no technological activities which are strictly feminine. One can, of course, name activities that are strictly feminine, e.g., nursing and infant care, but they fall outside the range of technological pursuits" (Murdock and Provost 1973, 210). The ethnologist doing a book on cradles (Mason 1889) would no longer be an oddity; and the inventions of the digging stick, child- and food-carriers, methods of food-processing, detoxification, cooking, and preserving, menstrual absorbers and other aspects of menstrual technology, infant formulas, trail foods, herbal preparations to ease (or prevent) childbirth would receive their proper share of attention and be discussed as technology (see, for example, Cowan 1979).

In the second place, *the definition of significant technology would change.* In prehistory, for example, the main focus would shift from hunting and its weapons to gathering and its tools (Tanner 1981)—gathering provided 60 to 80 percent by weight, and the only reliable part, of foraging peoples' diet (Lee and DeVore

1968, 7)—and eventually to horticulture and its tools and processes. In later times, the focus would shift from war and its weapons, industry and its machines, to healing and its remedies, fertility and antifertility technology, advances in food production and preservation, child care, and inventions to preserve and keep us in tune with our environment. Again, the change would be from what men do to what people do, with the added dimension of a shift in priorities.

To the degree that these major changes were slow in coming, two further or interim changes would take place. First, the *classification of many women's inventions would change*. For example, the digging stick would be classed as a simple machine, the first lever; the spindle whorl, the rotary quern, and the potter's wheel would be credited with the radical breakthrough of introducing continuous rotary motion to human technology; and women's querns (hand-operated grain mills) would be better known as bearing the world's first cranks. Herbal and other remedies would no longer be classified as "domestic inventions" when invented by women and as medicines or drugs when invented by men. Cosmetics would be classed as the chemical inventions they are, and built-in, multi-purpose furniture, moveable storage walls or room dividers, and the like would no longer be classed as architectural when invented by a man and as domestic when invented by a woman. The nineteenth century's inventions inspired by the Dress Reform Movement could be classed not as wearing apparel but as health and medicinal inventions; and food-processing in all its aspects, including cooking, would fall under agriculture.

Second, *women's creation of or contributions to many inventions significant by either or both definitions would be acknowledged*. In prehistory, women's early achievements in horticulture and agriculture, such as the hoe, the scratch plow, grafting, hand pollination, and early irrigation, would be pointed out. Architecture would grow out of weaving, chemistry out of cooking and perfumery, and metallurgy out of pottery. In more modern times, Julia Hall's collaboration with her brother in his process for extracting aluminum from its ore (Trescott 1979), Emily Davenport's collaboration with her husband on the small electric motor (Davenport 1929), Bertha Lammé's contribution to early Westinghouse generators and other great machines (Matthews n.d.), and Annie C-Y. Chang's contribution to genetic engineering (Patent 1981) would all be recognized.

As a result, we would almost certainly see females as primary technologists in proto-and early human societies, especially in any groups whose division of labor resembled that of the Kurnai ("Man's work is to hunt and fish and then sit down; women's work is all else," Reed 1975, 106); as at least equal technologists in such societies as those of the North American Indians and the African !Kung; and as highly important technologists in much of the so-called developing world today. Even in recent Western culture, when women's technological areas regain their true status and significance, and "Anonymous" is no longer so often a woman, women's contributions to technology emerge as much greater than previously imagined. In short, if we consider both history and prehistory, women hold up at least two-thirds of the technological sky.

To see how such a new view of technology might work out in practice, let us look at three areas of human technological endeavor—two that have traditionally been considered significant, but male, preserves; and one originally female preserve that began to be considered significant (i.e., worth including in histories of technology) only when males began to dominate it.

## SIGNIFICANT, ASSUMED MALE, TECHNOLOGIES: FIRE AND MACHINES

The taming of fire is one of the most important technological advances of prehistory. Coming as it did (in Europe) in the midst of an ice age, between 75,000 and 50,000 years ago, it enabled Neanderthals to compete with large animals for cave dwellings, and in those dwellings to survive the ice-age winters. It also transformed early human technology. Food could for the first time be cooked, softening it for toothless elders and allowing them to survive longer to transmit more of their culture. Foods could be created out of toxic or otherwise inedible plants, opening up entire new food supplies, and so on. Although this revolutionary advance is usually ascribed to men, Elise Boulding suggests "it seems far more likely that the women, the keepers of home base and the protectors of the young from wild animals, would be the ones whose need for [fire] would overcome the fear of it" (1976, 80).

Mythological evidence connects women strongly with the taming of fire. The deities and guardians of the hearth and of fire are often female, from Isis and Hestia, Unči Ahči (Ainu), Chalchinchinatl, and Manuiki (Marquesas) to the Vestal Virgins and the keepers of Brigit's sacred flame in Ireland (Corson 1894, 714–15; Frazer 1930, 83; Graves 1955, I, 43, 75; Ohnuki-Tierney 1973, 15). The ancient aniconic image of the Great Goddess herself was a mound of charcoal covered with white ash, forming the center of the clan gatherings. A hymn to Artemis tells how she cut her first pine torch on Mysian Olympus and lit it at the cinders of a lightning-struck tree (Graves 1955, I, 75, 84). In Yahi (American Indian) myth, an old woman stole a few coals of fire from the Fire People and brought them home hidden in her ear (Kroeber 1964, 79). In Congo myth, a woman named Favorite brought fire from Cloud Land to Earth (Feldman 1963, 102–03).

Several myths show women, particularly old women, as the first possessors of fire, and men stealing fire not from the gods but from women. Examples come from Australia, the Torres Straits, mainland New Guinea, Papua, Dobu Island, the Admiralty Islands, the Trobriand Islands, from the Maori, the Fakaofo or Bowditch Islands north of Samoa, Yap, and Northern Siberia. In a Wagifa myth (Melanesia) the woman, Kukuya, gives fire willingly to the people (Frazer 1930, 5, 15, 18, 23–28, 40, 43–45, 48–49, 50, 55–57, 74, 90–91, 104).

Other myths connect women directly with the making of fire, of course coming later than the taming of existing fire. A rather confusing and probably transitional Guiana Indian explanatory tale begins with an old woman who could vomit fire. At her death, "the fire which used to be within her passed into the surrounding fagots. These fagots happened to be hima-heru wood, and whenever we rub together two sticks of this same timber we can get fire" (Roth 1908–1909, 133). From the Taulipang of northern Brazil comes a very similar myth representing perhaps a more nearly complete transition: An old woman named Pelenosamo had fire in her body and baked her manioc cakes with it, whereas other people had to bake their cakes in the sun. When she refused to share fire with the people, they seized and tied her, collected fuel, then set Pelenosamo against it and squeezed her body till the fire spurted out. "But the fire changed into the stones called *wato,* which, on being struck, give forth fire" (Frazer 1930, 131).

Among the Sea Dyaks of Borneo, a lone woman survived a Great Flood. Finding a creeper whose root felt warm, she took two pieces of this wood, rubbed them together, and thus kindled fire. "Such was the origin of the fire-drill" (apparatus for making fire by friction). Biliku, ancestress of the Andaman Islanders and a creator figure, made fire by striking together a red stone and a pearl shell. In this case, a dove stole fire for the people. Among the Nagas of Assam, two women invented the fire-thong (another fire-making device where the friction comes from pulling) by watching a tiger (or an ape) pull a thong under its claw. The ape, having lost fire, is all hairy, whereas people, having fire to keep them warm, have lost their hairy covering. In the New Hebrides, a woman discovered how to make fire while amusing her little boy by rubbing a stick on a piece of dry wood. When the stick smoked and smoldered and finally burst into flame, she laid the food on the fire and found it tasted better because of it. From that time on, all her people began to use fire (Frazer 1930, 51, 94–95, 99, 105–06).

In the Torres Straits, the very operation of fire-making is called "Mother gives fire," the board from which the fire is extracted by the turning of the stick or drill upon it seen as "mother," and the drill as "child" (Frazer 1930, 26–27). More common is a sexual analogy. Commenting on some of these myths, Frazer (1930) says:

> The same analogy may possibly also explain why in the myths women are sometimes represented as in possession of fire before men. For the fire which is extracted from the board by the revolution of the drill is naturally interpreted by the savage as existing in the board before its extraction...or, in mythical language, as inherent in the female before it is drawn out by the male....(220–21)

This of course would not explain myths ascribing fire first to women in cultures using other fire-making methods.

Whatever the origins of fire in various cultures, women put fire to more uses in their work than men did: protecting infants from animals, warming their living area, fire-hardening the point of their digging sticks, cooking, detoxifying and preserving food, hollowing out wooden bowls and other vessels, making pottery, burning vegetation for gardens. Our familiar Prometheus myth may need a footnote.[1]

When Prometheus was on Olympus stealing fire from Hephaestus, he also stole "mechanical skill." Significantly enough, he stole it from a goddess, Athena, who shared a workshop with Hephaestus (Frazer 1930, 194). This seldom-cited Platonic version of the myth takes on new meaning when we reflect that women almost certainly invented the first lever, the digging stick (Stanley 1981, 291–92; Tanner and Zihlman 1976, 599), that the crank may have appeared first in the West on women's querns (Lynn White 1978, 18; Mason 1894, 23), and that at least one historian of technology rates the crank second only to the wheel in importance (Lynn White 1978, 17). Women also invented a cassava-processing device called a *mapiti* or *tipiti,* combining the principles of press, screw, and sieve (Mason 1902, 60–61; Sokolov 1978, 34, 38).

In the development of many mechanical processes, the first stages imitated human limb action in using reciprocal (back-and-forth) motion, as for instance in an ordinary handsaw. Real advances came with continuous or rotary motion, as in the wheel and the circular saw (Singer et al. 1954, chap. 9; Smith 1978, 6).[2] Women seem to have introduced rotary motion to human technology; at least three important early examples of rotary motion pertain unmistakably to women's work: the spindle whorl, the rotary quern, and the potter's wheel. In the spindle whorl, women invented the flywheel (Mason 1894, 57–58, 279–80; Lynn White 1978, 18n). These early examples of axial rotary motion would certainly have influenced the invention of the vehicular wheel—which may have been women's doing in some cultures. In Meso-America (Mexico and Central America) wheeled vehicles appeared only as miniatures that may be either children's toys or religious objects (Doster et al. 1978, 55; Halsbury 1971, 13). If toys, they could easily have been made by women.

As we move into the industrial era, we find further evidence refuting stereotypes about women and machines. Women invented or contributed to the invention of such crucial machines as the cotton gin, the sewing machine, the small electric motor, the McCormick reaper, the printing press, and the Jacquard loom. Catherine Greene's much-debated contribution to the cotton gin may never be proven conclusively; but note that Whitney did arrange to pay her royalties and, according to a Shaker writer, once publicly admitted her help (*Shaker Manifesto* 1890, Stanley 1984). In his most famous lecture, "Acres of Diamonds," nineteenth- and early-twentieth-century journalist and lecturer Russell H. Conwell has Mrs. Elias Howe completing in two hours the sewing machine her husband had struggled with for fourteen years. Conwell's source was impressive—Elias Howe himself (Conwell 1968, 46; Boulding 1976, 686). It was also a woman, Helen Augusta Blanchard (1839–1922) of Portland, Maine, who invented zigzag sewing and the machine to do it (Willard and Livermore 1893, 97). The nineteenth-century patent records show literally dozens of sewing machine improvements by women.

Emily Goss Davenport's role in the invention of the small electric motor usually ascribed to her husband Thomas—her continuous collaboration with him and her crucial suggestion that he use mercury as a conductor—is best described by Walter Davenport (1929, 47, 55, 62). Other sources merely sentimentally praise her for sacrificing her silk wedding dress to wind the coils of Thomas' first homemade electromagnet.

In the case of the reaper, Conwell (1968) cites "a recently published interview with Mr. McCormick," in which the inventor admitted that after he and his father had tried and failed, "a West Virginia woman...took a lot of shears and nailed them together on the edge of a board. Then she wired them so that when she pulled the wire one way it closed them, and...the other way it opened them. And there she had the principle of the mowing machine" (45–46). Another American woman, Ann Harned Manning of Plainfield, New Jersey, invented a mower-reaper in 1817–1818. This was apparently a joint invention with her husband William, who patented and is usually credited with inventing it. Ann and William also invented (and he patented) a clover-cleaner (U.S. Patents of Nov. 24, 1830 and May 3, 1831; Hanaford 1883, 623; Mozans 1913, 362; Rayne 1893, 116–117). The Manning Reaper, predating McCormick's by several years, was important enough to be mentioned in several histories of farm machinery.

Russell Conwell (1968) and Jessie Hayden Conwell (1962) state baldly that farm women invented the printing press and that Mme. Jacquard invented the loom usually credited to her husband.

## SIGNIFICANT TECHNOLOGY WHEN TAKEN OVER BY MALES: MEDICINE

As keepers of home base and then of the home, as preeminent gatherers and then propagators of plants, and as caretakers of children until puberty, women traditionally cared for the sick, creating the earliest form of medicine—herbal medicine. The original deities of healing were probably female. Many such deities and reports of their attributes survive, from Neith, Isis, and Gula in the Middle East to Panacea in Greece and Brigit in Ireland. Minerva Medica parallels Athena Hygeia—Great Goddesses worshipped in their healing aspect (Graves 1955, I, 80–81; Hurd-Mead 1938, 11, 32–33; Jayne 1925, 64–68, 71–72, 121, 513; Rohrlich 1980, 88–89).

Except for contraceptives, abortifacients, preparations to ease labor, and other elements of women's or children's medicine, it is difficult to state unequivocally that women invented or discovered any specific remedy or procedure. However, in general, the more ancient any given remedy, the likelier it is to be a woman's invention; and, of course, if a remedy occurs in a group where the healers are women, the presumption is strong.

Many plants are both foods and medicines: asparagus, whose species name *officinalis* means that it once stood on apothecary shelves; clover, a styptic (Weiner 1972, 144) and heart stimulant and also a food; rhubarb, both a stewed dessert and an effective laxative. Plants may be both foods and contraceptives: wild yams, Queensland matchbox bean (Himes 1970, 28–29; cf. Goodale 1971, 180–81). Or they may serve as food, medicine, and contraceptive, depending on method of preparation, dosage, and concomitant regimen. An example of this triple usage is the Indian turnip or jack-in-the-pulpit, *Arisaema triphyllum*. The root or corm of this North American wildflower contains needlelike crystals of calcium oxalate (oxalate of lime). After proper preparation, however—drying and cooking or pounding the roots to a pulp with water and allowing the mass to dry for several weeks—the Iroquois and other Indians used it as food. The Pawnee also powdered the root and applied it to the head and temples to cure headache, and the Hopi used it to induce temporary or permanent sterility, depending on the dosage (Jack-in-the-pulpit 1958, 851; Weiner 1972, 41, 64–65).

The most ancient medical document yet discovered, a Sumerian stone tablet, dates from the late third millennium B.C., when Gula was Goddess of healing and medicine, and most healers were probably still women. The tablet's several prescriptions call for plants and other natural curatives, mentioning not a single deity or demon, and giving no spells or incantations. A tablet from the time of Hammurabi (around 1750 B.C.) by contrast—when medicine had become a male profession serving mainly elites—blames diseases on demons and suggests incantations as cures. But women healers still ministered to the lower classes, probably continuing to use herbal remedies (Kramer 1963, 93–98; Rohrlich 1980, 88–89).

Precisely parallel developments occurred centuries later in Greece (see, e.g., Graves 1955, I, 174–ff.) and still later in Northern Europe, where male doctors trained mostly in theology in Church-run universities wrested control of medicine from their herbally trained female counterparts, some of whom still practiced the old religion. These male usurpers were aided, intentionally or unintentionally, by the Christian Church, which threatened the wise women they called witches with both hell- and earthly fire. Innumerable precious medical secrets no doubt burned with these women at the stake.[3]

To get some idea of what may have been lost in that medieval holocaust, we need only reflect that European peasant women bound moldy bread over wounds centuries before Alexander Fleming "discovered" that a *Penicillium* mold killed bacteria; that medieval wise women had ergot for labor pains and belladonna to prevent miscarriage; and that an English witch discovered the uses of digitalis for heart ailments (Ehrenreich and English 1973, 14; Raper 1952, 1). Ergot derivatives are the main drugs used today to hasten labor and recovery from childbirth; belladonna is still used as an antispasmodic, and digitalis is still important in treating heart patients (Ehrenreich and English 1973, 14). Medieval wise women knew all this at a time when male practitioners knew little to prescribe except bleeding and incantations. Edward II's physician, for example, boasting a bachelor's degree in theology and a doctorate in medicine from Oxford, recommended writing on a toothache patient's jaw "In the name of the Father, the Son, and the Holy Ghost, Amen," or touching a needle first to a caterpillar and then to the tooth (Ehrenreich and English 1973, 17). Ladies of medieval epic poetry, repeatedly called upon to treat the ghastly wounds of errant knights, worked their miraculous-seeming cures not through prayer but through deep herbal knowledge and careful nursing (Hughes 1943).

Indeed, these women were the repositories of medical knowledge coming to them in a line of women healers from the days when Hecate the Moon Goddess invented aconite teas for teething and children's fevers, when Rhea invented liniments for the pains of children, when the Egyptian Polydamna gave her pupil Helen the secret of Nepenthe, and when Artemisia of Caria—famed for knowing every herb used in medicine—discovered the uses of artemisia to cause (or in other combinations to prevent) abortion and to expel a retained placenta, the value of wormwood, and the delights of absinthe (Hurd-Mead 1938, 32, 37n, 40; Jayne 1925, 345). Thus, we should not be surprised to hear that in the sixteenth century Paracelsus burnt his text on pharmaceuticals because everything in it he had learned from "the Sorceress," i.e., from a wise woman or women he had known (Ehrenreich and English 1973, 17).

Although women are connected most intimately with herbal medicine, ancient women healers had accomplishments in other areas. The only surgery mentioned in the Bible is gynecological or obstetrical surgery—or circumcisions—done by women with their flint knives (Hurd-Mead 1938, 19). Flint in pre-Hellenic Greek myth was the gift of the Goddess (Spretnak 1978, 42). Ancient Scandinavian women's graves contain surgical instruments not found in men's graves. And California Indian medicine women used a technique only now being rediscovered, and still controversial in modern medicine—visualization, for focusing the body's own mental and physical powers of healing on the illness, tumor, or pain (Hurd-Mead 1938, 6, 14).

In late medieval and early modern Europe, women healers continued to work unofficially. The most outstanding of them, such as Trotula, Jacoba or Jacobina, Felicie, and Marie Colinet, sometimes were given more or less recognition by the male medical profession or protected by wealthy patients. Marie Colinet learned surgery from her husband, the renowned surgeon Fabricius of Hilden, but by his own admission she excelled him. For shattered ribs she opened the chest and wired together the fragments of bone—this in the seventeenth century. Her complex herbal plasters prevented infection and promoted healing. She also regulated the postoperative diet and used padded splints. Marie Colinet was first to use a magnet to remove fragments of iron or steel from the eye. Though most sources credit Fabricius with this invention, he credits her (Boulding 1976, 472–75; Hurd-Mead 1938, 361, 433).

During the American colonial period, it is thought that more women practiced medicine than did men (Hymowitz and Weissman 1978, 7). Even in the nineteenth century, women healers still ministered to a great many American families, especially in rural areas. Some operated as informally as the Misses Roxy and Ruey Toothacre portrayed in Harriet Beecher Stowe's *Pearl of Orr's Island* (1862, 17–18), and some more formally or professionally. But they relied on time-tested herbal knowledge brought from the Old World and enriched by contact with Indian women healers, while male practitioners relied heavily on bleeding and the poisonous calomel (containing mercury). Moreover, like Lady Aashild in *Kristin Lavransdatter,* like the medieval ladies in the epics, like the German Mother Seigel (b. ca. 1793), and like Sister Kenny in the Australian Outback in the twentieth century, they not merely made house calls, but stayed with their seriously ill patients for weeks at a time, personally conducting or supervising their care, medication, and diet (Kenny 1943, e.g., 21–29, 71–72; Stage 1979, chap. 2; 45–63; Anna J. White 1866). Women who observed their patients day and night, watching every symptom and the effect of every remedy, quite naturally gained more practical knowledge than the doctor who spent just a few moments with a patient.

Although the twentieth century finds male practitioners firmly in control of formal Western medicine, women doctors and healers still have important inventions and innovations to their credit.

For example, although three men received the Nobel Prize for penicillin, women participated significantly in the team effort that brought the drug to medical usefulness. Women had discovered the mold's usefulness centuries or perhaps millennia earlier (Halsbury 1971, 19; Raper 1952, 1), and one nineteenth-century Wisconsin woman, Elizabeth Stone, an early antibiotic therapist, specialized in treating lumberjacks' wounds with poultices of moldy bread in warm milk or water: she never lost an injury patient (Stellman 1977, 87). In the twentieth-century development of the drug, it was a woman bacteriologist, Dr. Elizabeth McCoy of the University of Wisconsin, who created the ultraviolet-mutant strain of *Penicillium* used for all further production, since it yielded *900 times as much penicillin* as Fleming's strain (Bickel 1972, 185; O'Neill 1979, 219).[4] And as Howard Florey, leader of the British penicillin team, was quick to point out, it was Dr. Ethel Florey's precise clinical trials that transformed penicillin from a crude sometimes-miracle-worker into a reliable drug. It was also a woman, Nobel laureate and x-ray crystallographer Dr. Dorothy Crowfoot Hodgkin, who finally determined the precise structure of the elusive penicillin molecule (Bickel 1972, 216; Opfell 1978, 211, 219).

Women were also involved in developing the sulfa drugs that preceded penicillin. For instance, it was a married pair of chemists, Prof. and Mme. Tréfouël, and their colleagues at the Pasteur Institute in Paris who split red azo dye to create sulfanilamide (Bickel 1972, 50).

At least two women have invented new antibiotics for which they receive sole credit. Dr. Odette Shotwell of Denver, Colorado, came up with two new antibiotics—duramycin and azacolutin—during her first assignment as a research chemist at the Agriculture Department laboratories in Peoria, Illinois. She has also invented new methods for separating antibiotics from fermentation by-products, and in doing so has played an important role in the development of two other antibiotics: cinnamycin and hydroxystreptomycin. Dr. Marina Glinkina of the U.S.S.R. directed the laboratory effort that produced a new antigangrene antibiotic during World War II. Her postwar work as a senior scientist has been theoretical (Dodge 1966, 226; O'Neill 1979, 32; Ribando 1980).

Follies-girl-turned-scientist Justine Johnstone Wanger (1895– ) was the laboratory part of the team that developed the slow-intravenous-drip method of administering drugs and other substances to the human body. She then joined a different medical team in applying this new method to the treatment of early syphilis, in an advance that was called the "greatest step since Ehrlich" (Hobson 1941, 298).

The DPT vaccine that protects virtually all infants in the developed world against three of their former mass killers (diphtheria; pertussis, or whooping cough; and typhus) was invented by Dr. Pearl Kendrick (1890–1980) and Dr. Grace Eldering (1900– ) in the early 1940s. In 1939 they had invented a whooping cough vaccine. Unlike Drs. Salk and Sabin, they refused to allow their vaccines to be named for them. In the 1920s, Dr. Gladys Henry Dick (1881–1963) and her husband conquered another great childhood killer, scarlet fever. They not only isolated the streptococcus causing the disease, but created the toxin and the antitoxin that prevent and cure it, respectively. They then went on to develop the Dick test, a skin test showing susceptibility to the disease. They were recommended for the Nobel Prize in 1925, but no prize was given in medicine for that year. They did, however, receive the Mickle Prize of the University of Toronto, the Cameron Prize of the University of Edinburgh, and several honorary degrees

(Dr. Kendrick...1980–81, 12; Kendrick 1942; O'Neill 1979, 217; *Notable American Women* 1980, 191–92; *Time* 1980, 105).

Although two male doctors are credited with developing the vaccines that conquered polio in the developed world, it was a woman, Sister Elizabeth Kenny (1886–1952) of Australia, who invented the only treatment useful once the disease had struck. Whereas the doctors of her day were splinting the affected limbs to prevent spasm—but also causing the damaged muscles to waste away and become useless for life—Sister Kenny used moist hot packs, massage, and daily gentle exercise, plus muscle reeducation. About 87 percent of her patients escaped paralysis, while about 85 percent of the doctors' patients were paralyzed for life. In spite of these results, the established medical profession long rejected her treatment. It was in the United States that she finally found acceptance. By the early 1940s, the National Infantile Paralysis Foundation had officially endorsed her treatment, and she saw the opening of the Elizabeth Kenny Institute in Minneapolis. Awarding her an honorary doctor of science degree, the president of the University of Rochester said, "In the dark world of suffering you have lit a candle that will never be put out" (Kenny 1943, passim and 267; Marlow 1979, 259–265).

In still more recent times, women have contributed significant inventions or innovations in the battle against cancer, on many fronts. Outstanding examples are Drs. Charlotte Friend and Ariel Hollinshead. While working as a virologist at the Sloan-Kettering Institute for Cancer Research in the 1950s, Dr. Friend (1921– ) not only demonstrated the viral origins of leukemia (the Friend mouse-leukemia virus), but developed the first successful anticancer vaccine for mammals. She won a *Mademoiselle* magazine achievement award for her work in 1957, and has since then won many other honors, including the Alfred P. Sloan Award (1954, 1957, and 1962), the American Cancer Society Award (1962), and the Virus-Cancer Progress Award from the National Institutes of Health (1974). In 1966, she became director of the Center for Experimental Cell Biology at New York's Mt. Sinai School of Medicine, where her work continues at this writing (Achievement awards 1958, 68; *American Men and Women of Science* 1979, 1602; O'Neill 1979, 224).

Dr. Hollinshead (1929– ) is professor of medicine at George Washington Medical Center in Washington, D.C., and director of its Laboratory for Virus and Cancer Research. Doing both basic and clinical research on cancer, she has helped develop immunotherapy for breast, lung, and gastric cancers as well as melanomas. But she may be best remembered for her lung-cancer vaccines. In the process of inventing these vaccines, which are made from antigens in cancer-cell membranes and are specific for their cancer of origin, she also invented a method of getting antigens out of membranes without destroying their structure, using low-frequency sound. Completed clinical tests show an 80 percent survival rate among those receiving the antigens as opposed to a 49 percent survival rate among the controls. Dr. Hollinshead's work, which opens possibilities for preventing as well as treating cancer, has been called brilliant, "the most advanced and exciting in the world" (*American Men and Women of Science* 1979, 2238; Arehart-Treichel 1980; Cancer vaccines...1979, 248).

Severely neglected by medical research is the field of menstrual disorders. Although this health condition affects 35 million people in the United States alone, and not just once but every month, in 1974 only eight articles on menstrual pain appeared in the entire world medical literature. Quipped Thomas Clayton, vice-president for medical affairs at Tampax in 1979, "If men had cramps, we'd have had a National Institute of Dysmenorrhea for years" (Thorpe 1980, 36; Twin 1979, 8).

Dr. Penny Wise Budoff, a family practitioner and medical school professor at the State University of New York (Stony Brook), undertook some dysmenorrhea research. Beginning in the 1970s with new findings on antiprostaglandins (drugs resembling aspirin but much stronger), she experimented first on herself. Most effective was mefenamic acid, and she next recommended mefenamic acid to a few women in her practice. When these patients also reported some relief, Dr. Budoff set up further experiments. By 1980 the U.S. Food and Drug Administration had approved mefenamic acid for treating menstrual pain. Eighty-five percent of the women tested so far have reported significant relief not only from pain but from nausea, vomiting, dizziness, and weakness. Dr. Budoff has also studied premenstrual tension, and recommends a simple dietary change that may give relief without drugs (Budoff 1980; Thorpe 1980, 36).

It seems fitting to climax and close this brief review of women inventors and innovators in health and medicine with Nobel laureate Dr. Rosalyn Sussman Yalow. Born in the Bronx in 1921, a brilliant and strong-willed child, she took Marie Curie for a role model at age 17, and seems to have moved with unswerving purpose ever since. She graduated from Hunter College with a physics major at age 19. If she was discouraged at being refused a graduate assistantship at Purdue because she was a New Yorker, Jewish, and a woman—or at being told at Columbia that she must start as secretary to a medical school professor—she did not reveal it. She received her Ph.D. in physics from the University of Illinois in 1945.

After working briefly as an electrical engineer, and teaching physics at Hunter College, she became interested in nuclear medicine and took a research position at the Bronx Veterans' Administration Hospital. Thus began the collaboration with Solomon Berson, M.D., that lasted until Berson's death and produced one of the most powerful research techniques, and one of the most powerful diagnostic tools, of the twentieth century: radioimmunoassay (RIA). RIA is a measurement technique so sensitive that it could detect a teaspoon of sugar in a lake 62 miles long, 62 miles wide, and 30 feet deep. In more practical terms, it has allowed doctors for the first time to measure the circulating insulin in a diabetic's blood.

Physicians and researchers continually find new and exciting uses for RIA. Pediatricians can prevent one kind of mental retardation by detecting and treating an infant thyroid deficiency. The RIA test uses only a single drop of the baby's blood, and costs only about a dollar. Thousands of blood banks now screen their blood with RIA to prevent transfusion hepatitis (the test can detect Hepatitis-B virus). RIA can also detect deficiencies or surpluses in human growth hormone in children so that they can be treated to prevent certain kinds of dwarfism and gigantism; can help explain high blood pressure and infertility; can detect hormone-secreting cancers and other endocrine-related disorders. It can detect heroin, methadone, and LSD in the bloodstream; it can gauge circulating vitamins and enzymes to shed light on human nutrition. It can make antibiotic treatment more precise and even help catch murderers, by revealing minute traces of poison in their victims' bodies. RIA was recently used to diagnose Legionnaires' Disease at an early stage, by detecting *Legionella* antigen in the urine.

Had Yalow and Berson decided to patent RIA, they could have been millionaires. Laboratories selling RIA kits do some $30 million in business each year. Thinking like scientists, however, instead of like entrepreneurs, the two freely published their work.

In awarding Rosalyn Yalow the Nobel Prize for physiology and medicine in 1977, the Prize Committee specifically recognized RIA as "the most valuable advance in basic research directly applicable to clinical medicine made in the past two decades" (Levin 1980, 135). In her acceptance speech she said:

> We still live in a world in which a significant fraction of people, including women, believe that a woman belongs and wants to belong exclusively in the home; that a woman should not aspire to achieve more than her male counterparts, and particularly not more than her husband.... But if women are to start moving toward [our] goal, we must believe in ourselves, or no one else will believe in us; we must match our aspirations with the competence, courage, and determination to succeed, and we must feel a personal responsibility to ease the path for those who come afterward (Stone 1978, 34).

Rosalyn Yalow has come a long way from the South Bronx to the chair of Distinguished Professor of Medicine at Mt. Sinai School of Medicine in New York City, where her outstanding work continues (Levin 1980, 133–37; Opfell 1978; Rapid diagnois...1981, 358; Stone 1978, 29–34ff; Yalow 1979).

Through examples from the taming and making of fire to the development of machines and medicine, we have glimpsed the history and prehistory of technology as they would be if women's contributions were included.[5] Most historians of technology to date, looking backward through the distorted glass of a prevailing cultural stereotype that women do not invent, have found, not surprisingly, that women never did invent. If, instead, we examine the evidence—from mythology, anthropology, and history—we find that, as H. J. Mozans wrote nearly seventy years ago:

> More conclusive information respecting woman as an inventor is...afforded by a systematic study of the various races of mankind which are still in a state of savagery [sic]. Such a study discloses the interesting fact that woman has...—*pace* Voltaire—been the inventor of all the peaceful arts of life, and the inventor, too, of the earliest forms of nearly all the mechanical devices now in use in the world of industry (1913, 338).

Our task now is to carry that systematic study of woman's achievement through to the present so that we can, once again, let her own works praise her in the gates.[6]

## NOTES

1 Or indeed, a full-scale companion myth. The Prometheus myth is late and literary. The name Prometheus apparently comes from the Sanskrit word for the fire-making drill, or for the process of making fire by friction. Thus it may be saying only that by inventing fire-making devices, humans stole fire from the realm of the gods and brought it down to the realm of the human (Corson 1894, 714).

2 In a striking nineteenth-century recapitulation of this ancient breakthrough, Sister Tabitha Babbitt (d. 1858) of the Harvard, Massachusetts, Shakers independently invented the circular saw about 1810. After watching the brothers sawing, she concluded that their back-and-forth motion wasted half their effort, and mounted a notched metal disk on her spinning wheel to demonstrate her proposed improvement (Deming and Andrews 1974, 153, 156, 157; Anna J. White and Taylor 1904, 312). Joseph and Frances Gies reveal that Sister Tabitha intended the blade to be turned by water power (1976, 255–256).

3 Mary Daly (1978) presents the most radical view of this tragic event in human history, accepting the highest reported figure for the almost entirely female deaths: 9,000,000. More conservative scholars have estimated as high as 3,000,000; and calmly rational Elise Boulding (1976) says:

> One could argue that there never was any overt decision to "get the women out," that it all happened by default. On the other hand, given the number of instances in which the church combined with various economic groups from doctors to lawyers to merchant guilds, not only to make pronouncements about the incapacities of women, but often to accomplish the physical liquidation of women through witchcraft and heresy trials, one can hardly say that it all happened without anyone intending it. The exclusion of women was a result of impersonal and intentional forces. (p. 505)

4 Had she done this today, she could have patented the organism (*Diamond* v. *Chakrabarty* 1980).

5 For further discussion, see Stanley (1984).

6 "Give her of the fruit of her hands, and let her own works praise her in the gates" (Proverbs 31:31).

## NOTES

Achievement awards, Dr. Charlotte Friend. 1958. *Mademoiselle* (January):68.

*American Men and Women of Science*. 1979. ed. Jacques Cattell Press. 14th ed. New York: Bowker.

Arehart-Treichel, Joan. 1980. Tumor-associated antigens: Attacking lung cancer. *Science News* 118 (July 12):26–28.

Bickel, Lennard. 1972. *Rise up to life: A biography of W. H. Florey*. New York: Scribner's.

Boulding, Elise. 1976. *The underside of history: A view of women through time*. Boulder, Colo.: Westview Press.

Budoff, Penny Wise. 1980. *No more menstrual cramps and other good news*. New York: Putnam.

Cancer vaccines in the works. 1979. *Science News* 117 (April 14):248.

Conwell, Jessie Hayden. 1926 (c. 1865). Inaugural editorial, Ladies' Department, *Minneapolis Daily Chronicle,* weekly ed., *Conwell's Star of the North. Russell H. Conwell and his work,* ed. Agnes Rush Burr. Philadelphia: J. C. Winston: 141–144.

Conwell, Russell H. 1968 (1877). *Acres of Diamonds*. Kansas City, Mo.: Hallmark.

Corson, Juliet. 1894. The evolution of home. *Congress of Women*. ed. Mary K. Eagle. Chicago: Conkey, Vol. II:714–18.

Cowan, Ruth Schwartz. 1979. From Virginia Dare to Virginia Slims: Women and technology in American life. *Technology and Culture* 20, 1 (January):51–63.

Daly, Mary. 1978. *Gynecology: The metaethics of radical feminism*. Boston: Beacon Press.

Davenport, Walter Rice. 1929. *Biography of Thomas Davenport, the "Brandon blacksmith." inventor of the electric motor*. Montpelier, Vt.: The Vermont Historical Society.

Deming, Edward; and Andrews, Faith. 1974. *Work and worship: The economic order of the Shakers*. Greenwich, Conn.: New York Graphic Society.

*Diamond* v. *Chakrabarty*. 100 US 2204 (1980).

Dr. Kendrick dies. 1980–81. *National NOW Times* (Dec./Jan.):12.

Dodge, Norton T. 1966. *Women in the Soviet economy: Their role in economic, scientific, and technical development*. Baltimore: Johns Hopkins.

Doster, Alexis III; Goodwin, Joe; and Ross, Jane M. 1978. *The Smithsonian book of invention*. Washington, D.C.: Smithsonian Exposition Books.

Ehrenreich, Barbara; and English, Deirdre. 1973. *Witches, midwives, and nurses: A history of women healers*. Old Westbury, N.Y.: Feminist Press.

Feldmann, Susan, ed. 1963. *African myths and tales*. New York: Dell.

Frazer, Sir James George. 1930. *Myths on the origin of fire*. London: Macmillan.

Fuller, Edmund. 1955. *Tinkers and genius*. New York: Hastings House.

Gies, Joseph; and Gies, Frances. 1976. *The ingenious yankees*. New York: Crowell.

Goodale, Jane C. 1971. *Tiwi wives: A study of the women of Melville Island, North Australia*. Seattle: University of Washington Press.

Graves, Robert. 1955. *The Greek myths*. Baltimore: Penguin. 2 vols.

Halsbury, Earl of. 1971. Invention and technological progress (fourth annual Spooner Lecture). *The Inventor* (London). (June):10–34.

Hanaford, Phebe A. 1883. *Daughters of America*. Augusta, Maine: True and Company.

Himes, Norman E. 1970 (1936). *Medical history of contraception*. New York: Schocken.

Hobson, Laura Z. 1941. Follies girl to scientist. *Independent Woman* 20 (October):297–298ff.

Hughes, Muriel Joy. 1943. *Women healers in medieval life and literature*. New York: King's Crown Press.

Hurd-Mead, Kate Campbell. 1938. *A history of women in medicine*. Haddam, Conn.: Haddam Press. [1973. Boston: Milford House.]

Hymowitz, Carol; and Weissman, Michaele. 1978. *A history of women in America*. New York: Bantam.

Jack-in-the-pulpit. 1958. *Encyclopedia Britannica* 12.

Jayne, Walter Addison. 1925. *The healing gods of ancient civilizations*. New Haven, Conn.: Yale University Press.

Kendrick, Pearl L. 1942. Use of alum-treated pertussis vaccine, and of alum-precipitated combined pertussis vaccine and diphtheria toxoid for active immunization. *American Journal of Public Health* 32 (June):615–626.

Kenny, Elizabeth. 1943. *And they shall walk*. New York: Dodd Mead.

Kramer, Samuel N. 1963. *The Sumerians: Their history, culture, and character*. Chicago: University of Chicago Press.

Kroeber, Theodora. 1964. *Ishi, last of his tribe*. Berkeley, Calif.: Parnassus.

Lee, Richard B.; and DeVore, Irven, eds. 1968. *Man the hunter*. Chicago: Aldine.

Levin, Beatrice S. 1980. *Women and medicine*. Metuchen, N.J.: Scarecrow.

Marlow, Joan. 1979. *The great women*. New York: A & W Publishers.

Mason, Otis T. 1889. *Cradles of the North American Indians*. Seattle: Shorey.
Mason, Otis. T. 1894. *Woman's share in primitive culture*. New York: D. Appleton.
Mason, Otis T. 1902 (1895). *Origin of inventions*. London: Scott.
Matthews, Alva. n.d. Some Pioneers, unpub. speech delivered to the Society of Women Engineers, n.p.
Mozans, H. J. 1913. *Woman in science*. New York: D. Appleton.
Murdock, George P.; and Provost, Caterina. 1973. Factors in the division of labor by sex. *Ethnology* 12 (April):203–225.
*Notable American Women: The modern period*. 1980. eds., Barbara Sicherman and Carol Hurd Green. Cambridge, MA: Harvard Univ. Press.
Ohnuki-Tierney, Emiko. 1973. The shamanism of the Ainu. *Ethnology* 12 (March):15ff.
O'Neill, Lois Decker, ed. 1979. *The women's book of world records and achievements*. Garden City, N.Y.: Doubleday/Anchor.
Opfell, Olga S. 1978. *The lady laureates; Women who have won the Nobel prize*. Metuchen, N.J.: Scarecrow Press.
Patent for gene-splicing, cloning, awarded Stanford. 1981. *Stanford Observer* (January):1f.
Raper, Kenneth B. 1952. A decade of antibiotics in America. *Mycologia* 45 (Jan.-Feb.):1–59.
Rapid diagnosis of legionellosis. 1981. *Science News* 6 (June):358.
Rayne, Martha Louise. 1893. *What can a woman do: Or, her position in the business and literary world*. Petersburgh, N.Y.: Eagle.
Reed, Evelyn. 1975. *Woman's evolution: From matriarchal clan to patriarchal family*. New York: Pathfinder.
Ribando, Curtis P. 1980. Personal communication, March 26; Patent Advisor, United States Department of Agriculture Northern Regional Research Center, Peoria, Illinois.
Rohrlich, Ruby. 1980. State formation in Sumer and the subjugation of women. *Feminist Studies* 6 (Spring):76–102.
Roth, Walter E. 1908–1909. An inquiry into the animism and folk-lore of the Guiana Indians. 30th Annual Report, Bureau of American Ethnology, Washington, DC.
*Shaker Manifesto*. 1890. American women receiving patents. Vol. 2, no. 7, July, n.p.
Singer, Charles; Holmyard, E. J.; and Hall, A. R. 1954. *A history of technology*. Oxford, Eng.: Oxford University Press.
Smith, Denis. 1978. Lessons from the history of invention. *The Inventor* (London). (January):6ff.
Sokolov, Raymond. 1978. A root awakening. *Natural History* 87 (November):34ff.
Spretnak, Charlene. 1978. *Lost goddesses of ancient Greece: A collection of pre-Hellenic mythology*. Berkeley, Calif.: Moon Books.
Stage, Sarah. 1979. *Female complaints: Lydia Pinkham and the business of women's medicine*. New York: W. W. Norton.
Stanley, Autumn. 1981. Daughters of Isis, daughters of Demeter: When women sowed and reaped. *Women's Studies International Quarterly* 4:3:289–304.
Stanley, Autumn. 1984. *Mothers of invention: Women inventors and innovators through the ages*. Metuchen, N.J.: Scarecrow.
Stellman, Jeanne M. 1977. *Women's work, women's health: Myths and realities*. New York: Pantheon.
Stone, Elizabeth. 1978. A Madame Curie from the Bronx. *The New York Times Magazine*, April 9, pp. 29–34ff.
Stowe, Harriet Beecher. 1862. *The pearl of Orr's Island*. Boston: Houghton Mifflin.

Tanner, Nancy M. 1981. *On becoming human: A model of the transition from ape to human and the reconstruction of early human social life*. Cambridge, Eng.: Cambridge University Press.

Tanner, Nancy; and Zihlman, Adrienne. 1976. Women in evolution. Part I: Innovation and selection in human origins. *Signs* 1 (Spring):585–608.

Thorpe, Susan. 1980. The cure for cramps: It took a woman doctor. *Ms*. (November):36.

*Time*. 1980. [Kendrick obit.] (Oct 20):105.

Trescott, Martha Moore. 1979. Julia B. Hall and aluminum. *Dynamos and virgins revisited,* ed. M. M. Trescott.: 149–179.

Trescott, Martha Moore, ed. 1979. *Dynamos and virgins revisited: Women and technological change in history*. Metuchen, N.J.: Scarecrow Press.

Twin, Stephanie L., ed. 1979. *Out of the bleachers: Writings on women and sport*. Old Westbury, N.Y.: Feminist Press.

Voltaire. 1764. *Dictionnaire philosophique*. Paris: Garnier.

Weiner, Michael A. 1972. *Earth medicine—earth foods: Plant remedies, drugs, and natural foods of the North American Indians*. London: Collier Macmillan.

White, Anna J. 1866. The mystery explained. *Shaker Almanac:* 6ff.

White, Anna J.; and Taylor, Leila S. 1904. *Shakerism, its meaning and message*. Columbus, Ohio: Fred J. Heer.

White, Lynn, Jr. 1978. *Medieval religion and technology*. Berkeley: University of California Press.

Willard, Frances E.; and Livermore, Mary A., eds. 1893. *A woman of the century*. Buffalo: Moulton.

Yalow, Rosalyn. 1979. Speech delivered to the American Physical Society Convention, January, New York City.

He reminds us that the invention of the steam engine led to a preoccupation in the 1850s and 1860s with the "harnessing of vapors, steam and gasses to unwonted ends." The infusion of carbon dioxide into bread dough to secure its instantaneous rise, soda water, the Bessemer method of steel production, even a hot air balloon designed to move a rail car up the side of a mountain: all these were the children of the steam engine.[2]

What then, according to Ellul, is the status of the individual human being in a society where technology has become autonomous? "The human being is delivered helpless, in respect to life's most important and trivial affairs, to a power which is in no sense under his control. For there can be no question of a man's controlling the milk he drinks or the bread he eats, any more than of his controlling his government."[3]

For Ellul, it is efficiency that offers the key to the success of the program of domination that technology pursues by means of these immutable laws. "Every rejection of a technique judged to be bad entails the application of a new technique, the value of which is estimated from the point of view of efficiency alone."[4] Of Ellul's thesis, the sociologist Robert Merton has written: "The essential point, according to Ellul, is that technique produces all this without plan; no one wills or arranges it to be so."[5]

Ellul tempers some 500 pages of pessimism in *The Technological Society* with a paragraph of qualification in the foreword to the American edition: "We must look at it dialectically, and say that man is indeed determined, but that it is open to him to overcome necessity, and that this is freedom. Freedom is not static, but dynamic; not a vested interest, but a prize continually to be won."

The twenty-three years between the original publication of *The Technological Society* and the publication in French in 1977 of *The Technological System* (an English translation became available in 1980) did little to soften Ellul's hard-line position. Whereas formerly he had attempted to describe a society in which technology had become dominant, his revised position was that all the important things of the life of a society, including work, leisure, religion, and other institutions, have become "technicized, homogenized, and integrated in a new whole, which *is not* the society. No more meaningful social or political organization is possible for this ensemble, every part of which is subordinate to the technologies and linked to other parts by the technologies."[6]

Ellul's basic argument, then, is that technique has become everywhere autonomous. It is beyond the control of ordinary men and women, beyond the control of the most powerful technocrats: even their power functions only within the narrow confines of the system. Human freedom becomes narrowed and conditioned by the technological gridlock. Choice is doubly conditioned by the technological system. First, men and women have been indoctrinated by the technological system to the extent that they have certain desires and no others; second, the range of choices is itself limited by the technological system.

And what of his remark that "it is open to [man] to overcome necessity"? For Ellul, true freedom can exist only outside the boundaries of the technological system. But what lies outside those ever-expanding boundaries? Ellul's somewhat vague answer is contained in his theological works. He writes of Christian hope in a time of God's silence, of the possibility of God's intervention into human history, and, above all, of the necessity of allowing the dialectic to do its work.

This last point is an important one which helps explain the surprising fact that the roots of Ellul's thought are, by his own account, in the historical dialectic of Karl Marx no less than in the Protestant theology of John Calvin. First, Ellul regards himself as the model of the lonely critic of the technological system, just as Marx was the model of the solitary critic of industrial capitalism. Second, Marx perceived his work as the exploration of the contradictions of industrial capitalism, and he was hopeful that once the working classes felt those contradictions with sufficient force then they themselves would find a way out. It is the same with Ellul, who seeks to display the contradictions within the technological society and to prod the oppressed to find a way out. In the work of both men there is a strong commitment to the idea that the dialectic, the passage from opposing contradictory forces into the resolution of a new synthesis, takes place on a broader scale than can be anticipated by any one author, whether that author be Marx or Ellul.[7]

Ellul's Marxist analysis of the autonomy of technology places his work in close proximity to that of more single-minded Marxists such as Herbert Marcuse. Marcuse's book *One Dimensional Man* was one of the primary texts of leftist students during the countercultural movement of the 1960s. His essay "The New Forms of Control," reprinted in this section, served as the introductory chapter to that book.

Marcuse's argument wends its way between two contradictory hypotheses. The first is that "advanced industrial society is capable of containing qualitative change for the foreseeable future," and the second is that "forces and tendencies exist which may break this containment and explode the society."[8] Like Ellul, Marcuse views the technological system as having absorbed culture, politics, and the economy. And like Ellul, he thinks that technique, or what he calls "technological rationality," grows according to its own internal laws in ways that are for the most part irresistible.

For Marcuse, the technological system of the "free" world is but a more subtle form of totalitarianism than the crude variety which he experienced firsthand in fascist Europe of the 1930s. Media manipulation, diminishing choices, and the subversion of language (in much the same sense as envisioned by George Orwell in *1984*) have in Marcuse's view led to a type of domination that is virtually incorrigible because it is freely accepted by individuals who are no longer able to see the totalitarian structures within their culture. Some of the contradictions within the technologically repressive society include what he terms "free competition at administered prices, a free press which censors itself, free choice between brands and gadgets." The mass media, for Marcuse, have ceased to be the means to information and have instead become the tools for superficial entertainment, subtle manipulation, and outright indoctrination.

Two key words in Marcuse's account are "alienation" and "transcendence." The alienated individual has uncritically accepted the messages and the products of the technological society. The alienated individual has sacrificed his or her freedom to the demands of technological expedience. Marcuse, like Ellul, thinks that there is little hope that the technological system can be improved from within because its laws and forces have become so nearly autonomous. He nevertheless suggests that it is still possible, though difficult, to "transcend" the system, that is, to view it critically from without. But unlike Ellul, there is no room in Marcuse's thought for an epiphany. Those who are outside the system are not Ellul's hopeful Christians but rather the outcasts, the homeless, the persecuted of other races, and the unemployed. "The fact that they start refusing to play the game," writes Marcuse, "may be the fact which marks the beginning of the end of a period."[9] Both Ellul and Marcuse harbor deep hope that the dialectic of history will do its work, but they differ greatly regarding what manifestation its next stage is likely to take.

In the work of Ellul and Marcuse, it is as if contemporary technology has generated an ideology which human beings have not created and cannot control. For Ellul and Marcuse, technology "embeds" an ideology in its physical structures, including the social structures which are consequent upon it. But as Daniel Lawrence O'Keefe reminds us, "Putting the core of the value system down into the 'ground' of technology (or biological character) where it cannot be discussed is a curious modern variation on the traditional process of putting these values 'up in the sky' of the sacred cosmos."[10]

An interesting alternative to the view of Ellul and Marcuse has been advanced by the sociologist Theodore Roszak. Although Roszak's work has not been included in this anthology for lack of space, a short discussion of his position may nevertheless help to bring the views of Ellul and Marcuse into sharper focus.

Roszak's basic argument is that technology, although each day more dominant, is still in large measure amenable to control by technocrats. For Ellul, of course, even the technocrats have lost control: technology goes barreling down the track of history like a runaway locomotive, unresponsive to the attempts of its engineer to direct it. For Roszak, however, the problem is a different one. The engineer is driving the train faster than can safely be done: the passengers are put in peril, and there is constant danger that the engineer will lose control. For Roszak, then, the problem is not technology in the abstract sense in which Ellul and Marcuse use the term but the technocrats who "talk of facts and probabilities and practical solutions. Their politics *is* the technocracy: the relentless quest for efficiency, for ever more extensive rational control. Parties and governments may come and go, but the experts stay on forever. Because without them the system does not work."[11]

But the difference between Ellul's view of technology and that of Roszak can be brought into even sharper focus. For Ellul, the only freedom we have is the freedom to act in the absence of reasonable expectation, to hope in the absence of reasonable evidence of success. His view is thus reminiscent of the discussion of human freedom by Jean-Paul Sartre in *Being and Nothingness*. Freedom is there described as an act of "choosing oneself." Freedom is not for Sartre made of dreams or wishes, but of the very act of choosing. "Thus we shall not say that a prisoner is always free to go out of prison, which would be absurd, not that he is free to long for release, which would be an irrelevant truism, but that he is always free to try to escape (or to get himself liberated); that is, that whatever his condition may be, he can project his escape and learn the value of his project by undertaking some action."[12]

There are important similarities between the prison of which Sartre writes and the technological society which is the subject of Ellul's argument. In both cases, the individual is "caught" in a system over which he or she has little or no control. The individual in the technological society, like the prisoner, may wish or long to be freed from his or her bondage, but this wishing or longing does not constitute or bring about emancipation. In both cases, the only freedom may be the possibility of planning an escape.

Alternatively, the men and women in Ellul's technological society may be more like the denizens of Plato's cave than the inmates of Sartre's modern prison. In Plato's allegory, the inhabitants of the cave were chained from birth. What they took to be real things were actually no more than shadows cast on the wall of the cave. But they had no desire to escape because they knew nothing of alternatives. The cave to them, as technology to many contemporary men and women, was an entirely natural situation from which escape or rescue appeared more a threat than a liberation.

Whereas for Sartre and Ellul it is a radical act of will, or choice, that constitutes human freedom in a context of incarceration, Roszak believes that men and women in a technological society have freedom of a much more far-ranging variety. And if Ellul and Marcuse argue that "transcendence" of the technological culture is required, Roszak thinks that he has found such transcendence in the countercultural movement of the 1960s. It was by establishing communities based on the primacy of myth and wonder in human experience that the cultural revolutionaries sought the kind of interpersonal interaction with one another that the technological society had denied them.

Ellul views autonomous technology as a kind of straitjacket that isolates human beings from one another and that allows only individual acts of courage by way of countering it. But in Roszak's book it is possible to free oneself from technological constraints by constructing bridges of personal relations to others in the same predicament. This type of interaction, in Roszak's view, is precisely what the technological society tends to inhibit. He thinks that the genius of the revolt of the 1960s was that it was not primarily political, but cultural. His view is that political movements tend to be thin and unidimensional forces. They rest on ideological positions that are rationally derived and propagated. They admit of orthodoxy and heterodoxy. Cultural movements, on the other hand, tend to be multidimensional, to maximize alternatives, and to increase participation.

For Ellul, technology is a vast system of regularities. It is in fact very much like the system of regularities we call "nature," although it competes with nature and attempts to complete and supplant natural sequences of events. Although technique is by definition artificial, its laws are just as "real"—that is, they exercise a control over us which is as unavoidable as a "law of nature."

For Roszak, the situation offers more room to move. Men and women in a technological society are in his view like people who have become addicted to a hard drug. The pushers are in this case the technocrats who continually devise new ways to hook their victims at the same time they continually raise the price of addiction. One can escape the tyranny of the drug if one chooses to do so, but such an escape is not easy and it is often possible only when a community of individuals works in concert, offering mutual support.

For Ellul, technology is beyond control. Political action is ultimately ineffective, and God's will is revealed only to individuals in specific situations. If there is to be salvation from technology, therefore, it will come from a God who transcends the human sphere. For Roszak, salvation from technology's autonomy rests firmly within the here and now of community endeavor. He thinks that amelioration is possible because technology is not *beyond* control but simply *out of* control and because appropriate actions can effect personal and communal release from its grip. But what is even more important is that Roszak rejects Ellul's "all or nothing" characterization of technology; he thinks that we can change the course of technology by forming and maintaining enlightened communities. We can recall technology when it has taken a wrong turn. We can put it back on course. One of the ways this can happen is by demythologizing science, that is, by taking aim at the outworn myths of scientific objectivity. Another of the ways this can happen is by engaging in the scientific analysis of myth. Roszak argues that one of the places where science and myth meet is in the realm of concern about the quality of the environment.

Certainly not all those who write about contemporary affairs agree with Ellul that technology is beyond control, or even with Roszak that it is out of control. Arnold Pacey, for example, has utilized case studies from the history of technology to demonstrate that the determinist thesis, "which presents technical advance as a process of steady development dragging human society along in its train," is insupportable. Because of its rich historical material, the essay by Pacey which I have included here might well have found a place in Part Five. And because of its detailed treatment of concepts of human progress, it might also have been placed in Part Seven, along with essays by Ayres and Habermas. I have located it in this chapter, however, because it represents an excellent counterargument to Ellul's view, which Pacey calls "machine mysticism."

By means of his analyses of the development of grain yields, improvements in lathe design, and advancements in steam engines, Pacey demonstrates three important points. The first is that the history of technology, at least in these areas, does not exhibit Ellul's scenario of a geometrical forward drive of technical rationality. Technological change in these areas has not been a smooth pattern of development from less to more complex and from lower to higher efficiency. It has instead been an affair of intermittent development, with plateaus and even declines. In other words, Pacey contends that the history of technology exhibits not so much a pattern of smooth development as a pattern of fitful evolution in which there are failures as well as successes.

Pacey's second point is that machines do not always make history, though they may at times do so. Against Marx's contention that the steam engine leads to industrial capitalism, he argues that the heart of industrial capitalism, the factory system, was an invention that had an earlier origin than most of the machines it contained.

Third, Pacey argues that if technological determinism is untenable, so is its complete opposite. Innovation, in his view, must be seen as an outcome of a number of interactive factors, among which are cultural, social, and technical forces. He suggests that the persistence of the myth of technological determinism may in fact be due to political purposes: People are more likely to accept the advice of technical "experts" if there is an intrinsic dynamic of scientific technology that must be managed.

In sum, Pacey argues against narrow "linear" views of technological development. He believes that these attempts to "understand in depth rather than to broaden awareness" and to measure whatever can be measured may neglect the stubborn fact that people are free to change their views about what is important, and consequently what type of world they wish to construct and inhabit.

The authors represented in this section are participants in a debate that extends far beyond the halls of academia. Each of us has felt the heady delights as well as the distressing strictures which accompany new forms of technology. At their worst, new technologies appear monolithic and unyielding: they seem to curtail choice and determine future action. At their best, they open up whole new ways of thinking and acting and are the source of exhilarating new freedoms. In these essays Ellul, Marcuse, and Pacey provide a wide variety of tools to help us take the measure of these facilities and constraints.

## NOTES

1 Jacques Ellul, *The Technological Society,* John Wilkinson (trans.), Alfred A. Knopf, New York, 1964, p. 87.
2 Siegfried Giedion, *Mechanization Takes Command,* W. W. Norton & Co., New York, 1969, p. 185.
3 Ellul (1964), p. 107.
4 Ellul (1964), p. 110.
5 Ellul (1964), p. vii.
6 Jacques Ellul, *The Technological System,* Joachim Neugroschel (trans.), Continuum, New York, 1980, p. 15.
7 For a good discussion of Ellul's Marxism, see David C. Menninger, "Marx in the Social Thought of Jacques Ellul," in Clifford G. Christians and Jay M. Van Hook (eds.), *Jacques Ellul: Interpretive Essays,* University of Illinois Press, Urbana, 1981. For a good discussion of the religious components of Ellul's work, see Jay M. Van Hook, "The Politics of Man, the Politics of God, and the Politics of Freedom," in the same volume.
8 Herbert Marcuse, *One Dimensional Man,* Beacon Press, Boston, 1964, p. xv.
9 Ibid., p. 257.
10 Daniel Lawrence O'Keefe, *Stolen Lightning: The Social History of Magic,* Continuum, New York, 1982, p. 82.

11 Theodore Roszak, *The Making of a Counter Culture,* Anchor Books, Garden City, N.Y., 1969, p. 21.

12 Jean-Paul Sartre, *Being and Nothingness,* Hazel Barnes (trans.), Philosophical Library, New York, 1956, p. 484. It should be added that Sartre's notion of freedom underwent a significant modification in *Critique of Dialectical Reason,* published in French in 1960 and in English in 1976. In that work freedom is presented in much more modest terms. It is limited by what Sartre calls the "practico-inert," his term for the network of constraints which have been bequeathed to us as a result of decisions made by our ancestors and by our past selves.

## The Autonomy of Technology

**Jacques Ellul**

The primary aspect of autonomy is perfectly expressed by Frederick Winslow Taylor, a leading technician. He takes, as his point of departure, the view that the industrial plant is a whole in itself, a "closed organism," an end in itself. Giedion adds: "What is fabricated in this plant and what is the goal of its labor—these are questions outside its design." The complete separation of the goal from the mechanism, the limitation of the problem to the means, and the refusal to interfere in any way with efficiency; all this is clearly expressed by Taylor and lies at the basis of technical autonomy.

Autonomy is the essential condition for the development of technique, as Ernst Kohn-Bramstedt's study of the police clearly indicates. The police must be independent if they are to become efficient. They must form a closed, autonomous organization in order to operate by the most direct and efficient means and not be shackled by subsidiary considerations. And in this autonomy, they must be self-confident in respect to the law. It matters little whether police action is legal, if it is efficient. The rules obeyed by a technical organization are no longer rules of justice or injustice. They are "laws" in a purely technical sense. As far as the police are concerned, the highest stage is reached when the legislature legalizes their independence of the legislature itself and recognizes the primacy of technical laws. This is the opinion of Best, a leading German specialist in police matters.

The autonomy of technique must be examined in different perspectives on the basis of the different spheres in relation to which it has this characteristic. First, technique is autonomous with respect to economics and politics. We have already seen that, at the present, neither economic nor political evolution conditions technical progress. Its progress is likewise independent of the social situation. The converse is actually the case, a point I shall develop at length. Technique elicits and conditions social, political, and economic change. It is the prime mover of all the rest, in spite of any appearance to the contrary and in spite of human pride, which pretends that man's philosophical theories are still determining influences and man's political regimes decisive factors in technical evolution. External necessities no longer determine technique. Technique's own internal necessities are determinative. Technique has become a reality in itself, self-sufficient, with its special laws and its own determinations.

Let us not deceive ourselves on this point. Suppose that the state, for example, intervenes in a technical domain. Either it intervenes for sentimental, theoretical, or intellectual reasons, and the effect of its intervention will be negative or nil; or it intervenes for reasons of political technique, and we have the combined effect of two techniques. There is no other possibility. The historical experience of the last years shows this fully.

To go one step further, technical autonomy is apparent in respect to morality and spiritual values. Technique tolerates no judgment from without and accepts no limitation. It is by virtue of technique rather than science that the great principle has become established: *chacun chez soi.* Morality judges moral problems; as far as technical problems are concerned, it has nothing to say. Only technical criteria are relevant. Technique, in sitting in judgment on itself, is clearly freed from this principal obstacle to human action. (Whether the obstacle is valid is not the question here. For the moment we merely record that it is an obstacle.) Thus, technique theoretically and systematically assures to itself that liberty which it has been able to win practically. Since it has put itself beyond good and evil, it need fear no limitation whatever. It was long claimed that technique was neutral. Today this is no longer a useful distinction. The power and autonomy of technique are so well secured that it, in its turn, has become the judge of what is moral, the creator of a new morality. Thus, it plays the role of creator of a new civilization as well. This morality—internal to technique—is assured of not having to suffer from technique. In any case, in respect to traditional morality, technique affirms itself as an independent power. Man alone is subject, it would seem, to moral judgment. We no longer live in that primitive epoch in which things were good or bad in themselves. Technique in itself is neither, and can therefore do what it will. It is truly autonomous.

However, technique cannot assert its autonomy in respect to physical or biological laws. Instead, it puts them to work; it seeks to dominate them.

Giedion, in his probing study of mechanization and the manufacture of bread, shows that "wherever mechanization encounters a living substance, bacterial or animal, the organic substance determines the laws." For this reason, the mechanization of bakeries was a failure. More subdivisions, intervals, and precautions of various kinds were required in the mechanized bakery than in the nonmechanized bakery. The size of the machines did not save time; it merely gave work to larger numbers of people. Giedion shows how the attempt was made to change the nature of the bread in order to adapt it to mechanical manipulations. In the last resort, the ultimate success of mechanization turned on the transformation of human taste. Whenever technique collides with a natural obstacle, it tends to get around it either by replacing the living organism by a machine, or by modifying the organism so that it no longer presents any specifically organic reaction.

The same phenomenon is evident in yet another area in which technical autonomy asserts itself: the relations between techniques and man. We have already seen, in connection with technical self-augmentation, that technique pursues its own course more and more independently of man. This means that man participates less and less actively in technical creation, which, by the automatic combination of prior elements, becomes a kind of fate. Man is reduced to the level of a catalyst. Better still, he resembles a slug inserted into a slot machine: he starts the operation without participating in it.

But this autonomy with respect to man goes much further. To the degree that technique must attain its result with mathematical precision, it has for its object the elimination of all human variability and elasticity. It is a commonplace to say that the machine replaces the human being. But it replaces him to a greater degree than has been believed.

Industrial technique will soon succeed in completely replacing the effort of the worker, and it would do so even sooner if capitalism were not an obstacle. The worker, no longer needed to guide or move the machine to action, will be required merely to watch it and to repair it when it breaks down. He will not participate in the work any more than a boxer's manager participates in a prize fight. This is no dream. The automated factory has already been realized for a great number of operations, and it is realizable for a far greater number. Examples multiply from day to day in all areas. Man indicates how this automation and its attendant exclusion of men operates in business offices; for example, in the case of the so-called tabulating machine. The machine itself interprets the data, the elementary bits of information fed into it. It arranges them in texts and distinct numbers. It adds them together and classifies the results in groups and subgroups, and so on. We have here an administrative circuit accomplished by a single, self-controlled machine. It is scarcely necessary to dwell on the astounding growth of automation in the last ten years. The multiple applications of the automatic assembly line, of automatic control of production operations (so-called cybernetics) are well known. Another case in point is the automatic pilot. Until recently the automatic pilot was used only in rectilinear flight; the finer operations were carried out by the living pilot. As early as 1952 the automatic pilot effected the operations of takeoff and landing for certain supersonic aircraft. The same kind of feat is performed by automatic direction finders in antiaircraft defense. Man's role is limited to inspection. This automation results from the development of servomechanisms which act as substitutes for human beings in more and more subtle operations by virtue of their "feedback" capacity.

This progressive elimination of man from the circuit must inexorably continue. Is the elimination of man so unavoidably necessary? Certainly! Freeing man from toil is in itself an ideal. Beyond this, every intervention of man, however educated or used to machinery he may be, is a source of error and unpredictability. The combination of man and technique is a happy one only if man has no responsibility. Otherwise, he is ceaselessly tempted to make unpredictable choices and is susceptible to emotional motivations which invalidate the mathematical precision of the machinery. He is also susceptible to fatigue and discouragement. All this disturbs the forward thrust of technique.

Man must have nothing decisive to perform in the course of technical operations; after all, he is the source of error. Political technique is still troubled by certain unpredictable phenomena, in spite of all the precision of the apparatus and the skill of those involved. (But this technique is still in its childhood.) In human reactions, howsoever well calculated they may be, a "coefficient of elasticity" causes imprecision, and imprecision is intolerable to technique. As far as possible, this source of error must be eliminated. Eliminate the individual, and excellent results ensue. Any technical man who is aware of this fact is forced to support the opinions voiced by Robert Jungk, which can be summed up thus: "The individual is a brake on progress." Or: "Considered from the modern technical point of view, man is a useless appendage." For instance, ten per cent of all telephone calls are wrong numbers, due to human error. An excellent use by man of so perfect an apparatus!

Now that statistical operations are carried out by perforated-card machines instead of human beings, they have become exact. Machines no longer perform merely gross operations. They perform a whole complex of subtle ones as well. And before long—what with the electronic brain—they will attain an intellectual power of which man is incapable.

Thus, the "great changing of the guard" is occurring much more extensively than Jacques Duboin envisaged some decades ago. Gaston Bouthoul, a leading sociologist of the phenomena of war, concludes that war breaks out in a social group when there is a "plethora of young men surpassing the indispensable tasks of the economy." When for one reason or another these men are not employed, they become ready for war. It is the multiplication of men who are excluded from working which provokes war. We ought at least to bear this in mind when we boast of the continual decrease in human participation in technical operations.

However, there are spheres in which it is impossible to eliminate human influence. The autonomy of technique then develops in another direction. Technique is not, for example, autonomous in respect of clock time. Machines, like abstract technical laws, are subject to the law of speed, and coordination presupposes time adjustment. In his description of the assembly line, Giedion writes: "Extremely precise time tables guide the automatic cooperation of the instruments, which, like the atoms in a planetary system, consist of separate units but gravitate with respect to each other in obedience to their inherent laws." This image shows in a remarkable way how technique became simultaneously independent of man and obedient to the chronometer. Technique obeys its own specific laws, as every machine obeys laws. Each element of the technical complex follows certain laws determined by its relations with the other elements, and these laws are internal to the system and in no way influenced by external factors. It is not a question of causing the human being to disappear, but of making him capitulate, of inducing him to accommodate himself to techniques and not to experience personal feelings and reactions.

No technique is possible when men are free. When technique enters into the realm of social life, it collides ceaselessly with the human being to the degree that the combination of man and technique is unavoidable, and that technical action necessarily results in a determined result. Technique requires predictability and, no less, exactness of prediction. It is necessary, then, that technique prevail over the human being. For technique, this is a matter of life or death. Technique must reduce man to a technical animal, the king of the slaves of technique. Human caprice crumbles before this necessity; there can be no human autonomy in the face of technical autonomy. The individual must be fashioned by techniques, either negatively (by the techniques of understanding man) or positively (by the adaptation of man to the technical framework), in order to wipe out the blots his personal determination introduces into the perfect design of the organization.

But it is requisite that man have certain precise inner characteristics. An extreme example is the atomic worker or the jet pilot. He must be of calm temperament, and even temper, he must be phlegmatic, he must not have too much initiative, and he must be devoid of egotism. The ideal jet pilot is already along in years (perhaps thirty-five) and has a settled direction in life. He flies his jet in the way a good civil servant goes to his office. Human joys and sorrows are fetters on technical aptitude. Jungk cites the case of a test pilot who had to abandon his profession because "his wife behaved in such a way as to lessen his capacity to fly. Every day, when he returned home, he found her shedding tears of joy. Having become in this way accident conscious, he dreaded catastrophe when he had to face a delicate situation." The individual who is a servant of technique must be completely unconscious of himself. Without this quality, his reflexes and his inclinations are not properly adapted to technique.

Moreover, the physiological condition of the individual must answer to technical demands. Jungk gives an impressive picture of the experiments in training and control that jet pilots have to undergo. The pilot is whirled on centrifuges until he "blacks out" (in order to measure his toleration of acceleration). There are catapults, ultrasonic chambers, etc., in which the candidate is forced to undergo unheard-of tortures in order to determine whether he has adequate resistance and whether he is capable of piloting the new machines. That the human organism is, technically speaking, an imperfect one is demonstrated by the experiments. The sufferings the individual endures in these "laboratories" are considered to be due to "biological weaknesses," which must be eliminated. New experiments have pushed even further to determine the reactions of "space pilots" and to prepare these heroes for their roles of tomorrow. This has given birth to new sciences, biometry for example; their one aim is to create the new man, the man adapted to technical functions.

It will be objected that these examples are extreme. This is certainly the case, but to a greater or lesser degree the same problem exists everywhere. And the more technique evolves, the more extreme its character becomes. The object of all the modern "human sciences" (which I will examine later on) is to find answers to these problems.

The enormous effort required to put this technical civilization into motion supposes that all individual effort is directed toward this goal alone and that all social forces are mobilized to attain the mathematically perfect structure of the edifice. ("Mathematically" does not mean "rigidly." The perfect technique is the most adaptable and, consequently, the most plastic one. True technique will know how to maintain the illusion of liberty, choice, and individuality; but these will have been carefully calculated so that they will be integrated into the mathematical reality merely as appearances!) Henceforth it will be wrong for a man to escape this universal effort. It will be inadmissible for any part of the individual not to be integrated in the drive toward technicization; it will be inadmissible that any man even aspire to escape this necessity of the whole society. The individual will no longer be able, materially or spiritually, to disengage himself from society. Materially, he will not be able to release himself because the technical means are so numerous that they invade his whole life and make it impossible for him to escape the collective phenomena. There is no longer an uninhabited place, or any other geographical locale, for the would-be solitary. It is no longer possible to refuse entrance into a community to a highway, a high-tension line, or a dam. It is vain to aspire to live alone when one is obliged to participate in all collective phenomena and to use all the collective's tools, without which it is impossible to earn a bare subsistence. Nothing is gratis any longer in our society; and to live on charity is less and less possible. "Social advantages" are for the workers alone, not for "useless mouths." The solitary is a useless mouth and will have no ration card—up to the day he is transported to a penal colony. (An attempt was made to institute this procedure during the French Revolution, with deportations to Cayenne.)

Spiritually, it will be impossible for the individual to disassociate himself from society. This is due not to the existence of spiritual techniques which have increasing force in our society, but rather to our situation. We are constrained to be "engaged," as the existentialists say, with technique. Positively or negatively, our spiritual attitude is constantly urged, if not determined, by this situation. Only bestiality, because it is unconscious, would seem to escape this situation, and it is itself only a product of the machine.

Every conscious being today is walking the narrow ridge of a decision with regard to technique. He who maintains that he can escape it is either a hypocrite or unconscious. The autonomy of technique forbids the man of today to choose his destiny. Doubtless, someone will ask if it has not always been the case that social conditions, environment, manorial oppression, and the family conditioned man's fate. The answer is, of course, yes. But there is no common denominator between the suppression of ration cards in an authoritarian state and the family pressure of two centuries ago. In the past, when an individual entered into conflict with society, he led a harsh and miserable life that required a vigor which either hardened or broke him. Today the concentration camp and death await him; technique cannot tolerate aberrant activities.

Because of the autonomy of technique, modern man cannot choose his means any more than his ends. In spite of variability and flexibility according to place and circumstance (which are characteristic of technique) there is still only a single employable technique in the given place and time in which an individual is situated. We have already examined the reasons for this.

At this point, we must consider the major consequences of the autonomy of technique. This will bring us to the climax of this analysis.

Technical autonomy explains the "specific weight" with which technique is endowed. It is not a kind of neutral matter, with no direction, quality, or structure. It is a power endowed with its own peculiar force. It refracts in its own specific sense the wills which make use of it and the ends proposed for it. Indeed, independently of the objectives that man pretends to assign to any given technical means, that means always conceals in itself a finality which cannot be evaded. And if there is a competition between this intrinsic finality and an extrinsic end proposed by man, it is always the intrinsic finality which carries the day. If the technique in question is not exactly adapted to a proposed human end, and if an individual pretends that he is adapting the technique to this end, it is generally quickly evident that it is the end which is being modified, not the technique. Of course, this statement must be qualified by what has already been said concerning the endless refinement of techniques and their adaptation. But this adaptation is effected with reference to the techniques concerned and to the conditions of their applicability. It does not depend on external ends. Perrot has demonstrated this in the case of judicial techniques, and Giedion in the case of mechanical techniques. Concerning the overall problem of the relation between the ends and the means, I take the liberty of referring to my own work, *Présence au monde moderne*.

Once again we are faced with a choice of "all or nothing." If we make use of technique, we must accept the specificity and autonomy of its ends, and the totality of its rules. Our own desires and aspirations can change nothing.

The second consequence of technical autonomy is that it renders technique at once sacrilegious and sacred. (*Sacrilegious* is not used here in the theological but in the sociological sense.) Sociologists have recognized that the world in which man lives is for him not only a material but also a spiritual world; that forces act in it which are unknown and perhaps unknowable; that there are phenomena in it which man interprets as magical; that there are relations and correspondences between things and beings in which material connections are of little consequence. This whole area is mysterious. Mystery (but not in the Catholic sense) is an element of man's life. Jung has shown that it is catastrophic to make superficially clear what is hidden in man's innermost depths. Man must make allowance for a background, a great deep above which lie his reason and his clear consciousness. The mystery of man perhaps creates the mystery of the world he inhabits. Or perhaps this mystery is a reality in itself. There is no way to decide between these two alternatives. But, one way or the other, mystery is a necessity of human life.

Man cannot live without a sense of the secret. The psychoanalysts agree on this point. But the invasion of technique desacralizes the world in which man is called upon to live. For technique nothing is sacred, there is no mystery, no taboo. Autonomy makes this so. Technique does not accept the existence of rules outside itself, or of any norm. Still less will it accept any judgment upon it. As a consequence, no matter where it penetrates, what it does is permitted, lawful, justified.

To a great extent, mystery is desired by man. It is not that he cannot understand, or enter into, or grasp mystery, but that he does not desire to do so. The sacred is what man decides unconsciously to respect. The taboo becomes compelling from a social standpoint, but there is always a factor of adoration and respect which does not derive from compulsion and fear.

Technique worships nothing, respects nothing. It has a single role: to strip off externals, to bring everything to light, and by rational use to transform everything into means. More than science, which limits itself to explaining the "how," technique desacralizes because it demonstrates (by evidence and not by reason, through use and not through books) that mystery does not exist. Science brings to the light of day everything man had believed sacred. Technique takes possession of it and enslaves it. The sacred cannot resist. Science penetrates to the great depths of the sea to photograph the unknown fish of the deep. Technique captures them, hauls them up to see if they are edible—but before they arrive on deck they burst. And why should technique not act thus? It is autonomous and recognizes as barriers only the temporary limits of its action. In its eyes, this terrain, which is for the moment unknown but not mysterious, must be attacked. Far from being restrained by any scruples before the sacred, technique con-

lieu in which we have always lived into a technological milieu. They are fundamental, I think, in that we must call upon a theory which is entirely new, the theory of the three milieus. For it is not true that we have passed directly from the natural to the technological milieu. In reality, we have known not two, but three successive milieus: the natural milieu; the milieu of society; and now, the milieu of technology.

The natural milieu was that of the prehistoric period, when there was no organized society as yet and when immediate contact with nature was absolutely permanent. This was really an immediate contact; nothing mediated, nothing served as an intermediary between the human group and nature in the traditional sense of the term. Nature provided a sustenance for human beings, who lived by hunting, by gathering; and nature also provided their principal danger—the danger of poisons, the danger of wild animals, certainly, but also the danger of barrenness, the danger of shortages. This was the first milieu, the one we think of quite spontaneously.

However, humans found a way of defending themselves against this natural milieu, getting the best from it and protecting themselves against it—something that would mediate between themselves and nature. This new means was society. The creation of human society appeared with the times traditionally known as historical. History is tied not to the existence of a natural milieu, but to the existence of a social environment. Society allowed humans to grow strong. The human group became an organized group, a group that has gradually dominated the natural environment, using it as best it can.

What I call the ''social period'' is the historical period beginning some seven thousand years ago, when human beings succeeded in more or less protecting themselves against nature and taming it, in grouping into a society and in utilizing technologies. During this period, society was the natural milieu for human beings, who remained in close contact with nature (there was a balance between town and country). Technologies were only means, instruments. They were not all-invasive. The great problems were those in the organization of society, the political form to choose, the distribution of labor and wealth, the circulation of information, and the maintenance of cohesion among groups. Thus, society was the environment which allowed human beings to live, and also caused problems.

But while becoming a human milieu, society also turned into something that allowed us to live and then imperiled us. For the chief dangers were now wars, which are an invention of societies. The social milieu still seems like a ''natural milieu,'' because people to some extent remained in nature. Throughout the historical period, this social milieu marks the intermediary period between the natural milieu and the one we know today, the technological milieu.

The third milieu, this technological one, has actually replaced society. Not only are natural data and natural facts utilized by technology, mediated by technology; not only are people alienated from nature by technology; but also social relations are mediated and shaped by technology. In short, the weight of society is far lighter now than the weight of technology.

Of course, when I speak of these three successive milieus for humanity, I am certainly not saying that the appearance of a new milieu eliminates and destroys the preceding one. I have just mentioned that when human beings organize themselves into a society, they still remain in contact with the natural milieu. Society is a means for best utilizing the means of nature and avoiding the disadvantages of the natural environment. By the same token, it is obvious that technology does not suppress nature or society; rather, it mediates them. Nature was mediated by society, with people living in the social group and beyond nature. Now, technology mediates both society and, on a secondary level, nature.

Each preexisting element—nature or society—is to some extent obsolete. But it still exists in regard to dangers. For instance, the dangers of natural epidemics were always imminent in the social environment. However, epidemics were a relatively less serious matter than the dangers inherent in society. Likewise, there are natural dangers and societal dangers that still survive, even though the environment we now live in is a technological one. There are still typhoons and earthquakes; there are still wars and dictatorships. Yet in reality, all things are already rendered obsolete and placed on a secondary level by the emergence of a new environment. In other words, the problems raised by a former, obsolete milieu are no longer the essential or fundamental problems.

When human beings were organized in a society, their fundamental problems—and this was the whole question of politics—were the very organization of society, the relations between various societies, the growth of political power, and the control of political power. These issues were far more important than those concerning natural phenomena.

Similarly, today, the technological phenomena, that is, the positive and the negative aspects of technology, the things that both endanger us and increase our power, are far more important than the problems caused by society itself. Hence, we ultimately come to the following conclusion: most of the problems we face today—especially the purely political ones, which relate to the foregoing historical period when the essential milieu was society—are all obsolete by now. These are ancient problems, if you will. During the historical period, it was more important to solve political problems than a certain number of purely natural problems. Likewise, today, it is more important, more decisive, to solve the difficulties raised by technology, the dangers coming from technology, than to solve purely political issues, the problems of elections, the question of whether a system should be democratic or not.

Of course, just as society employed the means of nature, so too technology employs the means of society. Hence, technology aggravates political problems. Political power is now in the hands of technological structures that far surpass any power ever held by older political authorities. However, this is no longer a political problem. Whatever the regime, it has its structures in hand. The problem is actually a technological one.

Thus, technology has become an environment. Beyond that, however, it has also become a system. I am using the term "system" in a sense that has now become customary since Ludwig von Berthalanffy: an ensemble of mutually integrated elements, situated in terms of one another and reacting to one another. On the one hand, every element in the system is understood only in terms of the whole, in terms of the system. Any variation in the whole has consequences for the integrated parts. And reciprocally, any change in the elements affects the whole.

This, I feel, is a new view of technology, with a difficulty that I have already pointed out: when I speak of technology as a system, I mean two different things. The first is that technology has in fact become a system. This means that each individual technology is actually integrated in a totality; each datum of technology must be understood in terms of this totality. Hence, there is actually a system of many technologies. Secondly, when I say that technology is a system, I mean that the concept of "system," used both philosophically and sociologically, is a means of interpreting what is happening technologically. It is virtually an epistemological instrument allowing us to know and understand technology better. Hence, the term "system" designates both the fact and the work instrument, the instrument of comprehension.

This interpretation of technology as a system has enormous consequences. I will mention only two.

First of all, technology as a system obeys its own law, its own logic. In other words, we are dealing with an autonomy of technology, a closure of technology in itself. There is a very small margin of possibility for intervention, for outside action—economic, political, or whatever—on technology. Furthermore, technology is autonomous in regard to morality, politics, and so on.

On the other hand, it is involved in a process of self-augmentation. Technology increases itself for its own reason and with its own causalities. We would obviously have to go into a long explanation of how the person who interferes with the technological environment and the technological system intervenes to some extent as an instrument of technology and not as its master. Technology has its augmentation power, which is intrinsic to it.

We encounter an apparent difficulty here. The technological system progresses by virtue of its intrinsic laws, and there is an autonomous process of organization. But at the same time, this can occur only by means of constant human decisions and interventions. By describing the system as autonomous, I do not mean an autonomy capable of directing itself and reproducing itself without human intervention. What happens is that the system determines the one who must make the decisions and who must act. The *sole* actions and decisions to be allowed are the ones that promote the growth of technology. The rest are rejected and quickly forgotten. Those who make the decisions are neither aesthetes, skeptics, critics, nor people free of obligation. Since childhood, they have been accustomed to technology: they feel that only technology is important, that only progressive thinking is valid; and they have learned technologies for their work and their leisure. In this way their decisions always support the autonomy of technology.

Here we have a problem. Like any system, technology ought to have its self-regulation, its feedback. Yet it has nothing of the sort. For instance, if one observes a set of negative effects by a group of technicians, one should not only repair the damage, but go back to the origin of the technologies involved and modify their application *at the source*—for fertilizers, say, or certain work methods, or chemical products. But this action is *never* taken. We prefer to let the drawbacks and dangers develop (on the pretext that they are not fully demonstrated) and to create new technologies to "repair" the problems. In fact this actually entails a positive feedback. There is no self-regulation of any kind in the technological system. This does not mean that it is not a system. It does, however, mean that we are confronted with a system that has gotten out of hand—a system incapable of controlling itself. Hence, we cannot expect any rationality, contrary to what we may believe. And this, I may say, is going to be the chief danger, the chief question when we think of ourselves within the system. That is a first set of consequences.

A second set of consequences is that, contrary to what we usually do, we can no longer understand technology per se. No technology can be understood in itself because it exists only in terms of the whole. Yet that is what we always do when, for instance, we consider television. We ask: "What are the effects of television? Can one escape the impact of television? Can we master television?" And the reaction is always totally elementary: "But I'm not the least bit addicted to television. I can switch off my set whenever I like. I'm completely free." We respond as if television were a separate phenomenon, likewise independent of the system. The same is true for the automobile. Observers are investigating, for example, the effects of the car on an individual or on an entire populace, as though the car were not located within the technological system, a part of an extremely complex set of technologies.

However, if we wish to understand television, we have to place it within the technological system, that is, television in relation to advertising, in relation to the fact that the world I live in is turning more and more into a visual world, or that I am constantly learning that only the image that I see corresponds to a reality or that the world in which I am likewise makes constant demands by way of a growing consumerism. This is the same world in which I am obliged by the group in which I live to keep up to date on whatever takes place. I am by no means free to watch my television set or not to watch it, because tomorrow morning the people I meet will talk to me about such and such a program and I do not want to put myself on the fringes of the group.

Likewise, I am part of a world in which the technological operation requires a certain amount of knowledge. I cannot enter a milieu or a job if I do not possess a certain quantity of knowledge, and a good portion of this knowledge is transmitted to me by television. Hence, in reality, I am not independent of my television set. With the set belonging to me, I am integrated in a totality that is the technological society, of which television is a part, and I am absolutely not free in my choices, in my decisions.

Naturally, I can decide not to watch a certain movie or program. But am I really sure that I can decide? For I am also a person who spends my day working at a generally technological job that is quite uninteresting, repetitive, and anything but absorbing. In the evening, what do I have for relaxing, for relieving the buildup of nervous tension that I have experienced all day long? Television. Hence, in a sense I watch television as a reward at the end of the day, and this too is caused by my living in this milieu.

Therefore, I am absolutely not independent in regard to television; and it is no use trying to understand the effect of television as an isolated phenomenon. The true problem is the situation of human beings in the totality of the technological society.

I already mentioned the absence of regulation in the system. This non-self-regulation and another feature of technology, its ambivalence, prohibit any accurate forecasting of what may happen. We are always left with two hypotheses: Huxley's brave new world; or else the "disasters" foreseen by science fiction or the Club of Rome. Neither possibility is predictable.

In fact, Huxley's brave new world, where everything is normalized, is, as I will explain later, absolutely impossible. On the other hand, the disasters predicted by the Club of Rome strike me as equally improbable since all precise scientific forecasts about the technological world seem false to me. They are false because the system has no self-regulation, and we are incapable of foretelling the actual developments.

Then there is the ambivalence of technology, the fact that each emerging technology brings either positive effects or negative effects mixed in with the others. It is extremely simplistic and elementary to think that one can separate, or to claim that one can suppress negative effects and retain the positive ones. Unhappily, this is never the case. I recall that when nuclear energy was launched, people simplistically said: "All we have to do is stop making atomic bombs and produce nuclear energy, and everything will be all right; we'll be pacifists." Alas, we know that the development of nuclear plants presents yet another danger and that ultimately every such plant is a potential atomic bomb.

Hence, the effects are by no means clearly separated. When we think of chemical products, we must bear in mind that a chemist comes up with a product of which we know certain effects. The secondary effects are only revealed a long time later; we are unable to discern them in advance. The same is true of fertilizers, medicines, and so on.

Thus, the positive effects and the negative effects of technology are closely, strangely interrelated. We may say that each technological advance increases both the positive and the negative effects, of which we generally know very little. I would therefore say that I cannot endorse either Huxley or the Club of Rome because of the margin of unpredictability. No scientific forecast seems certain to me. Nor can we now say that technology will keep progressing from innovation to innovation at the rate it has moved during the past thirty years, or that, on the contrary, we are veering toward a period of stoppage, of technological stagnation, which would obviously give us a certain amount of time, a delay. Clearly, a work like Huxley's or a cry of alarm like that of the Club of Rome's is meant to alert us, to warn us of certain possibilities that lie ahead, but there is no way that we can tell which possibility is bound to come true.

Still, one thing seems absolutely certain: the difference and opposition between the development of the technological system on one hand and society and human beings on the other.

People have said, and I myself have written, that our society is a technological society. But this does not mean that it is entirely modeled on or entirely organized in terms of technology. What it does mean is that *technology is the dominant factor, the determining factor within society,* which is altogether different from Huxley's brave new world.

Society is made up of many different factors. There are economic factors, there are political factors. Human beings, as I have said, have an irrational element. Hence, being irrational and spontaneous, they are not fit for technology, and society, being habituated to ideologies, being historical and a result of the past, and existing in an emotional world of nationalisms, is as irrational as humanity and as unfit for technology.

The result is a shock, a contradiction, a conflict between the technological system, which augments according to its own laws, and the technological society. To follow a comparison that I employed to shed light on the relationship between the technological system and technology, it is almost like cancer developing in a live organism. But I do not mean to say that technology is a cancer; this is just an analogy to present the problem more effectively. Cancer, the cancerous cells, proliferate according to their own law. Cancer increases with its own specific dynamism; and it does so within a live organism, within a different set of cells, which obey different laws and which will be disturbed, sometimes completely unbalanced and disrupted, by the development of cancer.

The technological system is rather similar in that it is located inside the technological society. Hence, one may say that wherever the technological system increases, there is a greater disturbance of the social environment and the human groups. In other words, there is a growth of what might be called a certain disorder, a certain chaos. Hence, contrary to what we might imagine, technology is quite rational, the technological system is quite rational; but it does not subordinate everything to this rationality. There continue to be areas that are absolutely not subject to the technological system; hence, some kind of crisis occurs. That is why I simply do not believe in the possibility of Huxley's brave new world. What we actually observe is a technological order, but *within* a growing chaos.

Will this state of affairs continue? Does this situation have no solution? As a matter of fact, we do not see any possible historical solution. It is quite simplistic, quite elementary to say that "we have only to adjust to technology" or "society has to be organized according to technological means." What this actually signifies is that during the five hundred thousand years of our existence, we have developed in a specific direction, and now we are suddenly being asked to change. Well, I am simply saying that we cannot suppress half a million years of evolution in a few short years. What we can predict for sure is that if technological growth continues, there will also be a growth of chaos. This does not at all mean a void or a crumbling of societies, but difficulties *will* increase.

Let us apply what we just learned about the milieu. We have developed (and here I might allude to Toynbee's theory of challenge) only when encountering challenges, only when meeting new circumstances to overcome. In a sense, the new challenge to us is our own invention of technology. But this is not necessarily negative. We are called upon to surmount technology just as we have surmounted the difficulties of society or the difficulties of primitive nature. In short, this is an expression of life, for life is a series of imbalances successfully restored to a state of equilibrium. Life is not something static that has been organized once and for all. Hence, this challenge of technology may be positive so long as we fully understand that it is a challenge to be overcome and that it is a fundamentally serious issue.

If, for instance, human beings had not taken seriously (and no intellectual interpretation was necessary) the challenge posed by the development of cave bears, then they would very simply not have survived. Then, it was an immediate challenge, which they experienced constantly. At present we are obliged to travel a long intellectual road in order to understand the crux of the matter.

Given the current depth of relations between the technological system and the technological society, and the depth of the true problems raised by technological development, no political action in the normal, strict sense of the term is adequate today. On the one hand, the politician and the political institutions are totally incapable of mastering technology. They are incapable of normalizing the techno-social phenomena and steering them. Our institutions were invented between the seventeenth and the eighteenth centuries, and they are adapted to situations that have nothing to do with what we now know. One need merely recall the total impotence of the legal system in fighting pollution. Obviously, we can always issue decrees and pass laws, which sufficed for the problems of society one hundred years ago. But none of this is effective against pollution, and I could multiply the examples along these lines.

Likewise, as we have already said, the politician is totally unfit for technological problems. But just as we cannot master technology, so too the politician cannot rationalize behavior or find a new organization for society. For this would require the most totalitarian and the most technological government that has ever been imagined. However, we are not about to create such totalitarian or technological governments. At most, we have political authorities that are gradually and with difficulty adapting a few old governmental methods to new instruments. Indeed, when politicians realize the full scope of the problem, they become totally impotent. Hence, I believe that politicians can change neither technology nor human beings and society. In any case, for the challenge now facing us, we cannot expect any response along the road of traditional politics.

Politics is in no way acting upon technology and its problems. It is actually providing a framework for the events and trying to respond to the circumstances. In short, there is no such thing anymore as large-scale politics. It is quite astonishing to see the extent to which the great ideological systems—for instance the Communist systems in both the Soviet Union and China—have vanished, giving way to step-by-step policies. The USSR and China are completely normalizing in terms of a development of technology and are therefore in the same situation as the Western world. Indeed, I believe that modern society has two entirely different, entirely distinct levels: the level of appearances and phenomena; and the level of structures.

In appearance, there are many movements, changes and events. Not so long ago the World Council of Churches investigated the question: "What is

Christianity becoming in a changing society?" As if—and we all believe this—as if change were the fundamental trait of our society! The only thing that is really changing is appearances. It is obvious that the Soviet influence, say, in Africa, is tending to replace the Chinese influence of the nineteen sixties. Granted, this is not unimportant. But ultimately, the Chinese and the Soviets are more or less doing the same thing. Hence, we witness a large number of events which always boil down to a certain number of rather simple elements. The surface may seem very agitated, but the depth remains extremely stable. One can draw a well-known comparison to the ocean: the surface may be extraordinarily whipped up with waves and a tempest; but if one descends fifty meters, everything is calm, nothing is stirring.

Sociologically, I would say, we actually have three levels: the level of events and circumstance, which is always the level of politics; the level of far-reaching changes, for instance economic phenomena, which are longer-lasting and less circumstantial; and the level of stable structures, which, I believe, are given us by technology.

Technology fundamentally structures modern society. It is not that technology does not change. When I say it is stable, I am not saying it does not change. But it obeys its own law of evolution, and it is only very slightly influenced by events. It can be limited in its own development. Clearly, in the world we live in, we do not know everything that technology would allow. Blockages crop up—for instance economic ones. In France, we know the contradictions in the National Health Service. The costs are so high that a choice must be made between an extremely sophisticated medical technology and an increase in hospital beds for the most common illnesses and operations; we cannot have both. Hence, in the basic structures, there are blockages coming from the two other levels.

However, there is no fundamental change. Technology does not obey events in any way. Yet obviously, what interests us, as people taken with information, with the news, with everything exciting and fascinating, is the events. However, the more fascinated we are by political circumstances, speeches, and ideologies, the more we leave the structures free to function as they do. We can focus on an important political discussion about the Third World, but in reality, the power of technology expands in regard to the Third World too—and this we do not see. We are so excited by events, by circumstances, by the latest news, that in regard to fundamentals, we always feel we have time. Even if we do not understand the stakes of the game in regard to technology, we always feel we have a great deal of time ahead of us. But this is not true. If technology keeps growing, then disorder will keep growing; and the more disorder increases, the greater our fundamental danger.

May one say that there is no help, no hope, that all is lost and we can only let things happen? By no means! I think that humanity—as I have already said—has frequently been challenged and endangered in an equally fundamental way, and at first sight, people saw no way out. In 1935, we saw no way out from the Hitlerian dictatorship. It was something terrifying, on which we seemed to have no grip. Likewise, those who were critical of Stalinism saw no way out. We were convinced, myself included, that things would go on in exactly the same way after Stalin's death. All the same, there were a certain number of changes. Hence, we may not see any way out for now, but we should not claim that none exists.

I feel that, in any case, there are groups who hold out some hope. On the one hand, the groups from certain milieus that express the chaos in the midst of which we go forward. That is, the groups, the milieus of a certain age—youths, for instance—who feel the shock of this society most strongly and most harshly and who tend to reject it, even if, for the moment, no solution can be found.

Then there are the groups who are beginning to be conscious of what is happening. I will limit myself to discussing the antinuclear movements, because all this is very well known. The technological validity or nonvalidity of their arguments does not matter. The important thing is to be capable of posing the problem on the most basic level. Even if one can affirm that the nuclear plants are totally harmless, the real question is one of society's choice; and the antinuclear groups would therefore be right. Likewise, the ecological movements, the consumer movements, the neighborhood associations. The latter are citizens' groups who feel that we don't get rid of problems just by electing a local government. After all, the city council can only run the municipality. Thus, we have groups who feel that everything concerning their neighborhood life is of interest to them, and they ask to receive all documents, they discuss all the decisions of the municipal council. They are capable of arousing public opinion in certain cases. Generally, they form a mechanism that I might call a spontaneous referendum. I find this a new phenomenon and a very important one in the political world.

Then, we have to take the women's movements into account. They strike me as extremely serious and fundamental—so long as their objective is not to become masculine! That is, so long as women understand their specific role and do not wish to play the same role as men in the same work, the same framework, and for the same technologies. If women become men, it is of little interest. On the contrary, what strikes me as fundamental is that in a society in which the masculine extreme is crystalized in technology, the feminine part, which, I would say, is focused on sensitivity, spontaneity and intuition is starting to rally again. In other words, I feel that women are now far more capable than men of restoring a meaning to the world we live in, of restoring goals for living and possibilities for surviving in this technological world. Hence, the women's movements strike me as extraordinarily positive.

In this list of groups, I have not mentioned the proletariat or the Third World. In European countries, thoroughly permeated with Marxist thought, the proletariat was the bearer of hope for the world because, even without precise knowledge of Marx, people saw the proletariat as the most wretched, the most "alienated," people who would be forced to revolt in order to wipe out their own inhuman condition. The proletariat is, in general, thoroughly integrated in the technological world by organisms like trade unions or political parties having a purely industrial view and goal, and by situations that involve the proletarian in technology. Hence, the proletariat still thinks about issues in terms of the social and economic situation of the nineteenth or early twentieth centuries. Movements like trade unions do not see the new problems at all. For now, at least, and until a new consciousness is reached, I do not believe that the proletariat offers a future for humanity, anymore than the Third World does.

We have already indicated that the Third World has progressively lost its specificity as the technologies introduced in those countries upset whatever was unique and singular about their cultures. I feel that it is a mistake to investigate the transfers of technology. It is not enough, as is all too often said, to act with great care, to seek ways of adjustment. The transfer of technology can take place, and individuals and even certain groups in the Third World can be psychologically adapted. But in reality, the shock of technology causes a total breakup of the society. Hence, new studies on the transfers of technology will not solve this problem. The question is whether the civilizations of the Third World—India, Islam, and so on—being totally different from the Western world, are capable of absorbing Western technologies and integrating them into a totality of culture and civilization that is utterly new.

The shock of absorbing technologies has apparently destroyed the specific character of most of these societies. When one tries to rediscover the cultural roots, they seem so backward and impossible that, in the eyes of all humanity, one is dealing with an absurdity. I am thinking of what has happened in Iran with the Ayatollah Khomeini—his desire to return to a pure, hard Islam, indeed to the Middle Ages, with a rejection of all technologies which is unthinkable and unacceptable. There is no integration of technologies into a society with a different culture. It is an either/or situation: either technology or our Islamic society. That is the conflict of Iran today. Obviously the Ayatollah Khomeini's position is absolutely untenable. He is bound to be defeated because one can no longer live without accepting technologies. Iran will have to renounce the specific nature of an Islamic society.

There is, however, a further element which makes me feel that the Third World is no longer a resource in regard to the challenge facing us. You see, the very mentality of the inhabitants of the Third World has been transformed. On the one hand, the elite have only one idea: to develop technology, to enter the mainstream of technologies. Both intellectuals and politicians are fascinated with this notion, just as the rich of the Third World are interested—in the most banal sense of the term—in developing Western technologies. In both cases the goal is to enter the circuit of Western technologies.

On the other hand, for the poor in the Third World, technology clearly seems like a hope, the hope of overcoming poverty. In the mythology of the Third World, technology has succeeded in making the West rise from its own poverty. Therefore, they believe all they have to do is adopt Western technologies, and they too will profit from this development. One cannot contradict this notion, in the light of how poor and wretched the people of the Third World are. But they fail to realize that they are launching the twofold process of destroying their culture and entering into a universe that is totally alien to them, a universe that will bring disruptions on a psychological level and that will in fact cause in all areas far more serious disruptions than in the Western world. The West has adjusted gradually to its development of technologies—and we know how badly and with how much difficulty. It has taken us two hundred years. How then can the Third

World endure the shock, psychologically and sociologically, when it is asked to absorb this technological apparatus and this technological system in just a few years?

Within this international framework, and especially considering what we have just said about the Third World and the gradual destruction of its unique cultures (despite the ideologies of, for example, Africanism), we must, I believe, realize that the true powers in our time are no longer the rich countries or the populous ones, but those possessing the technologies. The term "rich nation" instantly brings to mind the Arab countries with their oil. Of course, these countries do impress us greatly with their influence on all economic and political life. In fact, however, the accumulation of their wealth is not bringing any true interior development or any sort of independence from the West.

It is, I feel, very important to realize that these riches do not permit the emergence of a new type of society. They simply allow the adoption, the purchase of what the West has already done. One need only think of the very characteristic example of buying ready-made factories, delivered "key in hand," so to speak, and set up in the Arab countries. What is this? In fact, it is the implantation of Western technologies in the Arab world. Likewise, in the terrible war between Iraq and Iran, everything is Western, including the materials and the strategy. Nothing remains of Arab military culture.

Hence, the wealth of Arab countries does not give them real power. The countries with real power are those that have the technological instruments, that are capable of the technological progress that is confused with *development*. It is not real development, but simply growth, a growth of power. We ought to recall the difference that many sociologists and economists make between *growth* and *development*. Schematically, we may say that growth is chiefly quantitative and development qualitative. In economy, aiming at growth means trying to produce more cement, more iron, or more wheat. Aiming at development means looking for the most balanced and least harmful economic structure, recognizing the value of the statement "small is beautiful," and achieving higher quality in consumption.

This distinction between growth and development obtains equally for politics and societal organization as well as economics. So far, however, technology has always emphasized growth and the growth of power. And this power is both economic and political, of course.

By the same token, a large population does not imply real power. (This is the problem of the Third World.) People never stop emphasizing the dreadful injustice that exists because of the difference in standards of living between the Western world and the Third World. But this difference is accentuated by the very rapid advances of technologies in the Western world. It is not simply the dynamics of capitalism, but rather the development of technologies. Hence, the axis of power is determined by the progression of technologies. At the same time, however, these technologies entail certain similarities. In order to exploit and to utilize technologies as much as possible and to maximize their yield, we must be able to organize society in a certain way, we must be able to put people to work in a certain way, and we must get them to consume in a certain way. Hence, the ideological oppositions are growing less and less important. The ideological and political conflicts in the strict sense of the term are rendered obsolete by the identical nature of the technologies.

Technologies are pretty much the same in the Soviet Union, the United States, and Europe, with only slightly different rates of growth. China is now moving in the same direction, evolving in the same manner, and also to technologize progressively. As a result, political structures are growing more and more alike, as are economic structures. It is no coincidence that the Soviet world is beginning to talk about a market economy, a natural formation of prices through competition. Not that the capitalist system is better; rather, both sides are looking for the best forms, the most effective ways of using the technologies. Likewise, the Western world is talking more and more about economic planning. Hence, an obvious convergence, with identical objectives, namely technological power, and the domination and utilization of raw materials for technology. Ideologies no longer count. Whether the discourse is Communist or capitalist, liberal or Socialist, in fact, everyone is obliged to do more or less the same thing.

I could give countless examples of these facts. For instance, when the Swedish Socialist Party was beaten by an antinuclear platform, the Liberal Party, on coming into power, realized it simply could not carry out the electoral promises it had made. Technology won out, and Sweden was forced to begin constructing nuclear plants.

This example shows the consequence of the technologically powerful nations; however, this convergence does not automatically guarantee peace. All we can say is that ideological politics is now secondary, and that the conflict between the powers comes from an excess of power, an excess that extends beyond the national boundaries. In the past, people offered long explanations for the conflicts between capitalist nations, saying that capitalist production had to conquer new markets throughout the world. Hence, it was economic output that caused wars. Now, however, the risk is obviously the excessive power of the three and soon four great creators of technology. They will soon find themselves facing one another in such a way that a conflict will be inevitable—the conflict over the use of raw materials, for example. It is a question of life and death. This, ultimately, is what endangers world peace, and nothing else.

My interpretation of the technological phenomenon as a milieu, as a system, has led me to get involved in society, as I tried to explain earlier. However, it was never my goal to go back, to declare that technology must be eliminated. I was looking for a new direction. So I tried to reach what is known in France as "the base" of society. "The base" is the average person, the one who simply lives his or her own life, who has no great ambitions or special intellectual development; but who still has something like spontaneity, openness, often allowing him or her to understand the things that are happening, so long as they are shown, and to understand them in such a way that he or she is relatively better prepared than intellectuals, technical experts, and executives to take the values of life seriously. All this led me to concentrate on local initiatives—that is, to rely on direct and close relationships to form groups for investigating the issues that require people to take a stand on technology and the technological system, but which are also very concrete.

Let me give you an example of ecological action in the region of Aquitaine. I tried to get intellectuals to develop a critical attitude so that they would question the very technologies they are studying. These intellectuals included scientists, lawyers and administrators. The point was not to reject administrative technology or juridical technology, but to clearly know what we were doing by employing them; to know the visible, immediate results and the secondary and less visible drawbacks. In other words, very close attention must be paid to any technological interference in the social or psychological domain. It was a great consolation for me to see people not going backward but realizing that the most highly developed technological means are not necessarily the best, even though they are the most efficient. I am thinking of insights by the doctors I worked with. They saw that many tests, although highly developed from a technological point of view, are ultimately no more certain than the diagnoses that were once made by more elementary procedures, but demanded greater personal commitment from the physician. In other words, a very large number of laboratory tests and clinical examinations are absolutely useless. They are technologically highly developed, but often very dangerous and sometimes very painful. Ultimately physicians and surgeons (I am speaking of the most highly qualified) recognize that the results and the knowledge attained are no greater. This is an example of a critical stance in regard to the very technologies we use.

At the same time, I was obliged to remain on the fringes with all my activities. Again and again, people tried to draw me into political circles, saying that something was happening politically that might lead to the acceptance of my analyses! This is a trap for the ecological movements. I feel that any action pertaining to the technological milieu must remain on the fringes because this milieu is extraordinarily enveloping and, I might say, extraordinarily seductive. My work, therefore, is obviously on a small scale; it requires much effort for apparently meager results. While crowds of people adopt all the technological developments, we can act only on individual levels. Hence, this is a true artisan's work. Nonetheless I am fully convinced that my slow labor, involving small numbers of people, is actually a point of departure for an internal change in society. To use big words, confronted with the technological phenomenon and the new milieu we live in, we must have "mutants." Not the mutants of science fiction—the technological human being with a robot's brain—but quite the opposite. To be a mutant a person needs to become someone who can use the technologies and at the same time not be used by, assimilated by, or subordinated to them. This implies a development of the intellect and a development of consciousness which can come about only for individuals, but it is the only development possible.

This leads, obviously, to the problem of educating children. For a longer or shorter period, our children and grandchildren, we must realize, will be living in a technological milieu, and we cannot even for one second imagine that we can raise them without some contact with it. Once again, the point is not to refuse to admit that technology exists, because it does exist; it is our milieu.

This goes back to what I was saying about the milieu. I know that it has in fact happened that when historical societies organized, small groups or sometimes individual people absolutely refused, saying: "We want to keep living like monkeys in the forest." Of course, they could do so, rejecting the development of society. But this was no solution. Those who continued living in the forest became extinct.

In the same way, one cannot claim to go on living as in the nineteenth century. We cannot bring up our children as though they were ignorant of technology, as though they had not been introduced from the first into the technological world. If we tried to do that, we would make total misfits of our children, and their lives would be impossible. They would then be highly vulnerable to the powers of technology. Yet we cannot wish them to be pure technical experts, making them so well fit for the technological society that they are totally devoid of what has until now been considered human.

Hence, I think that on the one hand we must teach them, prepare them to live *in* technology and at the same time *against* technology. We must teach them whatever is necessary to live in this world and, at the same time, to develop a critical awareness of the modern world. This is a very delicate balance, and we should not delude ourselves. We are preparing a world that will be even harder to live in for our children than it is for us. For us it is already complicated. And our children will be forced to deal with even more difficult situations.

Let me tell you of an experience that strikes me as dreadfully enlightening in its cynicism. I am rather well acquainted with the president of *Electricité de France* (the French national utilities company which is also responsible for the nuclear power plants). I was talking to him, discussing the dangers of nuclear plants point by point. Finally, in regard to two items in particular, he acknowledged that there were indeed some insoluble problems. And then he made the following extraordinary comment: "After all, we have to leave some problems for our children to solve."

That is the cynical attitude of the technical expert who knows his limits; it reveals that our children are indeed going to have difficult problems. Hence, in the immediate future, I feel that our children should be like all the others, go to the same schools as everyone else. But, at the same time, we should try to set up an alternative school, as it were, a parallel institution, where children learn to live differently and, on an existential level, learn to question the certitudes taught them in regular schools. Of course, this can be done only in communities of parents. One simply cannot provide such an orientation for life in a purely familial framework; and one cannot do work of this sort all alone with one's own children.

## The New Forms of Control

**Herbert Marcuse**

A comfortable, smooth, reasonable, democratic unfreedom prevails in advanced industrial civilization, a token of technical progress. Indeed, what could be more rational than the suppression of individuality in the mechanization of socially necessary but painful performances; the concentration of individual enterprises in more effective, more productive corporations; the regulation of free competition among unequally equipped economic subjects; the curtailment of prerogatives and national sovereignties which impede the international organization of resources. That this technological order also involves a political and intellectual coordination may be a regrettable and yet promising development.

The rights and liberties which were such vital factors in the origins and earlier stages of industrial society yield to a higher stage of this society: they are losing their traditional rationale and content. Freedom of thought, speech, and conscience were—just as free enterprise, which they served to promote and protect—essentially *critical* ideas, designed to replace an obsolescent material and intellectual culture by a more productive and rational one. Once institutionalized, these rights and liberties shared the fate of the society of which they had become an integral part. The achievement cancels the premises.

To the degree to which freedom from want, the concrete substance of all freedom, is becoming a real possibility, the liberties which pertain to a state of lower productivity are losing their former content. Independence of thought, autonomy, and the right to political opposition are being deprived of their basic critical function in a society which seems increasingly capable of satisfying the needs of the individuals through the way in which it is organized. Such a society may justly demand acceptance of its principles and institutions, and reduce the opposition to the discussion and promotion of alternative policies *within* the status quo. In this respect, it seems to make little difference whether the increasing satisfaction of needs is accomplished by an authoritarian or a nonauthoritarian system. Under the conditions of a rising standard of living, nonconformity with the system itself appears to be socially useless, and the more so when it entails tangible economic and political disadvantages and threatens the smooth operation of the whole. Indeed, at least in so far as the necessities of life are involved, there seems to be no reason why the production and distribution of goods and services should proceed through the competitive concurrence of individual liberties.

Freedom of enterprise was from the beginning not altogether a blessing. As the liberty to work or to starve, it spelled toil, insecurity, and fear for the vast majority of the population. If the individual were no longer compelled to prove himself on the market, as a free economic subject, the disappearance of this kind of freedom would be one of the greatest achievements of civilization. The technological processes of mechanization and standardization might release individual energy into a yet uncharted realm of freedom beyond necessity. The very structure of human existence would be altered; the individual would be liberated from the work world's imposing upon him alien needs and alien possibilities. The individual would be free to exert autonomy over a life that would be his own. If the productive apparatus could be organized and directed toward the satisfaction of the vital needs, its control might well be centralized; such control would not prevent individual autonomy, but render it possible.

This is a goal within the capabilities of advanced industrial civilization, the "end" of technological rationality. In actual fact, however, the contrary trend operates: the apparatus imposes its economic and political requirements for defense and expansion on labor time and free time, on the material and intellectual culture. By virtue of the way it has organized its technological base, contemporary industrial society tends to be totalitarian. For "totalitarian" is not only a terroristic political coordination of society, but also a nonterroristic economic-technical coordination which operates through the manipulation of needs by vested interests. It thus precludes the emergence of an effective opposition against the whole. Not only a specific form of government or party rule makes for totalitarianism, but also a specific system of production and distribution which may well be compatible with a "pluralism" of parties, newspapers, "countervailing powers," etc.

Today political power asserts itself through its power over the machine process and over the technical organization of the apparatus. The government of advanced and advancing industrial societies can maintain and secure itself only when it succeeds in mobilizing, organizing, and exploiting the technical, scientific, and mechanical productivity available to industrial civilization. And this productivity mobilizes society as a whole, above and beyond any particular individual or group interests. The brute fact that the machine's physical (only physical?) power surpasses that of the individual, and of any particular group of individuals, makes the machine the most effective political instrument in any society whose basic organization is that of the machine process. But the political trend may be reversed; essentially the power of the machine is only the stored-up and projected power of man. To the extent to which the work world is conceived of as a machine and mechanized accordingly, it becomes the *potential* basis of a new freedom for man.

Contemporary industrial civilization demonstrates that it has reached the stage at which "the free society" can no longer be adequately defined in the traditional terms of economic, political, and intellectual liberties, not because these liberties have become insignificant, but because they are too significant to be confined within the traditional forms. New modes of realization are needed, corresponding to the new capabilities of society.

Such new modes can be indicated only in negative terms because they would amount to the negation of the prevailing modes. Thus economic freedom would mean freedom *from* the economy—from being controlled by economic forces and relationships; freedom from the daily struggle for existence, from earning a living. Political freedom would mean liberation of the individuals *from* politics over which they have no effective control. Similarly, intellectual freedom would mean the restoration of individual thought now absorbed by mass communication and indoctrination, abolition of "public opinion" together with its makers. The unrealistic sound of these propositions is indicative, not of their utopian character, but of the strength of the forces which prevent their realization. The most effective and enduring form of warfare against liberation is the implanting of material and intellectual needs that perpetuate obsolete forms of the struggle for existence.

The intensity, the satisfaction and even the character of human needs, beyond the biological level, have always been preconditioned. Whether or not the possibility of doing or leaving, enjoying or destroying, possessing or rejecting something is seized as a *need* depends on whether or not it can be seen as desirable and necessary for the prevailing societal institutions and interests. In this sense, human needs are historical needs and, to the extent to which the society demands the repressive development of the individual, his needs themselves and their claim for satisfaction are subject to overriding critical standards.

We may distinguish both true and false needs. "False" are those which are superimposed upon the individual by particular social interests in his repression: the needs which perpetuate toil, aggressiveness, misery, and injustice. Their satisfaction might be most gratifying to the individual, but this happiness is not a condition which has to be maintained and protected if it serves to arrest the development of the ability (his own and others) to recognize the disease of the whole and grasp the chances of curing the disease. The result then is euphoria in unhappiness. Most of the prevailing needs to relax, to have fun, to behave and consume in accordance with the advertisements, to love and hate what others love and hate, belong to this category of false needs.

Such needs have a societal content and function which are determined by external powers over which the individual has no control; the development and satisfaction of these needs is heteronomous. No matter how much such needs may have become the individual's own, reproduced and fortified by the conditions of his existence; no matter how much he identifies himself with them and finds himself in their satisfaction, they continue to be what they were from the beginning—products of a society whose dominant interest demands repression.

The prevalence of repressive needs is an accomplished fact, accepted in ignorance and defeat, but a fact that must be undone in the interest of the happy individual as well as all those whose misery is the price of his satisfaction. The only needs that have an unqualified claim for satisfaction are the vital ones—nourishment, clothing, lodging at the attainable level of culture. The satisfaction of these needs is the prerequisite for the realization of *all* needs, of the unsublimated as well as the sublimated ones.

For any consciousness and conscience, for any experience which does not accept the prevailing societal interest as the supreme law of thought and behavior, the established universe of needs and satisfactions is a fact to be questioned—questioned in terms of truth and falsehood. These terms are historical throughout, and their objectivity is historical. The judgment of needs and their satisfaction, under the given conditions, involves standards of *priority*—standards which refer to the optimal development of the individual, of all individuals, under the optimal utilization of the material and intellectual resources available to man. The resources are calculable. "Truth" and "falsehood" of needs designate objective conditions to the extent to which the universal satisfaction of vital needs and, beyond it, the progressive alleviation of toil and poverty, are universally valid standards. But as historical standards, they do not only vary according to area and stage of development, they also can be defined only in (greater or lesser) *contradiction* to the prevailing ones. What tribunal can possibly claim the authority of decision?

In the last analysis, the question of what are true and false needs must be answered by the individuals themselves, but only in the last analysis; that is, if and when they are free to give their own answer. As long as they are kept incapable of being autonomous, as long as they are indoctrinated and manipulated (down to their very instincts), their answer to this question cannot be taken as their own. By the same token, however, no tribunal can justly arrogate to itself the right to decide which needs should be developed and satisfied. Any such tribunal is reprehensible, although our revulsion does not do away with the question: how can the people who have been the object of effective and productive domination by themselves create the conditions of freedom?

The more rational, productive, technical, and total the repressive administration of society becomes, the more unimaginable the means and ways by which the administered individuals might break their servitude and seize their own liberation. To be sure, to impose Reason upon an entire society is a paradoxical and scandalous idea—although one might dispute the righteousness of a society which ridicules this idea while making its own population into objects of total administration. All liberation depends on the consciousness of servitude, and the emergence of this consciousness is always hampered by the predominance of needs and satisfactions which, to a great extent, have become the individual's own. The process always replaces one system of preconditioning by another; the optimal goal is the replacement of false needs by true ones, the abandonment of repressive satisfaction.

The distinguishing feature of advanced industrial society is its effective suffocation of those needs which demand liberation—liberation also from that which is tolerable and rewarding and comfortable—while it sustains and absolves the destructive power and repressive function of the affluent society. Here, the social controls exact the overwhelming need for the production and consumption of waste; the need for stupefying work where it is no longer a real necessity; the need for modes of relaxation which soothe and prolong this stupefication; the need for maintaining such deceptive liberties as free competition at administered prices, a free press which censors itself, free choice between brands and gadgets.

Under the rule of a repressive whole, liberty can be made into a powerful instrument of domination. The range of choice open to the individual is not the decisive factor in determining the degree of human freedom, but *what* can be chosen and what *is* chosen by the individual. The criterion for free choice can never be an absolute one, but neither is it entirely relative. Free election of masters does not abolish the masters or the slaves. Free choice among a wide variety of goods and services does not signify freedom if these goods and services sustain social controls over a life of toil and fear—that is, if they sustain alienation. And the spontaneous reproduction of superimposed needs by the individual does not establish autonomy; it only testifies to the efficacy of the controls.

Our insistence on the depth and efficacy of these controls is open to the objection that we overrate greatly the indoctrinating power of the "media," and that by themselves the people would feel and satisfy the needs which are now imposed upon them. The objection misses the point. The preconditioning does not start with the mass production of radio and television and with the centralization of their control. The people enter this stage as preconditioned receptacles of long standing; the decisive difference is in the flattening out of the contrast (or conflict) between the given and the possible, between the satisfied and the unsatisfied needs. Here, the so-called equalization of class distinctions reveals its ideological function. If the worker and his boss enjoy the same television program and visit the same resort places, if the typist is as attractively made up as the daughter of her employer, if the Negro owns a Cadillac, if they all read the same newspaper, then this assimilation indicates not the disappearance of classes, but the extent to which the needs and satisfactions that serve the preservation of the Establishment are shared by the underlying population.

Indeed, in the most highly developed areas of contemporary society, the transplantation of social into individual needs is so effective that the difference between them seems to be purely theoretical. Can one really distinguish between the mass media as instruments of information and entertainment, and as agents of manipulation and indoctrination? Between the automobile as nuisance and as convenience? Between the horrors and the comforts of functional architecture? Between the work for national defense and the work for corporate gain? Between the private pleasure and the commercial and political utility involved in increasing the birth rate?

We are again confronted with one of the most vexing aspects of advanced industrial civilization: the rational character of its irrationality. Its productivity and efficiency, its capacity to increase and spread comforts, to turn waste into need, and destruction into construction, the extent to which this civilization transforms the object world into an extension of man's mind and body makes the very notion of alienation questionable. The people recognize themselves in their commodities; they find their soul in their automobile, hi-fi set, split-level home, kitchen equipment. The very mechanism which ties the individual to his society has changed, and social control is anchored in the new needs which it has produced.

The prevailing forms of social control are technological in a new sense. To be sure, the technical structure and efficacy of the productive and destructive apparatus has been a major instrumentality for subjecting the population to the established social division of labor throughout the modern period. Moreover, such integration has always been accompanied by more obvious forms of compulsion: loss of livelihood, the administration of justice, the police, the armed forces. It still is. But in the contemporary period, the technological controls appear to be the very embodiment of Reason for the benefit of all social groups and interests—to such an extent that all contradiction seems irrational and all counteraction impossible.

No wonder then that, in the most advanced areas of this civilization, the social controls have been introjected to the point where even individual protest is affected at its roots. The intellectual and emotional refusal "to go along" appears neurotic and impotent. This is the socio-psychological aspect of the political event that marks the contemporary period: the passing of the historical forces which, at the preceding stage of industrial society, seemed to represent the possibility of new forms of existence.

But the term "introjection" perhaps no longer describes the way in which the individual by himself reproduces and perpetuates the external controls exercised by his society. Introjection suggests a variety of relatively spontaneous processes by which a Self (Ego) transposes the "outer" into the "inner." Thus introjection implies the existence of an inner dimension distinguished from and even antagonistic to the external exigencies—an individual consciousness and an individual unconscious *apart from* public opinion and behavior.[1] The idea of "inner freedom" here has its reality: it designates the private space in which man may become and remain "himself."

Today this private space has been invaded and whittled down by technological reality. Mass production and mass distribution claim the *entire* individual, and industrial psychology has long since ceased to be confined to the factory. The manifold processes of introjection seem to be ossified in almost mechanical reactions. The result is, not adjustment but *mimesis:* an immediate identification of the individual with *his* society and, through it, with the society as a whole.

This immediate, automatic identification (which may have been characteristic of primitive forms of association) reappears in high industrial civilization; its new "immediacy," however, is the product of a sophisticated, scientific management and organization. In this process, the "inner" dimension of the mind in which opposition to the status quo can take root is whittled down. The loss of this dimension, in which the power of negative thinking—the critical power of Reason—is at home, is the ideological counterpart to the very material process in which advanced industrial society silences and reconciles the opposition. The impact of progress turns Reason into submission to the facts of life, and to the dynamic capability of producing more and bigger facts of the same sort of life. The efficiency of the system blunts the individuals' recognition that it contains no facts which do not communicate the repressive power of the whole. If the individuals find themselves in the things which shape their life, they do so, not by giving, but by accepting the law of things—not the law of physics but the law of their society.

I have just suggested that the concept of alienation seems to become questionable when the individuals identify themselves with the existence which is imposed upon them and have in it their own development and satisfaction. This identification is not illusion but reality. However, the reality constitutes a more progressive stage of alienation. The latter has become entirely objective; the subject which is alienated is swallowed up by its alienated existence. There is only one dimension, and it is everywhere and in all forms. The achievements of progress defy ideological indictment as well as justification; before their tribunal, the "false consciousness" of their rationality becomes the true consciousness.

This absorption of ideology into reality does not, however, signify the "end of ideology." On the contrary, in a specific sense advanced industrial culture is *more* ideological than its predecessor, inasmuch as today the ideology is in the process of production itself.[2] In a provocative form, this proposition reveals the political aspects of the prevailing technological rationality. The productive apparatus and the goods and services which it produces "sell" or impose the social system as a whole. The means of mass transportation and communication, the commodities of lodging, food, and clothing, the irresistible output of the entertainment and information industry carry with them prescribed attitudes and habits, certain intellectual and emotional reactions which bind the consumers more or less pleasantly to the producers and, through the latter, to the whole. The products indoctrinate and manipulate; they promote a false consciousness which is immune against its falsehood. And as these beneficial products become available to more individuals in more social classes, the indoctrination they carry ceases to be publicity; it becomes a way of life. It is a good way of life—much better than before—and as a good way of life, it militates against qualitative change. Thus emerges a pattern of *one-dimensional thought and behavior* in which ideas, aspirations, and objectives that, by their content, transcend the established universe of discourse and action are either repelled or reduced to terms of this universe. They are redefined by the rationality of the given system and of its quantitative extension.

The trend may be related to a development in scientific method: operationalism in the physical, behaviorism in the social sciences. The common feature is a total empiricism in the treatment of concepts; their meaning is restricted to the representation of particular operations and behavior. The operational point of view is well illustrated by P. W. Bridgman's analysis of the concept of length:

> We evidently know what we mean by length if we can tell what the length of any and every object is, and for the physicist nothing more is required. To find the length of an object, we have to perform certain physical operations. The concept of length is therefore fixed when the operations by which length is measured are fixed: that is, the concept of length involves as much and nothing more than the set of operations by which length is determined. In general, we mean by any concept nothing more than a set of operations; *the concept is synonymous with the corresponding set of operations.*[3]

Bridgman has seen the wide implications of this mode of thought for the society at large:

> To adopt the operational point of view involves much more than a mere restriction of the sense in which we understand "concept," but means a far-reaching change in all our habits of thought, in that we shall no longer permit ourselves to use as tools in our thinking concepts of which we cannot give an adequate account in terms of operations.[4]

Bridgman's prediction has come true. The new mode of thought is today the predominant tendency in philosophy, psychology, sociology, and other fields. Many of the most seriously troublesome concepts are being "eliminated" by showing that no adequate account of them in terms of operations or behavior can be given. The radical empiricist onslaught thus provides the methodological justification for the debunking of the mind by the intellectuals—a positivism which, in its denial of the transcending elements of Reason, forms the academic counterpart of the socially required behavior.

Outside the academic establishment, the "far-reaching change in all our habits of thought" is more serious. It serves to coordinate ideas and goals with those exacted by the prevailing system, to enclose them in the system, and to repel those which are irreconcilable with the system. The reign of such a one-dimensional reality does not mean that materialism rules, and that the spiritual, metaphysical, and bohemian occupations are petering out. On the contrary, there is a great deal of "Worship together this week," "Why not try God," Zen, existentialism, and beat ways of life, etc. But such modes of protest and transcendence are no longer contradictory to the status quo and no longer negative. They are rather the ceremonial part of practical behaviorism, its harmless negation, and are quickly digested by the status quo as part of its healthy diet.

One-dimensional thought is systematically promoted by the makers of politics and their purveyors of mass information. Their universe of discourse is populated by self-validating hypotheses which, incessantly and monopolistically repeated, become hypnotic definitions or dictations. For example, "free" are the institutions which operate (and are operated on) in the countries of the Free World; other transcending modes of freedom are by definition either anarchism, communism, or propaganda. "Socialistic" are all encroachments on private enterprises not undertaken by private enterprise itself (or by government contracts), such as universal and comprehensive health insurance, or the protection of nature from all too sweeping commercialization, or the establishment of public services which may hurt private profit. This totalitarian logic of accomplished facts has its Eastern counterpart. There, freedom is the way of life instituted by a communist regime, and all other transcending modes of freedom are either capitalistic, or revisionist, or leftist sectarianism. In both camps, nonoperational ideas are nonbehavioral and subversive. The movement of thought is stopped at barriers which appear as the limits of Reason itself.

Such limitation of thought is certainly not new. Ascending modern rationalism, in its speculative as well as empirical form, shows a striking contrast between extreme critical radicalism in scientific and philosophic method on the one hand, and an uncritical quietism in the attitude toward established and functioning social institutions. Thus Descartes' *ego cogitans* was to leave the "great public bodies" untouched, and Hobbes held that "the present ought always to be preferred, maintained, and accounted best." Kant agreed with Locke in justifying revolution *if and when* it has succeeded in organizing the whole and in preventing subversion.

However, these accommodating concepts of Reason were always contradicted by the evident misery and injustice of the "great public bodies" and the effective, more or less conscious rebellion against them. Societal conditions existed which provoked and permitted real dissociation from the established state of affairs; a private as well as political dimension was present in which dissociation could develop into effective opposition, testing its strength and the validity of its objectives.

With the gradual closing of this dimension by the society, the self-limitation of thought assumes a larger significance. The interrelation between scientific-philosophical and societal processes, between theoretical and practical Reason, asserts itself "behind the back" of the scientists and philosophers. The society bars a whole type of oppositional operations and behavior; consequently, the concepts pertaining to them are rendered illusory or meaningless. Historical transcendence appears as metaphysical transcendence, not acceptable to science and scientific thought. The operational and behavioral point of view, practiced as a "habit of thought" at large, becomes the view of the established universe of discourse and action, needs and aspirations. The "cunning of Reason" works, as it so often did, in the interest of the powers that be. The insistence on operational and behavioral concepts turns against the efforts to free thought and behavior *from* the given reality and *for* the suppressed alternatives. Theoretical and practical Reason, academic and social behaviorism meet on common ground: that of an advanced society which makes scientific and technical progress into an instrument of domination.

"Progress" is not a neutral term; it moves toward specific ends, and these ends are defined by the possibilities of ameliorating the human condition. Advanced industrial society is approaching the stage where continued progress would demand the radical subversion of the prevailing direction and organization of progress. This stage would be reached when material production (including the necessary services) becomes automated to the extent that all vital needs can be satisfied while necessary labor time is reduced to marginal time. From this point on, technical progress would transcend the realm of necessity, where it served as the instrument of domination and exploitation which thereby limited its rationality; technology would become subject to the free play of faculties in the struggle for the pacification of nature and of society.

Such a state is envisioned in Marx's notion of the "abolition of labor." The term "pacification of existence" seems better suited to designate the historical alternative of a world which—through an international conflict which transforms and suspends the contradictions within the established societies—advances on the brink of a global war. "Pacification of existence" means the development of man's struggle with man and with nature, under conditions where the competing needs, desires, and aspirations are no longer organized by vested interests in domination and scarcity—an organization which perpetuates the destructive forms of this struggle.

Today's fight against this historical alternative finds a firm mass basis in the underlying population, and finds its ideology in the rigid orientation of thought and behavior to the given universe of facts. Validated by the accomplishments of science and technology, justified by its growing productivity, the status quo defies all transcendence. Faced with the possibility of pacification on the grounds of its technical and intellectual achievements, the mature industrial society closes itself against this alternative. Operationalism, in theory and practice, becomes the theory and practice of *containment*. Underneath its obvious dynamics, this society is a thoroughly static system of life: self-propelling in its oppressive productivity and in its beneficial coordination. Containment of technical progress goes hand in hand with its growth in the established direction. In spite of the political fetters imposed by the status quo, the more technology appears capable of creating the conditions for pacification, the more are the minds and bodies of man organized against this alternative.

The most advanced areas of industrial society exhibit throughout these two features: a trend toward consummation of technological rationality, and intensive efforts to contain this trend within the established institutions. Here is the internal contradiction of this civilization: the irrational element in its rationality. It is the token of its achievements. The industrial society which makes technology and science its own is organized for the ever-more-effective domination of man and nature, for the ever-more-effective utilization of its resources. It becomes irrational when the success of these efforts opens new dimensions of human realization. Organization for peace is different from organization for war; the institutions which served the struggle for existence cannot serve the pacification of existence. Life as an end is qualitatively different from life as a means.

Such a qualitatively new mode of existence can never be envisaged as the mere by-product of economic and political changes, as the more or less spontaneous effect of the new institutions which constitute the necessary prerequisite. Qualitative change also involves a change in the *technical* basis on which this society rests—one which sustains the economic and political institutions through which the "second nature" of man as an aggressive object of administration is stabilized. The techniques of industrialization are political techniques; as such, they prejudge the possibilities of Reason and Freedom.

To be sure, labor must precede the reduction of labor, and industrialization must precede the development of human needs and satisfactions. But as all freedom depends on the conquest of alien necessity, the realization of freedom depends on the *techniques* of this conquest. The highest productivity of labor can be used for the perpetuation of labor, and the most efficient industrialization can serve the restriction and manipulation of needs.

When this point is reached, domination—in the guise of affluence and liberty—extends to all spheres of private and public existence, integrates all authentic opposition, absorbs all alternatives. Technological rationality reveals its political character as it becomes the great vehicle of better domination, creating a truly totalitarian universe in which society and nature, mind and body are kept in a state of permanent mobilization for the defense of this universe.

**1** The change in the function of the family here plays a decisive role: its "socializing" functions are increasingly taken over by outside groups and media. See my *Eros and Civilization* (Boston: Beacon Press, 1955), p. 96 ff.

**2** Theodor W. Adorno, *Prismen, Kulturkritik und Gesellschaft* (Frankfurt: Suhrkamp, 1955), p. 24 f.

**3** P. W. Bridgman, *The Logic of Modern Physics* (New York: Macmillan, 1928), p. 5. The operational doctrine has since been refined and qualified. Bridgman himself has extended the concept of "operation" to include the "paper-and-pencil" operations of the theorist (in Philipp J. Frank, *The Validation of Scientific Theories* [Boston: Beacon Press, 1954], Chap. II). The main impetus remains the same: it is "desirable" that the paper-and-pencil operations "be capable of eventual contact, although perhaps indirectly, with instrumental operations."

**4** P. W. Bridgman, *The Logic of Modern Physics,* loc. cit., p. 31.

# From *The Culture of Technology*

**Arnold Pacey**

## MEASURING PROGRESS

It is understandable that in thinking about particular machines—hand-pumps or VDUs—we habitually focus on hardware rather than human activity. It makes less sense, though, to think like this about more general concepts such as 'progress'. Yet there is a long history of identifying the overall progress of technology with specific inventions or with other strictly technical advances. As early as 1600, inventions such as the printing press, the magnetic compass and firearms were being quoted as evidence of technical progress; more recently, the steam engine and electric light have been added to the list.

There has also been a growing interest in measurable factors which allow key aspects of progress to be expressed numerically and displayed by means of graphs. Some authors have used this method to show how scientific knowledge is cumulative in time,[1] or have plotted the improving performance of specific types of machine. Others have tried to form a composite picture of how technology develops by superimposing data that relate to a number of different techniques. Thus Wedgwood Benn presents a schematic graph showing how computing, transport, communications and weapons have developed, describing this as 'a citizen's guide to the history of technology'.[2] Chauncey Starr goes further and quotes a technological index which combines factors relating to energy efficiency, steel output, communications and skilled manpower in science and engineering.[3]

One striking feature of these studies is that many types of machine appear to show steady development over very long periods of time. The improving accuracy of clocks is a favourite example with many commentators, because a consistent improvement in timekeeping can be plotted over nearly four centuries. Nicholas Rescher, a keen exponent of statistics and diagrams as representations of progress, comments that this kind of consistency in technical improvement is a common finding. 'Again and again—in one science-correlative branch of technology after another', a smooth curve is observed when 'performance-improvement' is plotted as a graph.[4]

These ways of thinking about progress have very serious weaknesses, however. They tend to be overselective, and lead us to overlook the fact that improvements in one dimension are sometimes accompanied by less desirable developments elsewhere. In agriculture, for example, the amount of food produced can be judged in relation to land, labour or energy. In Britain, as in other western countries, grain output has increased enormously in the present century, especially in relation to the area of land cultivated (Figure 1) and the number of people employed. But grain output per unit of energy consumed on farms has *decreased*.[5]

How progress is evaluated depends on circumstances and, in particular, on whether there is a shortage of land or of energy. In small, heavily populated

greater skills from operators. However, development did not stop there. As lathes became more nearly automatic they displaced not only muscular and manual skills, but the operator's judgment as well.

It is impossible to say precisely when lathe development passed from work enhancement to deskilling, though Braverman makes suggestions. But clearly, from the operator's point of view, the impression of consistent progress implied by figure 2 is extremely misleading. If one could check job satisfaction, it might be found to have reached a peak at some date in the later nineteenth century, after which the negative effects of deskilling became predominant.

In recent years, the division of labour and the deskilling of work have extended to many more occupations, aided very often by the computer. Mike Cooley describes how the engineering draughtsman in the 1930s was at the centre of design work in industry. 'He could design a component, draw it, stress it out, specify the material for it and the lubrication required.' Nowadays the work is divided between several specialists: the 'draughtsman draws, the metallurgist specifies the material, the stress analyst analyses the structure and the tribologist specifies the lubrication'.[21]

The process of deskilling has been taken much further recently, however. A computer can now generate the drawings on which many draughtsmen would once have had to work, and the designer himself, using a VDU as an electronic drawing board, can produce drawings and detailed designs much faster. But even the professional in charge of a computer design facility may find his work partly deskilled as systematized design procedures are programmed into the computer to form what Cooley calls an 'automated design manual'. Thus the work of the designer has been undergoing exactly the same process of deskilling as manual work. Sometimes he is reduced to making a series of routine choices between fixed alternatives, in which case 'his skill as a designer is not used, and decays'.

Obviously, computer-aided design has the potential to be used far more creatively. In more modest ways, word processors applied to office work present a similar choice between systems that are planned merely to speed up work and increase management control, and systems which enhance job interest. On the shop floor, when numerically controlled machine tools are used, the advent of microprocessors provides an opportunity for building in 'more opportunity for shop-floor intervention to improve performance'. In each of these instances, though, managements are often reluctant to take advantage of the more creative option, 'because it would lessen *their* opportunities for controlling output'.[22]

While it is acknowledged that much tedious work is displaced by this type of technology, claims are made about the new skilled jobs being created, which relate both to software and to the maintenance of equipment. However, the philosophy of maintenance has altered so that the work involved has also been deskilled. Braverman notes that even householders have observed a deterioration in the repair skills of the men who service their washing machines; modern equipment is designed as a system of standard modules which can be replaced without much knowledge.

Undoubtedly, computerization can help us cope with the complexity of the modern world and the pressure of resource shortages, but the problems associated with it should not be disguised. Microprocessors allow many kinds of equipment to be more compact and energy-efficient. Computers and modern communications may allow a trade in information to grow at a time when trade in material goods could become restricted. This would make knowledge more important as a resource in its own right; one consequence already evident is that computers change power relationships within firms and within the community as knowledge-based power is reduced for some people and increased for others. The result is that computerization is tending to 'strengthen rather than weaken... centralization and hierarchy' in modern organization.[23]

Harry Braverman draws an instructive comparison with the first industrial revolution. That was not primarily a technical revolution; there was no change in the nature of many processes, which were merely reorganized on the basis of the division of labour. Craft production was dismembered and subdivided so that it was no longer 'the province of any individual worker'. In the modern 'revolution' the whole system is transformed. New materials, techniques and machines are used in an effort 'to dissolve the labour process as a process conducted by the worker and reconstitute it as a process conducted by management'. The individual workman or operative is analysed almost as a piece of machinery; he or she is seen as a 'sensory device', linked to a 'computing mechanism' and 'mechanical linkages'. This, says Braverman, is what modern industry 'makes of humanity'; labour is 'used as an interchangeable part' and progress is seen as a matter of indefinitely increasing the number of tasks that can be carried out by machine. The final triumph is achieved when all the human components have been exchanged for mechanical or electronic ones.

## DETERMINIST DEDUCTIONS

Beyond value judgements about the desirability of all this, two contrasting beliefs about progress are evident. There is the linear view, well expressed by the graphs showing smooth, steady, upward development. But in sharp contrast, there is a view in which the context of innovation—including its organizational aspect—is more broadly considered.

Beliefs based on both approaches have some validity, but the linear view oversimplifies matters in a way which encourages a false optimism among those who approve the kind of progress it portrays, and a deep, despairing pessimism among others. The reason is that technical progress, as the linear view presents it, may seem inevitable and inescapable. There is a consistency in it which appears to imply an inflexible logic. When graphs conceal ambiguities and smooth out irregularities, they may seem to indicate an unvarying pattern in the unfolding of technological rationality—a pattern which is independent of the ups and downs of human affairs. This regularity, based on the internal logic of technology, has sometimes encouraged efforts to discover the 'laws' which supposedly govern progress. For example, one of two laws put forward by Jacques Ellul is that technical progress tends to act according to a geometrical progression because, 'the preceding technical situation alone is determinative'. He argues that, 'Technical progress today is no longer conditioned by anything other than its own calculus of efficiency',[24] and this leads him to a bleakly pessimistic view. Others have claimed that because technology 'carries its own culture with it', it also 'determines the ownership structure of industry'.[25]

All these views are variants of an attitude often referred to as technological determinism, which presents technical advance as a process of steady development dragging human society along in its train. Then many social problems are regarded as being due to 'culture lag', which arises when social norms and institutions fail to adapt to the latest developments in, say, automation or cable television.

This idea of technical advance as the leading edge of progress is widely held; it constitutes what some have called 'machine mysticism'. Thus we see ourselves as living in the computer age or the nuclear age which has succeeded the nineteenth-century age of steam. Each era is thought of in terms of its dominant technology, extending back to the early history of man. There, we think of development from stone tools to bronze ones and the later emergence of an iron age, as a logical technical progression bringing social evolution in its wake. And we think about each era in terms of the impact of technique on human affairs, rarely enquiring about the converse.

The Chicago World's Fair of 1933 was a particularly strong expression of this view of technology and progress in the way its exhibits were arranged and presented. The guidebook described how 'science discovers, genius invents, industry applies, and man adapts himself to, or is moulded by, new things'. It went on to assert that individuals, groups, 'entire races of men' are compelled to 'fall into step with... science and industry... Science finds—Industry applies—Man conforms.'

Such views are not often stated so starkly today, but they still seem to lie at the back of people's minds. Yet quite often, new patterns of organization had to be invented or evolved before innovations in technique could arise. The idea of television, for example, could hardly have arisen in a society without mass entertainment and organized news media. Thus Raymond Williams has described how radio and television evolved from an urbanized institutional background in which there was a growing need for communications of all kinds.[26] It is not sufficient to consider only the technical logic of the first photoelectric cells and cathode ray tubes, invented just before 1900. These, perhaps, were the beginnings of a capability to create television, but there had to be a conscious intention as well, and this was to a large extent 'an effect of a particular social order'.

In a similar way, it is possible to turn the conventional history of almost any invention on its head, and instead of showing how technical developments grew one on another, influencing social change, we may demonstrate how organizational development led to new technology. As already noted, instead of saying that James Watt's steam engine led to the industrial revolution, it is possible to argue that the prior development of factory organization gave Watt the opportunity to perfect his inventions.

As Williams says, technological determinism is untenable—but so is its complete opposite. Most inventions have been made with a specific social purpose in mind, but many also have an influence which nobody has expected or intended. The reality is perhaps easier to comprehend by thinking about the concept of technology-practice with its integral social components. Innovation may then be seen as the outcome of a cycle of mutual adjustments between social, cultural and technical factors. The cycle may begin with a technical idea, or a radical change in organization, but either way, there will be interaction with the other factors as the innovation comes to fruition.

This applies as much to the stone and bronze ages through which early man passed as to the industrial revolution. Explanations are insufficient if they focus only on the development of tools; there is also need to recognize the 'whole complex of mutually enhancing agencies arising from ecological circumstances... tool using and making, symbolic communication... group conduct'.[27]

So in the development of modern technology, it is not just the influence of tools and techniques on society that needs to be understood, but the whole array of 'mutually enhancing agencies' which has led to the spectacular advances of our own time. As another student of human evolution has put it, 'technology has always been with us. It is not something outside society, some external force by which we are pushed around... society and technology are... reflections of one another.'[28] Equally, it is a myth that cultural lag occurs in every community as people try to keep up with their progressive technology. In the interactions which take place between various aspects of human activity 'it is often technology that is lagging'.[29]

With such views so clearly expressed, why are conventional beliefs about the inevitability of technological progress and its leading role in social development still so widely held? The answer seems to be partly that the conventional beliefs serve a political purpose. When people think that the development of technology follows a smooth path of advance predetermined by the logic of science and technique, they are more willing to accept the advice of 'experts' and less likely to expect public participation in decisions about technology policy. Thus Leslie Sklair remarks that, 'the argument about the intrinsic dynamic of science and technology seems to me... the defence of those who find the idea of the democratization of science and technology unpalatable'.[30]

Such people argue that technical logic 'determines a unique progression from one stage of development to the next'. The implication is that although we may not like the idea of nuclear power, microelectronics, or heart transplant surgery, we have to solve the technical problems connected with these things if engineering and medicine are to develop. Such attitudes were reflected by the comments of a British government minister in December 1978, when he launched an initiative to promote the manufacture of 'silicon chips' and microprocessors. He remarked that we cannot 'stop technology'; it would have been no use trying 'to stop the steam engine, or... electric light', and it is no use now 'trying to stop the silicon chip revolution'.

That kind of statement is clearly designed to defuse political dissent—but it puts the issues the wrong way round. Not many people want to stop microelectronics, but they may want to state preferences about how it is used. The minister's remarks present a view of progress which implies only one dimension of choice: either you accept innovation unreservedly, or you opt out. Silicon microchips have potential for many kinds of development, though; it is choices between these that matter.

The development of the steam engine was by no means inevitable and 'unstoppable'. On the continent of Europe its adoption was halting and slow. Rapid development in Britain reflected the success of the new ways of organizing industry, and the freedom of mine owners and factory masters to pursue their ends without much social or political restraint. If the advance of microelectronics now seems inevitable, we ought to ask what kind of organizational pressure lies behind it, and what restraints might be appropriate in order to give proper effect to the wider public choice? We need beliefs about progress which help us recognize the real choices available. Existing views about how technology develops seem mostly to hinder perception of choice and allow experts and the industrialists they serve to get their own way.

Explanations of the rapid development of the steam engine in Britain and its slow progress on the continent include another point of modern concern. From 1712 until Watt extended the range of their application, most steam engines were used to pump water from coal mines. In any other situation, the early, inefficient engine used far too much fuel to be of much economic worth. And Britain was far in advance of most of Europe in mining coal, because deforestation had gone much further in Britain than elsewhere, and firewood was scarce and expensive. In much of the rest of Europe, as in America, wood fuel continued to be relatively abundant for another century; so there was less need to mine coal, less need for engines at mines, and little other opportunity for the economic application of steam. Thus there were strong ecological or environmental reasons for the adoption of steam power in Britain which were less important elsewhere. But there is no ecological determinism, and the environmental condition of Britain in 1712 did not dictate the development of the steam engine. A vigorous policy of reafforestation and fuel-wood coppicing could have solved the problem.

So also in our own immediate future, shortage of conventional energy will not eliminate choice; there is no determinism making it inescapably necessary to extend nuclear power systems, or to develop solar energy techniques. The issue, rather, is whether those who make decisions will prefer that we pay the environmental and social cost of one option or others—or whether we shall reduce these costs by conserving energy and by recycling materials. The choices available, in any case, are much wider than decisions about energy. As we have seen, the growing importance of knowledge as a resource means that industry may already be developing in directions that depend less on materials and energy.[31] Other options are possible, some relating to social policy, and others influenced by voluntary changes in lifestyle such as individuals have sometimes already adopted. The future is rich with choice, and the most balanced view of the environmental problem is not that it 'destroys the notion of... progress',[32] nor that it dictates the adoption of any particular technique, but that it requires, 'a new evaluation of what does and does not constitute "progress"'.[33]

## MOVEMENTS IN PROGRESS

One way of rethinking our concept of progress may be to take an altogether broader view of the many factors which interact in 'mutually enhancing' ways at especially creative moments. At such times, the various technical, organizational and cultural workings of technology-practice seem all at once to start meshing together in new and more harmonious, effective ways. A new pattern emerges, and people experience a new awareness of practical possibility. The age of Columbus and the discovery of America was such a time. People had long realized that the earth is a sphere—Columbus did not have to teach them that. What happened was the dawning of recognition that this familiar academic fact had an unrealized practical potential.

Much the same can be said about the invention of the factory system. Many of the ideas on which it was based were commonplace. Italian merchants had operated the division of labour and a degree of mechanization in textile manufacture three or four centuries earlier, and the division of labour had been discussed in Britain since the 1650s. What happened at the start of the industrial revolution was a recognition of how to fit these ideas to economic opportunity in effective ways. And in the enthusiasms of the time, voiced particularly by Adam Smith, one may catch the same sense of a reaching toward a new and enormous potential that one feels with the age of Columbus.

There is clearly no determinism about these experiences of progress. A human, not a mechanistic process is at work, in which there is certainly an element of choice. But choice here is not the simple weighing of known options—it involves, rather, different ways of approaching the unknown. It is a decision between different attitudes of mind. We may cultivate an exploratory, open view of the world in which awareness can grow; or we can maintain a fixed, inflexible view in which new possibilities are not recognized. The magic of European culture from the age of Columbus until after the time of Watt was perhaps chiefly in its openness and growing awareness. The dominance more recently of linear views of progress, which restrict expectations to narrow patterns of development, may be symptomatic of how that openness has been lost.

Rather than speculate on such general matters, however, it is more realistic here to think again about the way improvements in agriculture and in steam engines have clustered together, leading to a step-wise pattern of advance rather than smoothly continuous progress. Each step, whether representing Cornish engines or modern farming, was characterized by specific organizational arrangements as well as by new techniques; it thus seemed right to describe these distinct phases of development as movements in technology-practice. What we can now add is that innovation is not simply the outcome of rational logic. It involves purpose and intention, and reflects awareness of possibility and economic opportunity. So in these minor technological movements, as well as in larger developments, there are crucial moments of recognition when a varied collection of different factors fit together and a new form of practice takes off.

Examination of diagrams, such as those presented in this chapter, enables a few of these moments of takeoff to be identified. Figure 3, for example, illustrates the effects of several innovative movements in the use of steam power. The two upward steps in performance associated with James Watt and with Cornish engine development are clearly seen.

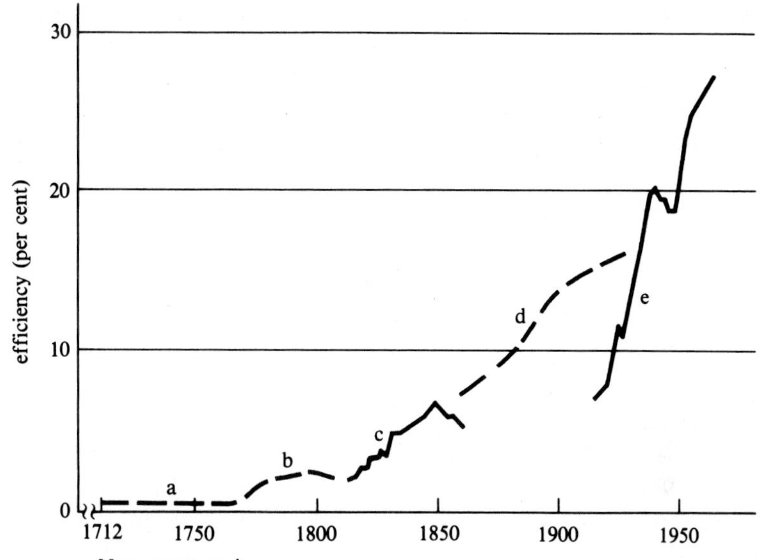


a Newcomen engines
b Watt engines
c Cornish engines
d Compound, triple expansion and uniflow engines
e Steam turbines in British power stations

Data have been recalculated to a common base, in which a full line is used only for consistently recorded series. Untypical efficiencies recorded during special engine tests are not shown.

**FIGURE 3**
Average efficiencies of representative British steam engines and turbines (with boilers) from the invention of the Newcomen engine.

In more recent times, the outstanding innovative movement involving steam power has been the development of electricity generation using steam turbines. This movement acquired a very particular self-awareness in Britain during the 1920s and 1930s when efficiency figures for individual power plant were regularly published by the Electricity Commissioners. Leslie Hannah[34] quotes the data and describes how engineers vied with each other to be at the top of the league table for their type of plant. The development of electricity supply was a movement in a much wider sense, though. It evoked considerable idealism about the creation of clean, pollution-free cities and factories. In England, it brought the vision of the garden city within reach. Women actively campaigned for it, seeing the use of electricity in the home as a step toward emancipation, and so did the Labour movement. It is ironic that in the 1980s, this progressive mood has largely dissipated, and the electricity industry is now the target of criticism about its over-ambitious expansion, and its role as a source of chemical, nuclear and thermal pollution.

There is a further point to notice: innovative movements are not usually restricted to specialist fields. There were links between the achievements of Cornish engineers and the development of steam power for railways. There are strong and obvious links between modern agriculture, with its remarkable crop yields, and the chemical industries that produce fertilizers and pesticides. New electrical appliances were invented, and older industries were modernized as electricity supplies developed.

Because of these links, a clustering of innovations in one branch of technology may lead to inventions in a wide variety of others; indeed, detailed analysis shows a distinct bunching of inventions at key points in the overall history of industry. This has led to the suggestion that we should not picture the last two centuries of technological progress as a smooth, continuous advance, but more as a succession of waves of innovation. Christopher Freeman puts this point by saying that invention is not characterized by a 'linear trend', but by peaks and troughs, with peaks occurring as linked developments in 'new technology systems'.[35]

Thus when a large number of important developments occur within a couple of decades—as they did during the 1870s and 1880s with chemicals, steel, electricity and automobiles—then the economy may enter a phase of rapid expansion stimulated by several different technologies simultaneously. Freeman points out, though, that a major economic upswing of this kind not only depends on technical innovation, but also, very often, on changes in organization; every new phase of economic expansion depends on 'a whole cluster of related... innovations and institutional changes.'

Freeman and his predecessors have meticulously charted the varying number of patents taken out from one year to the next in order to gain firm evidence for the clustering of inventions, and mainly agree with earlier work by Kuznets[36] about dates of the chief phases of technological development. This leads to a perspective in which four 'long waves' of industrialization are distinguished (and some shorter cycles also), the first being the classic industrial revolution in Britain. With each wave, inevitably, there is a time lag between inventions being patented, their coming into limited use, and their later impact on the economy; table 1 quotes two sets of dates to indicate this.

In a time of economic recession, such cyclical theories are understandably more attractive than linear ones. They offer the hope that, after the cyclic downturn, there will eventually be recovery and renewed growth.

There is danger in this model, however, as much as in linear views. Any historical analysis which seeks to identify patterns and rhythms in development may tend to become deterministic. It may seem to imply that processes are at work which no human intervention can decisively alter.[37] Because there have been four long waves of industrialization, some people are ready to expect a fifth. They see it as based on microelectronics and biotechnology, and centred on the 'Pacific rim' linking California, Japan and south-east Asia. But there is no inevitability about this. There are also those—economists as well as environmental-

**TABLE 1**
LONG WAVES OF INDUSTRIALIZATION

The four historical waves are, with minor modification, those identified by Christopher Freeman and his colleagues, and before them by Simon Kuznets and N. Kondratieff.

| Dates for clustering of innovations | Key innovative technologies | Geographical base | Period of rapid economic growth |
|---|---|---|---|
| *1st long wave* | | | |
| 1760s | textile manufacture | Britain | 1780–1815 |
| 1770s | (also steam engines, figure 3), chemistry, civil engineering | France | |
| *2nd long wave* | | | |
| 1820s | railways, mechanical engineering | Britain, Europe | 1840–70 |
| *3rd long wave* | | | |
| 1870s | chemistry, electricity, | Germany, United | 1890–1914 |
| 1880s | internal combustion engines | States | |
| *4th long wave* | | | |
| 1930s | electronics, | United States | 1945–70 |
| 1940s | aerospace, chemistry (e.g. chemical farming, figure 1) | | |
| POSITIVE PROSPECTS AND OPTIONS | | | |
| *5th long wave* | | | |
| 1970s | microelectronics, biotechnology | Japan, California | 1985–? |
| *Other waves?* | | | |
| social development and improvement in quality of life with little economic growth | public health, nutrition | South Asia | |
| | renewable energy, conservation, agriculture, reafforestation | China, United States? | |

ists—who think that industrialization is ending. They point out that all previous civilizations have had limited duration, and even if there is neither an environmental nor a nuclear disaster, we should expect that one day industrial civilization will go into decline.

There is also a third prospect, indicated speculatively on table 1 in order to stress that many options still lie open. This is the possibility of human development and progress involving technology, but independent of growth.

Interpretation of the long waves of industrialization may easily become too narrow, because conventional analysis of economic and technical development must of necessity focus on facts and figures, graphs and statistics. Whatever individual authors intend, this inevitably gives little weight to human experience and choice, and leads to a mechanistic, if not a deterministic picture. To get beyond that point, we need to leave graphs and statistics behind and use other ideas. One social thinker who has done this offers an image which may help to show what innovative movements and waves may mean in human terms. With interests focused less on technology than on the broad social consequences of an energy crisis and an end to economic growth, he uses the analogy of people's habits in conversation. When one subject is exhausted, they talk about something else. Similarly, the goals of human development can change, be they technological or social, because, 'history proceeds by changing the subject rather than by progressing from one stage to the next'. Innovative movements and new waves of industrialization can be seen in human terms as changes of subject; and Ralf Dahrendorf, whose analogy this is, goes on to point out that what makes the difference is altered awareness:

> One day, people wake up to the experience that what was important yesterday, what preoccupied and divided them, no longer matters in the same way. We rub our eyes and discover that the way to solve the problem that kept us awake last night is not to do more or even to do better about it but to turn to something different which may be more relevant...we are in the process of one such historical change of subject...
>
> In the advanced societies of the world, with their market economies, open societies and democratic politics, a dominant theme appears to be spent, the theme of progress in a certain, one-dimensional sense, of linear development, of the implicit and often explicit belief in the unlimited possibilities of quantitative expansion. The new theme which might take its place...is not a negation of growth...but what I shall call improvement, qualitative, rather than quantitative development.[38]

One does not need to agree in detail with the ideas which Dahrendorf goes on to develop in order to appreciate his starting point. The need is to keep open the possibility of waking up to the experience that there are new, radically different ways of dealing with economic problems, and that there are unexplored options for human benefit from technology. There is still the difficulty, however, that our habitual style of writing and analysis, whether in sociology, economics or technology, is itself basically linear. Its aim is usually to understand in depth rather than to broaden awareness. It is a style based on following logical connections, pursuing meticulous detail, and measuring whatever can be measured. Unless it is skillfully used, the very literary form of such discussion can itself trap one into a narrow, linear view.

## NOTES

1 Derek J. de S. Price, *Little Science, Big Science,* New York: Columbia University Press, 1963, pp. 10, 29.
2 Anthony Wedgwood Benn, 'Introduction', *The Man-Made World: The Book of the Course,* Milton Keynes: The Open University Press, 1971, p. 13.
3 Chauncey Starr, *Current Issues in Energy,* Oxford and New York: Pergamon Press, 1979, pp. 77, 87, 91.
4 Nicholas Rescher, *Scientific Progress,* Oxford: Basil Blackwell, 1978, p. 178.

**5** Gerald Leach, 'Energy and food production', *Food Policy,* **1** (1975), especially p. 64.
**6** For use of these terms, see David Dickson, *Alternative Technology and the Politics of Technical Change,* London: Fontana/Collins, 1974, pp. 43–4; Ralf Dahrendorf, *The New Liberty,* London: Routledge, and Stanford (California): Stanford University Press, 1975, p. 14.
**7** J. R. Jensma, 'The silent revolution in agriculture', *Progress: The Unilever Quarterly,* **53** (4), 1969, pp. 162–5.
**8** W. G. Hoskins, 'Harvest fluctuations and English economic history', *Agricultural History Review,* **16** (1968), pp. 15–45; Carlo M. Cipolla (ed.), *The Fontana Economic History of Europe,* vol. 2, London: Collins/Fontana, 1974, Statistical Tables, pp. 612–15. In plotting the graph in Figure 1, all figures have been reduced to a common base by taking a yield of 10 bushels of wheat per acre as equivalent to 690 kg per hectare, or to a yield ratio of around 4 to 6.
**9** The Lawes and Gilbert data are quoted and assessed by Susan Fairlie, 'The Corn Laws and British wheat production', *Economic History Review,* ser. 2, **22** (1969), pp. 109–16. Some untypical and very high yields are quoted by M. J. R. Healey and E. L. Jones, 'Wheat yields in England, 1815–1859', *Journal of the Royal Statistical Society,* ser. A, **125** (1962); these latter are not used in Figure 1, except to establish the direction of trends. Recent data are from Ministry of Agriculture sources published as *Agricultural Statistics, United Kingdom,* 1974, 1978 and other years, London: HMSO.
**10** On steam engine efficiency, see Starr, *Current Issues,* p. 78; also Carlo M. Cipolla, *The Economic History of World Population,* Harmondsworth: Penguin Books, revised edn. 1964, p. 57; Richard G. Wilkinson, *Poverty and Progress,* London: Methuen, 1973, p. 144.
**11** D. S. L. Cardwell, *From Watt to Clausius: The Rise of Thermodynamics in the early Industrial Age,* London: William Heinemann, 1974, pp. 158, 179.
**12** Anthony Wedgwood Benn, possibly echoing Engels, 'Introduction', *The Man-Made World,* p. 11; compare Dickson *Alternative Technology,* p. 46.
**13** D. S. Landes, *The Unbound Prometheus: Technological Change and Industrial Development in Western Europe from 1750,* Cambridge: Cambridge University Press, 1969, p. 65.
**14** Charles Babbage's preface in Peter Barlow, *A Treatise on the Manufactures and Machinery of Great Britain,* London: 1836, pp. 50–5.
**15** Andrew Ure, *The Philosophy of Manufactures,* London: Charles Knight, 1835, p. 15.
**16** D. S. Landes, quoted by Stephen A. Marglin, 'What do bosses do?', in *The Division of Labour,* ed. André Gorz, Hassocks (Sussex): Harvester Press, 1977.
**17** Arnold Pacey, *The Maze of Ingenuity,* Cambridge (Mass): MIT Press, 1976, pp. 223–6, 272, 277.
**18** Ure, *The Philosophy of Manufactures,* p. 23.
**19** Ian Crockett, 'An intermediate technology approach to the design of lathes for small workshops', unpublished third year report, Mechanical Engineering Department, University of Manchester Institute of Science and Technology, 1971.
**20** Harry Braverman, *Labour and Monopoly Capital,* New York: Monthly Review Press, 1974, pp. 213–20: pp. 169–70, 180, 182 and 223 are also drawn on in this section.
**21** Mike Cooley, *Architect or Bee? The Human/Technology Relationship,* Slough: Langley Technical Services, 1980, pp. 2 and 76.
**22** Roy Rothwell and Walter Zegveld, *Technical Change and Employment,* London: Frances Pinter, 1979, p. 117.
**23** Albert Cherns, 'Automation...How it may affect the Quality of Life', *New Scientist,* **78** (8 June 1978), pp. 653–5.

**24** Jacques Ellul, *The Technological Society,* trans. John Wilkinson, New York: Vintage Books, 1964, chapter 2, pp. 74, 89.

**25** George McRobie, *Small is Possible,* London: Jonathan Cape, 1981, p. 192.

**26** Raymond Williams, *Television: Technology and Cultural Form,* London: Fontana/Collins, 1974, pp. 14–21, 128–9.

**27** J. S. Weiner, *The Natural History of Man,* London: Thames and Hudson, 1971, pp. 77–8.

**28** Solly Zuckerman, *Beyond the Ivory Tower,* London: Weidenfeld & Nicolson, 1970, p. 129.

**29** C. Wright Mills, *The Sociological Imagination,* Harmondsworth: Penguin Books, 1970, p. 101.

**30** Leslie Sklair, *Organized Knowledge,* London: Hart-Davis MacGibbon, 1973, pp. 237–8.

**31** This argument is put by Nathan Rosenberg, *Perspectives on Technology,* Cambridge: Cambridge University Press, 1976, pp. 240–2.

**32** Jeremy Rifkin, *Entropy: A New World View,* New York: Viking Press, 1980, pp. 6, 30.

**33** Peter Chapman, *Fuel's Paradise: Energy Options for Britain,* Harmondsworth: Penguin Books, 1979 edn., p. 219.

**34** Leslie Hannah, *Electricity before Nationalization,* London and Basingstoke: Macmillan, 1979, p. 136. Other data on the efficiency of steam plant used in plotting Figure 3 comes from Thomas Lean, *Historical Statement of the . . . Steam Engines in Cornwall,* London: 1839; D. B. Barton, *The Cornish Beam Engine,* Truro: Bradford Barton, 1969; D. S. L. Cardwell, *From Watt to Clausius: The Rise of Thermodynamics in the Early Industrial Age,* London: William Heinemann, 1974; H. W. Dickinson, *A Short History of the Steam Engine,* new edn., London: Frank Cass, 1963; A. J. Pacey, 'Some early heat engine concepts', *British Journal for the History of Science,* 7 (1974), pp. 135–45.

**35** Christopher Freeman, John Clarke and Luc Soete, *Unemployment and Technical Innovation,* London: Frances Pinter, 1982, pp. 63, etc. Freeman is also quoted at length on this subject by Rothwell and Zegveld, *Technical Change,* pp. 28–34.

**36** Simon Kuznets, *Economic Change: Selected Essays,* London: William Heinemann, 1974, pp. 109–118.

**37** R. A. Buchanan, *History and Industrial Civilization,* London and Basingstoke: Macmillan, 1979, p. 151.

**38** Dahrendorf, *The New Liberty,* pp. 13–14.

PART **SEVEN**

# ORDER AND CHANGE IN HUMAN SOCIETIES

## INTRODUCTION

The premise of most of the essays in this anthology is that technologies are forms of human society. Thompson's essay on nuclear weapons, Ortega's on the stages of technology, and Pacey's on the ways in which technological hardware interacts with social relations would all be at home in this section. In order to focus matters more sharply, however, the essays collected in this section address three specific questions within the broader area of the social and political implications of technology: Which models offer better characterizations for contemporary social relations—those which are political, or those which are technological? How do technological revolutions alter the balance of power between political decision makers and those whose base is technological? Finally, what constitutes technological progress and to what degree is it possible?

Put in somewhat different terms, these three questions set the stage for the two controversies around which this section is organized: the role of the technocrat in societies that purport or aspire to be democratic and the nature of technological progress.

In "Technē and Politeia: The Technical Constitution of Society," Langdon Winner argues that the traditional relationship between technology and politics has been reversed. It was once thought that politics, or what Plato called "statecraft," was the loftiest form of human production. Plato borrowed the metaphors of the technological arts for his discussion of the leadership of his ideal republic. And modern political thinkers such as Rousseau, Locke, and Jefferson drew upon metaphors of the reigning scientific paradigm, namely, that the universe is a machine and that societies and individuals are lesser machines inside the greater one. Even as late as the early years of the nineteenth century, Jefferson was still

use of such contradictions was a part of the introduction to the previous section. In his essay on "Paradoxes of the Information Age," the media analyst Herbert I. Schiller addresses a set of concrete problems that exhibit such tragic motifs or contradictions. Schiller focuses his attention on the astounding growth of the transnational corporation (TNC) and its heavy dependence on new information technologies. To some, there seems to be ample evidence that every significant form of human social organization, for better or worse, will soon be dominated by TNCs. But Schiller argues that when matters are subjected to close scrutiny it is not autonomy or inevitability that becomes apparent but paradox and contradiction.

First, although it appears that the global corporate system has once and for all thwarted its adversaries, such as labor movements and nation-states, TNCs have in fact worked themselves into a situation which is increasingly tenuous. Their goal of self-enlargement tends to undercut the "relatively secure world system of stable nation-states" that makes their continued growth possible. Their success contains the seeds of prodigious difficulty.

Second, TNCs have apparently been quite effective in selling the idea that widened consumer choice is equivalent to wider political-economic choice. But when this message reaches third and fourth world economies, many of which tend to be based on corruption and inequality and are infamous for their squandering of natural resources, it tends to generate not more but less political stability. The political instability that has accompanied the rise of Islamic fundamentalism, for example, can be viewed as a direct reaction to attempts by TNCs to expand into new markets and to dominate them.

Third, the tendency of western democracies to privatize the production of social goods and services, that is, to turn them over to TNCs, appears to portend a gradual eclipse of the power of elected public officials by the power of the directors of the TNCs, whose own accountability is narrowly circumscribed. But here, too, there are seeds of great difficulty. As nation states provide fewer and fewer social services, they shift instead toward social control, surveillance, and coercion. Schiller reminds us that "in the U.K. and in the U.S.A., as national educational and health expenditures contract, military, law enforcement, intelligence and police outlays expand." The implication is the one that the leaders of the Soviet Union have apparently begun to heed, namely, that money spent on social control is not available for consumer goods.

Fourth, there currently exists a veritable explosion of the ability to generate, process, store, and disseminate information. On the surface, this situation would tend to indicate wider and more informed participation in the political process. But information and the means of access to it are becoming increasingly privatized. One effect of this situation is the increasing gap between the information haves and have-nots. Moreover, as access to information becomes increasingly commercialized, some important information will simply no longer be produced because it lacks profitable markets. These and other countertrends within the growth of information technologies portend a deep-seated political instability.

Schiller's fifth point is related to his first one: the dominance of TNC management over traditional labor movements has removed an important stabilizing factor within industrial democracies.

Sixth, the concentration of information technologies in a few hands, which if viewed superficially would indicate greater efficiency, actually tends to destabilize the political situation in two ways. First, the "right to know," which has been an integral part of western democracies, is now being eroded. Greater homogenization of information means decreased public awareness of real dangers and impoverished public input into policy formulation. Second, there has come to be considerable resentment among the populations of third and fourth world nations against the domination by the industrialized nations of their national media.

Finally, Schiller points to one paradox that overshadows each of the preceding six in terms of its long-term implications. It is what he refers to as the "gate keeping" which insulates the American people against knowledge of conditions that exist in the rest of the world. What is regarded as a desirable means of political control by the managers of the American military-industrial complex may well contain the seeds of the decay of the nation that it purports to serve, and this because Americans are often so poorly informed about external affairs that they are incapable of realistic political judgment.

Whereas Lowrance and Schiller are concerned with the tragic motifs or contradictions inherent in high-tech societies, economist C. E. Ayres attempts to determine what constitutes technological progress and to ascertain to what extent it is possible.

In his 1943 essay "The Path of Progress," Ayres argues that doubt about the possibility of progress disguises a deeper pessimism regarding the nature and function of values in general. Like John Dewey, whose work greatly influenced his, Ayres maintains that far from being either recondite because transcendent, or arbitrary because invented, values are not only knowable, but attainable in greater measure than they are commonly thought to be. He offers a definition of progress: it is "the continuous development of the technological arts and crafts and the accompanying recession of superstition and ceremonially invested status. . . ."

One of the central tenets of capitalism, Ayres contends, is that the industrial system only functions as long as business enterprise operates in a laissez-faire and unhindered fashion. But this thesis in fact assumes what it sets out to prove and it is in any case little more than a superstition. He rejects the standard laissez-faire view that consumption is a transcendent end, that is, that it is a final cause valuable "in itself," to which another state, namely, "production," is a means. He argues instead that values, including the value we call "consumption," arise out of and are testable within what he terms "the technological continuum."

Ayres insists that once values are grounded in human technological (including conceptual or ideational) activities rather than in ideals that are transcendent or supernatural, "progress" will once again be a respectable word. His notion of progress involves advances in what is popularly called scientific technology, to be sure; but he also asserts that the fine arts are an essential part of the technological continuum.

Finally, Ayres seeks to undercut the position of classical liberals (who are now known as "conservatives" and "libertarians") as well as the stance of those who would later be called "welfare liberals." He argues that if consumption could be regarded as one element of a continuous industrial effort within the context of technical progress and not as a be-all and end-all, the homeless would be housed and the poor would be fed not because they have the "right" to be housed and fed or even because housing and feeding them would be "just," but rather because their capacity to contribute to the progress of the entire society is impaired by their debilitated situation.

Ayres's remarks were prescient: the decade of the 1980s in the United States has seen an enormous squandering of human resources. A study undertaken at the Harvard University School of Public Health indicated that between 1980 and 1984, the first term of the Reagan administration, poverty in America increased 20.5 percent and that "the cumulative impact of budget and tax cuts enacted during [the period] 1981–1985 was to transfer some $23 billion from poorer households, and to give an additional $35 billion to those already earning over $80,000 each."[1] The report called government policy during those years "mean spirited." The same period also saw a rise in the number of the homeless, and nutritional diseases usually associated with third world countries increasingly became a part of life in the inner city.

But Ayres argues that intelligent technical judgment militates against consigning a large segment of the population to the status of an "underclass" whose members are underfed, undereducated, and underemployed, whose native talents are underdeveloped and underutilized, and who have a decreasing stake in the present and future of the society in which they find themselves. The implication of Ayres's argument is that to cut funds for school lunches and to designate catsup as a "vegetable" for those which remained, as the Reagan administration did during the early days of its first term, is worse than bad judgment: it is an egregious mismanagement of human resources that portends disaster because it substitutes greed for good technology.

It is Ayres's conclusion that progress is possible only to the extent that the public decides that it is desirable and decides to work for it. Progress is not automatic. It does not arrive as a gift of God. Nor will it occur as long as we follow the suggestions of the laissez-faire economists to just allow things to take their "natural" course, to allow the "invisible hand" of the marketplace to do its work. Progress is invariably the result of the hard work and hard choices that characterize responsible human behavior.

Jürgen Habermas, who is a leading member of the Marxist "Frankfurt School" of social criticism, also addresses the question of technological progress in his essay "Technical Progress and the Social Life-World." The model of progress against which Habermas argues is one that had been put forward by Aldous Huxley in his book *Literature and Science*. Huxley argued that whereas in the domain of literature real human worlds of love and hate, triumph and humiliation, hope and despair are reflected and examined, the domain of science is "worldless" in the sense that it deals with quantified regularities that lie outside the "real" worlds of the human condition. Huxley concluded that progress would occur when literature began to assimilate the results of science and when science began to take on the "flesh and blood" of literature.

But Habermas rejects Huxley's model in favor of one that is more complex. It is his view that science can never inform literature until it first enters the lives of men and women through the technologies that determine their quotidian consciousness. In short, "poems arise from the consideration of Hiroshima and not from the elaboration of hypotheses about the transformation of mass into energy."

As a part of his wider thesis, Habermas argues that the traditional distinctions between theory and practice, as well as between "scientific" and "technical" education, have broken down. These distinctions, which were such an important part of the idealism of the German educational system of the nineteenth century, have proved to be bankrupt. His conclusion is a recapitulation of Dewey's a half century earlier: technological progress can no longer be thought of as a matter of the private cultivation of a personal consciousness but must instead be identified with the bringing of the "power of technical control within the range of the consensus of acting and transacting citizens."

Habermas explicitly rejects models of technological progress that treat it as automatic or inevitable. Because of this, however, he must also reject the view advanced by Marx in certain portions of his work that the technical control which leads to greater production of material and intellectual artifacts is the sufficient condition for a society in which democratic decision making is dominant. In short, Habermas argues that technical rationality does not guarantee human progress, since technical rationality sometimes means that technocrats have found ways of dominating entire cultures. What is required for real progress is open dialogue among politically aware men and women who have thrown off the domination of the technocrats. Though his language is different, Habermas's position is at root quite similar to the one advanced by Winner in the first essay in this section: both argue that political virtue must once again be developed and treated as an important technological artifact. The equally unhappy alternatives to the development of political virtue are anarchy on one side and domination by antidemocratic technocrats on the other.

Finally, in his essay "The Ethics of Order and Changes: An Analytical Framework," Godfrey Gunatilleke examines the assumption that technological progress means the same for societies that are developed as it does for those that are developing. He is especially concerned with the impact of technology on societies that have what he calls an "order-oriented" ethos, as distinguished from the "rapid-change" ethos that characterizes developed countries. Order-oriented societies, such as many of those in southeast Asia, have long histories of treating values as transcendent and stable. Developed countries, on the other hand, from which developing countries now tend to import technologies and their implicit and explicit accompanying social models, tend to accept change and discontinuity as essential features of personal and social life. This disparity often precipitates major cultural clashes in developing countries.

Gunatilleke calls for increased understanding of the historical context in which developing countries confront new technologies, and he points out that the demographic and economic changes experienced during the past few decades by many such societies are considerably greater than those experienced by societies that are already industrialized.

He argues that the two major competing models for third world development—neoclassical capitalism and Marxism—are both inappropriate because they are reductionistic. Neoclassical capitalism is a logical order, based on the assumption that progress is essentially technoeconomic and involves the development of productive capacity. The competing Marxist model is a dialectical process that assumes that social restructuring can release productive capacities; it therefore places political processes at the center of its plans for development. Despite their different methods, he argues, both models are inappropriately reductive because both focus single-mindedly upon transformation of the system of production and distribution. In addition, there is in both models a fundamental assumption that "the historical evolution of humanity leads to progressively higher stages." In short, both reductive models sequester a faulty notion of progress.

Against these models Gunatilleke proposes one that is more complex and, he believes, more appropriate to the cultures of the third world. It replaces the single-minded goal of increased production and distribution with clusters of equally desirable goals. His model is inclusive of the logic of capitalism and the dialectic of Marxism but adds a third feature, which he calls "the paradoxical." The paradoxical approach is nonreductive because it establishes clusters of goals and gives them equal weight. It aims for fair distribution of wealth, income, and power. It calls for decreasing dependence upon industrialized countries. It requires that the ecosystem be respected and that growth be long-term and sustainable. It also demands human rights and political freedoms, as well as increasing participation by those affected by technological decisions.

Gunatilleke is careful to stress that each of these clusters should be treated as of equal value and that none should be traded off for the benefit of another. Social needs must be balanced with demands for growth, and the needs of the present must be balanced with respect to the needs of the future. In other words, there must be a holistic vision of development.

But how is this vision to be secured and maintained? Gunatilleke believes that the transcendental paradigm of the world's major religions offers the key. He argues that the transcendental paradigm is superior because it is nonreductionistic, that is, because it treats ethical decision not as simple choice between acceptable and unacceptable consequences, but as a real and tragic dilemma in which real values are sacrificed in order to secure other values. Technological choice is thus a real dilemma, because it "implies humility in the presence of life's complexity, an acceptance of the limitations of human knowledge, an anguish at the possibility of losing one value when choosing another value. The unfolding of reality confounds the simple principles of logical order and sequence where every object, condition, and value has a separate identity and is not to be confused with another." Further, the transcendental model treats social and moral obligations as a part of a wider cosmic order.

His argument, then, is that technological change in developing countries must work in tandem with existing cultural values, including existing religious values. This means that western notions of change and progress will increasingly be viewed as inappropriate for the developing world, for there is no simple historical progression in the transcendental paradigm that carries human beings outside of the realm of the paradoxical. Solutions can never be fully anticipated; instead, they grow, change, and mature as experience opens up the consequences of action.

Gunatilleke's argument is in many of its aspects both cogent and forceful. His characterization of appropriate technology is reminiscent of that of Alan Drengson in Part One, and his treatment of the paradoxical elements in knowledge is very much like that of John Dewey, who argued that knowledge is never complete and that solutions can never be completely anticipated. Dewey contended that the quest for certainty is both illusive and unnecessary. But Gunatilleke's argument that developing countries must begin where they are and build upon transcendental, order-oriented value systems also seems to have serious difficulties. Scientific technology prizes objective measurement, democratic debate, and testable results. Transcendental value systems have rarely prized any of these things. They have tended to be authoritarian and to view history as in the hands of supernatural powers. Where they have been otherwise, as in the case of some forms of Buddhism, there has often been an emphasis on technologies of the self at the cost of wider social concern. Some transcendental models, such as the "social gospel" of Reinhold Niebuhr in the 1930s and "liberation theology" in the 1970s and 1980s, have in fact evinced a commitment to wider social concerns, but neither of these movements could be described as "order-oriented."

Gunatilleke is manifestly correct in his criticism of those who have held that the methods of scientific technology lead to absolute certainty. But it is essential to place the blame where it properly lies. The difficulty is not with scientific inquiry, but with some of its putative practitioners. Dewey and Ayres, among others, also criticized those who claimed absolute certainty, but they were able to do so in a way that did not demand reliance upon the transcendental model.

It would therefore seem that Gunatilleke is both right and wrong in this regard. He is correct in arguing that developing cultures must begin where they are and that most of them are deeply immersed in transcendental models. But if he wishes to claim that scientific technology is incapable of approaching problems in a holistic, multivalent manner, then he has a very narrow conception of it.

There is little room to doubt Gunatilleke's claim that "progress" will continue to mean something very different in developing countries than it does in those which are industrialized. Historical and cultural conditions in Sri Lanka are very different from those in the United States. But the larger question is whether the patterns of inquiry associated with the successes of scientific-technological societies can be improved upon by incorporating them into a transcendental model.

Both Dewey and Ayres argued that the methods of scientific technology are not just the best methods so far invented for solving problems but are the only methods, short of luck. Scientific methods, unlike transcendental ones, are self-corrective; improvements in the methods of scientific inquiry are also scientific. Dewey maintained that those who argue against the methods of scientific technology usually make one of two mistakes. Some of them assume that other methods of inquiry such as faith, appeal to authority, or luck work better than the methods of scientific technology. But their view is false because those other methods are undemocratic and not self-corrective. Others have mistakenly seen a failure of scientific methods where such methods have never been applied, where they have in fact been supplanted by quests for personal wealth or power, allegiances to class interests, or applications of intransigent ideologies.

Each of the essays in this section has addressed the ways in which human societies are organized and governed. Each essayist faults the present balance between the political and the technological, and each offers suggestions for redressing that balance. Dewey and Ayres argue that individuals and their governments have not yet taken advantage of the methods of scientific technology. Winner, Dewey, Lowrance, and Habermas each maintains in his own unique way that political virtue is a kind of technology and that it must once again take priority over narrower forms of technological expertise. Schiller envisions a coming crisis in which significant social rearrangements will have to be made, regardless of our intent. Gunatilleke tries to find a way of wedding scientific technology and the transcendentalism of traditional order-oriented cultures. Beyond their differences, however, all our authors share the common perception that technologies are laden with values and that every form of technology represents a form of life that is at best deliberately crafted, at worst just gotten by default.

## NOTES

1 The Physician Task Force on Hunger in America, *Hunger in America: The Growing Epidemic*, Harvard University School of Public Health, Boston, 1985, p. 94.

# Technē and Politeia: The Technical Constitution of Society

**Langdon Winner**

To achieve a political understanding of technology requires that we examine the realm of tools and instruments from a fresh point of view. We have already begun to recognize some of the ways in which conditions of power, authority, freedom, and social justice are deeply embedded in technical structures. From this standpoint no part of modern technology can be judged neutral a priori. All varieties of hardware and their corresponding forms of social life must be scrutinized to see whether they are friendly or unfriendly to the idea of a just society.

But what does the study of politics have to contribute to our thinking about the realm of instrumental things? Where can one turn to look for a political theory of technology?

## A CLASSIC ANALOGY

In the previous chapter I noted that rooted in Western political thought is a powerful analogy linking the practice of technology to that of politics. In his *Republic, Laws, Statesman,* and other dialogues, Plato asserts that statecraft is a *technē,* one of the practical arts. Much like architecture, weaving, shipbuilding, and other arts and crafts, politics is a field of practice that has its own distinctive knowledge, its own special skills. As we have seen, one purpose of Plato's argument was to discredit those who believed that the affairs of public life could be left to mere amateurs, the democratic masses. But beyond that it is clear he thought the art of politics could be useful in the same way as any other *technē,* that it could produce well-crafted works of lasting value.

The works he had in mind were good constitutions, supremely well-crafted products of political architecture. *Politeia,* the title of the *Republic* in Greek, means the constitution of a polis, the proper order of human relationships within a city-state. The dialogue describes and justifies what Plato holds to be the institutional arrangements appropriate to the best *politeia*. He returns to this theme in the *Laws,* a discussion of the "second best" constitution, comparing his work to that of a well-established craft. "The shipwright, you know, begins his work by laying down the keel of the vessel and indicating her outlines, and I feel myself to be doing the same thing in my attempt to present you with outlines of human lives....I am really laying the keels of the vessels by due consideration of the question by what means or manner of life we shall make our voyage over the sea of time to the best purpose."[1] There is evidence that Plato actually sought to realize his skills as a designer/builder of political societies. He traveled from Athens to live at the court of Dionysius the Elder, tyrant of Syracuse, hoping to transform his host into a genuine philosopher-king, a person willing to apply the true principles of political *technē*. The attempt did not succeed.

In Plato's interpretation the analogy between technology and politics works in one direction only; *technē* serves as a model for politics and not the other way around. Although respectful of the power of the material arts, he remained deeply suspicious of them. Thus, in the *Laws* he excludes craftsmen from positions of citizenship, explaining that they already have an art that requires their full attention. At the same time he forbids citizens to engage in any material craft whatsoever because citizenship makes full demands on them. Plato's discomfort with technology has remained characteristic of moral and political philosophers to this day. Most of them have politely ignored the substance of technical life, hoping perhaps that it would remain segregated in a narrowly defined corner of human life. Evidently, it did not occur to anyone that Plato's pregnant analogy would at some point qualify in reverse, that *technē* itself might become *politeia,* that technical forms of life might in themselves play a powerful role in shaping society. When that finally happened, political theory would find itself totally unprepared.

The one-sided comparison of technical and political creativity appears again in modern political thought. Writing in *The Social Contract,* Jean Jacques Rousseau employs a mechanical metaphor to illuminate the art of constitution making. "A prince," he says, "has only to follow a model which the lawgiver provides. The lawgiver is the engineer who invents the machine; the prince is merely the mechanic who sets it up and operates it."[2] At another point in the book, Rousseau compares the work of the lawgiver to that of an architect. With a frustrated ambition reminiscent of Plato's, Rousseau offered himself as a political engineer or architect of exactly this kind, writing treatises on the constitutions of Corsica and Poland in the hope that his ideas might influence the founding of new states.

A practical opportunity of exactly that kind later became available to the founders of modern nation states, among them the leaders of the American Revolution. From the earliest rumblings of rebellion in the seventeenth century to the adoption of the U.S. Constitution in 1787, the nation was alive with disputes about the application of political principles to the design of public institutions. Once again the ancient analogy between politics and technology became an expressive idea. Taking what they found useful from previous history and existing theories, thinkers like Madison, Hamilton, Adams, and Jefferson tried to devise a "science of politics," a science specifically aimed at providing knowledge for a collective act of architectonic skill. Thus, in *The Federalist Papers,* to take one example, we find a sustained discussion of how to move from abstract political notions such as power, liberty, and public good to their tangible manifestations in the divisions, functions, powers, relationships, and limits of the Constitution. "The science of politics," Hamilton explains in "Federalist No. 9," "like most other sciences, has received great improvement. The efficacy of various principles is now well understood, which were either not known at all, or imperfectly known to the ancients. The regular distribution of power into distinct departments; the introduction of legislative balances and checks; the institution of courts composed of judges holding their offices during good behavior; the representation of the people in the legislature by deputies of their own election: these are wholly new discoveries, or have made their principal progress towards perfection in modern times." Metaphors from eighteenth-century science and mechanical invention—for example, "the ENLARGEMENT of the ORBIT within which such systems are to revolve"[3] and references to the idea of checks and balances—pervade *The Federalist Papers* and indicate the extent to which its writers saw the founding as the creation of an ingenious political/mechanical device.

But even as the eighteenth century was reviving the comparison between technology and politics, even as philosopher statesmen were restoring the *technē* of constitution making, another extremely powerful mode of institutionalization was taking shape in the United States and Europe. The industrial revolution with its distinctive ways of arranging people, machines, and materials for production very soon began to compete with strictly political institutions for power, authority, and the loyalties of men and women. Writing in 1781 in his *Notes on Virginia,* Thomas Jefferson noted the new force abroad in the world and commented upon its probable meaning for political society. The system of manufacturing emerging at the time would, he argued, be incompatible with the life of a stable, virtuous republic. Manufacturing would create a thoroughly dependent rather than a self-sufficient populace. "Dependence," he warned, "begets subservience and venality, suffocates the germ of virtue, and prepares fit tools for the designs of ambition." In his view the industrial mode of production threatened "the manners and spirit of a people which preserve a republic in vigor. A degeneracy in these is a canker which soon eats to the heart of its laws and constitution."[4] For that reason he advised, in this book at least, that Americans agree to leave the workshops in Europe.

### ABUNDANCE AND FREEDOM

Jefferson's plea echoes a belief common in writings of ancient Greece and Rome that civic virtue and material prosperity are antithetical. Human nature, according to this view, is easily corrupted by wealth. The indolent, pleasure-seeking habits of luxurious living tend to subvert qualities of frugality, self-restraint, and self-sacrifice needed to maintain a free society. By implication, any society that wishes to maintain civic virtue ought to approach technical innovation and economic growth with the utmost caution. At the time of the founding of the American republic, the country did not depend upon high levels of material production and consumption. In fact, during political discussions of the 1770s and 1780s, the quest for material wealth was sometimes mentioned as a danger, a source of corruption. A speaker before the Continental Congress in 1775 called upon the citizenry to "banish the syren LUXURY with all her train of fascinating pleasures, idle dissipation, and expensive amusements from our borders," and to institute "honest industry, sober frugality, simplicity of manners, and plain hospitality and christian benevolence."[5]

There are signs that a desire to shape industrial development to accord with the ideals of the republican political tradition continued to interest some Americans well into the 1830s. Attempts to include elements of a republican community in the building of the factory town in Lowell, Massachusetts, show this impulse at work.[6] But these efforts were neither prominent in the economic patterns then taking shape nor successful in their own right. In the 1840s and decades since the notion that industrial development might be shaped or limited by republican virtues dropped out of common discourse, echoed only in the woeful lamentations of Henry David Thoreau, Henry Adams, Lewis Mumford, Paul Goodman, and a host of others now flippantly dismissed as "romantics" and "pastoralists."

In fact, the republican tradition of political thought had long since made its peace with the primary carrier of technical change, entrepreneurial capitalism. Moral and political thinkers from Machiavelli to Montesquieu and Adam Smith had argued, contrary to the ancient wisdom, that the pursuit of economic advantage is actually a civilizing, moderating influence in society, the very basis of stable government. Rather than engage the fierce passion for glory that often leads to conflict, it is better, so the argument goes, to convince people to pursue their self-interst, an interest that inclines them toward rational behavior.[7] The framers of the American Constitution were, by and large, convinced of the wisdom of this formula. They expected that Americans would act in a self-interested manner, employing whatever instruments they needed to generate wealth. The competition of these interests in society would, they believed, provide a check upon the concentration of power in the hands of any one faction. Thus, in one important sense republicanism and capitalism were fully reconciled at the time of the founding.

By the middle of the nineteenth century this point of view had been strongly augmented by another idea, one that to this day forms the basic self-image of Americans—a notion that equates abundance and freedom. The country was rich in land and resources; people liberated from the social hierarchies and status definitions of traditional societies were given the opportunity to exploit that material bounty in whatever ways they could muster. In this context new technologies were seen as an undeniable blessing because they enabled the treasures to be extracted more quickly, because they vastly increased the product of labor. Factories, railroads, steamboats, telegraphs, and the like were greeted as the very essence of democratic freedom for the ways they rendered, as one mid-nineteenth-century writer explained, "the conveniences and elegancies of life accessible to the many instead of the few."[8]

American society encouraged people to be self-determining, to pursue their own economic goals. That policy would work, it was commonly believed, only if there were a surplus that guaranteed enough to go around. Class conflict, the scourge of democracy in the ancient world, could be avoided in the United States because the inequalities present in society would not matter very much. Material abundance would make it possible for everybody to have enough to be perfectly happy. Eventually, Americans took this notion to be a generally applicable theory: economic enterprise driven by the engine of technical improvement was the very essence of human freedom. Franklin D. Roosevelt reportedly remarked that if he could put one American book in the hands of every Russian, it would be the Sears, Roebuck catalogue.

In this way of looking at things the form of the technology you adopt does not matter. If you have cornucopia in your grasp, you do not worry about its shape. Insofar as it is a powerful thing, more power to it. Anything that history, literature, philosophy, or long-standing traditions might have to suggest about the prudence one ought to employ in the shaping of new institutions can be thrown in the trash bin. Describing the industrial revolution in Britain, historian Karl Polanyi drew an accurate picture of this attitude. "Fired by an emotional faith in spontaneity, the common-sense attitude toward change was discarded in favor of a mystical readiness to accept the social consequences of economic improvement, whatever they might be. The elementary truths of political science and statecraft were first discarded, then forgotten. It should need no elaboration that a process of undirected change, the pace of which is deemed too fast, should be slowed down, if possible, so as to safeguard the welfare of the community. Such household truths of traditional statesmanship, often merely reflecting the teachings of a social philosophy inherited from the ancients, were in the nineteenth century erased from the thoughts of the educated by the corrosive of a crude utilitarianism combined with an uncritical reliance on the alleged self-healing virtues of unconscious growth."[9] Indeed, by the late nineteenth century, an impressive array of scientific discoveries, technical inventions, and industrial innovations seemed to make the mastery of nature an accomplished fact rather than an idle dream. Many took this as a sign that all ancient wisdom, like all the old-fashioned machines and techniques, had simply been rendered obsolete. As one chronicler of the new technology wrote in *Scientific American:* "The speculative philosophy of the past is but a too empty consolation for short-lived, busy man, and, seeing with the eye of science the possibilities of matter, he has touched it with the divine breath of thought and made a new world."[10]

According to this view, everything one might desire of the relationship between expanding industrial technology and the building of a good society will happen automatically. All that is necessary is to make sure the machinery is up to date, well maintained, and well oiled. The only truly urgent questions that remain are ones of technical and economic efficiency. For unless a society keeps pace with the most efficient means available anywhere in the world, it will lag behind its competitors, a precondition of cultural decline.

A fascination with efficiency is a venerable tradition in American life. It is announced early on, for example, in Benjamin Franklin's maxims that economizing on time, effort, and money is a virtue. With the advance of industrialism in the late nineteenth and early twentieth centuries, this concern grew to something of an obsession among the well-educated in the United States. Understood to be a criterion applicable to personal and social life as well as to mechanical and economic systems, efficiency was upheld as a goal supremely valuable in its own right, one strongly linked to the progress of science, the growth of industry, the rise of professionalism, and the conservation of natural resources. During the Progressive Era the rule of efficient, well-trained professionals was upheld as a way of sanitizing government of the corruption of party machines and eliminating the influence of selfish interest groups. An eagerness to define important public issues as questions of efficiency has continued to be a favorite strategy in American politics over many decades; adherence to this norm has been (and still is) welcomed as the best way to achieve the ends of democracy without having to deal with democracy as a living political process. Demonstrating the efficiency of a course of action conveys an aura of scientific truth, social consensus, and compelling moral urgency. And Americans do not even worry much about the specific content of numerators and denominators used in efficiency measurements. As long as they are getting more for less, all is well.[11]

## THE TECHNICAL CONSTITUTION OF SOCIETY

With the passage of time the cornucopia of modern industrial production began to generate some distinctive institutional patterns. Today we can examine the interconnected systems of manufacturing, communications, transportation, and the like that have arisen during the past two centuries and appreciate how they form de facto a constitution of sorts, the constitution of a sociotechnical order. This way of arranging people and things, of course, did not develop as the result of the application of any particular plan or political theory. It grew gradually and in separate increments, invention by invention, industry by industry, engineering project by engineering project, system by system. From a contemporary vantage point, nevertheless, one can notice some of its characteristics and begin to see how they embody answers to age-old political questions—questions about membership, power, authority, order, freedom, and justice. Several of the characteristics that matter in this way of seeing things—characteristics that would certainly have interested Plato, Rousseau, Madison, Hamilton, and Jefferson—can be summarized as follows.

First is the ability of technologies of transportation and communication to facilitate control over events from a single center or small number of centers. Largely unchecked by effective countervailing influences, there has been an extraordinary centralization of social control in large business corporations, bureaucracies, and the military. It has seemed an expedient, rational way of doing things. Without anyone having explicitly chosen it, dependency upon highly centralized organizations has gradually become a dominant social form.

Second is a tendency for new devices and techniques to increase the most efficient or effective size of organized human associations. Over the past century more and more people have found themselves living and working within technology-based institutions that previous generations would have called gigantic. Justified by impressive economies of scale and, economies or not, always an expression of the power that accrues to very large organizations, this gigantism has become an accustomed feature in the material and social settings of everyday life.

Third is the way in which the rational arrangement of sociotechnical systems has tended to produce its own distinctive forms of hierarchical authority. Legitimized by the felt need to do things in what seems to be the most efficient, productive way, human roles and relationships are structured in rule-guided patterns that involve taking orders and giving orders along an elaborate chain of command. Thus, far from being a place of democratic freedom, the reality of the workplace tends to be undisguisedly authoritarian. At higher levels in the hierarchy, of course, professionals claim their special authority and relative freedom by virtue of their command of scientific and technical expertise. At the point in history in which forms of hierarchy based on religion and tradition had begun to crumble, the need to build and maintain technical systems offered a way to restore pyramidal social relations. It was a godsend for inequality.

Fourth is the tendency of large, centralized, hierarchically arranged sociotechnical entities to crowd out and eliminate other varieties of human activity. Hence, industrial techniques eclipsed craftwork; technologies of modern agribusiness made small-scale farming all but impossible; high-speed transportation crowded out slower means of getting about. It is not merely that useful devices and techniques of earlier periods have been rendered extinct, but also that patterns of social existence and individual experience that employed these tools have vanished as living realities.

Fifth are the various ways that large sociotechnical organizations exercise power to control the social and political influences that ostensibly control them. Human needs, markets, and political institutions that might regulate technology-based systems are often subject to manipulation by those very systems. Thus, to take one example, psychologically sophisticated techniques of advertising have become a customary way of altering people's ends to suit the structure of available means, a practice that now affects political campaigns no less than campaigns to sell underarm deodorant or Coca-Cola (with similar results).

There are many other characteristics of today's technological systems that can accurately be read as political phenomena. And it is certainly true that there are factors other than technology that strongly influence the developments I have mentioned. But it is important to note that as our society adopts one sociotechnical system after another it answers some of the most important questions that political philosophers have ever asked about the proper order of human affairs. Should power be centralized or dispersed? What is the best size for units of social organization? What constitutes justifiable authority in human associations? Does a free society depend upon social uniformity or diversity? What are appropriate structures and processes of public deliberation and decision making? For the past century or longer our responses to such questions have often been instrumental ones, expressed in an instrumental language of efficiency and productivity, physically embodied in human/machine systems that seem to be nothing more than ways of providing goods and services.

If we compare the process through which today's sociotechnical constitution evolved to the process employed by the framers of the U.S. Constitution, the contrast is striking. Clearly, the founding fathers considered all of the crucial questions in classical political thought. When they included a particular feature in the structure of government, it was because they had studied the matter, debated it, and deliberately chosen the result. They understood that if all went well the structures they were building would last a good long time. Because the rules, roles, and relationships they agreed upon would shape the future life of a whole nation, the framers acknowledged a special responsibility—a responsibility of wise political craftsmanship. To realize this responsibility required a depth of knowledge about political institutions and sensitivity to human motives altogether rare in human history. The results of their work include two centuries of relatively stable government in the United States, a sign that they practiced their craft well.

Of course, there are founding fathers of our sociotechnical constitution as well—the inventors, entrepreneurs, financiers, engineers, and managers who have fashioned the material and social dimensions of new technologies. Some of their names are well known to the public—Thomas Edison, Henry Ford, J. P. Morgan, John D. Rockefeller, Alfred P. Sloan, Thomas Watson, and the like. The names of such figures as Theodore Vail, Samuel Insull, and William Mullholland are not household words, but their accomplishments as builders of technological infrastructures are equally impressive. In one sense the founders of technological systems are no strangers to politics; many of them have engaged in fierce political struggles to realize their aims. William Mullholland's ruthless machinations to bring Owens Valley water to Los Angeles' desert climate is a classic case in point.[12] But the qualities of political wisdom we find in the founders of the U.S. Constitution are, by and large, missing in those who design, engineer, and promote vast systems. Here the founding fathers have been concerned with such matters as the quest for profits, organizational control, and the pleasures of innovation. They have seldom been interested in the significance of their work on the overall structure of society or its justice.

For those who have embraced the formula of freedom through abundance, however, questions about the proper order of society do not matter very much. Over many decades technological optimists have been sustained by the belief that whatever happened to be created in the sphere of material/instrumental culture would certainly be compatible with freedom, democracy, and social justice. This amounts to a conviction that all technology—whatever its size, shape, or complexion—is inherently liberating. For reasons noted in the previous chapter, that is a very peculiar faith indeed.

It is true that on occasion agencies of the modern state have attempted to "regulate" business enterprises and technological applications of various kinds. On balance, however, the extent of that regulation has been modest. In the United States absolute monopolies are sometimes outlawed only to be replaced by enormous semimonopolies no less powerful in their ability to influence social outcomes. The history of regulation shows abundant instances in which the rules and procedures that govern production or trade were actually demanded or later captured by the industries they supposedly regulate. In general, the rule of thumb has been if a business makes goods and services widely available, at low cost with due regard for public health and safety, and with a reasonable return on investment, the republic is well served.[13]

In recent times the idea of recognizing limits upon the growth of certain technologies has experienced something of a revival. Many people are prepared to entertain the notion of limiting a given technology if:

1 Its application threatens public health or safety
2 Its use threatens to exhaust some vital resource
3 It degrades the quality of the environment (air, land, and water)
4 It threatens natural species and wilderness areas that ought to be preserved
5 Its application causes social stresses and strains of an exaggerated kind.

Along with ongoing discussions about ways to sustain economic growth, national competitiveness, and prosperity, these are the only matters of technology assessment that the general public, decision makers, and academicians are prepared to take seriously.

While such concerns are valid, they severely restrict the range of moral and political criteria that are permissible in public deliberations about technological change. Several years ago I tried to register my discomfort on this score with some colleagues in computer science and sociology who were doing a study of the then-novel systems of electronic funds transfer (EFT). They had concluded that such systems contained the potential for redistributing financial power in the world of banking. Electronic money would make possible a shift of power from smaller banks to large national and international financial institutions. Beyond that it appeared such systems posed serious problems about data protection and individual privacy. They asked me to suggest an effective way of presenting the possible dangers of this development to their audience of scholars and policy makers. I recommended that their research try to show that under conditions of heavy, continued exposure, EFT causes cancer in laboratory animals. Surely, that finding would be cause for concern. My ironic suggestion acknowledged what I take to be the central characteristic of socially acceptable criticism of technology in our time. Unless one can demonstrate conclusively that a particular technical practice will generate some physically evident catastrophe—cancer, birth defects, destruction of the ozone layer, or some other—one might as well remain silent.

The conversation about technology and society has continued to a point at which an obvious question needs to be addressed: Are there no shared ends that matter to us any longer other than the desire to be affluent while avoiding the risk of cancer? It may be that the answer is no. The prevailing consensus seems to be that people love a life of high consumption, tremble at the thought that it might end, and are displeased about having to clean up the messes that modern technologies sometimes bring. To argue a moral position convincingly these days requires that one speak to (and not depart from) people's love of material well-being, their fascination with efficiency, or their fear of death. The moral sentiments that hold force can be arrayed on a spectrum ranging from Adam Smith to Frederick W. Taylor to Thomas Hobbes. I do not wish to deny the validity of these sentiments, only to point out that they represent an extremely narrow mindset. Concerns about particular technological hazards are sometimes the beginning of a much broader political awareness. But for the most part we continue to disregard a problem that has been brewing since the earliest days of the industrial revolution—whether our society can establish forms and limits for technological change, forms and limits that derive from a positively articulated idea of what society ought to be.

As a way of beginning that project, I would suggest a simple heuristic exercise. Let us suppose that every political philosophy in a given time implies a technology or set of technologies in a particular pattern for its realization. And let us recognize that every technology of significance to us implies a set of political commitments that can be identified if one looks carefully enough. What appear to be merely instrumental choices are better seen as choices about the form of social and political life a society builds, choices about the kinds of people we want to become. Plato's metaphor, especially his reference to the shipwright, is one that an age of high technology ought to ponder carefully: we ought to lay out the keels of our vessels with due consideration to what means or manner of life best serves our purpose in our voyage over the sea of time. The vessels that matter now are such things as communications systems, transit systems, energy supply and distribution systems, information networks, household instruments, biomedical technologies, and of course systems of industrial and agricultural production. Just as Plato and Aristotle posed the question, What is the best form of political society? so also an age of high technology ought to ask, What forms of technology are compatible with the kind of society we want to build?

Answers to that question often appear as subliminal themes or concealed agendas in policy discussions that seem to be about productivity, efficiency, and economic growth. A perfect set of examples can be found among the dozens of sophisticated energy studies conducted during the 1970s in response to what was then called "the energy crisis." A careful reader can survey the various reports and interpret the political and social structures their analyses and recommendations imply.[14] Would it be nuclear power administered by a benign priesthood of scientists? Would it be coal and oil brought to you by large, multinational corporations? Would it be synthetic fuels subsidized and administered by the state? Or would it be the soft energy path brought to you by you and your neighbors?

Whatever one's position might be, the prevailing consensus required that all parties base their arguments on a familiar premise: efficiency. Regardless of how a particular energy solution would affect the distribution of wealth and social power, the case for or against it had to be stated as a practical necessity deriving from demonstrable conditions of technical or economic efficiency. As the Ford Foundation's Nuclear Energy Policy Study Group explained: "When analyzing energy, one must first decide whether ordinary rules of economics can be applied." The group decided that, yes, energy should be considered "an economic variable, rather than something requiring special analysis."[15] After that decision had been made, of course, the rest was simply a matter of putting Btus or kilowatt-hours in the numerator and dollars in the denominator and worshiping the resulting ratio as gospel.

Even those who held unorthodox viewpoints in this debate found it necessary to uphold the supreme importance of this criterion. Thus, Amory B. Lovins, a leading proponent of soft energy paths, wrote of his method: "While not under the illusion that facts are separable from values, I have tried...to separate my personal preferences from my analytic assumptions and to rely not on modes of discourse that might be viewed as overtly ideological, but rather on classical arguments of economic and engineering efficiency (which are only tacitly ideological)."[16] To Lovins's credit, he consistently argued that the social consequences of energy choices were, in the last analysis, the most important aspect of energy policy making. In his widely read *Soft Energy Paths,* Lovins called attention to "centrism, vulnerability, technocracy, repression, alienation" and other grave problems that afflict conventional energy solutions. Lovins compares "two energy paths that are distinguished by their antithetical social implications." He notes that basing energy decisions on social criteria may appear to involve a "heroic decision," that is, "doing something the more expensive way because it is desirable on other more important grounds than internal cost."

But Lovins is careful not to appeal to his readers' sense of courage or altruism. "Surprisingly," he writes, "a heroic decision does not seem to be necessary in this case, because the energy system that seems socially more attractive is also cheaper and easier."[17] But what if the analysis had shown the contrary? Would Lovins have been prepared to give up the social advantages believed to exist along the soft energy path? Would he have accepted "centrism, vulnerability, technocracy, repression, alienation," and the like? Here Lovins yielded ground that in recent history has again and again been abandoned as lost territory. It raises the question of whether even the best intentioned, best qualified analysts in technological decision making are anything more than mere efficiency worshipers.

Much the same strategy often appears in the arguments of those who favor democratic self-management, decentralization, and human-scale technology. As Paul Goodman once noted, "Now, if lecturing at a college, I happen to mention that some function of society which is highly centralized could be much decentralized without loss of efficiency, or perhaps with a gain in efficiency, at once the students want to talk about nothing else."[18] That approach is, indeed, one way of catching people's attention; if you can get away with it, it is certainly a most convincing kind of argument. Because the idea of efficiency attracts a wide consensus, it is sometimes used as a conceptual Trojan horse by those who have more challenging political agendas they hope to smuggle in. But victories won in this way are in other respects great losses. For they affirm in our words and in our methodologies that there are certain human ends that no longer dare be spoken in public. Lingering in that stuffy Trojan horse too long, even soldiers of virtue eventually suffocate.

## REGIMES OF INSTRUMENTALITY

In our time *technē* has at last become *politeia*—our instruments are institutions in the making. The idea that a society might try to guide its sociotechnical development according to self-conscious, critically evaluated standards of form and limit can no longer be considered a "heroic decision"; it is simply good sense. Because technological innovation is inextricably linked to processes of social reconstruction, any society that hopes to control its own structural evolution must confront each significant set of technological possibilities with scrupulous care.

Applied in this setting, political theory can help reveal strategic decisions in technological design. From its perspective, each significant area of technical/functional organization in modern society can be seen as a kind of regime, a regime of instrumentality, under which we are obliged to live. Thus, there are a number of regimes of mass production, each with a structure that may be interpreted as a technopolitical phenomenon. There are a number of regimes in energy production and distribution, in petroleum, coal, hydroelectricity, nuclear power, etc., each with a form that can be scrutinized for the politics of its structural properties. There is, of course, the regime of broadcast television and that of the automobile. If we were to identify and characterize all of the sociotechnical systems of in our society, all of our regimes of instrumentality and their complex interconnections, we would have a clear picture of the second constitution I mentioned earlier, one that stands parallel to and occasionally overlaps the constitution of political society as such.

The important task becomes, therefore, not that of studying the "effects" and "impacts" of technical change, but one of evaluating the material and social infrastructures specific technologies create for our life's activity. We should try to imagine and seek to build technical regimes compatible with freedom, social justice, and other key political ends. Insofar as the possibilities present in a given technology allow it, the thing ought to be designed in both its hardware and social components to accord with a deliberately articulated, widely shared notion of a society worthy of our care and loyalty. If it is clear that the social contract implicitly created by implementing a particular generic variety of technology is incompatible with the kind of society we deliberately choose—that is, if we are confronted with an inherently political technology of an unfriendly sort—then that kind of device or system ought to be excluded from society altogether.

What I am suggesting is a process of technological change disciplined by the political wisdom of democracy. It would require qualities of judiciousness in the populace that have rarely been applied to the judgment of instrumental/functional affairs. It would, presumably, produce results sometimes much different from those recommended by the rules of technical and economic efficiency. Other social and political norms, articulated by a democratic process, would gain renewed prominence. Faced with any proposal for a new technological system, citizens or their representatives would examine the social contract implied by building that system in a particular form. They would ask, How well do the proposed conditions match our best sense of who we are and what we want this society to be? Who gains and who loses power in the proposed change? Are the conditions produced by the change compatible with equality, social justice, and the common good? To nurture this process would require building institutions in which the claims of technical expertise and those of a democratic citizenry would regularly meet face to face. Here the crucial deliberations would take place, revealing the substance of each person's arguments and interests. The heretofore concealed importance of technological choices would become a matter for explicit study and debate.

There are any number of ways in which the structural features of instrumental regimes might become a focus for democratic decision making. Technologies introduced in the workplace are the ones most often mentioned; in such cases it is usually fairly clear, as in developments in automation and robotization, whose interests are immediately helped and harmed in building a new system. There are, however, a wide variety of areas in which the political complexion of technological systems could fruitfully be explored. The following illustration is one in which there is no "crisis" or obvious social problem at hand, but in which the shape of an evolving system has interesting political dimensions.

The field of solar electricity, photovoltaic energy, is one in which the crucial choices are, to a certain extent, still open for discussion. We can expect to see events unfold in our lifetimes with outcomes that could have many different dimensions. If solar cells become feasible to mass produce, if their price in installed systems comes down to a reasonable level, solar electricity could make a contribution to our society's aggregate energy needs. If the day arrives that photovoltaic systems are technically and economically feasible (and many who work with solar electric prototypes believe they will be), there will be—at least in principle—a choice about how society will structure these systems. One could, for example, build centralized photovoltaic farms that hook directly into the existing electrical grid like any other form of centrally generated electric power. It might also be possible to produce a great number of stand-alone systems placed on the rooftops of homes, schools, factories, and the like. Or one could design and build medium-sized ensembles, perhaps at a neighborhood level. When it comes time to choose which model of photovoltaic development our society will have, a number of implicit questions will somehow be answered. How large should such systems be? How many will be built? Who should own them? How should they be managed? Should they be fully automatic? Or should the producer/consumer of solar power be actively involved in activities of load management?

All of these are questions about the shape of a new regime of instrumentality. What kind of regime do we wish to build? What material and social structure should it have? In light of the patterns of technological development I mentioned earlier—patterns of centralization, gigantism, hierarchical authority, and so forth—perhaps it would be desirable to choose a model of photovoltaic development based on a more flexible, more democratic principle. Here is an opportunity to extend responsibility and control to a greater number of people, an opportunity to create diversity rather than uniformity in our sociotechnical constitution. Is this not an opportunity we should welcome and seek to realize? Suggesting this, I would not ask anyone to make exorbitant economic sacrifices. But rather than pursue the lemminglike course of choosing only that system design which provides the least expensive kilowatt, perhaps we ought to consider which system might play the more positive role in the technical infrastructure of freedom.

It goes without saying that the agencies now actually developing photovoltaics have no such questions in mind. Government-subsidized research, such as it is, focuses upon finding the most efficient and effective form of solar electricity and seeing it marketed in the "private sector."[19] Huge multinational petroleum corporations have bought up the smaller companies at work in this field; their motive seems to be desire to control the configuration of whatever mix of energy sources and technologies we will eventually have. As we have done so often in the past, our society has, in effect, delegated decision-making power to those whose plans are narrowly self-interested. One can predict, therefore, that when photovoltaic systems are introduced they will carry the same qualities of institutional and physical centralization that characterize so many modern technologies.

A crucial failure in modern political thought and political practice has been an inability or unwillingness even to begin the project I am suggesting here: the critical evaluation and control of our society's technical constitution. The silence of liberalism on this issue is matched by an equally obvious neglect in Marxist theory. Both persuasions have enthusiastically sought freedom in sheer material plenitude, welcoming whatever technological means (or monstrosities) seemed to produce abundance the fastest. It is, however, a serious mistake to construct one sociotechnical system after another in the blind faith that each will turn out to be politically benign. Many crucial choices about the forms and limits of our regimes of instrumentality must be enforced at the founding, at the genesis of each new technology. It is here that our best purposes must be heard.

## NOTES

1 Plato, *Laws,* 7.803b, translated by A. E. Taylor, in *The Collected Dialogues of Plato,* Edith Hamilton and Huntington Cairns (eds.), Princeton, N.J., Princeton University Press, 1961, p. 1374.

2 Jean-Jacques Rousseau, *The Social Contract,* translated and introduced by Maurice Cranston, New York, Penguin Books, 1968, p. 84.

3 Alexander Hamilton, "Federalist No. 9," *The Federalist Papers,* with an introduction by Clinton Rossiter, New York, Mentor Books, 1961, pp. 72–73.

4 Thomas Jefferson, "Notes on Virginia," *The Life and Selected Writings of Thomas Jefferson,* Adrienne Koch and William Peden (eds.), New York, Modern Library, 1944, pp. 28–281.

5 Quoted in Gordon S. Wood, *The Creation of the American Republic, 1776–1787,* New York, W. W. Norton, 1972, p. 114.

6 John Kasson, *Civilizing the Machine: Technology and Republican Values in America, 1776–1900,* New York, Grossman, 1976.

7 Albert O. Hirschman, *The Passions and the Interests: Political Arguments for Capitalism Before Its Triumph,* Princeton, Princeton University Press, 1977.

8 Denison Olmsted, "On the Democratic Tendencies of Science," *Barnard's Journal of Education,* vol. 1, 1855–1856, reprinted in *Changing Attitudes Toward American Technology,* Thomas Parke Hughes (ed.), New York, Harper and Row, 1975, p. 148.

9 Karl Polanyi, *The Great Transformation,* Boston, Beacon Press, 1957, p. 33.

10 "Beta" (Edward W. Byrn), "The Progress of Invention During the Past Fifty Years," *Scientific American,* vol. 75, July 25, 1896, reprinted in Hughes (see note 8), pp. 158–159.

11 For a discussion of the wide-ranging influence of the idea of efficiency in American thought, see Samuel Haber, *Efficiency and Uplift: Scientific Management in the Progressive Era, 1890–1920,* Chicago, Chicago University Press, 1964; and Samuel P. Hays, *Conservation and the Gospel of Efficiency: The Progressive Conservation Movement, 1890–1920,* New York, Atheneum, 1959.

12 William L. Kahrl, *Water and Power: The Conflict Over Los Angeles' Water Supply in the Owens Valley,* Berkeley, University of California Press, 1982.

13 An analysis of the efficacy of various kinds of government regulation is contained in *The Politics of Regulation,* James Q. Wilson (ed.), New York, Basic Books, 1980.

14 A representative selection of the ambitious energy studies done during the 1970s would include the following: *A Time to Choose,* The Energy Policy Project of the Ford Foundation, Cambridge, Ballinger, 1974; *Nuclear Power: Issues and Choices,* Report of the Nuclear Energy Policy Study Group, Cambridge, Ballinger, 1977; *Energy in America's Future: The Choices Before Us,* National Energy Studies Project, Baltimore, Johns

Hopkins University Press, 1979; *Energy: The Next Twenty Years,* Report of the Study Group Sponsored by the Ford Foundation and Administered by Resources for the Future, Hans Landsberg et al., Cambridge, Ballinger, 1979; *Energy in Transition, 1985–2010,* National Research Council Committee on Nuclear and Alternative Energy Systems, San Francisco, W. H. Freeman, 1980.

**15** *Nuclear Power: Issues and Choices,* pp. 44, 41.

**16** Amory B. Lovins, "Technology Is the Answer! (But What Was the Question?): Energy as a Case Study of Inappropriate Technology," discussion paper for the Symposium on Social Values and Technological Change in an International Context, Racine, Wis., June 1978, p. 1.

**17** Amory B. Lovins, *Soft Energy Paths: Toward a Durable Peace,* Cambridge, Ballinger, 1977, pp. 6–7.

**18** Paul Goodman, *People or Personnel and Like a Conquered Province,* New York, Vintage Press, 1968, p. 4.

**19** See, for example, J. L. Smith, "Photovoltaics," *Science* 212:1472–1478, 1981.

## Science and Society

**John Dewey**

The significant outward forms of the civilization of the western world are the product of the machine and its technology. Indirectly, they are the product of the scientific revolution which took place in the seventeenth century. In its effect upon men's external habits, dominant interests, the conditions under which they work and associate, whether in the family, the factory, the state, or internationally, science is by far the most potent social factor in the modern world. It operates, however, through its undesigned effects rather than as a transforming influence of men's thoughts and purposes. This contrast between outer and inner operation is the great contradiction in our lives. Habits of thought and desire remain in substance what they were before the rise of science, while the conditions under which they take effect have been radically altered by science.

When we look at the external social consequences of science, we find it impossible to apprehend the extent or gauge the rapidity of their occurrence. Alfred North Whitehead has recently called attention to the progressive shortening of the time-span of social change. That due to basic conditions seems to be of the order of half a million years; that due to lesser physical conditions, like alterations in climate, to be of the order of five thousand years. Until almost our own day the time-span of sporadic technological changes was of the order of five hundred years; according to him, no great technological changes took place between, say, 100 A.D. and 1400 A.D. With the introduction of steam-power, the fifty years from 1780 to 1830 were marked by more changes than are found in any previous thousand years. The advance of chemical techniques and in use of electricity and radio-energy in the last forty years makes even this last change seem slow and awkward.

Domestic life, political institutions, international relations and personal contacts are shifting with kaleidoscopic rapidity before our eyes. We cannot appreciate and weigh the changes; they occur too swiftly. We do not have time to take them in. No sooner do we begin to understand the meaning of one such change than another comes and displaces the former. Our minds are dulled by the sudden and repeated impacts. Externally, science through its applications is manufacturing the conditions of our institutions at such a speed that we are too bewildered to know what sort of civilization is in process of making.

Because of this confusion, we cannot even draw up a ledger account of social gains and losses due to the operation of science. But at least we know that the earlier optimism which thought that the advance of natural science was to dispel superstition, ignorance, and oppression, by placing reason on the throne, was unjustified. Some superstitions have given way, but the mechanical devices due to science have made it possible to spread new kinds of error and delusion among a larger multitude. The fact is that it is foolish to try to draw up a debit and credit account for science. To do so is to mythologize; it is to personify science and impute to it a will and an energy on its own account. In truth science is strictly impersonal; a method and a body of knowledge. It owes its operation and its consequences to the human beings who use it. It adapts itself passively to the purposes and desires which animate these human beings. It lends itself with equal impartiality to the kindly offices of medicine and hygiene and the destructive deeds of war. It elevates some through opening new horizons; it depresses others by making them slaves of machines operated for the pecuniary gain of owners.

The neutrality of science to the uses made of it renders it silly to talk about its bankruptcy, or to worship it as the usherer in of a new age. In the degree in which we realize this fact, we shall devote our attention to the human purposes and motives which control its application. Science is an instrument, a method, a body of technique. While it is an end for those inquirers who are engaged in its pursuit, in the large human sense it is a means, a tool. For what ends shall it be used? Shall it be used deliberately, systematically, for the promotion of social well-being, or shall it be employed primarily for private aggrandizement, leaving its larger social results to chance? Shall the scientific attitude be used to create new mental and moral attitudes, or shall it continue to be subordinated to service of desires, purposes and institutions which were formed before science came into existence? Can the attitudes which control the use of science be themselves so influenced by scientific technique that they will harmonize with its spirit?

The beginning of wisdom is, I repeat, the realization that science itself is an instrument which is indifferent to the external uses to which it is put. Steam and electricity remain natural forces when they operate through mechanisms; the only problem is the purposes for which men set the mechanisms to work. The essential technique of gunpowder is the same whether it be used to blast rocks from the quarry to build better human habitations, or to hurl death upon men at war with one another. The airplane binds men at a distance in closer bonds of intercourse and understanding, or it rains missiles of death upon hapless populations. We are forced to consider the relation of human ideas and ideals to the social consequences which are produced by science as an instrument.

The problem involved is the greatest which civilization has ever had to face. It is, without exaggeration, the most serious issue of contemporary life. Here is the instrumentality, the most powerful, for good and evil, the world has ever known. What are we going to do with it? Shall we leave our underlying aims unaffected by it, treating it merely as a means by which uncooperative individuals may advance their own fortunes? Shall we try to improve the hearts of men without regard to the new methods which science puts at our disposal? There are those, men in high position in church and state, who urge this course. They trust to a transforming influence of a morals and religion which have not been affected by science to change human desire and purpose, so that they will employ science and machine technology for beneficent social ends. The recent Encyclical of the Pope is a classic document in expression of a point of view which would rely wholly upon inner regeneration to protect society from the injurious uses to which science may be put. Quite apart from any ecclesiastical connection, there are many "intellectuals" who appeal to inner "spiritual" concepts, totally divorced from scientific intelligence, to effect the needed work. But there is another alternative: to take the method of science home into our own controlling attitudes and dispositions, to employ the new techniques as means of directing our thoughts and efforts to a planned control of social forces.

Science and machine technology are young from the standpoint of human history. Though vast in stature, they are infants in age. Three hundred years are but a moment in comparison with thousands of centuries man has lived on the earth. In view of the inertia of institutions and of the mental habits they breed, it is not surprising that the new technique of apparatus and calculation, which is the essence of science, has made so little impression on underlying human attitudes. The momentum of traditions and purposes that preceded its rise took possession of the new instrument and turned it to their ends. Moreover, science had to struggle for existence. It had powerful enemies in church and state. It needed friends and it welcomed alliance with the rising capitalism which it so effectively promoted. If it tended to foster secularism and to create predominantly material interests, it could still be argued that it was in essential harmony with traditional morals and religion. But there were lacking the conditions which are indispensable to the serious application of scientific method in reconstruction of fundamental beliefs and attitudes. In addition, the development of the new science was attended with so many internal difficulties that energy had to go to perfecting the instrument just as an instrument. Because of all these circumstances the fact that science was used in behalf of old interests is nothing to be wondered at.

The conditions have now changed, radically so. The claims of natural science in the physical field are undisputed. Indeed, its prestige is so great that an almost superstitious aura gathers about its name and work. Its progress is no longer dependent upon the adventurous inquiry of a few untrammeled souls. Not only are universities organized to promote scientific research and learning, but one may almost imagine the university laboratories abolished and still feel confident of the continued advance of science. The development of industry has compelled the inclusion of scientific inquiry within the processes of production and distribution. We find in the public prints as many demonstrations of the benefits of science from a business point of view as there are proofs of its harmony with religion.

It is not possible that, under such conditions, the subordination of scientific techniques to purposes and institutions that flourished before its rise can indefinitely continue. In all affairs there comes a time when a cycle of growth reaches maturity. When this stage is reached, the period of protective nursing comes to an end. The problem of securing proper use succeeds to that of securing conditions of growth. Now that science has established itself and has created a new social environment, it has (if I may for the moment personify it) to face the issue of its social responsibilities. Speaking without personification, we who have a powerful and perfected instrument in our hands, one which is determining the quality of social changes, must ask what changes we want to see achieved and what we want to see averted. We must, in short, plan its social effects with the same care with which in the past we have planned its physical operation and consequences. Till now we have employed science absentmindedly as far as its effects upon human beings are concerned. The present situation with its extraordinary control of natural energies and its totally unplanned and haphazard social economy is a dire demonstration of the folly of continuing this course.

The social effects of the application of science have been accidental, even though they are intrinsic to the private and unorganized motives which we have permitted to control that application. It would be hard to find a better proof that such is the fact than the vogue of the theory that such unregulated use of science is in accord with "natural law," and that all effort at planned control of its social effects is an interference with nature. The use which has been made of a peculiar idea of personal liberty to justify the dominion of accident in social affairs is another convincing proof. The doctrine that the most potent instrument of widespread, enduring, and objective social changes must be left at the mercy of purely private desires for purely personal gain is a doctrine of anarchy. Our present insecurity of life is the fruit of the adoption in practice of this anarchic doctrine.

The technologies of industry have flowed from the intrinsic nature of science. For that is itself essentially a technology of apparatus, materials and numbers. But the pecuniary aims which have decided the social results of the use of these technologies have not flowed from the inherent nature of science. They have been derived from institutions and attendant mental and moral habits which were entrenched before there was any such thing as science and the machine. In consequence, science has operated as a means for extending the influence of the institution of private property and connected legal relations far beyond their former limits. It has operated as a device to carry an enormous load of stocks and bonds and to make the reward of investment in the way of profit and power one out of all proportion to that accruing from actual work and service.

Here lies the heart of our present social problem. Science has hardly been used to modify men's fundamental acts and attitudes in social matters. It has been used to extend enormously the scope and power of interests and values which anteceded its rise. Here is the contradiction in our civilization. The potentiality of science as the most powerful instrument of control which has ever existed puts to mankind its one outstanding present challenge.

There is one field in which science has been somewhat systematically employed as an agent of social control. Condorcet, writing during the French Revolution in the prison from which he went to the guillotine, hailed the invention of the calculus of probabilities as the opening of a new era. He saw in this new mathematical technique the promise of methods of insurance which should distribute evenly and widely the impact of the disasters to which humanity is subject. Insurance against death, fire, hurricanes and so on have in a measure confirmed his prediction. Nevertheless, in large and important social areas, we have only made the merest beginning of the method of insurance against the hazards of life and death. Insurance against the risks of maternity, of sickness, old age, unemployment, is still rudimentary; its idea is fought by all reactionary forces. Witness the obstacles against which social insurance with respect to accidents incurred in industrial employment had to contend. The anarchy called natural law and personal liberty still operates with success against a planned social use of the resources of scientific knowledge.

Yet insurance against perils and hazards is the place where the application of science has gone the furthest, not the least, distance in present society. The fact that motor cars kill and maim more persons yearly than all factories, shops, and farms is a fair symbol of how backward we are in that province where we have done most. Here, however, is one field in which at least the idea of planned use of scientific knowledge for social welfare has received recognition. We no longer regard plagues, famine and disease as visitations of necessary "natural law" or of a power beyond nature. By preventive means of medicine and public hygiene as well as by various remedial measures we have in idea, if not in fact, placed technique in the stead of magic and chance and uncontrollable necessity in this one area of life. And yet, as I have said, here is where the socially planned use of science has made the most, not least, progress. Were it not for the youth of science and the historically demonstrated slowness of all basic mental and moral change, we could hardly find language to express astonishment at the situation in which we have an extensive and precise control of physical energies and conditions, and in which we leave the social consequences of their operation to chance, laissez-faire, privileged pecuniary status, and the inertia of tradition and old institutions.

Condorcet thought and worked in the Baconian strain. But the Baconian ideal of the systematic organization of all knowledge, the planned control of discovery and invention, for the relief and advancement of the human estate, remains almost as purely an ideal as when Francis Bacon put it forward centuries ago. And this is true in spite of the fact that the physical and mathematical technique upon which a planned control of social results depends has made in the meantime incalculable progress. The conclusion is inevitable. The outer arena of life has been transformed by science. The effectively working mind and character of man have hardly been touched.

Consider that phase of social action where science might theoretically be supposed to have taken effect most rapidly, namely, education. In dealing with the young, it would seem as if scientific methods might at once take effect in transformation of mental attitudes, without meeting the obstacles which have to be overcome in dealing with adults. In higher education, in universities and technical schools, a great amount of research is done and much scientific knowledge is imparted. But it is a principle of modern psychology that the basic attitudes of mind are formed in the earlier years. And I venture the assertion that for the most part the formation of intellectual habits in elementary education, in the home and school, is hardly affected by scientific method. Even in our so-called progressive schools, science is usually treated as a side line, an ornamental extra, not as the chief means of developing the right mental attitudes. It is treated generally as one more body of ready-made information to be acquired by traditional methods, or else as an occasional diversion. That it is the method of all effective mental approach and attack in all subjects has not gained even a foothold. Yet if scientific method is not something esoteric but is a realization of the most effective operation of intelligence, it should be axiomatic that the development of scientific attitudes of thought, observation, and inquiry is the chief business of study and learning.

Two phases of the contradiction inhering in our civilization may be especially mentioned. We have long been committed in theory and words to the principle of democracy. But criticism of democracy, assertions that it is failing to work and even to exist are everywhere rife. In the last few months we have become accustomed to similar assertions regarding our economic and industrial system. Mr. Ivy Lee, for example, in a recent commencement address, entitled "This Hour of Bewilderment," quoted from a representative clergyman, a railway president, and a publicist, to the effect that our capitalistic system is on trial. And yet the statements had to do with only one feature of that system: the prevalence of unemployment and attendant insecurity. It is not necessary for me to invade the territory of economics and politics. The essential fact is that if both democracy and capitalism are on trial, it is in reality our collective intelligence which is on trial. We have displayed enough intelligence in the physical field to create the new and powerful instrument of science and technology. We have not as yet had enough intelligence to use this instrument deliberately and systematically to control its social operations and consequences.

The first lesson which the use of scientific method teaches is that control is coordinate with knowledge and understanding. Where there is technique there is the possibility of administering forces and conditions in the region where the technique applies. Our lack of control in the sphere of human relations, national, domestic, international, requires no emphasis of notice. It is proof that we have not begun to operate scientifically in such matters. The public press is full of discussion of the five-year plan and the ten-year plan in Russia. But the fact that the plan is being tried by a country which has a dictatorship foreign to all our beliefs tends to divert attention from the fundamental consideration. The point for us is not this political setting nor its communistic context. It is that by the use of all available resources of knowledge and experts an attempt is being made at organized social planning and control. Were we to forget for the moment the special Russian political setting, we should see here an effort to use coordinated knowledge and technical skill to direct economic resources toward social order and stability.

To hold that such organized planning is possible only in a communistic society is to surrender the case to communism. Upon any other basis, the effort of Russia is a challenge and a warning to those who live under another political and economic regime. It is a call to use our more advanced knowledge and technology in scientific thinking about our own needs, problems, evils, and possibilities so as to achieve some degree of control of the social consequences which the application of science is, willy-nilly, bringing about. What stands in the way is a lot of outworn traditions, moth-eaten slogans and catchwords, that do substitute duty for thought, as well as our entrenched predatory self-interest. We shall only make a real beginning in intelligent thought when we cease mouthing platitudes; stop confining our idea to antitheses of individualism and socialism, capitalism and communism, and realize that the issue is between chaos and order, chance and control: the haphazard use and the planned use of scientific techniques.

Thus the statement with which we began, namely, that we are living in a world of change extraordinary in range and speed, is only half true. It holds of the outward applications of science. It does not hold of our intellectual and moral attitudes. About physical conditions and energies we think scientifically; at least, some men do, and the results of their thinking enter into the experiences of all of us. But the entrenched and stubborn institutions of the past stand in the way of our thinking scientifically about human relations and social issues. Our mental habits in these respects are dominated by institutions of family, state, church, and business that were formed long before men had an effective technique of inquiry and validation. It is this contradiction from which we suffer to-day.

Disaster follows in its wake. It is impossible to overstate the mental confusion and the practical disorder which are bound to result when external and physical effects are planned and regulated, while the attitudes of mind upon which the direction of external results depends are left to the medley of chance, tradition, and dogma. It is a common saying that our physical science has far outrun our social knowledge; that our physical skill has become exact and comprehensive while our humane arts are vague, opinionated, and narrow. The fundamental trouble, however, is not lack of sufficient information about social facts, but unwillingness to adopt the scientific attitude in what we do know. Men floundered in a morass of opinion about physical matters for thousands of years. It was when they began to use their ideas experimentally and to create a technique or direction of experimentation that physical science advanced with system and surety. No amount of mere fact-finding develops science nor the scientific attitude in either physics or social affairs. Facts merely amassed and piled up are dead; a burden which only adds to confusion. When ideas, hypotheses, begin to play upon facts, when they are methods for experimental use in action, then light dawns; then it becomes possible to discriminate significant from trivial facts, and relations take the place of isolated scraps. Just as soon as we begin to use the knowledge and skills we have to control social consequences in the interest of shared abundant and secured life, we shall cease to complain of the backwardness of our social knowledge. We shall take the road which leads to the assured building up of social science just as men built up physical science when they actively used the techniques of tools and numbers in physical experimentation.

In spite, then, of all the record of the past, the great scientific revolution is still to come. It will ensue when men collectively and cooperatively organize their knowledge for application to achieve and make secure social values; when they systematically use scientific procedures for the control of human relationships and the direction of the social effects of our vast technological machinery. Great as have been the social changes of the last century, they are not to be compared with those which will emerge when our faith in scientific method is made manifest in social works. We are living in a period of depression. The intellectual function of trouble is to lead men to think. The depression is a small price to pay if it induces us to think about the cause of the disorder, confusion, and insecurity which are the outstanding traits of our social life. If we do not go back to their cause, namely our half-way and accidental use of science, mankind will pass through depressions, for they are the graphic record of our unplanned social life. The story of the achievement of science in physical control is evidence of the possibility of control in social affairs. It is our human intelligence and human courage which are on trial; it is incredible that men who have brought the technique of physical discovery, invention, and use to such a pitch of perfection will abdicate in the face of the infinitely more important human problem.

## The Relation of Science and Technology to Human Values

**W. H. Lowrance**

Admiration of the extraordinary powers of science often tempts people to hope that the laboratory and clinic will hand down social *oughts,* "Thou shalts." Convictions on the issue range from the view that science and technology can and must be used to generate human values, to the view that science and technology are just as value-neutral as banking or playing soccer are and have little moral or political character.

Neither extreme, I will argue, is correct. Although scientific knowledge, once attained, may be considered *ambi-potent* for good or evil, the work of pursuing new science and developing technologies is by no means value-neutral (as, in the context of international politics, banking and playing soccer may not be, either). And although they don't dictate values, technical analyses and accomplishments profoundly influence social philosophies and choices.

### VALUES AND FACTS: THE BASIC RELATIONS

Does science generate moral oughts? Almost, on occasion, but usually not, and never without reference to embedded social values. Observations of unfitness in offspring of incestuous human mating, reinforced by analogous observations in nonhuman species and long proscription by most societies as a "crime against nature," support the almost universal taboo against incest. Technical estimates of the physical, biological, and social consequences of nuclear war make us dread it. But even in these extreme cases our responses still depend on value judgments that lie outside science: abhorrence of giving birth to defective children, abhorrence of genocide.[1]

Toward desired ends, enabling oughts can be formulated in the light of knowledge from many sources, including—powerfully—science. As we become aware of environmental connections and consequences, we weave ecological sensibilities into our values fabric. As we accumulate evidence on how life-habits affect personal health, we reassign social responsibilities for health promotion.

Social values simply cannot be derived from science qua science alone. To speak, as some cavalierly do, of "the values of science" may mislead. To be sure, over the past several centuries scientists have developed tenets of method, evidence, and proof, and they have cultivated an ethos of intellectual openness, truthfulness, and international fraternity. Some writers have been so impressed with the ethics and etiquette through which scientific work proceeds that they have urged that these mores be adopted as the foundation of social ethics. In *Science and the Social Order* Bernard Barber suggested that the "rationality, universalism, individualism, 'communality,' and 'disinterestedness'" that serve science so effectively "could even some day become the dominant moral values for the whole society" (1952, 90). Anatol Rapoport went so far as to say that "the ethics of science must become *the* ethics of humanity" (1957, 798). Jacob Bronowski argued that since science flourishes in societies fostering such "values of science" as "independence and originality, dissent and freedom and tolerance," such norms should be adopted for other social endeavors as well (1965, 62). But science has no monopoly on creativity, truthtelling, or tolerance, nor is it uniquely the definer of these traits. Science's precedent is hardly a sufficient model for the redesign of social ethics.

Technical people contribute richly to the alleviation of suffering and the enhancement of culture. Like everyone else, scientists hold deeply cherished personal convictions, which they express often and articulately. Groups of scientists, very large groups even, vigorously pursue social goals. Coalitions may go so far as to engage in partisan politicking, as the Scientists, Engineers, and Physicians for Johnson/Humphrey did in the 1964 presidential campaign. But such actions derive no more from the methods or scientific knowledge of Pasteur or Bohr than the 1983 Artists Call Against U.S. Intervention in Central America derived from the aesthetic tenets or oeuvre of Turner or Cézanne.

On the other hand, any assertion that scientific activity is value-free or value-neutral is disingenuous. Disclaimers have been made at least since Robert Hooke's 1663 proposed charter for the Royal Society: "The business and design of the Royal Society is—To improve the knowledge of naturall things, and of all useful Arts, Manufactures, Mechanick Practises, Engynes and Inventions by Experiments—(not meddling with Divinity, Metaphysics, Moralls, Politicks, Grammar, Rhetorick or Logick)" (Lyons 1944, 41). But even as Hooke drafted that antiseptic mandate, interpretations of Divinity and Metaphysics were brought under severe challenge by science; Mechanick Practises, Engynes, and Inventions were pursued that would profoundly affect the Moralls of warfare and Politicks of labor; studies of the Grammar of the world's languages grew through philology toward modern semantics; and lines of Rhetorick and Logick were explored that would lead straight to G. E. Moore and Bertrand Russell.

When, in London in 1838, William Whewell, Charles Babbage, and their colleagues founded the *Journal of the Statistical Society* they chose as its symbol the wheatsheaf, to stand for the facts the journal would gather that "alone can form the basis of correct conclusions with respect to social and political government." On a band around the sheaf these canny masters emblazoned the motto "*Aliis Exterendum,*" "It must be threshed by others," as though the facts threshed aren't conditioned by the gathering and sheaving. But even the titles of the journal's first papers gave them away: "Social and moral statistics of criminal offenders," "Vicious extent and heavy expense of advertisements in England," "On the accumulation of capital by the different classes of society." They kept the wheatsheaf but dropped the motto in 1857.[2]

Occasionally even today a neutral gray waistcoat is donned against suggestions that the work of science and technology is value-freighted. But, just as in Hooke's day, and in Whewell and Babbage's, no major creative activities in society—especially those that are pragmatically or symbolically powerful—should be allowed to claim valuative or moral immunity.

Jacob Bronowski's distinction is key: "Those who think that science is ethnically neutral confuse the findings of science, which are, with the activity of science, which is not" (1965, 63). Research, once accomplished, must be considered in the long run ambi-potent, usuable for either good or evil. The anticholinesterase chemicals developed as nerve gases between the World Wars later turned out to be elegant research weapons in the protein biochemistry revolution, and botanical research on how trees drop their leaves in autumn led to development of the military defoliant Agent Orange.[3] In the short run, of course, facts and know-how can be kept secret, or applied under close control, but they are likely to be revealed or discovered independently elsewhere, eventually.

But: We must vigorously resist any notion that researchers are helpless to make choices among envisionable future lines of research and development, or among possible conditions of pursuit. Although it may not be useful to regard already published knowledge as having any particular moral or ethical cast, surely it is wrong to view not-yet-accomplished research, which cannot be undertaken without commitment of will and resources, as being anything other than value-laden. And, because timing and pacing always are important, it would be naïve not to recognize that new knowledge may at the moment of its emergence have maleficent or beneficent potency that demands attention. Questions of practical ethics always lie in *what to do next.*

*Technical activity must be considered value-laden in two senses: technical people's social values and value perceptions affect their research and service; and that work, in turn, affects the value-situation of others in the public.*

Thus I consider to be value-laden: the undertaking and supporting of research (for example, committal of funds to research on acquired immune deficiency syndrome (AIDS), or other disease); the choosing of conditions of experimentation (diplomatic auspices of hurricane-seeding experiments in the South Pacific

Basin); the marshaling of science to analyze, assess, and help decide on socially important problems (agricultural policymaking); the investigating of people's values, using social-scientific methods (studying jail guards' attitudes); the applying of technology and medicine to practical problems (treatment of breast cancer); and the incorporating of scientific knowledge into the fabric of social philosophies and policies (taking the findings of child psychology and of income-maintenance economics into account when revising child welfare programs).

Technical experts make crucial decisions for, and in the name of, the public. Although geology students usually don't view themselves as moving into a value-charged realm—what could be less social than rocks, after all?—as their careers progress geologists find themselves making seismicity assessments for hydroelectric or nuclear power plant siting decisions, advising on beach protection, municipal building codes, transnational water resources, seabed mining, and strategic minerals supply, and leading projects on causes of acidification of lakes and on underground disposal of radioactive waste. Teams of automotive engineers design cars and sell the designs through corporate management to the public. Pharmaceutical experts develop, test, and push drugs toward the market. Nuclear power plant designers weight the ratio of instant to delayed (cancer) death risks they design into reactors. Nuclear managers decide, in cleaning up after an accident, between exposing a few workers to radiation for relatively long times and exposing more workers for shorter times. Much research by social scientists—on the effect of school busing on educational achievement, on the effect of incarceration on criminal recidivism, on the influence of wage incentives on acceptance of occupational hazard—is so integral to policymaking that analysis can hardly be distinguished from advocacy. And of course some scientists and physicians themselves become high official decisionmakers in industry, labor, and government.

Experts' overt value stances can be argued about. Much harder to dig out and deal with are inarticulate premises. Genetic counselors' counsel is bound to be tempered by their attitude toward contraception and abortion. Marine ecologists' advice on the dumping of wastes into the ocean hinges on whether they think of the oceanic environment as being fragile or resilient, and on whether they prefer a pristine ocean to a "working" ocean. In psychotherapy, as Anne Seiden has argued, "the assumption that dependency, masochism, and passivity are normal for women and the tendency to treat assertiveness and aggression differently for women than for men" leads to "different standards of health for women and men"; therapists' practice thus is conditioned by their intuited, schooled, and inferred interpretation of gender (1976, 1116).

Later chapters will address such questions that arise on this account as: How should advisors and advisory committees, in their procedures and reports, deal with their factual biases and value preferences? (As with bias in a textile, "bias" simply means inclination, and isn't necessarily pejorative.) Since most of the work individual scientists do is guided by personal motives and morals rather than by grand ethical schemes, and since each researcher contributes only small increments to the overall technical enterprise, how should individual scientists' actions be oriented to "society's" values? How should scientific research freedoms be balanced against societal constraints?

Serious trouble arises when the distinction between facts and values is blurred or not recognized, or when disputants engage in mislabeling. Nothing has illustrated this more dramatically than the congressional abortion battle of 1981. Senate Bill 158 was introduced which would extend to unborn fetuses the rights of due process guaranteed by the Constitution's Fourteenth Amendment, circumventing Supreme Court decisions preserving women's right to abortion. In hearings, five medical researchers and physicians were drafted into testifying that a human being is formed at the moment sperm fuses with egg, and that this is a "scientific fact." Professor Jérôme Lejeune of the Medical College of Paris asserted, "To accept the fact that after fertilization has taken place a new human has come into being is no longer a matter of taste or of opinion. The human nature of the human being from conception to old age is not a metaphysical contention, it is plain experimental evidence." Boston physician Micheline Mathews-Roth insisted that "one is being scientifically accurate if one says that an individual human life begins at fertilization or conception." Entering the fray, Yale geneticist Leon Rosenberg protested that "the notion embodied in the phrase 'actual human life' is not a scientific one, but rather a philosophic and religious one." He quoted geneticist Joshua Lederberg: "'Modern man knows too much to pretend that life is merely the beating of the heart or the tide of breathing. Nevertheless he would like to ask biology to draw an absolute line that might relieve his confusion. The plea is in vain. There is no single, simple answer to 'When does life begin?''" Rosenberg emphasized, "I have no quarrel with anyone's ideas on this matter, so long as it is clearly understood that they are personal beliefs...and not scientific truths" (U.S. Senate, Subcommittee on Separation of Powers, 1981). The transgression was important enough to move the National Academy of Sciences to pass one of its rare resolutions (April 28, 1981):

> It is the view of the National Academy of Sciences that the statement in Chapter 101, Section 1, of U.S. Senate Bill S158, 1981, cannot stand up to the scrutiny of science. This section reads "The Congress finds that present-day scientific evidence indicates a significant likelihood that actual human life exists from conception." This statement purports to derive its conclusions from science, but it deals with a question to which science can provide no answer. The proposal in S158 that the term "person" shall include "all human life" has no basis within our scientific understanding. Defining the time at which the developing embryo becomes a "person" must remain a matter of moral or religious values.

## FACT/VALUE INTERPLAY, FROM SOCIAL DARWINISM THROUGH WILSONIAN SOCIOBIOLOGY

Some of the most extreme exploitations of science in moral debate occurred around Social Darwinism late in the nineteenth century. During this clamor every major social movement struggled to accommodate the unsettling revelations of Wallace and Darwin (or popular interpretations of their theories), while at the same time defending its own ingrained views of class, wealth, race, sex, progress, and justice. The "moral economy of nature" and the "vital order of society" were analyzed in terms of specialization of function and social equilibrium. Spencer justified laissez-faire capitalism by equating economic competition with natural selection, and he drew upon a revised Malthusianism to explain why poverty was unavoidable. Engels fought the trend by pointing out how economic activity intervenes in selection, and how, as Darwin himself had recognized, cooperative behavior can enhance survivability. Comte represented himself as siring sociology out of the biological sciences. And later, on an openly Darwinian theme, Galton founded the eugenics movement. Reference to evolutionary science often was simply sham scientism dragged in for justification, but many eminent scholars' excesses stemmed from sincere internal struggles.[4]

Many of the utopian visions that have come along since then have yearned vaguely for a society founded on somehow-scientific principles. In medicine, for instance, Frederick T. Gates, the Baptist minister and chairman of the board of the Rockefeller Institute for Medical Research, on that institution's tenth anniversary in 1911 promised (Corner 1964, 4):

> As medical research goes on, it will find out and promulgate, as an unforeseen byproduct of its work, new moral laws and new social laws—new definitions of what is right and wrong in our relations with each other. Medical research will educate the human conscience in new directions and point out new duties. It will make us sensitive to new moral distinctions.

Engineers similarly have reached for moral "rationality." In his presidential address to the American Institute of Electrical Engineers in 1919 Comfort Adams wondered (1919, 792):

> Are there no laws in this other realm of human relations which are just as inexorable as the physical laws with which we are so familiar? Is there no law of compensation which is the counterpart of our law of conservation of energy?

And in the heady 1910 of the University of Chicago, Albion Small foresaw social scientists serving as "sailing masters" for the nation (1910, 242):

> The most reliable criteria of human values which science can propose would be the consensus of councils of scientists representing the largest possible variety of human interests, and co-operating to reduce their special judgments to a scale which would render their due to each of the interests in the total calculation.

One stream of latter-day Social Darwinism began with the great English biologist T. H. Huxley at the turn of the century, was modified two wars later by his grandson Julian, and was elaborated upon at mid-century by C. H. Waddington. In *The Ethical Animal* Waddington hypothesized that "the function of ethical beliefs is to mediate human evolution, and that evolution exhibits some recognizable direction of progress" (1961, 59). Acerbically he noted, "The horrifying effects of social actions based on excessive beliefs of an allegedly ethical character, as they are exhibited in the wars and persecutions in the name of religion, politics, nationalism, racism and various other idealisms, is sufficient evidence that the human condition might well be improved" (p. 203), leading him to conclude that "the major ethical problems of today in the context of individual-to-individual behaviour would, I think, according to our criteria, have to be sought in those types of attitude and activity which facilitate or hinder the development of a healthy authority structure" (p. 205). But he didn't specify how the search should proceed, or what authority structures were to be considered healthy (T. H. Huxley and Julian Huxley 1947; Waddington 1961).

From time to time social scientists have tried to derive cultural guidance from what they see as historical "moral evolution," for example from sanctions observed in primitive cultures and in utopian communities, and from analogues to human morality, such as altruism and sexual fidelity, perceived in the behavior of lower animals. I don't hold much hope for most of these efforts. The analogical constructions seem quite shaky. Despite promises, these studies have not generated any specific guidance on real-world problems.[5]

The latest phase in this line of propositions began in 1975 when Harvard biologist Edward O. Wilson published *Sociobiology: The New Synthesis*. The book's central theoretical query was: "How can altruism, which by definition reduces personal fitness, possibly evolve by natural selection?" Wilson outlined how he would answer (1975a, 3):

> The hypothalamic-limbic complex of a highly social species, such as man, "knows," or more precisely it has been programmed to perform as if it knows, that its underlying genes will be proliferated maximally only if it orchestrates behavioral responses that bring into play an efficient mixture of personal survival, reproduction, and altruism. Consequently, the centers of the complex tax the conscious mind with ambivalences whenever the organisms encounter stressful situations. Love joins hate; aggression, fear; expansiveness, withdrawal; and so on—in blends designed not to promote the happiness and survival of the individual, but to favor the maximum transmission of the controlling genes.

Building upon his renowned lifelong research on insect societies and comparative biology, for twenty-six chapters Wilson conducted an enchanting tour, albeit with disturbing overtones, of African ant and termite nests, Scottish barnacle colonies, Montana sagebrush grouse leks, and Serengeti wildebeest herds, up through equatorial African chimpanzee societies. Then came the concluding, overreaching Chapter 27, "Man: From Sociobiology to Sociology," which carried a section on ethics that began: "Scientists and humanists should consider together the possibility that the time has come for ethics to be removed temporarily from the hands of the philosophers and biologicized."

Three almost baiting propositions (among the most egregious) from Wilson, from *Sociobiology* and related publications, will illustrate why he drew criticism. More contentious theses hardly can be imagined. On division of labor between genders (1975b, 48):

> In hunter-gatherer societies, men hunt and women stay at home. This strong bias persists in most agricultural and industrial societies and, on that ground alone, appears to have a genetic origin.

On religious indoctrinability (1975a, 561):

> The enduring paradox of religion is that so much of its substance is demonstrably false, yet it remains a driving force in all societies. Men would rather believe than know, have the void as purpose, as Nietzche said, than be void of purpose. . . . Human beings are absurdly easy to indoctrinate—they *seek* it.

On rights (1976, 189):

> To the extent that the biological interpretation noted here proves correct, men have rights that are innate, rooted in the ineradicable drives for survival and self-esteem, and these rights do not require the validation of ad hoc theoretical constructions produced by society.

Wilson came under attack in part because he made some strong assertions that he didn't (and I think couldn't) defend, in part because sporadically he lapsed into fuzzy language, and in part because he was presumed guilty of having Spencerian tendencies. Some critics then made sociobiology into a proxy debate over sexism, economic justice, and other perennial issues.[6]

Most reviewers have found Wilson's ethological analyses and interdisciplinary coverage stimulating, though not novel, but have objected strenuously to his more extreme speculations. On that twenty-seventh-chapter leap, Wilson's Harvard colleague Stephen Jay Gould remarked (1977, 252):

> We who have criticized this last chapter have been accused of denying altogether the relevance of biology to human behavior, of reviving an ancient superstition by placing ourselves outside the rest of "the creation." Are we pure "nurturists?" Do we permit a political vision of human perfectibility to blind us to evident constraints imposed by our biological nature? The answer to both statements is no. The issue is not universal biology vs. human uniqueness, but biological potentiality vs. biological determinism.

Gould then asked (1977, 257):

> Why imagine that specific genes for aggression, dominance, or spite have any importance when we know that the brain's enormous flexibility permits us to be aggressive or peaceful, dominant or submissive, spiteful or generous? Violence, sexism, and general nastiness *are* biological since they represent one subset of a possible range of behaviors. But peacefulness, equality, and kindness are just as biological—and we may see their influence increase if we can create social structures that permit them to flourish.

Especially harsh criticism came from a Boston Sociobiology Study Group, an affiliation of the leftist organization Science for the People. Acrimoniously the Group attacked *Sociobiology,* "the manifesto of a new, more complex, version of biological determinism," as drawing unfounded analogies between nonhuman and human societies, as engaging in "speculative reconstructions of human prehistory," as overemphasizing the genetic bases of behavior, and as making untestable assumptions about selection-adaptive drives in behavior. Wilson countercharged them with practicing "academic vigilantism" (Allen et al. 1975; Wilson 1976; Wade 1976).

My view is that although Wilson pulled together some fascinating themes in a provocative way, he overreached badly in his social speculations. To Wilson's claim that a "neurologically based learning rule" makes humans "absurdly easy to indoctrinate," literary critic Stuart Hampshire admonished, "Vast obscurities are concealed in that phrase 'neurologically based'" (1978, 65), and Stephen Gould shrugged, "I can only say that my own experience does not correspond with Wilson's" (1977, 254). Philosopher Ruth Mattern expressed my own reservations in saying, "A form of sociobiology attenuated enough to be plausible seems to be too weak to take ethics out of the hands of philosophers" (1978, 470).

Its unfortunate features aside, the sociobiology controversy has stimulated vigorous interdisciplinary discussion, and it continues to reward following as an example of is/ought exploration.

## MEANING OF "VALUES" AND "VALUE"

"Values" and "value" have been given so many connotations by the public and by philosophers, theologians, psychologists, economists, and other specialists that I cannot hope to refine them into neat definitions. What I will try to do is draw some commonsense perspective, illustrate how values are manifested, and show how values enter into analysis and decisionmaking.

In ordinary usage, *values* are taken to be abstract aspirations: freedom of speech, cohesive family life, national security. Such goals may be neither perfectly definable nor attainable, but, as with the Constitution of the United States, they can serve as ideals. It is in this spirit that the World Health Organization constitution declares that all people have a right to the "highest possible level of health," which it defines as "a state of complete physical, mental, and social well-being, not merely the absence of disease or infirmity."

Values can, in narrower usage, be taken as potentially attainable states of affairs, as objectives: assurance of infant survival, eradication of leprosy, achievement of energy independence, maintenance of forests for future generations. Values can govern means as well: protection of entrepreneurial access to seabed minerals, requirement that behavioral experiments not be carried out unless the subjects freely grant informed consent.

*Value* I take to be ascribed worth, as reflected in social preferences and transactions: market value of zinc, information value of a blueprint, political value of a senator's endorsement, social value of literacy, aesthetic value of a cityscape, symbolic value of a new medical clinic.

*Value-laden (-freighted, -charged, -oriented)* then connotes that an analysis, decision, or action is influenced by personal or institutional proclivities and prejudices, and that the analysis, decision, or action may affect people's value situation—their opportunities, status, wealth, happiness, or aspirations.

In 1752 David Hume made clear how valuation regresses to deep "sentiments":[7]

> Ask a man *why he uses exercise;* he will answer, *because he desires to keep his health.* If you then inquire *why he desires health,* he will readily reply, *because sickness is painful.* If you push your inquiries further and desire a reason *why he hates pain,* it is impossible he can ever give any. This is an ultimate end, and is never referred to any other object.
>
> Perhaps to your second question, *why he desires health,* he may also reply that *it is necessary for the exercise of his calling.* If you ask *why he is anxious on that head,* he will answer, *because he desires to get money.* If you demand, *Why? It is the instrument of pleasure,* says he.... Something must be desirable on its own account, and because of its immediate accord or agreement with human sentiment and affection.

Three fundamental questions have pervaded value-laden decisions and actions throughout history. How should choices be made when options cannot all be pursued or when they conflict? (If people are differently vulnerable to health hazards, and workplaces can never be perfectly risk-free, how should equal health protection be reconciled with equal employment opportunity?) How should collective societal goods be pursued with least erosion of the rights and goods of affected individuals? (What degree of vaccination efficacy for a population should be judged to outweigh side-effect risks to individuals?) And how should specific guidance on particular real actions be derived from abstract high precepts? (What does "right to personal privacy" mean in our present world of electronic financial transactions, international campaigns against terrorism and drug smuggling, and institutionalized health recordkeeping?)

People's strivings toward what they value—home and neighborhood lifestyles, opportunities for their children, assistance to citizens of less fortunate countries—hardly can be expressed precisely. Within any group, preferences will differ, and values will change over time. It is especially hard to resolve grand goals (national security) into the prosaic objectives (missile deployment plans, titanium stockpile policies, computer export restrictions), themselves value-laden, required for achieving those goals.

Public opinion polls can to some extent reveal value concerns. I remain a skeptic, and believe that polls that don't force respondents to choose among real options and confront trade-offs are worthless. Actions speak louder than answers to surveys. To me, actual manifestations of valuation—in political actions, budgets, laws, treaties, regulatory policies, military strategies; in court, corporation, and labor union decisions; in consumer purchasing behavior; in medical preferences; and in wage differentials and insurance schemes—are much more telling than casually expressed "opinions" are.

## INFLUENCE OF SCIENCE AND TECHNOLOGY ON SOCIAL PHILOSOPHIES AND CHOICES

Although they overlap and are not a formal taxonomy, the following modes of influence can be distinguished.

*Science deeply informs our cultural outlook.* Science has transmuted quite a few major cultural myths; negated many superstitions; left us living in a "disenchanted" world; imparted substance to a host of miasmas, humours, auras, scourges, and vital forces; recast the mind–body, nature–nurture, and other classic mysteries; and conspicuously revealed the hand of Man where none was seen before but Fortune's. Science reveals fundamentals about the occurrence and causes of mortality, genetic inheritance, and material wealth; gives us insights into where we have come from, and into our place in the universe; enables us to understand how we perceive what we see and mean what we say; and not only describes particular cultures, but helps us elaborate the very notions of "culture" and "society."

*Scientific and technological advance can create options for public consideration.* Choices from intimately personal to Malthusianly global have been opened up by the invention of the condom, diaphragm, spermacidal foam, Pill, and IUD, while other choices remain distant because of the practical unavailability of a male Pill, reversible vas deferens valve, and other contraceptive options. In addition to such options for *doing,* technology can create options for *knowing* (and then perhaps doing): until the recent development of amniocentesis and related techniques, never had it been possible to know with any certainty the gender, genetics, or pathology of a fetus in utero and thus be presented with informed choices over carrying to term, or seeking pre- or postnatal therapeutics, or terminating the pregnancy.

*Technology can strongly alter the relative attractiveness of competing social alternatives and can induce value changes.* The increasing practicality of solar energy and conservation methods will have implications, in the long run, for issues ranging from insulation installers' pulmonary health to diplomatic relations with Persian Gulf sheikdoms. Advance toward such dreams as rooftop solar electric cells, or vaccines against venereal diseases, depends not just on more efficient manufacture or adaptation, but on fundamental scientific discovery. Possibility changes tend to induce value changes, although they don't necessarily do so: the development of surgical anesthesias in the nineteenth century led to revision of the risks and costs patients were willing to bear in the surgical "calculus of suffering" (Pernick 1985).

*Science can identify and analyze consequences of choices and events, and help raise issues to public attention.* Science describes causal conditionals. Social attitudes and actions are altered by the knowledge that masturbation does not, contrary to earlier dogma, lead to madness; that pellagra, far from being an inherited inferiority, is the result of a dietary niacin deficiency that can easily be remedied; that some forms of schizophrenia can be attributed not to a mystically evil soul but to treatable physiological anomalies; and that soft, fluffy, seemingly harmless textile dust causes brown lung disease. Such knowledge allows causes to be assigned, opportunities and liabilities identified, and ethical issues altered.

*Science can anticipate and analyze perturbations in society itself, including impacts of technological change.* It can estimate the effects of entering alternative energy futures, of building a high-speed train system between Los Angeles and San Francisco, of mandating kindergarten attendance. It can project demographic changes, the relation of future populations to the resources they will have available, and the likely future interactions between people and technologies.

*The social sciences can observe and analyze expressed and implicit social values and valuation processes.* Reports on social attitudes and practices, such as surveys of sexual habits, can prime the way for changes in the way people evaluate their own attitudes. Although social scientists from Max Weber through Clyde Kluckhohn to the present have hoped to become able to analyze values and valuation psychologies, their methods are just now gaining enough acuity to warrant practical use. Social scientists are making progress in constructing value typologies, in surveying voter, medical client, and consumer attitudes, and in analyzing the preferences implicit in economic and legal actions. All of this information can be brought to bear on social decisions.

## TRACTATUS

The argument so far has been intended to establish these fundamentals.

- Social values cannot be derived from science qua science alone.
- Although scientific knowledge, once attained, may be considered ambi-potent for good or evil, the work of pursuing new science and developing technologies, which requires commitment of will and resources to undertake, is by no means value-neutral.
- Besides, at the moment of its emergence new knowledge may well have maleficent or beneficent potency that demands attention.
- Technical activity must be considered value-laden in two senses: technical people's social values and value perceptions affect their research and service; and that work, in turn, affects the value-situation of others in the public.
- Technical experts make crucial decisions for, and in the name of, the public.
- Science and technology affect our philosophies and choices by: deeply informing our cultural outlook; creating options for public consideration; altering the relative attractiveness of competing alternatives, and inducing value changes; identifying consequences of choices and events, and helping raise issues to public attention; anticipating and analyzing perturbations in society itself, including impacts of technological change; and observing and analyzing expressed and implicit social values and valuation processes.

## TECHNICAL PROGRESS AS DIRECTED TRAGEDY

Because it pervades this essay's outlook, I must now introduce my view of technical progress as tragedy. Not mere sadness or misfortune, but tragedy in a high sense. And not fatalistic tragedy, but, in what has become a characteristic of modern society, deliberately directed tragedy.

"The myths warn us that the wresting and exploitation of knowledge are perilous acts, but that man must and will know, and once knowing, will not forget," David Landes has reminded us (1969, 555).

> Adam and Eve lost Paradise for having eaten the fruit of the tree of knowledge; but they retained the knowledge. Prometheus was punished, and indeed all of mankind, for Zeus sent Pandora with her box of evils to compensate the advantages of fire; but Zeus never took back the fire. Daedalus lost his son, but he was the founder of a school of sculptors and craftsmen and passed much of his cunning on to posterity.

Man—*bestia cupidissima rerum novarum,* the "species cupidinous of new things"—must and will know. That ambitiousness has long been embodied in our Western tragic sense of ourselves. Alfred North Whitehead provocatively recast it (1925, 14):

> The pilgrim fathers of the scientific imagination as it exists today are the great tragedians of ancient Athens, Aeschylus, Sophocles, Euripides. Their vision of fate, remorseless and indifferent, urging a tragic incident to its inevitable issue, is the vision possessed by science. Fate in Greek Tragedy becomes the order of nature in modern thought.

Crucially, Whitehead continued: "Let me remind you that the essence of dramatic tragedy is not unhappiness. It resides in the solemnity of the remorseless working of things." And I would add the emphasis, "... especially as human agency intervenes."

For present purposes, I take tragedy to mean the deliberate confrontation of deeply important but nearly irresolvable life issues. Tragedy begins in our knowing of causalities, in our intervening in particular causes, and in our technical enlargement of interventional possibilities.

Robert Oppenheimer's confesso—"In some sort of crude sense which no vulgarity, no humor, no overstatement can quite extinguish, the physicists have known sin"—sounds only to be a starkly lame *mea culpa,* unless one realizes that the ages-old service of science and technology to war-making was too well known to Oppenheimer and his colleagues (1948, 66). No; surely the emphasis was on the verb: "Physicists have *known* sin." And not bad-boy sin, but Original sin. I think it not unlikely that the father of the A-bomb would approve of my transmutation: *scientists have known tragedy*. It is in that knowing that many of the issues of this book reside.

Nowhere has this "remorseless working of things" been more profoundly evident than in the development of nuclear weapons. In a *Discovery* editorial in September 1939, C. P. Snow, noting that the idea of explosive chain reaction had become accepted among leading physicists, grimly predicted that a project to make an atomic bomb would "certainly be carried out somewhere in the world." The Manhattan Project, of course, went forward, as did its Japanese and German counterparts. Looking back on the 1945 decision to drop the Hiroshima weapon, Robert Wilson recalled (1970, 32):

> Things and events were happening on a scale of weeks: the death of Roosevelt, the fall of Germany, the 100-ton TNT test of May 7, the bomb test of July 16, each seemed to follow on the heels of the other. A person cannot react that fast. Then too, there was an absolutely Faustian fascination about whether the bomb would really work.

Similarly Norbert Wiener's melancholia of 1948 over the development of cybernetics, although it strikes me as giving in too easily (1948, 28):

> Those of us who have contributed to the new science of cybernetics stand in a moral position which is, to say the least, not very comfortable. We have contributed to the initiation of a new science which embraces technical developments with great possibilities for good and for evil. We can only hand it over into the world that exists about us, and this is the world of Belsen and Hiroshima. We do not even have the choice of suppressing these new technical developments. They belong to the age....

Once arcane knowledge is generated somewhere, its transmission, or independent reconstruction, is almost, though not absolutely, inevitable. As Dürrenmatt had Möbius say in *The Physicists,* "What has once been thought can't be unthought." After basic nuclear information was released after World War II, the question of proliferation became not *whether* but *when* and *under what circumstances* other countries would pursue nuclear options—and, in Whitehead's phrase, "tragic incident moved to inevitable issue."

Three fundamental tragic motifs can be recognized.

*First, by describing flatly the way things are, science raises tragic awarenesses: that certain things are happening, others may happen, others will happen, others cannot happen; that events are determined by causes; that causes may reflect willful human agency and decision.*

Even as scientists achieve long-sought humanitarian breakthroughs, the timeless lament of Ecclesiastes (1:18) resonates: "He that increaseth knowledge increaseth sorrow" ("sorrow," or "mental anguish," is a standard translation of the Hebrew, *mak'ôbâh*).

In absence of knowledge, people may resign themselves to Fate, to "blind chance" (mongoloid birth just happens). With knowledge (the chance of Down's syndrome increases sharply with maternal age above thirty-five), intentionality issues arise; chances still have to be taken, but odds may be altered or stakes adjusted deliberately. Tragedy lies not in resigned fatalism, but in considered confrontation of the near-irreconcilables (wanting to have a child, but wanting to pursue other early-life goals first, but of course wishing that the child not be born infirm).

Biomedical science is forcing us to confront basic facts about differences among people, such as differences in allergic vulnerability, color perception, reflex quickness, and lower-back resiliency, that can bring occupational health protection squarely into conflict with equal employment opportunity. Differences we have pretended don't exist will have to be recognized.

*Second, in their inventions and in the systems they weave our lives into, technology and medicine confront us with tragic choices among life-extending options whose consequences we have at least some foreknowledge of.*[8] In a trend that can only lead, eventually, to anguishing decisions, improvements in neonatal care, coupled with a desire to save all infants no matter what, are preserving ever-more-premature babies (down to 500 grams, or a little over a pound, birthweight, now), at ever-increasing costs, with higher incidences of permanent medical deficiencies.[9] As society invests in life-extending medical technologies, sorts out endangered-species protection priorities, and debates the future of fourth-world nations, exceedingly traumatic choices will have to be confronted— not just made (we do that already, often by defaulting), but *confronted.* Weighed. Debated. Faced.

The answers do not reside in knowledge by itself. Again we hear a classic voice, Tiresias moaning when Oedipus commanded him to consult "bird-flight or any art of divination" to guide Thebes from the plague: "How dreadful knowledge of the truth can be when there is no help in truth."

*Third, technological advance challenges society with tragic commitments to consequences.* Warning that commercial nuclear power comes as a "Faustian bargain," in 1972 Alvin Weinberg, the director of Oak Ridge National Laboratory, said "the price that we demand of society for this magical energy source is both a vigilance and a longevity of our social institutions that we are quite unaccustomed to" (1972a, 33). Regardless of future decisions about nuclear weapons or nuclear power, the high-level radioactive waste already accumulated in many countries will demand curatorship for thousands of years; the stuff will not go away. The great system of dikes that creates and protects one-third of the

Netherlands requires similar massive perpetual commitment. Now that we have eradicated smallpox worldwide as a clinical entity, we shall forever have to monitor our increasingly unvaccinated populace, to watch against recrudescence of the virus from who-knows-what lurking source.

These tragic awarenesses, choices, and commitments caused or mediated by technical ventures may leave us happy or not. But of their solemnity there can be no question.

Think for a moment about our progress in dealing with public health risks, in which all of these themes are so evident. Health in the industrial West surely is, in general, more robust than ever before. Many of the most dangerous infectious diseases, such as tuberculosis, diphtheria, smallpox, cholera, typhus, and polio, have been conquered, and progress has been made against many others. Scurvy, pellagra, iron deficiency anemia, and other nutritional diseases have been mastered. Many illnesses that have not yet been eliminated, such as diabetes, have at least been brought under control. Exposure to mercury, lead, arsenic, chromium, and other heavy-metal poisons has been substantially reduced, as has exposure to asbestos, halocarbon solvents, and many other chemicals. Through prediction and protection much damage from hurricanes, floods, and earthquakes has been mitigated. All over the world infant mortality continues to decrease, and life expectancy to increase. More people are living longer, healthier, more vigorous lives.

Nonetheless, almost ruefully, we have progressed to an inherently discomfiting state, a state in which we must expect to remain from now on. Why inherently discomfiting? Because steadily we have broadened our apprehensions to include not only natural catastrophes, infectious diseases, everyday mechanical accidents, and acute poisons, but also large-scale technological accidents, chronic low-level hazards from chemicals, radiation, and noise, and lifestyle vices, such as addictions to tobacco, alcohol, barbiturates, narcotics, caffeine, and rich foods. To our struggle against cancer and other classical illnesses we have added concern about reproductive, genetic, immunological, behavioral, and other debilitations. And of course we continue to create new hazards, to identify risks that have existed without being recognized, and to resolve to reduce previously tolerated risks.

Now, about many hazards we know enough, scientifically, to "worry," but not enough to know *how much* to worry—or how much protective action to invest. Knowledge has grown enormously, and we even have the luxury of going around searching for possible trouble. But many scientific disciplines are still in their adolescence and are unable to evaluate risks precisely. We can detect minuscule traces of manmade pesticides in mothers' milk all over the world, which is vaguely disturbing; but, with rare exceptions, we don't have a clue as to whether the chemicals exert any effect on mother or infant (Jensen 1983; Wolff 1983). Toxicology, epidemiology, and medicine have taken us out to their borders, but it's unruly territory.

At the same time that we are learning more, we are heightening our societal aspirations beyond all previous limits. We intend to help all infants get a vigorous start in life. And we strive to afford first-rate, broadly defined health protection to all citizens and noncitizens, even immigrant aliens, through an enormous range of risks, throughout their lives. No civilization ever before has had these ambitions.

The crux: In our knowing so much more, though imperfectly, and aspiring to so much more, we have passed beyond the sheltering blissfulness of ignorance and risk-enduring resignation. This has given rise to considerable social apprehensiveness, which is affecting both the outlook of individuals and the functioning of institutions.

Similar phase-changes have occurred historically when people became aware of specific causes of disease and deformity, as when it became clear that moral turpitude alone was not the cause of syphilis, and when societal aspirations, such as commitment to worker protection, rose. We are going through both kinds of change at the same time.

Risks and aspirations will continue to evolve. In his 1803 revised *Essay on Population* Thomas Malthus observed of Jenner's new vaccine, "I have not the slightest doubt that if the introduction of cowpox should extirpate the smallpox, we shall find...increased mortality of some other disease" (1803, 522). Malthus was right. As any risk is reduced, others inevitably increase in the mortality and morbidity tables—though perhaps setting in at later ages. Risks in the industrial West are evolving now as rural and agrarian risks are succeeded by urban, industrial, and medical-care risks. A similar progression is occurring, displaced in time, in less developed countries: cholera and fatal infant diarrhea are being succeeded by cancer, heart disease, and drug side effects. I can't imagine that our societal aspirations won't expand even further, both for the industrial West and for the rest of the world.

"We are in for a sequentiality of improbable possibles," *Finnegans Wake*'s Shem knew to expect. Some possibles are, of course, more predictable than others. Recent years' debates over such matters as food additives, contraceptives, pesticides, and energy sources have brought broad public recognition that nothing can be risk-free, that there are no rewards without risks, and that risktaking for benefit is the essence of human striving.

Our analyses and decisions are making us face risks ever more explicitly and comparatively. The Occupational Safety and Health Administration's 1983 standard for worker exposure to airborne inorganic arsenic was a striking example. In tightening the standard from 500 to 10 micrograms arsenic per cubic meter of air, OSHA concluded "that inorganic arsenic is a carcinogen, that no safe level of exposure can be demonstrated, and that 10 micrograms per cubic meter is the lowest possible level to which employee exposure could be controlled." Further (U.S. Occupational Safety and Health Administration 1983, 1867):

> The level of risk from working a lifetime of exposure at 10 micrograms per cubic meter is estimated at approximately 8 excess lung cancer deaths per 1000 employees. OSHA believes that this level of risk does not appear to be insignificant. It is below risk levels in high risk occupations but it is above risk levels in occupations with average levels of risk.

and sometimes instantaneously. Capital now moves routinely on a global scale. Labor remains a national, if not a locally bound factor of production. As a consequence, national labor forces are, if not entirely, increasingly subject to capital's capability to relocate if its demands are not acceded to.

In Scotland, for example, this capability of transnational capital is called the "Hyster factor," and refers to the "take it or leave it" offer to the 500 workers at the Hyster company's forklift truck factory at Irvine, in Ayrshire. They were told recently that the company (a U.S. transnational headquartered in Portland, Oregon) would pull out of Scotland if they did not accept a 10 percent pay cut.[1] The workers, with only 11 workers voting "no," accepted.

How duplicitous as well as brutal is the exercise of this new power of transnational capital, and how helpless labor is before it, is indicated in still another account, this one in an American newspaper, on the behavior of the Hyster Co.:

> Hyster's workers in Irvine have no union. The vote might have been no different if they did. The most muscular union in Europe isn't a match for a company run out of a place like Portland. A union is concerned with the workers in its country; a multinational knows no bounds. While the chief executive can look to the far corners of his realm through the screen of a desk-top terminal, the unionist still hesitates to make an international phone call. While a shop steward dickers with a plant manager, the plant's destiny may be determined on another continent.

How one group of workers in one country (or locale) can be played off against another is described in the same report:

> Plant managers within a multinational company could easily tell their Swiss workers, as the managers of one reputedly did, that new work was going to the lower-paid British; tell the British that production was being passed to the more efficient French; and tell the French it was creating jobs for the highly cooperative Swiss. *Companies rarely tell their scattered workers everything. Workers of different nationalities rarely tell each other anything. In an age when information is power, they lack both.*[2]

And, as if these enhanced capabilities were not enough, the availability of powerful global communications networks enables the big firms to penetrate national and world markets formerly inaccessible.

For these and other reasons, many of the stagnation theories of capitalism, which abounded in the 1930s and early 1940s, quite justifiably, have been mothballed. Dynamism and growth along with some sectoral decay seem more appropriate descriptors of the current scene. But the picture is not a complete one.

While capitalism admittedly is enjoying a new burst of energy, its room for maneuver narrows steadily. The complete dependence of the central force in modern capitalism, the transnational corporation, on unimpeded international communication, requires a relatively secure world system of stable nation states. This is the least likely outcome of the forces now in motion, actively stimulated by the renewed energies of high-tech capitalism. A second paradox explains this more fully.

## CONSUMERISM AND THE (TEMPORARY) WITHERING OF RADICAL CONSCIOUSNESS

Capitalism has achieved remarkable popular support with its fostering of consumerism. It has sold successfully a way of life and a set of beliefs that tie human well-being to the individual possession of an ever-expanding array of purchasable goods and services. Acquiring material goods has either superseded, or been made the equivalent of, love, friendship, and community.

In the highly industrialized market economies, people have widely accepted this ethic, however much they may disavow it in public conversation. To a very considerable extent, the inability of radical movements in Western Europe to change decisively the political and economic structures of their societies is explained by the unwillingness of a majority of people to engage in activities that threaten either the possession of, or the hope of acquiring consumer goods and perquisites. It is evidently felt, whether articulated or not, that literally nothing is worth losing the opportunity to get, or to hold on to, consumer goods.

Accepted as well, by a dominant fraction of the population in the North Atlantic region, is the linkage of consumption to democracy. Consumer choice here is equated with meaningful politico-economic choice. Shopping in a supermarket, with its crowded shelves of products is, in this perspective, a democratic practice.

The intense drilling of the people to participate in the consumer society is facilitated greatly, if it is not directly motivated, by advertising, which saturates most information channels—especially in the United States. The newest information technologies, which have many, and sometimes contradictory applications, as they are now being utilized, substantially increase the penetrative power of the marketing system. Interactive, two-way television, for example, still in its early development, is seen largely as a home marketing technology.

From all this, it seems reasonable to conclude that the individualist, possessive ethic has triumphed, in the short run at least, in the heartlands of capitalist enterprise. The system itself is stronger than ever on account of this. Though not eliminated, radical, equalitarian impulses have been weakened greatly, and alternative conceptions of life in an industrial world are, for the moment, unable to attract support, or bleaker still, unable to be conceptualized.

Yet here, too, the situation is hardly stabilized. In the developed market economies, the strength of a "democracy of consumption" rests almost entirely on economic growth and, at a minimum, on shares for all in an enlarging economic pie. If growth cannot be maintained, the consumerist ethos falters. Indeed, it is likely to become a source of growing dissatisfaction.

As the consumerist economies rely heavily on mass media advertising to keep demand high, there is a built-in, potentially disruptive force continuously at work, either, when the economic engine begins to stall, or if the resource availabilities begin to contract. A Club of Rome study touches on the second of these possibilities: "...economic policies based on quantitative growth through the stimulation of consumption will hardly prove effective in an era of resource limitation...."[3]

If this is so in the already industrialized states, the condition is still more aggravated in the Third World, where large numbers of people live close to, or sometimes fall below, the margin of survival. In these regions, now being integrated rapidly into the world market system of transnational enterprise by the new information technologies, the paradox is striking.

The information systems in most of the nations in the less industrialized world are being transformed into marketing networks for the resident transnational corporations. Increasingly, advertising-supported television carries the consumption messages of the world business system. Additionally, local banking, industrial, transport, and tourist sectors are being connected to the metropolitan informational circuits of the TNCs.

Local elites and the new professional classes quickly accept and embrace the consumerist message. They are also in a position to act on it. For the rest of the population, the overwhelming majority, the effects are less satisfactory. Unable to participate, the artifacts and stimuli of consumerism surround them. At the same time, the character of the economy is distorted to enable a relatively small number of people to enjoy Western consumption standards, while actually diminishing the output of vitally required goods.

Natural resources, on a global scale, are being plundered for a mode of consumption that embodies wastefulness and inequality. The consumerist model, carried into the world at large is, therefore, a radicalizing force, swallowing up irreplaceable natural resources and simultaneously feeding and thwarting human expectations. Political stability is diminished therefore in proportion to the speed with which the marketing system and its advertising component are extended to the still-impoverished parts of the world.

Installing itself in all corners of the globe, and spreading the message of consumption through its advertising-supported media channels, the transnational corporation is promoting what it most fears, future massive political instability.

## THE NEW INFORMATION TECHNOLOGIES CONFRONT NATIONAL SOVEREIGNTY

Domestically and internationally, governance is being changed radically by the availability of the new information technologies. Internationally, the very basis of national sovereignty, for a majority of states, is threatened. A combination of developments, utilizing satellite communications and the linkage of computers, directly undercut national jurisdictions. Remote sensing, for example—scanning a territory with powerful sensors attached to orbiting satellites—routinely maps the globe, obtaining all sorts of resource information without requiring the permission of the scanned region's government. The recipients of this information, moreover, are generally the power centers in the few industrialized countries that possess the technical capability to interpret and to take advantage of the data.

Along with remote sensing, direct satellite broadcasting is now at hand, offering the capability of transmitting messages from the satellite to receivers across the earth, irrespective of national boundaries. Most important of all the new developments, electronic transborder data flows now move in great volume across frontiers, silently and invisibly, transferring data, mostly *within* the transnational corporation's many branches, without oversight or accountability to national authorities.

Unless counter measures are adopted, these developments, taken together, herald the demise of most nation-states, at least as effective control agents of their own national space.

As there is no world government waiting to assume global responsibilities, these developments also suggest the enhanced power of a few superstates who exercise these technical capacities, and a still greater influence of already powerful transnational corporations. In short, the erosion of national sovereignty appears to offer still further reinforcement to the world business system. Thus, the devastating impact of the new technologies on national political organization redounds to the benefit of capitalist enterprise.

But here, too, appearances may be deceptive. The anti-imperialist struggles of the twentieth century are too recent to have been expunged from popular consciousness. Genuine national independence and sovereignty, though hardly (fully) attained by most countries, despite ceremonial trappings of flags and airlines, remain powerful aspirations. The new technological threats to national sovereignty can only rally great oppositional force—already observable in international negotiations over issues such as access to the geostationary orbit, radio frequency allocation, data flow regulation, and the right of nations to control the messages coming into their national space ("prior consent").

American policymakers are not entirely unaware of this opposition. This is discreetly, if indirectly admitted, by George Shultz, Secretary of State: "...the evolution of communication and information technology and its international significance makes it a foreign policy concern of the first magnitude since the ultimate course of communication and information can affect every dimension of our foreign relations...."[4]

There is a domestic equivalent to this dialectic, one with more menacing implications. In the United States, where these developments are most advanced, and therefore most observable, the new information technologies are also deeply affecting the character and role of the national state. In this instance, the impacts also are contradictory.

The new information technologies, for one thing, have contributed greatly to the weakening of the public sector by increasing the profit-making potential of a large number of activities which formerly were nonprofitable. Education, health, welfare, and public service functions (and public utilities in Western Europe) fall into this category. The advent of computers and the growth of the information sector in general have transformed information into a saleable good and encourage many former public service activities to be contracted out for profit. To be sure, factors other than computerization and sale of information are involved in these developments, but nonetheless their contribution to privatization cannot be minimized.

In any case, the trends toward a diminishing public sector in the United States and the United Kingdom are very evident. Yet while public service activities increasingly are privatized, the state's power and role by no means declines correspondingly. It shifts instead toward social control, surveillance, and coercion. In the U.K. and the U.S.A., as national educational and health expenditures contract, military, law enforcement, intelligence, and police outlays expand.

In the high-tech economy, a weak public sector is "balanced" by a strong state, availing itself of the most up-to-date communication instrumentation to maintain social equilibrium alongside grievous economic deterioration for increasing numbers of working people. In the developed market economies, in this time of unrelenting crisis, the state is divested as much as possible of its welfare function, while it strengthens its coercive capability to handle potentially unruly domestic groupings and perceived (fabricated?) international adversaries.

Not everyone however, agrees with this strategy. Some point out that the problems created by the new technologies can only be met with *increased* government intervention to protect the weaker sections of the population. Charles Lecht, writing about the catastrophic layoffs made by AT&T, for example, states:

> This is no time for our government to remove itself from the scene, whether it does so in the name of some ideology or of exhaustion of its moral/material resources, while the contesting parties struggle to work things out on their own. Short of instituting its own welfare system, A.T.&T. is powerless to solve the problems of its legion of new ex-employees. It cannot even promise job security to many of its remaining white-collar personnel.[5]

This view goes unattended in the current U.K. and U.S. administrations.

In the poorer lands the situation is reversed. The State is called upon to protect the national patrimony against the TNC's greatly expanded powers provided by the new communications capability to bypass national authority. Yet in both instances, different as they are, the communication technologies are being employed *against* human needs and aspirations.

## INFORMATION IMPOVERISHMENT ACCOMPANIES INFORMATION ABUNDANCE

Satellites, computers, and cable make possible a qualitatively new level of information availability. The present capacity to generate, process, store, retrieve, and disseminate information now exceeds, whatever their extravagance, earlier Utopian visions. Viewed exclusively as a technological capability, it is hardly unrealistic to regard the present situation as one of potential unprecedented abundance and richness of information.

It seems all the more shocking therefore to acknowledge at the same time the deepening division of the society into informationally privileged and informationally impoverished sectors. What accounts for this?

Above all, the responsibility rests with the social arrangements that are governing the development, utilization, and distribution of the new information technologies and their products. The private firms and institutions that are organizing the new information age are, as a matter of course, making information a merchandisable good, a commodity produced for profit and sale. Specifically, there are companies which design the systems, manufacture the hardware and the software, process the data, create the data bases, and transmit the messages. These and other activities create a new and expanding information sector of the economy.

From the time of Gutenberg, and even before, information production has been controlled and has led to social stratification based on unequal access. What is of special significance about the current situation is the centrality of information in all spheres of material production, as well as its increasing prominence throughout the economy. Today, information increasingly serves as a primary factor in production, distribution, administration, work, and leisure.

For these reasons, how information itself is produced and made available become crucial determinants affecting the organization of the overall social system.

Largely as a consequence of its heightened value and importance to the entire economy, information is being privatized at an accelerating rate. As we have noted, public sector information functions are being curtailed to facilitate private expansion in this area. Where the public information functions continue to exist, they are brought into the commercial sphere and are compelled to adopt market principles in their operation.

Libraries, for example, though remaining public institutions, introduce computer services and then levy user charges, breaking with longstanding traditions of "free" service. Similarly, the U.S. public mail service is being dismembered though no public official will acknowledge this. However, as large corporations adopt private electronic communication systems and abandon the public mails, costs to *individual* users rise rapidly and services begin to be curtailed as demand falls in response to the increased costs.

Throughout the economy, the privatization and commoditization of information are being accompanied by commercial charges that separate and stratify users by their ability to pay. What hypothetically could be a truly information-rich society is on the way to becoming a community divided into information "haves" and "have-nots." Across the information spectrum, the evidence of inequality of access, determined by income differentials, multiplies.

The commercialization of information knows no bounds. No source is exempt from the privatization wave, especially as the state's authority is put heavily on the side of the privatizers. And so, there is the spectacle of information deprivation developing in the midst of information abundance.

Is this a reversible development? The answer rests partly in the political process. Sufficient popular mobilization/demand for information access and for protection against rising information costs—higher telephone bills, for instance—*might* arrest these trends. The problem is the familiar one. How can the issue be made understandable, in all its aspects, to impel popular involvement, when the channels of communication remain at the disposal, almost exclusively, of those benefiting most from current policies?

There is yet another dimension to the information gap now widening. This is that the overall pool of information will contract eventually as access and availability become almost entirely tied to an ability-to-pay standard. Increasingly, certain kinds of social information just won't be produced, stored, indexed, or available, for there will be no ready market for it. Yet in a highly interdependent industrial economy, the absence of such information can become a grave source of vulnerability to the functioning of the social order.

This outcome, however, will occur, if it does, only in the long-term future. It cannot be expected to influence near-term developments.

### THE DISAPPEARANCE OF ORGANIZED LABOR REMOVES A POWERFUL SYSTEM STABILIZER

The new information technologies that provide the transnational corporation with greatly enhanced operational flexibility, locally and globally, have recast as well the historic balance between capital and labor.

Though the labor movement at its most organized and disciplined has never actually enjoyed an equal position with capital, it has, in several countries, including the United States, won considerable leverage. Until recently, it was capable of defending *some* of its interests, and sometimes advancing them.

For the moment, and perhaps for a long time to come, labor's capability to act as a check on capital and as a defender of its material interests has been weakened to the point of marginality. The great corporations, domestically and internationally, now have the means to make labor accept their terms, in ways almost unparalleled since the early days of the industrial revolution.

The giant firms, in their ability to shift capital, production, and investment, globally and nationally, have overturned the already less than rough balance exerted by formerly strong, national labor movements. Capital, in its transnational corporate organizational form, is now structurally and operationally thoroughly international. Labor remains local, nationally effective at best.

In these changed circumstances, the era of rising real wages for the industrial work force has come to a halt. Workers now are expected to grant "give backs"—the return of benefits won through decades of social struggle. The very existence of unionism is threatened. In the United States and Britain and elsewhere, the older "smokestack" industries were the strongholds of labor organization. It is these industries that are being "exported" and their work forces left to scrounge for jobs or hand-outs.

At the same time, the new high-tech industries have been singularly successful in avoiding (preventing?) unionization. "Near Boston," according to one account, "where a proliferation of high technology companies has created tens of thousands of new jobs since 1975, not one of the 133 companies that belong to the Massachusetts High Technology Council is unionized."[6]

The white-collar worker and open-collar professional have been imbued from earliest schooling with an individualist, anti-labor organizational ethic. Added to this, the new technologies are being utilized, in some cases, to restore patterns of home work and piece work. These further atomize the labor force and interfere with its organization.

While the old industrial centers of unionization shut down, the new complexes of high-tech activity move to locales long resistant to labor organization. In these new centers, the awareness of capital's mobility serves as a continuous warning to the unorganized labor force that it develops solidarity at its peril.

As if these were not problems enough for labor, the new media technologies are being employed to provide an ideological assault on independent worker organization. Increasingly, the big companies, whatever their main industrial activities, engage in direct media production, taking advantage of video and film to manufacture anti-union messages for in-house and national audiences.

These efforts join with corporate advertising, which continues to blanket the commercial media system. Throughout the range of sound and visual imagery, with the limited exception of some popular (rock) music, the ideology of property, individualism, and consumerism prevail. The cultural sphere of the mass media is a realm in which the labor movement and values of human solidarity have no place.[7]

However lamentable, it appears that in the most advanced centers of industrialization, the historic working class movement is being ground down and worn away. Capital's dream of (either) a nonorganized labor force, or production entirely without workers, seems on the way to fulfillment. Marx seemingly has been confounded. It is not the "expropriation of the expropriators" that is happening but the liquidation of the proletariat.

Whether it will come to this is not yet totally clear. What is certain is that the extent to which these developments do occur, and labor as an organized social force largely disappears, capital's dream may well turn out to be its nightmare.

Though rarely acknowledged, the labor movements in the developed market economies have provided, against capital's will to be sure, a powerful stabilizer, economically and politically, for the market system. Economically, mass purchasing power has been sustained and broadened, assuring a market for a good part of the rising output of the industrial system. Though labor's share of the economic pie has remained remarkably steady over the decades, it has been maintained and somewhat expanded.

Without the ever present claim of organized labor, the intensity of cyclical crisis—never far away—might have been stronger and could have wrought great havoc in the system. On the political side, the existence of an organized labor force, intent on joining electoral coalitions in the United States and presenting its own reformist platform in the U.K. and elsewhere, went a long way in extinguishing radical fires and revolutionary changes during capitalism's most exploitative period of growth and maturation.

Throughout the twentieth century, with few exceptions, organized labor has been a conservative force, often, if that is imaginable, as opposed to radical restructuring of the social order as capital itself. Unions have diminished some of capital's prerogatives, but the ultimate costs to capital have been readily sustainable, because the end result was systemic stabilization.

It is the stabilizing function that now is being removed in the shift to high technologies and an information-based economy. It is quite possible that new forms of labor organization may develop, mobilized around new kinds of issues, not all of which may be totally work-site based. If this does in fact occur, the labor-capital struggle will move to new ground with unpredictable outcomes.

However, if the information-based economy continues to resist unionization successfully, it is difficult at this point to see what social grouping can be called upon to provide the income stabilizing and political defusing function of the old labor movement.

This then is the paradox! Capitalism without organized labor may become a capitalism of political gyrations and persistent and intense economic slump. In the economic sphere, one question alone emphasizes the growing dilemma. Where will the consumer purchasing power come from, as automation bites deeper in the work force and as rising productivity increases the capacity to turn out more goods and services with fewer hands? The authors of the Club of Rome study, *Micro-Electronics and Society,* face the issue squarely: "At the heart of the matter," they write, "is the question of employment." But, "there is the danger that against the background of economic crisis and high unemployment, the rationalization possibilities brought about by technological change will be pursued to the exclusion of all else...."[8]

In other words, left to its own dynamic, capitalism will pursue profits before full employment. In doing so, by means of cost-cutting, the number of jobless will increase.

Where then will the pressure come from to redistribute income, lacking a strongly organized work force? And, if there is no redistribution, or not enough to matter, can moderation continue to prevail in the political sphere?

## SHORT-RUN ECONOMIC DECISIONS AND LONG-RUN CULTURAL CONSEQUENCES

The introduction of cable television, the use of communication satellites, and the pellmell rush into computerization repeat a pattern of technological invention and utilization that appeared early in the history of communication technologies. Whatever differences that there may be today are accountable largely to the enormous concentration of capital that now characterizes the industrial system in general, and the high-tech communication industries in particular.

The construction and launching of a communication satellite, for example, are multi-million-dollar ventures, and these exclude the initial costs of research, development, and experimentation. The outlays for cabling a metropolitan area are even more substantial. The manufacture of computers and the creation of computer/telecommunication networks are the exclusive preserves of a handful of giant firms or governments.

This being the material reality of modern communications, the decision-making authority on whether to proceed with any or all of these new technologies is vested in few hands, all of which are mostly, if not exclusively, responsive to the calculus of immediate economic benefit: profits foremost, and industrial primacy not far behind. When governments are engaged in the process, their concerns are also economic and they center largely around economic growth rates and providing jobs.

In the United States, the United Kingdom, West Germany, France, Canada, Japan, and a few other market economies, the readiness to promote the new communication technologies arises from a combination of motives, varying from one economy to another, but all sharing a short-term economic emphasis. With the possible exception of France—and this must be seen only as a tentative qualification—the consequences of following near-term objectives—profits and jobs—promise to produce far-reaching though longer term destabilizing cultural consequences.

The United States must be seen as exceptional at the same time as it is perhaps the most susceptible to the longer term destabilization forces released by the new technologies. Yet in the time immediately ahead it is difficult not to believe that the U.S. economy, or at least some sectors of it, will derive significant benefits from the widespread adoption domestically and internationally of the new communications instrumentation and processes.

Already the center of the global information system, America's media products and informational goods cannot fail to gain still greater advantage in the world market. The economies of scale available to U.S. producers who have a huge internal market at their disposal make the prices of their outputs unmatchable, for the most part, in the international market. This is particularly observable in TV programming but it applies elsewhere.

One result of this phenomenon will be a further increase of American media messages and imagery in national communication circuits in Western Europe, and, more pronouncedly still, in the vast, less-industrialized peripheral areas of the world. So, while West European and other market economies are launching high-tech promotions to stimulate production, create jobs, and grab a chunk of the international market, they are, in fact, opening their societies still further to the advances of the transnational corporations. The TNCs rapidly utilize for their own purposes the new communications infrastructure ostensibly established to improve national authority. Accordingly, further economic distortions as well as popular frustration may be expected to intensify throughout these regions.

Domestically, the greatly expanded number of options for receiving information of necessity will require a continuing escalation of user charges and a consequent accelerated widening of the societal knowledge gap. What some describe as the "Technologies of Freedom"[9] are rapidly becoming the basis for a thoroughly stratified nation.

Increasingly employed for marketing and monitoring, the new communication facilities constitute the infrastructure for an aggressive program of "law and order," aimed at checking any signs of popular discontent. While personal computers are being sold to the public as freedom-enhancing and self-enriching instruments, the multiplying distress signals in the economy are making it less and less likely for solutions to be found on an individual basis.

All the same, national leaderships in the West, with unfailing media support, encourage the belief that the new information technologies are essential to deliver information abundance as well as escape the hardships of economic crisis.

Still, this good news is received with some skepticism by many people. For this reason, there is concern at policy-making levels that the new information technology may begin to suffer the fate of nuclear power, which now is greatly feared by wide sectors of the population. Care must be taken, therefore, to consolidate in the general public a favorable impression of information technology. An Organization of Economic Cooperation and Development (OECD) proposal for an international conference on information, trade, and communication services makes this concern explicit:

> The ultimate aim of the conference would be to... make sure that the social consensus on the benefits of information technology continues unabashed in the years to come.... There is a certain danger that information technology might increasingly be regarded as a scare technology like nuclear technology; the conference might contribute to avoiding this potentially negative feeling in various strata of the public, and, on the contrary, stress the positive aspects.[10]

The anxiety of the OECD Secretariat is not misplaced. There is indeed evidence that people are deeply concerned with the lack of social responsibility and accountability for some of the information technologies already installed and operating. How else to explain the sudden temporary halt in the Reagan administration's proposed policy of allowing individual census data to be passed around from one government agency to another? No guardian of the public's right to know or to its right to information privacy, the general outcry was sufficient to stop for the time being, at least, the administration taking another step in the direction of omnipotent information surveillance and control.[11]

Whether large-scale manifestation of public concern over the direction of the information technology can be expected to continue and deepen remains to be seen.

In sum, long-term cultural impacts may well overturn the short run economic advantages that are today the motivating force behind the communications "revolution." This should not be viewed, however, with unqualified satisfaction. The cultural impacts may be far from neat and their political expression may be less than rational. Yet if the future is clouded, what does seem clear is that the information age now looming carries little promise of fulfilling what its proponents proclaim: ease, well-being, and humanization.

## GROWING SEPARATION OF AMERICAN THINKING FROM INTERNATIONAL REALITIES

Finally, there is one paradox that overshadows all the others in its long-term implications. It is, actually, the summation, or perhaps, outcome, of those paradoxes already mentioned.

Americans are forever being congratulated by their leaders for being the beneficiaries of the most technologically advanced, complex, expensive, and adaptable communication facilities and processes in the world. This notwithstanding, and this is the paradox, people in the United States may be amongst the globe's least knowledgeable in comprehending the sentiments and changes of recent decades in the international arena.

Despite thousands of daily newspapers, hundreds of magazines, innumerable television channels, omnipresent radio, and instantaneous information delivery systems, Americans are sealed off surprisingly well from divergent, outside (or even domestic) opinion.

This comes not exclusively, but largely, from the vigilant "gate keeping" of private U.S. information controllers, i.e., the big international news agencies' filters, the network news strainers, and the editors of the major newspapers and magazine chains.

It comes as well, however, from the privileged material position of an extremely large middle class, as well as a good section of the skilled and professional classes, though often these categories overlap significantly. In each instance, there is a demonstrable inability to recognize, but much less empathize, with a huge, have-not world.

It originates also with the very different experiences of the United States population in the wars of the twentieth century. World War I and World War II devastated much of Europe and parts of Africa and Asia as well. For the United States, the wars were times of great economic growth and expansion and of corporate and individual enrichment.

For these and other reasons, Americans, enveloped in a rapidly forming information-based economy—saturated with messages and images—remain unable to comprehend or sympathize with the most elemental and powerful feelings and social movements of this era. Symptomatic, if not typical, of this insensitivity was the reaction of the President of the United States, when informed that more than a hundred nations protested the invasion of Grenada. Reagan exclaimed this news didn't disturb his breakfast.

The consequences of the already great and still growing divergence between American and world sentiment on fundamental issues of peace and social change can hardly be overstated. The least that can be said is that it creates a perilous atmosphere for the time ahead.

Efforts to achieve national sovereignty, economic autonomy, and cultural independence are accelerating globally. American reactions to these attempts at changing world power relations are (mis)informed and shaped by an information apparatus almost completely dominated by transnational capital and permeated with its perspectives. It is to be expected, and in fact happens, that most of the struggles for change become subjects of suspicion and hostility in America.

The American people are thrust, ever more frequently, into fateful and tragic positions of opposition to international social movements for human material improvement and individual liberation. They are led also to assess local disputes that have had ancient origins as issues of potential danger to American survival.

When the content of the messages and the quality of the news reporting come under scrutiny in international discussions, American representatives of the transnational information apparatus invoke "freedom of the press" and the U.S. Constitution's First Amendment. Though increasingly regarded with skepticism outside the country, this tactic still serves to persuade Americans that the world is filled with tyrants seeking to curtail their informational freedom.

At the same time the omission, distortion, and partial representation of global events and popular sentiments, in the American media, assume phenomenal levels.

What may be expected, eventually, from the combination of a misinformed and underinformed people, supporting a government, representative of transnational corporate interests, armed with nuclear weapons of unimaginable destructive capacity? It can only be hoped that the paradoxes that have been described above may succeed in unhinging or overturning some of the dangerous relationships and directions that now prevail.

## NOTES


1 P. Hetherington, "The Electronic Warning for the Trade Unions," *The Guardian* (England), April 23, 1983, p. 19.
2 B. Newman, "Single-Country Unions of Europe Try to Cope with Multinationals," *The Wall Street Journal,* November 30, 1983, p. 1.
3 Guenter Friedrichs and Adam Schaff, editors, *Micro-Electronics and Society,* Club of Rome report, Mentor, New York, 1983, p. 24.
4 George Shultz, *Chronicle of International Communication,* Vol. IV, No. 8, October, 1983, p. 2.
5 Charles P. Lecht, "The A.T.&T. Strike: Automate or Die," *Computerworld,* August 22, 1983, p. 35.
6 Robert Lindsey, "Unions Press Drive to Enlist High-Technology Workers," *The New York Times,* May 25, 1983.
7 See, for example, the study of commercial television's view of labor in, "Television: Voice of Corporate America," *The Machinist,* 1981.
8 Friedrichs and Schaff, op. cit., pp. 32 and 152.
9 Ithiel de Sola Pool, *The Technologies of Freedom,* Cambridge, Harvard, 1983.
10 "Beyond 1984: The Societal Challenge of Information Technology," OECD Secretariat, Proposal for a Conference in 1984. *Document Service,* IIC, October 1983, Vol. III, No. 2.
11 David Burnham, "White House Scraps Plan to Share Census Figures," *The New York Times,* November 24, 1983.


# The Path of Progress

**C. E. Ayres**

Economic thinking has always embodied some conception of progress and must always do so; for the concept of value is the chief concern of economic thinking, and progress is indissociable from value. Agnosticism with regard to value implies agnosticism with regard to progress. It may be a gay agnosticism like that of the old American folk song, "We don't know where we're going, but we're on our way!" As Professor Walton Hamilton once pointed out, this refrain is a remarkably apt characterization of the state of mind into which some contemporary economists have got themselves. But gay or not, the state of mind which is described by this characteristically Hamiltonian irony is one of complete and stultifying agnosticism. Value may also be conceived to be known but unattainable, in which case progress also is unattainable. But if value is knowable and attainable, then progress also is knowable and attainable. If the technological process is the locus of value, the continuous development of the technological arts and crafts and the accompanying recession of superstition and ceremonially invested status is progress.

If the industrial revolution is itself the vehicle of progress, then Condorcet and the other optimists of the "age of reason" were not so far wrong as subsequent generations have believed. This does not mean that perfection is "just around the corner." But the authors of the idea of "infinite perfectability" really made no such rash promise. In attributing the disorders and violence of the times to bad institutions, Condorcet was speaking the language of Veblen and Dewey more than a century before them; and in declaring that we are now entering a period of "neo-technics," he was only anticipating Patrick Geddes and Lewis Mumford. The fact that we have not yet fully realized the possibilities of science and technology—possibilities of emancipation from the follies of the past and of attainment of an "economy of abundance"—is of secondary importance. The primary consideration is the fact that we do now realize these possibilities more clearly and more generally than ever before. The disorders of the present age are more widespread and more cataclysmic than those even in which Condorcet himself was "liquidated." But no one any longer believes that disorder and destruction are inevitable or necessary. The "demonstration" that increase of population necessarily and inevitably nullifies all the achievements of advancing technology, by which the Reverend T. R. Malthus, avowed spokesman of the landed gentry, undertook the final refutation of Condorcet's revolutionary optimism, was abandoned even by its author in the second and subsequent editions of his celebrated *Essay* and is now completely discredited. No one any longer doubts the physical and technological possibility of a worldwide economy of abundance.

Far more than in the time of Condorcet the twentieth century has accepted the machine. No serious student attributes the evils of the age to its machines. Popular essayists sometimes write as though tanks and airplanes were responsible for the bloodshed which is now going on, and novelists occasionally draw pictures of the horrors of a future in which life will have become wholly mechanized, with babies germinating in test tubes, "scientifically" maimed for the "more efficient" performance of industrial tasks. But this of course is literary nonsense, two kinds of nonsense. One kind portrays the devices of the future as horrible perversions, just as traveling in stagecoaches at the vertiginous speed of fifteen miles an hour was once thought to be. Extra-corporal gestation might well be a great improvement on nature, just as the extraction of the mammary secretion of the cow is a great improvement and one to which we have been able to reconcile our sentiments of decency, though it must have seemed a horrible perversion to the stalwart moralists of primitive society. As Mr. J. B. S. Haldane,[1] pointed out many years ago, all biological inventions seem disgusting at first. But this is nonsense. If science can reduce infant mortality by establishing an "unnatural" relation between a human baby and a lactating quadruped, then by all means let it be done. Such, happily, is now the prevailing attitude.

To represent schemes of mutilation as the teaching of science for the attainment of efficiency is nonsense of quite another kind; it simply is not true. Mutilation is neither scientific nor efficient. If we can credit science at all, we must know that any community in which any sort of mutilation is practised is a mutilated community. Modern industry demands the full powers of all its participants. Its development has all along been coincident with the expansion of the powers of a continually larger part of the community. Any deviation from this procedure is contrary to science and to industrial efficiency. It is said that the control of subject populations has recently been attempted by the withholding of certain vitamins from their diet; but no one has ever claimed that such a procedure enhances the efficiency of its victims, and no one who knows anything about science has ever seriously supposed that it is the discovery of the vitamins which has brought about such practises. After all, this is not the first time that victors have maimed the vanquished, as every good Bible reader knows.

There is nothing wrong with the machines. Nevertheless many people whose minds are entirely free from nonsensical aversions are still unable to think of progress in terms of the advancement of science and the arts, chiefly for this reason. The traditional conception of progress is that of movement toward the attainment of an "end." Within the limits of day to day activity finite and provisional ends are of course set up. Thus one may speak of progress toward the attainment of an academic degree. In a much more general but still limited sense one may even speak of the advancement of science as progress toward knowledge, or something of the sort. But the idea still persists that the attainment of such limited objectives constitutes "real" progress only insofar as these limited objectives contain some particularization of the universal "end."

This is also true of value which has likewise been traditionally conceived in terms of ultimate value, the summum bonum of the philosophers. Thus the difficulty with regard to "ends" is a major obstacle to a technological (or instrumental) understanding of the whole value-progress complex. It is frequently expressed in simple and direct language such as this. A machine is neither good nor bad in itself. The question is, what is it for? What does it do? What end does it serve? A machine (or instrumental technique) may serve desirable ends. It may save life or enrich personality. But a machine may also serve the ends of destruction and debasement. Machines are used in war, and scientific knowledgment may be employed in the commission of crime. How then can we speak of machines, or even of the arts and crafts and instrumental procedures as a whole, as being good in themselves, irrespective of the ends for which they are employed? How can we speak of the advancement of science and technology as progress except with reference to some conception of the end to the attainment of which all human efforts are directed?

It is by virtue of this way of thinking that consumption plays its unique role in economic theory. Consumption is the "end" for which all other economic activities are carried on, by definition. Textbook writers have fallen into the habit of explaining consumption to their readers as the process in which goods are "used up"; and this involves them in difficulties, since many things—such as diamonds, or even books—are not used up by their consumers, whereas many other things—such as fuel—are used up in processes otherwise identified as production. The truth is, the other meaning of this root, by which it is linked to "consummatory'' and ''consummation,'' is the only one by which it can be clearly distinguished from production and is in fact the meaning which its earlier users definitely intended to invoke, as any student can demonstrate for himself by substituting the word ''consummation'' for ''consumption'' wherever it appears. This is why no one ever undertook to prove that consumption is the ''end'' for which all the rest is carried on. The distinction of ''consumption'' from ''production'' is synonymous with the distinction of ''end'' from ''means.''

So deeply is this distinction embedded in the thinking of the community that even avowed revolutionaries have been unable to eradicate it. No other revolutionary slogan has been more widely used and none has made a more effective appeal than the formula, ''Production for use.'' To most people these words seem to appeal to simple common sense. Nevertheless they are in fact a transliteration into economic terminology of Kant's ''categorical imperative,'' and their appeal is to metaphysical tradition. In proposing that we should ''treat every man as an end and never as a means,'' Kant assumed ''man'' to be a spiritual entity. He did so on the basis of the immemorial tradition according to which it has been believed throughout the ages that every man has direct intuitive knowledge of himself as a spiritual entity. For all their anticlericalism it is to this essentially religious belief that modern revolutionaries appeal when they advocate ''production for use,'' and it is this belief alone which sustains the conviction that machines, economic processes, and human life itself can have significance only in terms of the ''end'' to which all else is a ''means.''

What is the evidence by which man knows himself ''intuitively'' to be a ''mind'' or ''spirit''? It is ''intuitive'' in the sense that this is ''inner'' knowledge, ''inner'' in the sense that it is not based on the evidence of the senses. The ''knowledge'' of primitive man was derived from the evidence of dreams, the departure of ''life'' with a dying gasp, and the like. But for all these phenomena modern science has other explanations, explanations which cover not only the actual phenomena of dreams, respiration, and the like but also the social processes of legend-creation and transmission by virtue of which these phenomena have been so persistently misconceived, with the result that no evidence remains; and in destroying the last remaining vestige of supposed evidence of direct, intuitive, inner, self-knowledge of spiritual ''reality'' modern science has precipitated an intellectual revolution far more momentous than the one effected by Copernicus.

For what is at issue now is the ''common sense'' of the community. Copernican astronomy and Newtonian physics claimed the whole physical universe as the domain of science; but through the efforts of Descartes and his successors, of whom Kant was perhaps the greatest, an armistice was arranged between science and metaphysics. A boundary was established between the ''outer'' world of science and the ''inner'' world of metaphysics. According to the terms of this armistice the validity of the findings of science was conceded, subject only to this reservation. Such an arrangement was of course extremely favorable to science. Not only did it bring an end to the long struggle in which scientists had been engaged, permitting them to explore the moons of Jupiter and even the organs of the human body without further opposition; it also permitted scientists to be scientists and still be men, retaining with regard to the "inner" and "real" world the beliefs with which they no less than all their neighbors had been indoctrinated "at their mothers' knees."

The relief was more than personal. Many a troublesome problem could be solved by judicious application of the Cartesian compromise. Thus it was that classical political economy solved the troublesome problems of value and progress. Price is a physical phenomenon, a feature of the "outer" world, and therefore subject to scientific analysis. But the valuations which this mechanism of the market assembles and summarizes are the private experiences of individual souls and are therefore real and valid within the purview of Cartesian and Kantian metaphysics. The mechanism of production and the pecuniary organization of society is the "means" to which the satisfaction of the inner aspirations (wants) of mankind is the consummatory "end." In theory these two worlds are linked by price which is both a physical mechanism and a register of spiritual experience.

This happy compromise was upset by the Darwinian revolution. It was of course science which violated the terms of the Cartesian armistice, and not in the field of biology alone. The demonstration of the continuity of the human species with all other species was of climactic importance, but archaeological evidence of the continuity of present civilization with extreme antiquity, increasing knowledge of comparative cultures, analysis of social mechanisms in terms of "collective representations," "folkways," and "mores," greatly increased knowledge of the physical mechanisms of behavior and of the process by which behavior patterns are formed in individual and social experience, all contributed to the elimination of the last frontier between knowledge and belief. As a result of all these developments science no longer respects the frontier by which the universe was once thought to be divided into "outer" and "inner" worlds, and no longer credits the supposed "immediate" knowledge of "inner" spiritual reality or recognizes the so-called "individual" wants and satisfactions as having any unique validity or as being in any sense "consummatory."

The disrepute into which the idea of progress has fallen in recent years is a further consequence of the collapse of metaphysical dualism and a phase of the general moral nihilism of the times. As such it is historically explicable. Just as the identification of the mores, the recognition of the traditional character of the "eternal verities," has given rise to the assumption that there are no verities, so the nullification of the "inner" world of consummatory spiritual experience has given rise to the assumption that consummation is meaningless; and since progress itself is supposedly meaningless except in terms of such attainment, the idea of progress itself has fallen into disrepute.

But however explicable, this situation is a paradox. It is the validity of science which has supposedly destroyed the values of the modern world, and it is the progress of science which has rendered the idea of progress itself supposedly untenable. Clearly there is more here than meets the eye. Why do we say that machines must be "for" use? The meaning "use" is implicit in the meaning "machine." We know that every paradise is a projection of some community's actual social arrangements into infinity. For South Sea Island dwellers it is the

Ultimate Atoll, for Eskimoes the Infinite Snowbank, in each case ruled by the Perfect Chief, and so on. Such projections, we know, are without validity. Yet we still insist that progress must be conceived in this way or not at all. Why? What principle of logic, or of common sense, presents our thinking with this absolute disjunction: either progress must be traditionally conceived and therefore without general validity, or it cannot be conceived at all? Such a disjunction can be sustained by definition. We can agree to limit the use of the word "progress" to "progress-as-it-has-been-traditionally-conceived," and by doing so we can assert with confidence that progress-so-defined can be conceived only as-progress-has-been-traditionally-conceived. But this is only a restatement of the initial agreement. The question still remains, Why should we subject our thinking to such limits in the first place? Doubtless it would only add to the confusion if we were to agree to throw the meaning of the word "progress" wide open by making it synonymous with "change." On the basis of such a preconceived definition we might then declare that a chemical reaction is progress; but that would certainly not increase our understanding of social development. Surely there is some meaning which all the "collective representations" of human societies have had in common. What is it? What have they all been trying to do?

All human behavior exhibits a certain continuity of a technological, instrumental, or cause-and-effect character. It is with reference to these observed and instrumentally "controlled" continuities that we use such terms as "value" and "progress" in common speech. In speaking of his "progress" down the page a writer is thinking in terms of the instrumental continuity of each written line with the line which precedes and the line which is to follow it. Such continuities are clearly more significant the further they extend. Progress "toward" the "completion" of an essay is an extension of this character. Here also what the mind is grappling with is not a preconception of the finished essay but a continuity which exists in any given sentence or paragraph and extends to the paragraph, the sentence, and the final word to which this continuity extends. Meanings such as this are capable of a considerable degree of extension without confusion. Thus we speak quite easily of "the progress of science." It is the paradox of our present state of mind that in spite of the disrepute into which the whole conception of progress has fallen we do actually continue to employ such phrases as this quite without embarrassment. When a scientist speaks of the progress of science other scientists do not leap up to reproach him with having uttered nonsense, for the phrase "the progress of science" is not nonsense. Neither does it depend for its meaning on any preconceived idea of what "the total realization of all scientific knowledge" might be. The meaning to which such a phrase refers is not that of a quantity of knowledge—not a finite quantity any more than infinity. It is that of a process which is now going on and which may quite reasonably be conceived as continuing.

It is this meaning of process-continuity which has given rise to the conception of progress as a metaphysical projection. In the effort to extend our understanding of the continuities in which we are engaged we have inevitably raised even such extensive continuities as that of science to a larger scale. The question then becomes, in what fashion is science continuous with human activity generally?

At this point, however, the imagination of mankind is liable to that peculiar sort of stimulation which we have recently identified as "ceremonial." We become excited, and we begin to think in capital letters. The everyday thinking which has sufficed for an understanding of common continuities now gives way to our inveterate propensity for myth-making; group loyalties become obsessive; and so we find ourselves insisting that the progress of science is but a "means" to the far more sublime "end" which is the eventual triumph of the Republican Party, or something of the sort. Does this mean that human behavior is wholly without significance? Or does it mean that our problem is one of decontamination?

Is there no point of which we can say, "This is the point at which we went astray. Up to this point our thinking was sound; beyond this point it was unsound; and consequently it is to this point that we must return and renew the attempt to carry on from here by the same sound methods which had been employed hitherto?" Those who declare that the concept of progress "must" have reference to metaphysical ultimates, that metaphysical ultimates are without significance, and therefore that the concept of progress is itself without significance, seem to deny the existence of any such point. In doing so they seem to be making the same mistake into which we have been misled by the principle of "mores," that of asserting that all judgments are conventional observances and nothing more. Said the Cretan, all Cretans are liars. Since the effort to extend our understanding of the continuities of human behavior has resulted in metaphysical fatuities, they seem to say, all intellectual efforts must be of this character.

It is the progress of science which belies this judgment, and it does so not only by example but by precept. Not only is the progress of science and technology itself a significant reality; its inevitable extension to the study of human behavior has given us the means of distinguishing between technological and ceremonial activities. This is the point at which scientific generalization is securely tied to the everyday judgments of which common existence is composed. Speaking of the progress of science, for example, we can say with certainty that it is continuous with the technological practices in which men have engaged as far back as our knowledge goes, as it is also continuous with all present tool-using activities of the commonest and humblest sort. It is also continuous with all the "creative" activities which we designate as the arts.

This total activity, as we know, has undergone progressive development throughout human experience. All that we can now do is done by virtue of that progressive development. Progress is the continuation of this process. We speak with certainty of the progress of aviation, meaning that better planes are built now than formerly—better in the sense of larger, faster, stronger, lighter per horsepower, and so forth. This judgment is valid quite without reference to the "ends" for which planes may be used. The fact that some people are using planes to kill other people is quite as irrelevant as it would be for a hardware merchant to inquire whether a hammer is to be used to bash in someone's skull before venturing an opinion which is the better of two hammers. In the same sense the judgment that the progress of aviation is a part of general progress is a valid judgment. The continuity it asserts is between plane-building and building in general. Since the building of better planes is in fact contingent upon and contributory to better building generally, it is part of a general process, co-extensive with human existence, by virtue of which the human race has risen above the brutes and gives every indication of rising far higher than anyone can now foresee.

The fact of war is by no means irrelevant to this judgment. We sometimes hear it said that the only result of the invention, for example, of airplanes is that people are killing each other on a larger scale than ever before. If such a proposition were true, it would indeed nullify the technological conception of progress; for if people are indeed being killed on a larger scale than ever before, this circumstance must eventually operate to the disadvantage of further airplane building and of technological development generally. But is it true? To say that killing is the "only" result of the technical development of the airplane is patently false, but this is perhaps a rhetorical exaggeration. The essential question is whether advancing technology creates disorder, and whether the disorders so created are in fact increasing by a cumulative process such as might be conceived to nullify the progress of the arts and sciences.

There is a sense in which technological development might be said to give rise to disorder. It has been recognized all along that technological development alters the physical habitat of a community in such a way that a shift in the institutional balance of power becomes inevitable. This shift may well be accompanied by disorder. In this sense the perfection of the airplane may be said to have brought on the present war; since, if the supposed supremacy of the French army and the British navy had not been a technological illusion, doubtless the present war would not have occurred. Does this mean that German (and Italian and Japanese) aggression had no part in bringing on the conflict? To say so would be equivalent to attributing the increase of kidnapping in recent decades solely to the development of the automobile without any reference to pre-existing organized crime (especially of the prohibition era) or to police corruption and inefficiency, the confusion of legal jurisdictions from which law enforcement has always suffered in America, etc., etc. Doubtless it was the development of fast automobiles and motor highways which gave to crime this particular direction, and doubtless it was a change in the technology of war which gave international conflict this particular direction; but the forces of conflict are in every case institutional.

Even so, the question still remains whether conflict and disorder are in fact becoming more general and catastrophic. If they are, progress is nullified irrespective of the distinction between causes and directions. But on this point the evidence is conclusive. Current pessimism to the contrary notwithstanding, population has increased tremendously throughout modern times. To be sure, this is no positive guarantee that it will continue to do so throughout the indefinite future, but neither is there any conclusive evidence that it will cease to do so. If the present disorders were unique, the situation would be rather more terrifying than it is. The very fact that they are not unique suggests that we must judge future probabilities in terms of an experience in which disorders such as the present ones have nevertheless been accompanied by continuing increase of population. It has been said that wars have been increasing in frequency throughout modern times, but in that case they must have been decreasing in violence—appearances to the contrary notwithstanding—since throughout the same period population has unquestionably increased. If later wars had brought the same devastation throughout the areas involved which the Hundred Years' War and the Thirty Years' War brought to the areas most seriously affected, the case would be quite different. But such is not the case. To recognize these facts is not to condone war, nor even to accept it as "inevitable." The only question at issue is whether the current evidence shows that disorders are in fact increasing catastrophically; and the answer is that the evidence shows nothing of the sort—or rather, just the contrary.

What the evidence shows is that humbug, cruelty, and squalor have been decreasing for the population as a whole throughout modern times as they have been decreasing throughout the history of the race. No one seriously advocates turning back the clock to the day when Plato dispensed sweet wisdom to a few disciples while all the rest of the world lived in fear of evil spirits, or to the day when theology was most angelic and the clergy lived in open concubinage, lords enjoyed first night rights with every bride, and no man was safe from violent molestation or from smallpox, typhus, and starvation. In spite of all sentimentality and all the intellectual scruples of scientific caution, we are all committed by the whole continuous series of everyday judgments and activities to carrying on those achievements of tool and instrument, hand and brain, the genuineness of which no one really doubts.

It is from the pattern of this continuing activity that the idea of progress derives its meaning. Nevertheless this meaning can be projected into the future. If the progressive advance of technology means a similarly cumulative diminution of the extent and importance in the affairs of the community of superstition and ceremonial investiture, then the projection of this process into the infinitely remote future would seem to reveal an "ultimate" condition of complete enlightenment and efficiency wholly devoid of mystic potencies. Such a state of affairs is perhaps difficult to imagine, and yet these phrases have a familiar sound. This would be in effect a classless society, one in which as a consequence of the withering away of the state (that is, the whole institutional scheme of rank and privilege) all prerogatives of status would have disappeared. It would be a society in which men and women would go about their concerns with the simple innocence of little children, one in which the lion and the lamb would lie down together in common amity.

These are poetic expressions. They lack the precision and detail of scientific formulas. What they express is perhaps vision rather than analysis. Nevertheless, as scholars have often remarked, the visions of the great spiritual leaders, the visions by which mankind has been most profoundly moved, exhibit striking similarities. It has often been remarked that the teachings of Jesus and Buddha were both characterized by a gentleness, an abhorrence of every manifestation of coercion, which is more than a mere quality of temperament. For both, the injunction to turn the other cheek is accompanied by an equally fundamental abhorrence of phariseeism, of the mores of conformity, and of the institutionalization of human behavior. These ideas, or attitudes, are also found in the teachings of lesser men such as Marx and even Condorcet. Perhaps it is impious to couple the name of Condorcet with that of Gautama Buddha, but Condorcet's aversion to phariseeism and his conviction that emancipation comes only by enlightenment are singularly reminiscent of the teachings of Buddha. Scholars are still uncertain as to what "nirvana" meant to Buddha himself (as distinguished from the institutionalization of Buddhism in later centuries), and therefore we may perhaps be allowed to conjecture that the "nothingness" by the attainment of which man was to free himself from spiritual slavery was less metaphysical and more sociological than the priestcraft of organized Buddhism has supposed and was not altogether unrelated to the Marxian nothingness of the classless society which follows the withering away of the state. It is also worthy of remark that all these seers viewed the use of tools, the ordinary act of the common artisan, as a function of the profoundest import. The fact that Voltaire closed *Candide* by retiring to cultivate his garden means more than a mere shrug of ironic shoulders; it imputes a reality to the act of cultivation which is absent from the institutionalized humbug of the world of affairs. We must not overinterpret these poetical expressions. Certainly we must not impute to the teachers of the past—in some cases of many centuries past—all the analytical clarity which our generation owes to the sum of the scientific achievements of the race. But perhaps the difference is more one of terminology than of substance. Perhaps the knowledge we have attained by laborious analysis may be essentially the same as the insights of poetic vision, the vision of a world in which enlightenment would have replaced superstition, and efficiently organized teamwork institutional coercion.

But even such a vision is a projection of the current process into the indefinite future, not an independently conceived "end" by which present process is to be judged and guided. What it represents is insight into the current realities of human life. It is these current realities of which the vision is a poetical expression and from which it derives its meaning, not the other way about. In this sense perfection may be conceived to have an operational meaning like the mathematical concept of infinity. Doubtless mankind will achieve perfection only at infinity. Doubtless technological progress is an asymptotic function. There is no finite moment in the past at which human behavior is known to have been wholly ceremonial. As far back as our knowledge goes rudimentary tool-activities have been going on; and our knowledge of the present situation does not encourage any expectation of the total disappearance of superstition, status, and institutional coercion within the foreseeable future. This does not mean that our interpretation of current process as one of progressive enlightenment and efficiency is incorrect. It means that the reality of progress is implicit in the finite process of which visions of infinity are a projection, just as mathematical infinity is a projection of finite series.

Within the limits of current process it is true that mankind needs superstition and coercion. This fact is often cited as the climactic nullification of the "illusion" of progress. But such an interpretation is an expression of the metaphysical misconception of the idea of progress. To whatever degree superstition and institutionalized status may prevail at any given time, the habituation of the race to those forms of behavior does constitute a need, just as a cripple needs a crutch. But the fact that a person is habituated to the use of crutches does not establish that crutches are good in themselves or that the attainment of crutchless health is a fatuous illusion. It also means that needs conceived in weakness are not a sound criterion of possible achievement, for individuals or for societies. The supposition that the prevalence of institutionalized humbug and coercion at any given time proves the impossibility of progress is a special case of the paradox of Zeno. It was precisely by this method that Zeno was supposed to have "proved" that a moving object does not move, since at any given moment it is at a given point. This fact, as we have long since assured ourselves, does not prevent an object from passing through an infinite series of points during an infinite series of moments; and in the same sense the deplorable conditions which prevail in any community at any given time do not constitute a proof that such conditions must continue to prevail. Doubtless the immediate future will be not wholly different from the immediate past; but the fact that a given difference is infinitesimal does not mean that it is not profoundly significant.

The changes which have accompanied industrial revolution have been felt to be significant by the whole community throughout modern times. It is this judgment which has given rise to the idea of progress, an idea which is one of the most characteristic features of modern western civilization. The idea has of course been institutionalized. When dynastic power was paramount, that was the force to which the progress of opulence was prospectively attributed. When money power superseded dynasties, the attribution was to "Capital the Creator." Throughout both these periods the nature of the process was but dimly apprehended. It is much clearer now. But the identification of technological process and its dissociation from institutional obsessions has been at the expense of the idea of progress. What we now have to do is to de-institutionalize that idea itself—to recognize as a misconception the idea of progress as movement toward the attainment of some previsioned "end," and to reconstitute the criterion of progress in terms of the continuity of technological development. If we can do this—if we can now see that the path of progress is the advancement of the arts and sciences, tools, instruments, and the machine process, and not the apotheosis of any legendary power-system—we shall have consummated the revolution to which the great Copernican revolution was a mere preliminary skirmish.

## NOTES

1 Daedalus (1924), p. 44.

# Technical Progress and the Social Life-World

**Jürgen Habermas**

When C. P. Snow published *The Two Cultures* in 1959, he initiated a discussion of the relation of science and literature which has been going on in other countries as well as in England. Science in this connection has meant the strictly empirical sciences, while literature has been taken more broadly to include methods of interpretation in the cultural sciences. The treatise with which Aldous Huxley entered the controversy, however, *Literature and Science,* does limit itself to confronting the natural sciences with the belles lettres.

Huxley distinguishes the two cultures primarily according to the specific experiences with which they deal: literature makes statements mainly about private experiences, the sciences about intersubjectively accessible experiences. The latter can be expressed in a formalized language, which can be made universally valid by means of general definitions. In contrast, the language of literature must verbalize what is in principle unrepeatable and must generate an intersubjectivity of mutual understanding in each concrete case. But this distinction between private and public experience allows only a first approximation to the problem. The element of ineffability that literary expression must overcome derives less from a private experience encased in subjectivity than from the constitution of these experiences within the horizon of a life-historical environment. The events whose connection is the object of the lawlike hypotheses of the sciences can be described in a spatio-temporal coordinate system, but they do not make up a world:

> The world with which literature deals is the world in which human beings are born and live and finally die; the world in which they love and hate, in which they experience triumph and humiliation, hope and despair; the world of sufferings and enjoyments, of madness and common sense, of silliness, cunning and wisdom; the world of social pressures and individual impulses, of reason against passion, of instincts and conventions, of shared language and unsharable feelings and sensations.... [1]

In contrast, science does not concern itself with the contents of a life-world of this sort, which is culture-bound, ego-centered, and pre-interpreted in the ordinary language of social groups and socialized individuals:

> ...As a professional chemist, say, a professional physicist or physiologist, [the scientist] is the inhabitant of a radically different universe—not the universe of given appearances, but the world of inferred fine structures, not the experienced world of unique events and diverse qualities, but the world of quantified regularities.[2]

Huxley juxtaposes the *social life-world* and the *worldless universe of facts.* He also sees precisely the way in which the sciences transpose their information about this worldless universe into the life-world of social groups:

> Knowledge is power and, by a seeming paradox, it is through their knowledge of what happens in this unexperienced world of abstractions and inferences that scientists have acquired their enormous and growing power to control, direct, and modify the world of manifold appearances in which human beings are privileged and condemned to live.[3]

But Huxley does not take up the question of the relation of the two cultures at this juncture, where the sciences enter the social life-world through the technical exploitation of their information. Instead he postulates an immediate relation. Literature should assimilate scientific statements as such, so that science can take on "flesh and blood."

> ...Until some great artist comes along and tells us what to do, we shall not know how the muddled words of the tribe and the too precise words of the textbooks should be poetically purified, so as to make them capable of harmonizing our private and

unsharable experiences with the scientific hypotheses in terms of which they are explained.[4]

This postulate is based, I think, on a misunderstanding. Information provided by the strictly empirical sciences can be incorporated in the social life-world only through its technical utilization, as technological knowledge, serving the expansion of our power of technical control. Thus, such information is not on the same level as the action-orienting self-understanding of social groups. Hence, without mediation, the information content of the sciences cannot be relevant to that part of practical knowledge which gains expression in literature. It can only attain significance through the detour marked by the practical results of technical progress. Taken for itself, knowledge of atomic physics remains without consequence for the interpretation of our life-world, and to this extent the cleavage between the two cultures is inevitable. Only when with the aid of physical theories we can carry out nuclear fission, only when information is exploited for the development of productive or destructive forces, can its revolutionary practical results penetrate the literary consciousness of the life-world: poems arise from consideration of Hiroshima and not from the elaboration of hypotheses about the transformation of mass into energy.

The idea of an atomic poetry that would elaborate on hypotheses follows from false premises. In fact, the problematic relation of literature and science is only one segment of a much broader problem: *How is it possible to translate technically exploitable knowledge into the practical consciousness of a social life-world?* This question obviously sets a new task, not only or even primarily for literature. The skewed relation of the two cultures is so disquieting only because, in the seeming conflict between the two competing cultural traditions, a true life-problem of scientific civilization becomes apparent: namely, how can the relation between technical progress and the social life-world, which today is still clothed in a primitive, traditional, and unchosen form, be reflected upon and brought under the control of rational discussion?

To a certain extent practical questions of government, strategy, and administration had to be dealt with through the application of technical knowledge even at an earlier period. Yet today's problem of transposing technical knowledge into practical consciousness has changed not merely its order of magnitude. The mass of technical knowledge is no longer restricted to pragmatically acquired techniques of the classical crafts. It has taken the form of scientific information that can be exploited for technology. On the other hand, behavior-controlling traditions no longer naively define the self-understanding of modern societies. Historicism has broken the natural-traditional validity of action-orienting value systems. Today, the self-understanding of social groups and their worldview as articulated in ordinary language is mediated by the hermeneutic appropriation of traditions as traditions. In this situation questions of life conduct demand a rational discussion that is not focused exclusively either on technical means or on the application of traditional behavioral norms. The reflection that is required extends beyond the production of technical knowledge and the hermeneutical clarification of traditions to the employment of technical means in historical situations whose objective conditions (potentials, institutions, interests) have to be interpreted anew each time in the framework of a self-understanding determined by tradition.

This problem-complex has only entered consciousness within the last two or three generations. In the nineteenth century one could still maintain that the sciences entered the conduct of life through two separate channels: through the technical exploitation of scientific information and through the processes of individual education and culture during academic study. Indeed, in the German university system, which goes back to Humboldt's reform, we still maintain the fiction that the sciences develop their action-orienting power through educational processes within the life history of the individual student. I should like to show that the intention designated by Fichte as a "transformation of knowledge into works" can no longer be carried out in the private sphere of education, but rather can be realized only on the politically relevant level at which technically exploitable knowledge is translatable into the context of our life-world. Though literature participates in this, it is primarily a problem of the sciences themselves.

At the beginning of the nineteenth century, in Humboldt's time, it was still impossible, looking at Germany, to conceive of the scientific transformation of social life. Thus, the university reformers did not have to break seriously with the tradition of practical philosophy. Despite the profound ramifications of revolutions in the political order, the structures of the preindustrial work world persisted, permitting for the last time, as it were, the classical view of the relation of theory to practice. In this tradition, the technical capabilities employed in the sphere of social labor are not capable of immediate direction by theory. They must be pragmatically practiced according to traditional patterns of skill. Theory, which is concerned with the immutable essence of things beyond the mutable region of human affairs, can obtain practical validity only by molding the manner of life of men engaged in theory. Understanding the cosmos as a whole yields norms of individual human behavior, and it is through the actions of the philosophically educated that theory assumes a positive form. This was the only relation of theory to practice incorporated in the traditional idea of university education. Even where Schelling attempts to provide the physician's practice with a scientific basis in natural philosophy, the medical *craft* is unexpectedly transformed into a medical *praxiology*. The physician must orient himself to Ideas derived from natural philosophy in the same way that the subject of moral action orients itself through the Ideas of practical reason.

Since then it has become common knowledge that the scientific transformation of medicine succeeds only to the extent that the pragmatic doctrine of the medical art can be transformed into the control of isolated natural processes, checked by scientific method. The same holds for other areas of social labor. Whether it is a matter of rationalizing the production of goods, management and administration, construction of machine tools, roads, or airplanes, or the manipulation of electoral, consumer, or leisure-time behavior, the professional practice in question will always have to assume the form of technical control of objectified processes.

In the early nineteenth century, the maxim that scientific knowledge is a source of culture required a strict separation between the university and the technical school because the preindustrial forms of professional practice were impervious to theoretical guidance. Today, research processes are coupled with technical conversion and economic exploitation, and production and administration in the industrial system of labor generate feedback for science. The application of science in technology and the feedback of technical progress to research have become the substance of the world of work. In these circumstances, unyielding opposition to the decomposition of the university into specialized schools can no longer invoke the old argument. Today, the reason given for delimiting study on the university model from the professional sphere is not that the latter is still foreign to science, but conversely, that science—to the very extent that it has penetrated professional practice—has estranged itself from humanistic culture. The philosophical conviction of German idealism that scientific knowledge is a source of culture no longer holds for the strictly empirical scientist. It was once possible for theory, via humanistic culture, to become a practical force. Today, theories can become technical power while remaining unpractical, that is, without being expressly oriented to the interaction of a community of human beings. Of course, the sciences now transmit a specific capacity: but the capacity for control, which they teach, is not the same capacity for life and action that was to be expected of the scientifically educated and cultivated.

The cultured possessed orientation in action. Their culture was universal only in the sense of the universality of a culture-bound horizon of a world in which scientific experiences could be interpreted and turned into practical abilities, namely, into a reflected consciousness of the practically necessary. The only type of experience which is admitted as scientific today according to positivistic criteria is not capable of this transposition into practice. The capacity for *control* made possible by the empirical sciences is not to be confused with the capacity for *enlightened action*. But is science, therefore, completely discharged of this task of action-orientation, or does the question of academic education in the framework of a civilization transformed by scientific means arise again today as a problem of the sciences themselves?

First, production processes were revolutionized by scientific methods. Then expectations of technically correct functioning were also transferred to those areas of society that had become independent in the course of the industrialization of labor and thus supported planned organization. The power of technical control over nature made possible by science is extended today directly to society: for every isolatable social system, for every cultural area that has become a separate, closed system whose relations can be analyzed immanently in terms of presupposed system goals, a new discipline emerges in the social sciences. In the same measure, however, the problems of technical control solved by science are transformed into life problems. For the scientific control of natural and social processes—in a word, technology—does not release men from action. Just as before, conflicts must be decided, interests realized, interpretations found—through both action and transaction structured by ordinary language. Today, however, these practical problems are themselves in large measure determined by the system of our technical achievements.

But if technology proceeds from science, and I mean the technique of influencing human behavior no less than that of dominating nature, then the assimilation of this technology into the practical life-world, bringing the technical control of particular areas within the reaches of the communication of acting men, really requires scientific reflection. The prescientific horizon of experience becomes infantile when it naively incorporates contact with the products of the most intensive rationality.

Culture and education can then no longer indeed be restricted to the ethical dimension of personal attitude. Instead, in the political dimension at issue, the theoretical guidance of action must proceed from a scientifically explicated understanding of the world.

The relation of technical progress and social life-world and the translation of scientific information into practical consciousness is not an affair of private cultivation.

I should like to reformulate this problem with reference to political decision-making. In what follows we shall understand "technology" to mean scientifically rationalized control of objectified processes. It refers to the system in which research and technology are coupled with feedback from the economy and administration. We shall understand "democracy" to mean the institutionally secured forms of general and public communication that deal with the practical question of how men can and want to live under the objective conditions of their ever-expanding power of control. Our problem can then be stated as one of the relation of technology and democracy: how can the power of technical control be brought within the range of the consensus of acting and transacting citizens?

I should like first to discuss two antithetical answers. The first, stated in rough outline, is that of Marxian theory. Marx criticizes the system of capitalist production as a power that has taken on its own life in opposition to the interests of productive freedom, of the producers. Through the private form of appropriating socially produced goods, the technical process of producing use values falls under the alien law of an economic process that produces exchange values. Once we trace this self-regulating character of the accumulation of capital back to its origins in private property in the means of production, it becomes possible for mankind to comprehend economic compulsion as an alienated result of its own free productive activity and then abolish it. Finally, the reproduction of social life can be rationally planned as a process of producing use values; society places this process under its technical control. The latter is exercised democratically in accordance with the will and insight of the associated individuals. Here Marx equates the practical insight of a political public with successful technical control. Meanwhile we have learned that even a well-functioning planning bureaucracy with scientific control of the production of goods and services is not a sufficient condition for realizing the associated material and intellectual productive forces in the interest of the enjoyment and freedom of an emancipated society. For

Marx did not reckon with the possible emergence at every level of a discrepancy between scientific control of the material conditions of life and a democratic decision-making process. This is the philosophical reason why socialists never anticipated the authoritarian welfare state, where social wealth is relatively guaranteed while political freedom is excluded.

Even if technical control of physical and social conditions for preserving life and making it less burdensome had attained the level that Marx expected would characterize a communist stage of development, it does not follow that they would be linked automatically with social emancipation of the sort intended by the thinkers of the Enlightenment in the eighteenth century and the Young Hegelians in the nineteenth. For the techniques with which the development of a highly industrialized society could be brought under control can no longer be interpreted according to an instrumental model, as though appropriate means were being organized for the realization of goals that are either presupposed without discussion or clarified through communication.

Hans Freyer and Helmut Schelsky have outlined a counter-model which recognizes technology as an independent force. In contrast to the primitive state of technical development, the relation of the organization of means to given or pre-established goals today seems to have been reversed. The process of research and technology—which obeys immanent laws—precipitates in an unplanned fashion new methods for which we then have to find purposeful application. Through progress that has become automatic, Freyer argues, abstract potential continually accrues to us in renewed thrusts. Subsequently, both life interests and fantasy that generates meaning have to take this potential in hand and expend it on concrete goals. Schelsky refines and simplifies this thesis to the point of asserting that technical progress produces not only unforeseen methods but the unplanned goals and applications themselves: technical potentialities command their own practical realization. In particular, he puts forth this thesis with regard to the highly complicated objective exigencies that in political situations allegedly prescribed solutions without alternatives.

> Political norms and laws are replaced by objective exigencies of scientific-technical civilization, which are not posited as political decisions and cannot be understood as norms of conviction or weltanschauung. Hence, the idea of democracy loses its classical substance, so to speak. In place of the political will of the people emerges an objective exigency, which man himself produces as science and labor.

In the face of research, technology, the economy, and administration—integrated as a system that has become autonomous—the question prompted by the neohumanistic ideal of culture, namely, how can society possibly exercise sovereignty over the technical conditions of life and integrate them into the practice of the life-world, seems hopelessly obsolete. In the technical state such ideas are suited at best for "the manipulation of motives to help bring about what must happen anyway from the point of view of objective necessity."

It is clear that this thesis of the autonomous character of technical development is not correct. The pace and *direction* of technical development today depend to a great extent on public investments: in the United States the defense and space administrations are the largest sources of research contracts. I suspect that the situation is similar in the Soviet Union. The assertion that politically consequential decisions are reduced to carrying out the immanent exigencies of disposable techniques and that therefore they can no longer be made the theme of practical considerations, serves in the end merely to conceal preexisting, unreflected social interests and prescientific decisions. As little as we can accept the optimistic convergence of technology and democracy, the pessimistic assertion that technology excludes democracy is just as untenable.

These two answers to the question of how the force of technical control can be made subject to the consensus of acting and transacting citizens are inadequate. Neither of them can deal appropriately with the problem with which we are objectively confronted in the West and East, namely, how we can actually bring under control the preexisting, unplanned relations of technical progress and the social life-world. The tensions between productive forces and social intentions that Marx diagnosed and whose explosive character has intensified in an unforeseen manner in the age of thermonuclear weapons are the consequence of an ironic relation of theory to practice. The direction of technical progress is still largely determined today by social interests that arise autochthonously out of the compulsion of the reproduction of social life without being reflected upon and confronted with the declared political self-understanding of social groups. In consequence, new technical capacities erupt without preparation into existing forms of life-activity and conduct. New potentials for expanded power of technical control make obvious the disproportion between the results of the most organized rationality and unreflected goals, rigidified value systems, and obsolete ideologies.

Today, in the industrially most advanced systems, an energetic attempt must be made consciously to take in hand the mediation between technical progress and the conduct of life in the major industrial societies, a mediation that has previously taken place without direction, as a mere continuation of natural history. This is not the place to discuss the social, economic, and political conditions on which a long-term central research policy would have to depend. It is not enough for a social system to fulfill the conditions of technical rationality. Even if the cybernetic dream of a virtually instinctive self-stabilization could be realized, the value system would have contracted in the meantime to a set of rules for the maximization of power and comfort; it would be equivalent to the biological base value of survival at any cost, that is, ultrastability. Through the unplanned sociocultural consequences of technological progress, the human species has challenged itself to learn not merely to affect its social destiny, but to control it. This challenge of technology cannot be met with technology alone. It is rather a question of setting into motion a politically effective discussion that rationally brings the social potential constituted by technical knowledge and ability into a defined and controlled relation to our practical knowledge and will. On the one hand, such discussion could enlighten those who act politically about the tradition-bound self-understanding of their interests in relation to what is technically possible and feasible. On the other hand, they would be able to judge practically, in the light of their now articulated and newly interpreted needs, the direction and the extent to which they want to develop technical knowledge for the future.

This *dialectic of potential and will* takes place today without reflection in accordance with interests for which public justification is neither demanded nor permitted. Only if we could elaborate this dialectic with political consciousness could we succeed in directing the mediation of technical progress and the conduct of social life, which until now has occurred as an extension of natural history; its conditions being left outside the framework of discussion and planning. The fact that this is a matter for reflection means that it does not belong to the professional competence of specialists. The substance of domination is not dissolved by the power of technical control. To the contrary, the former can simply hide behind the latter. The irrationality of domination, which today has become a collective peril to life, could be mastered only by the development of a political decision-making process tied to the principle of general discussion free from domination. Our only hope for the rationalization of the power structure lies in conditions that favor political power for thought developing through dialogue. The redeeming power of reflection cannot be supplanted by the extension of technically exploitable knowledge.

### NOTES

1 Aldous Huxley, *Literature and Science,* New York, 1963, p. 8.
2 *Ibid.*
3 *Ibid.,* p. 9.
4 *Ibid.,* p. 107.

## The Ethics of Order and Change

**Godfrey Gunatilleke**

### THE ETHICS OF STRUCTURAL CHANGE

For the present, however, we need to go beyond the generic elements common to the ethical concern of all societies and ask the question with which we began this discussion: How are the ethical dilemmas of managing a society within an order-oriented ethos different from those of managing the rapid process of socioeconomic change that is the primary characteristic of development? Can rapid change be singled out as a distinguishing attribute of the current phase of human history in that part of the world we describe as *developing?* A simple affirmative answer represents a sweeping generalization that ignores the complexity of historical reality.

First, fundamental ideological changes have taken place in different phases of human history, even during the preindustrial era. These—like Christianity, Islam, and to a lesser extent Buddhism—often went together with far-reaching changes in the entire structure of social relations, replacing existing value systems with radically new ones. Christianity, for example, provided the ethical framework for the major social transitions from a tribal slave society to a feudal system and, in the period of the Reformation, from feudalism to capitalism. Without debating the relative roles of ideology and the productive forces in these major social transformations, we can safely conclude that the goals of structural change were an essential part of these ideological movements. After these changes were accomplished, however, the ideology moved away from its reforming role and became the major stabilizing agent for the new order. Its main task then became the achievement of a new stasis. We know too little of the inner processes that governed these transitions, how they were managed, how the dilemmas of conflicting value systems were resolved, and how the old and the new were reconciled to make any valid generalizations. We do know that the transitions were periods of great conflict, violence, and discontinuity, and that the human cost in the form of liquidation of social groups, religious persecution, and destruction of past achievements was enormous indeed.

Second, the social ethos of traditional societies was not necessarily order oriented. They were not exclusively devoted to the stabilization of existing social structures, totally bypassing the need for structural change. Within these traditions there were elements that struggled for systemic changes that would result in a spiritual transformation of the terrestrial order. Many traditions fostered the messianic hope, the search for the kingdom of heaven on earth, the longing for the rule of the *dharma-chakravarti.* These religious and social impulses moved toward the ideal of a fundamental reordering of the life of the individual and society. It is true that these aspirations did not take the form of a social agenda that the adherents would systematically strive to implement. The historical materialist would argue that the objective historical conditions were not yet ripe for the realization of such near-perfect, classless societies. Mainstream religious thinking, on the other hand, considers it misguided to aspire to the condition of a perfect or timeless *being* in the historical or temporal state of *becoming* something never to be realized in time. In many traditions these ideal transformations assumed the form of an apocalyptic intervention, but minority movements within the main tradition directed themselves toward less ambitious goals and strove to bring about major structural reforms, with occasional success.

The growth of capitalism and the breakdown of the feudal order, however, has witnessed a continuous process of far-reaching change that differs significantly from that in any previous period in human history. Past societies experienced a long period of relative stasis within a stable value system after the initial period of convulsion and change, but the industrial societies ushered in a period in which dynamic processes of change became the principal attribute and sustaining rationale of the system. This made the processes of change that came with the scientific and industrial transformation fundamentally different from the changes and discontinuities within the order-oriented ethos discussed earlier. It was this period that saw the evolution of the value system and the institutional framework that pose the major ethical dilemmas for the developing societies. Each value and institution had to be formed, molded, and established through sustained struggle and effort—the secular state, the abolition of serfdom and slavery, universal franchise and elected and democratic forms of government, religious tolerance, public welfare and state responsibility for the poor, a free press, civil rights, the independence of the judiciary. Free-enterprise systems provided their own strategies for achieving these goals through a mixed economy in which market forces were allowed to play a predominant role. Here structural changes were the result of a long incremental process that took place within structures of power that allowed for conflict resolution. The socialist system provided its own path to some of the goals through a socially owned, centrally planned system, often emerging out of a deepening of social cleavages and conflicts.

It is in this historical context that developing countries must manage processes of change that have their own special character and attributes, if only because change is much more rapid and intense than what societies faced in the past. Politically, the pressures for new forms of social decision making and the pace at which progressively larger sections of the population enter the political processes are of an unprecedented order. Demographically, the increases in population that have occurred in these societies over a few decades surpass what has occurred in centuries in any part of the world, including the developed societies during their period of development. Economically, growth has occurred at rates much higher than those that prevailed when developed societies were undergoing their socioeconomic transformation—5–6 percent on the average compared with 1–2 percent for most developed countries.[1] Technologically, societies are propelled from systems largely dependent on human and animal energy to the use of increasingly complex forms of inanimate mechanical energy; they are able to acquire, albeit only piecemeal, the advanced technology of the developed countries—even while large parts of their societies function at a preindustrial level.

Second, a new international value system has emerged that seeks to define the parameters within which change should occur. Growth and change of the order described would place an immense strain on the inner coherence of social systems and structures, on their capacity for constant adaptation, and on the sensitive living tissue of human relations within these structures. Transitions in the industrial societies, even proceeding much more slowly over longer periods, often burst the bounds of rationality and humanitarian constraints. These societies achieved their new equilibria through open conflicts, violent displacements, and a long record of social oppression and inhumanity. Through these processes, new value systems emerged. The developing countries, however, are in a different historical situation. The lessons and experience of the transition and the value systems that accompany it are already available. The value systems themselves act as a powerful conscientizing force within the developing countries. They have altered the fundamental structures of knowledge and perceptions among crucial elites and social groups in these societies and have transformed the rules of the game for the transition, setting the limits within which change should take place.

Therefore, an evaluation of how developing countries manage their transition is itself an exacting moral task. On the one hand, there must be a profound understanding of the historical processes in these societies, a capacity to see the ethical issue in relation to these processes, and a sensitivity to the inner struggle of a society in the midst of an unprecedented transition. The moral excesses and social deformities themselves must be understood in terms of this struggle before they are judged in terms of ethical norms. On the other hand, such an analysis should not imply either a tolerance for the violation of human values or a lowering of ethical norms and goals. It is important to recognize that only through sensitivity to the historical processes that underlie the present dilemmas is it possible for us to perceive the modes of action that enable these societies to mitigate and contain the excesses and create the right ethical framework to guide the development process.

Finally, another issue lies at the heart of our inquiry into the ethical dilemmas of development. We can state this in terms of the conflict between the order-centered ethos, which is the historical heritage of the developing societies, and the change-oriented ethos that came out of the historical revolution and that they must adopt for their transition. The oppositon, however, goes much deeper than the dualities of order and change, of traditional stability and future shock. Acceptance of discontinuity and change as essential attributes of social and personal life seems to alter the perception of values themselves. No longer do values function as the stable centers of meaning they had been in the past. Rather, they are subject to continuous displacement. They are pushed into a realm of relativity in which every value is time bound, is specific to a historical context, and has validity only in a relative sense. These contemporary processes of change break lose from the transcendental value centers around which all discontinuity and changes in the past had revolved. In doing so they disturb the unity that those value centers had strived to impose on both the personal and the social morality. Developing societies must grapple in the course of development with confrontation between different visions of life, different modes of knowing reality. The development process brings the transcendental model of man and society that has organized human experience in the traditional societies into a sharp encounter with the secular humanist model centered on the present life span and the satisfaction of material wants.[2] The empirical positivist structures of knowledge and reasoning that usually underpin the development process become the unrelenting adversary of the holistic, existential mode of knowing, the inner experiencing of reality that is part of the religious vision of life.

This fundamental clash of value systems and structures of knowledge has a pervasive impact on development goals and choices. It impinges on all major development goals, as well as on the strategies selected to achieve these goals. On the one hand, there is the need to preserve, renew, and redefine the holistic vision to enable these societies to respond critically and selectively to the processes of modernization and the massive intrusion of new values and life goals that destroy that vision. On the other hand, it is necessary that these societies remove the historical outgrowths of the traditional value systems that perpetuate structures of oppression and ignorance and, in doing so, make room for new processes of humanization and new sources of spirituality. Many elements in the system of production and the framework of social and political institutions that we associate with modernization will unquestionably form part of the inevitable changes in developing countries. For each of these elements, however, the question arises: How should developing countries absorb the new into their own indigenous experience and culture? For example, is the technology of the industrial societies separable from the entire value system, organization of human life, and vision of reality that seem to go with it.[3] Can developing societies acquire the former selectively while rejecting the entire value system in which it is embedded? How can these societies assimilate elements of the new humanism, with its emphasis on social equality and individual creativity, into their own concept of the intrinsic transcendental worth of the individual being? These tensions between past and future are particularly significant for Asian societies, where the presence of history is felt more profoundly than perhaps anywhere else. They are the societies with the longest human memory, together representing almost all the major civilizing traditions of humanity—the Hindu, the Islamic, the Buddhist, the Confucian, and the Christian.

## ETHICS AND TECHNOECONOMIC RATIONALITY

Let us now consider the second assumption implicit in the title of the project—that there is an intrinsic ethical component in the dilemmas of development that needs to be conceptualized clearly in any development strategy. The discussions held over the course of the project often reverted to this basic question. A variety of positions were taken during the initial defining of terms. They oscillated between two extremes—one that held that developmental dilemmas should be defined exclusively in technoeconomic, nonethical terms and that there was little purpose in introducing an ethical component, and the other arguing that all developmental dilemmas were ethical in character because all development outcomes were human outcomes about which we must eventually make value judgments involving ethical criteria. As the discussion unfolded, it became clear that these positions were themselves based on underlying concepts of development that in turn were derived from fundamental assumptions about people and society.

The first position is based on the premise that development is essentially a technoeconomic process involving the growth and expansion of productive capacity. This supposition implies that all noneconomic outcomes of development have their source in the technoeconomic component. If this were so, then the primary criterion governing development choices should be technoeconomic criteria and the basic principle guiding the development process should be technoeconomic rationality. In this framework, technoeconomic rationality is not merely one among several important principles; it is the key principle, superseding others and directing and managing the development process as a whole. The increase in human welfare in this framework is measured in terms of the increase in material output, which in turn leads to the improvement in material well-being, all of which is derived from the way in which the productive system is organized, developed, and utilized. This would imply that the best techno-economic solutions to development dilemmas are also the best decisions in terms of human welfare. The introduction of independent ethical criteria would thus become unnecessary.

Development ideologies that make material well-being the primary goal of development adopt a conceptual frame in which technoeconomic rationality emerges as the organizing center. This applies to ideologies based on the market system as well as to those based on socialist principles. In the former case social welfare can be best achieved through an efficient market system that gives full play to individual and collective preferences. No conscious ethical choices are necessary apart from technoeconomic ones. The basic presuppositions of this approach rely heavily on the model of people and society derived from the discipline of economics, particularly the neoclassical model.[4] In this model the political processes themselves recede into the background. Conflicts are moderated through the competitive system that provides for their resolution. The main role of the political process is to create the conditions for the full interplay of the market forces that enable the correct technoeconomic choices to be made.

In contrast to this model, the various approaches derived from socialism, especially those based on the Marxist interpretation of social change, move the political processes to the center. In the first instance, the framework conducive for development is set through conscious social action directed at structural changes that can release the productive forces that are being obstructed by the prevailing structure. These relate primarily to the structures that determine the ownership of the productive resources in a society and the relations among those engaged in production. When this agenda is completed, the solutions to development problems that arise thereafter would be essentially technoeconomic solutions, which are obtained mainly through centralized planning. In this conceptual framework, the rationale for the structural changes that are sought is perceived as historical rationality. What is ethical is ultimately what is necessary and inevitable in terms of historical forces and objective historical conditions. Therefore, the solutions and choices that are available are characterized as scientific even while they encompass political choices and modes of social action. In the final analysis, the ethical is subsumed in historical rationality.[5]

Although these ideologies confine the concept of development within the framework of material output and material well-being, however, they cannot avoid addressing themselves to the issue of distribution, as even within this framework well-being implies not only increases in output but also their distribution. One would expect, therefore, that with the distributive issue we move into the ethics of structural change; for here we are involved in political and social choices that lead to the right and equitable structure of production and distribution. Even these issues, however, ultimately tend to be reduced to technoeconomic and historical rationality. The choices and modes of action that move humanity along this path of structural change could be either the entrepreneurial neocapitalist transformation offered by the market system or the revolutionary strategies offered by the socialist approach. Two aspects of both ideological types are relevant to this discussion. First, in both, structural change will inevitably bring about the liquidation of social groups, often necessitate violence and destruction, and result in political systems that are authoritarian and oppressive. These tend to be perceived as the necessary human costs of achieving the collective good. Second, in both, the main goal of structural change is the transformation of the productive and distributive system, which is seen as the basic determinant of the nature of people and society. The central societal problem is the sharing of material output.

In each case, therefore, technoeconomic efficiency, productivity, and growth emerge as the dominant values. In the analysis of what is conducive to human welfare, the ethical component is progressively dissolved. For example, it is interesting to see how within the Western discipline of economics, attempts to correct market imperfections through interventionist measures such as public welfare remain controversial. The proponents of productivity and growth would contend that the effort to reach a better distribution of well-being through public welfare is self-defeating, as it constantly interferes with the incentive system that promotes productivity and growth. Distribution is better achieved on a sustained basis by permitting market forces to operate in terms of productivity and growth. A system that is excessively distributive slows growth and will eventually reduce the well-being of all, whereas an efficient growth-centered system will eventually lead to better distribution and greater well-being for all. The choice of instruments would then revert back to criteria that are exclusively technoeconomic.

Within the socialist framework the political processes are primarily concerned with the sharing of material output and the conflicts that consequently arise. It is significant that the measurement of growth in the socialist system is much more rigorously confined to the material product and excludes much of the output in the service sectors. The bias in this form of measurement and valuation leans much more heavily toward quantifiable material output. The rationale for the equitable structure is ultimately historical; that is, only this structure is capable of solving the contradictions that have arisen in the production system. Therefore, the ultimate criterion for choices relating to structural change is whether they accelerate the processes that are historically necessary. This criterion of historical rationality leads certain neo-Marxist schools to argue that, historically, capitalist relations of production have not outlived their capacity to carry through a major program of industrialization and structural change in certain parts of the developing world,[6] or to present a scenario wherein the next phase of capitalist expansion is transnationalization, which will solve some of the internal contradictions that have arisen and may thereby enable capitalism to survive for a long period in world history.[7] Whether these developments are good or bad, ethically desirable or undesirable, is irrelevant in the face of what appears to be the inescapable historical outcome. Any other approach that attempts to push toward an egalitarian society primarily on the ground that it is ethically desirable becomes unhistorical, unscientific, and utopian.[8]

In both the market and the socialist approaches, there is a fundamental assumption that the historical evolution of humanity leads to progressively higher stages. These higher stages, however, are measured primarily in terms of human technological capability, man's capacity to dominate nature and to transform natural resources for his own material uses. Both approaches are basically deterministic, severely restricting individual human choice and reducing it to insignificance in the context of external environmental and historical forces, on the one hand, and internal psychophysical determinants on the other. Both tend to produce their own forms of inhumanity, in which the means are justified by the end, whether it be technocratic efficiency, totalitarian rule, or larger concepts of military deterrence or restricted nuclear warfare. The human suffering involved is seen as irrelevant within a scientifically well designed and efficient program; human beings are simply another disposable input in a larger calculus directed toward goals of growth, power, and aggrandizement of the system.

The line of argument followed here is open to the criticism that it is an oversimplification of both the neoclassical and Marxist models. Proponents of both schools of thought deny vehemently that they reduce complex historical and social processes to the technoeconomic base. Marxists point to the writings that analyze the complex interrelationships between the superstructure of culture, ideology, and social institutions, on the one hand, and the base of production relations on the other. Members of the neoclassical school assert that their model is a partial model of reality, as is any scientific model, and that, in their concern with economic phenomena and the laws governing them, they do not deny the importance of other phenomena.

The main issues raised in such a refutation cannot be discussed fully in this brief chapter. The question posed here relates to the way in which the foundations of an intellectual system, the organizing centers of an ideology, determine the interpretation of reality as a whole. The belief that the "economic structure of society always furnishes the real basis starting from which alone we can work out the ultimate explanation of the whole superstructure and political institutions as well as of religious philosophical and other ideas of a given period," will necessarily lead to certain values and norms regarding the management of human affairs as a whole, if the ideology operates in an internally consistent manner.[9] Similarly, a system of thought that is postulated on the belief that the larger part of human behavior is governed by economic motives will produce a hierarchy of values in which the noneconomic values are assigned a low or indeterminate ranking. The discussion in this section has attempted to indicate only how in the development models that have come from industrialized societies, the principles of technoeconomic and historical rationality are paramount and tend to give the main directions to the development process.

This, however, does not mean that these models or the ideologies that support them refuse to recognize the importance of factors other than the technoeconomic ones, or that they ignore the ethical dimension when they apply themselves to the real world. Marshall, in the introductory chapter to *Principles of Economics,* states that "the two great forming agencies of the world's history have been the religious and the economic."[10] The Marxist model, on the other hand, is inspired by a vision of human freedom and fulfillment that is realized after the contradictions of the class struggle have been finally resolved. Revolutionary action toward this end is animated by a profound moral and ethical concern. This, in effect, seems to demonstrate that human reality in its total material and spiritual essence ultimately breaks through the bounds set by these ideologies and frames of thought. The ideologies themselves must come to terms with the reality, making the necessary accommodation to admit the part of that reality that they tend to exclude by virtue of their fundamental beliefs. Human nature must go to spiritual sources beyond these ideologies to draw moral sustenance for its actions.

## THE INDIVISIBILITY OF DEVELOPMENT

In a reductionist approach to development, one that ultimately perceives development as essentially a unidimensional technoeconomic phenomenon, there are no fundamental dilemmas. The dilemmas arise when we define development as a process that goes far beyond technoeconomic change, when we perceive it as a process in which technoeconomic changes are the means and accompaniment to other far-reaching societal changes in the political, social, and cultural dimensions that enhance the quality of life. In such a framework, dilemmas are ever present. They are irreducible to simple choices, whether we think within social-democratic limits or a revolutionary framework of fundamental structural change. These dilemmas become those of all political actors, whether they be the elites or the poor and disadvantaged. They will all have to grapple with the problem of choosing the development path and the strategies of change that lead to a fully human society.

Here it is useful to consider how the meaning of *development* has changed and been enlarged over the last three decades. We began by perceiving development in its technoeconomic dimension largely as the growth of productive capacity and output, the increase in national goods and services.[11] At the end of the 1960s we were already asking for a great deal more from development than a rapid increase in material output. Development had to manifest itself in social advances that improved standards of health and nutrition, gave people a longer life span, and raised the levels of education and knowledge. The strategy of development had to be such that the pattern of growth itself would yield these positive social outcomes.[12]

In the 1970s we moved further beyond the socioeconomic boundaries to redefine development in terms of other structural characteristics. Development also had to liberate developing countries from a system that reinforced and perpetuated their dependence on developed countries, and provide them with a capacity for self-reliance, both technologically and economically.[13] We began to emphasize the changes in the structure of society and in the distribution of wealth, income, and power that had to accompany changes in the structure of the economy, the size and composition of its product and work force. The elimination of poverty and the satisfaction of the basic needs of society as a whole became a major objective of development.[14] The international debate on development has added yet other dimensions. There is a need for development to take place within the harmonious balance between manmade structures and the total ecosystem that makes possible the management of resources on a long-term self-sustaining basis.[15] The desirable development process is also seen as one in which the structures of national decision making are established on the basis of civic freedoms and rights, in which dissent can be freely voiced, in which governments can be changed by popular choice, and in which there is increasing participation of the community in both defining development goals and acting on them.[16]

Genuine development therefore must succeed in all these areas. It must aim simultaneously to achieve the entire plurality of goals. This is what gives rise to the dilemmas of development. These goals cannot be arranged in a simple order of priority. The effort to order human needs and development priorities in relation to these needs within a well-defined preordained hierarchy oversimplifies and often falsifies the problem of development choices.[17] Development is a totality that is ultimately indivisible. In the development goals we enumerated, however, it is possible to identify three broad clusters. One is organized around growth and the structures that accelerate and sustain it. These include productive capacity, the technological capability for self-reliance, and ecologically sound management of resources. The second cluster focuses on equity: the distribution of growth, wealth, and income; the sharing of physical well-being, the eradication of poverty and ignorance. The third is organized around participation and freedom—the political processes that ensure human rights and provide for a wide sharing of power. Development, a total process of human growth for the individual and society, encompasses all three clusters.

The experience of development seems to demonstrate that these clusters are closely interrelated, that they are part of an interdependent structure. This requires us to strive toward all goals simultaneously if we are to achieve any significant fulfillment of any one cluster of goals. The equity goals cannot be achieved and sustained without adequate effort to achieve the growth cluster. The growth cluster requires rapid expansion of markets, broad-based purchasing power, and sustained growth of effective demand in progressively larger sections of the population. Both equity and growth require viable political systems that in turn can be sustained only through some degree of participation and sharing of power. The historical constraints may always result in lags and shortfalls in one cluster or another; in the short term, inevitable choices may place emphasis on one more than the other. These, however, will still have inevitable developmental costs in terms of the specific cluster that is neglected; they will have to be corrected and made good eventually in terms of that specific cluster. The interdependence of the development goals implies that no permanent trade-off is possible between any two of these clusters. This does not mean, however, that any one cluster can be conceived as a means for achieving another. In the reductionist technoeconomic approach, all nontechnoeconomic elements are means to the technoeconomic end. Redistribution is rational because it leads to growth; equity is a precondition for increasing effective demand. In contrast, what is suggested here is that each cluster must be recognized for its own intrinsic and autonomous value as an indispensable component in the total well-being.

## THE SIMULTANEOUS PURSUIT OF MULTIPLE GOALS

The analogy of the inorganic hierarchical column in which the fulfillment of development needs can follow in sequence does not represent or explain this process adequately. Development seen as an interdependent organic process is not a sequence but a simultaneity. For example, the approach that suggests growth first, equity later, and freedom to follow is not valid in either political, economic, or human terms. If the relationship between the various components of development is organic, then any part that is lost or neglected results in a process of mutilation. A cost must be borne for which there is no real compensation. The correct analogy is an organic structure in which every part has a living relation to the whole and removal or loss of any, although it may not endanger survival, yet results in an intrinsic loss, an impairment of a faculty or function of that part that is needed by the whole. The need for food, clothing, and shelter will always exist, together with the need for the nourishment of the inner person; for fulfillment in interpersonal relationships; for the possibility of spiritual inquiry.

Thus one cannot regard the choices between development goals as capable of trade-offs. The normal condition is the healthy functioning of the whole. Equity cannot be traded off for growth or stability and order for participation. The choices must always be perceived as part of an effort to sustain an organic process, to promote simultaneous movement toward the plurality of development goals. Development must be a process of harmonious growth. Approached in this manner, the choices must create and preserve a system that enables societies to return constantly to the simultaneity of effort, to the total configuration of development values and goals, even when inevitable choices result in the sacrifice of any one component. Such choices, which result in short-term costs in terms of one or another of the development clusters, are likely to be inevitable in the development process. The total framework guiding development choices, however, would need to make sure that they are indeed short term, that the loss of plurality, the excision of any component is transitional, and that the time span during which the loss must be borne is reduced to a minimum. There have to be social, economic, and political mechanisms built into the development process that will automatically steer it back to the right path and restore it to the plurality of development goals.

This approach to development goals implies that each cluster has an intrinsic value and that they are not substitutable for each other. They cannot be reduced to a common unit of value in which one cluster can be valued in terms of another and hence traded off for another, and by which, given a certain volume of resources, the various alternative proportions of the development goods from each cluster can be easily determined for alternative demand schedules that are equally desirable. There are no elasticities of substitution between the different development components.

The fact that development values are not substitutable for each other also requires a different approach to the ordering of priorities and the time frame for that ordering. Any given historical situation will demand an ordering of priorities, and the ordering of priorities itself will change in response to the urgency of the needs. All the time, however, the ordering itself will be subordinate to the overriding indivisibility of the development goal. It will have to discipline the choices and actions and set the time frame for them so that the living generations have the opportunity to participate in a process in which the full range of development goals is kept in sight.

The time frame for development goals raises an entire set of issues that have far-reaching human implications. Most of the global scenarios for development that project the future of mankind at the beginning of the twenty-first century shows a large part of the human race living in conditions that could still be described as poverty.[18] In the most recent projections made by the World Bank, the low-income countries, with more than 35 percent of the world population (excluding China), will probably still have annual levels of income below U.S. $275 per capita in 1990 even on the basis of relatively optimistic projections.[19] These projections signify the inescapable limits of the expectations of most developing societies. Within these limits they need to organize their development effort without illusions. Their development goals must be disciplined by a clear perception of the material limitations imposed on them by the condition of poverty. They have to recognize that within a time horizon that is meaningful for the present generation, the level of living they could reach is circumscribed by these limitations. This still does not mean, however, that within the limits of their poverty and the development path they choose, developing societies cannot organize their life and their upward movement in a manner that is qualitatively satisfying.

In the imagery that is latent in the development terminology used here, terms like "reaching a level of living," or "rates of growth" suggest that the process of growth is all important and that everything should be subordinated to the climb to affluence, which itself extends over several generations. In this imagery the imbalances that are inherent in the ascent and the human suffering they cause in the present receive little attention. A balanced social well-being, it seems, must be achieved in the distant future—after the ascent. Each generation, however, seeks a plateau, a point of rest in the movement where it tries to achieve a fullness of living in its own life span. All communities, at every stage of development, are as much concerned with material and spiritual fulfillment in their own lifetimes as with the effort to change their lives in order to enjoy a better life in the future. They are as much concerned with *being* in the sense of experiencing of life fully in the present, as with *becoming,* in the sense of growing into something better in the future.[20]

Although there is no question that the agenda of development requires speedy structural changes that can eradicate poverty and satisfy the basic needs of the population as a whole, it seems equally clear that developing societies are capable of reaching this condition long before they approach the state of material well-being enjoyed by developed countries. Further, the social and political framework, the human organization that promotes equity, participation, self-reliance, and a harmonious balance with the nonhuman environment, need not themselves be specific to any particular level of material and technological development. This framework can be achieved early in the process of technoeconomic change and should be sustained and adapted in that process if technoeconomic change itself is to become the means to full human development.

The human quality of the development that takes place would depend very much on the right dynamic balance between present social need and the drive for economic expansion and improvement—the correct tension between the present and the future. It is in the search for this balance that a society's social and political goals, its ideologies for development are defined. It will, for example, decide whether a country will pursue strategies of forced industrialization, uprooting of communities, suppression of human rights, and destruction of the environment—all in order to accelerate the pace of growth—or will seek a more humane course of development, with more equitable sharing of resources and growth. Here, too, the dilemma lies in the fact that the concept of such a balance could become a recipe for stagnation and backwardness if it were merely employed to make people content with their present poverty and the social system that perpetuates it. Also, like the growth-first strategy, it could become ecologically destructive and neglectful of the resource base that has to support future generations. The optimal balance between present and future is a balance between fulfillment and growth. In the normative framework of development that is defined here, the conflicts inherent in present consumption and future growth are resolved as far as possible within the limits of the temporal; for if development is harmonious growth—if technoeconomic change is disciplined within a fully human, social, and political organization—then the very striving for a better future and the austerities it may impose are also part of present fulfillment. This reiterates the previous statement that genuine development is an organic process arising out of the simultaneous pursuit of the full range of development goals. One essential element of this process is the way in which the time frame is set for the pursuit and achievement of development goals in order to achieve that most humane equilibrium for the present generation.[21]

### THE CENTERS OF VALUE AND THE SOCIETAL VISION

The full range of development goals cannot be kept in sight if development is treated as a large aggregate of various elements that must somehow be contained together in the process of change. The intensity and pace of change in developing societies will necessarily generate the corresponding intensity and recurrence of conflicts between the different components of development. What we might describe as the human entitlement to development encompasses all the goals of development. Orchestrating the different goals in response to the conflicts, however, and achieving the necessary balance in the numerous choices that developing societies have to make between competing alternatives, is a task that demands a capacity to envision development as a whole. It requires a framework of developmental values and criteria, and ideology relating to development, that can guide the choices and discipline the development process. It calls for a societal vision of the good life, which is the goal of development. Within such a vision, the entire planning of development goals discussed here is subject to organizing principles and value centers that are ultimately independent of the development process itself. They are independent because they are derived from a vision of people and society that in almost all ideologies goes beyond development goals themselves.

It is the continuing search for and the affirmation of these organizing principles and value centers that produce some of the most profound conflicts in the development process. Here the dichotomies of past and present, the struggle between value systems evolved by traditional cultures and those transmitted through development emerge most acutely. We return then to the theme discussed earlier—the conflicts between the secular model of man and society and the transcendental model, which are manifested in the changes that occur as development proceeds. At this level we witness a conflict between different visions of life, different modes of knowing reality, and different paradigms for organizing human experience.

The conflicts considered here relate to the relatively stable core of human values and life goals that are part of any human society and thus lie beyond development itself. The priorities of development ultimately arrange themselves around this human core. Their arrangement will depend on the questions we ask about the outcome of development as it relates to the core. The questions center on such issues as the way in which the new society or the condition of development for which we are striving provides a pattern of meaning for the individual's life. How does it organize human experience in its totality, teaching the individual how to live through his life cycle and preparing him for death? What does the pattern of development do to structures that promote community or to forms of social organization that protect the interpersonal world of voluntary relationships based on love, kinship, and the interior moral life? What is the equilibrium between material and spiritual well-being that development expects to produce? These are questions to which current development thinking rarely addresses itself. They are regarded as part of a subjective world that cannot be brought into the framework of such thinking and analysis. The reluctance to confront these questions is itself related to the underlying ideology of man and society on which such development thinking is based.

It is not possible here to examine in detail the nature of these conflicts or inquire into their validity. Such a task requires an extensive philosophical inquiry. We can, however, present the conflicts in terms of broad dichotomies between the two models. First, the ideology of development that comes to the developing countries out of the experience of developed countries—the ideology of the secular model—gives primacy to technoeconomic rationality, which becomes the organizing principle for individual and social action. Second, this ideology places the pursuit of material well-being at the center of human life goals. Third, it is based on a methodology and a structure of knowledge that is positivistic and empirical. The reality that is not capable of empirical verification, not capable of being identified and treated as an object of knowledge through the methodology of the empirical sciences, is not reality; it is a linguistic creation produced by the structures of language themselves. Fourth, the frame of reference in this ideology is anthropocentric; it is confined to man and his experience, and this experience itself has a limited life span, with no before or after. This centrality of man has perhaps been the driving force in the liberal humanist movements that redefined individual freedoms and human rights, that liberated man from many forms of social oppression and moved the social order toward equality. At the same time, by making man the ultimate source of meaning, it limited all meaning to man. The larger configuration that contained the human scheme became devoid of meaning, and man himself became a meaningless entity within a materialist, spiritless universe.

Fifth, all these elements combine to promote a way of life that is hedonistic and consumerist. The driving force in this system is the continuous expansion of demand, the ceaseless multiplication of new material wants and the flow of goods and services to satisfy them. Sixth, in the secular model the emphasis is on human rights and the institutionalization of these rights through juridical systems, contractual relationships, and systems of reward and punishment that are administered by bureaucracies and formal organizations. It has the virtue that these rights depend more on the rightness of the system than on individual power and the personal morality of individuals. Finally, implicit in this ideology is an underlying presupposition that leans it toward forms of determinism, whether we characterize it as economic or historical determinism. Historical economic and social forces become reified and assume specific identities as major actors in human affairs, whether they be "the market" or "history." They acquire an independent existence in our conceptual framework in relation to human choice. It is true that human will has a place in all these ideologies and that the dialectics of interaction between the individual and society attempt to allow for the freedom of human action. Even so, however, human volition and the capacity to choose between alternatives, the essence of ethical action by an individual or by society, are progressively diminished to the point of becoming immaterial as variables in the process of change. The individual exists only in society and in the vast collective reality of which he or she is a single psychophysical unit. Each person is subsumed within the material forces and the impersonal laws governing them. In the process, the concept of *the person* tends to get removed from our field of vision.

In contrast, the transcendental paradigm is posited on a different concept of rationality. Rationality is derived from life goals that link man to a transcendental reality, whatever conceptual form it assumes—Brahman, Nirvana, Tao, the Christian God, the Islamic Allah. Rationality in the transcendental model is therefore irreducible to any single principle that can be grasped through logical reasoning. The rationality of individual and social action in this condition is rooted in the perception of the contradictions and paradoxes of human existence. If we take one major transcendental tradition, rational action is related to the different stages in the life cycle of the individual as he moves through the many *asramas* or dwelling places in life—childhood and adolescence (the stage of the *sisya*), adulthood and the householder's life (the stage of the *grhasti*), contemplation and the spiritual search (the stage of the *vanaprasti*), and the journey on the path to deliverance (the stage of the *sanyasin*). The states of well-being are also numerous and are related to these stages. They include economic well-being *(artha),* sensual enjoyment *(kama),* responsibility in action in everyday life *(dharma),* and the final release or liberation from desire *(moksha).* Each has its rationality that might be contradicted by the other—*kama* by *dharma, artha* by *moksha*. The social and moral order provide for their coexistence, however, as at a given time some part of humanity is at each of these *asramas*.[20]

In opposition to the second attribute of the secular model, the transcendental model gives material well-being a limited value in the total well-being of the body and the spirit. *Artha* is constantly in tension with *moksha*. In the transcendental model the metaphysical is real; and the modes of cognition that go with it are intuitive, holistic, existential modes that illuminate reality in ways different from the analytical, empirical modes. Again, in this transcendental order man, despite his central place, exists within a reality that is more than human, to which the human order is subordinate and is related in a pattern of meaning that goes beyond the dimension of time. The concept of the *person* in the religious paradigm is different from the concept of the *individual* and the values of freedom and equality derived from it in the secular ideology. The person is nevertheless at the center; meaning is ultimately related to the person and his personal relationship with the transcendental reality. The consumerism and immoderate pursuit of well-being that characterize the secular model are incompatible with the equilibrium and moderation cultivated in the transcendental model, where the value clusters in each stage of life must be in harmony with the other clusters and the individual must move harmoniously, liberating himself from one stage as he enters the next, and must respond to the world of material well-being with the inner detachment provided by a spiritual center of gravity.

In the transcendental model human rights are inseparable from social and moral obligations that are part of a total order. The way they are upheld in society depends greatly on the voluntary human relationships and the internal moral discipline of the interpersonal world. By this process of internalizing moral codes, the transcendental model develops the capacity to regulate those components of human interaction—whether in communities, families, or organizations—that cannot be brought within formal structures of law and order and that depend on voluntary relationships based on love and compassion. On the other hand, by emphasizing individual moral conduct, it tends to make the protection of human rights a matter of personal choice and ethical behavior rather than an automatic outcome of the system.

Finally, in both the materialistic and the religious paradigm, there is a basically deterministic character in the relationship between human action and total reality. In both the individual is reduced to insignificance within the larger order; the personal act is altered beyond its original purpose; almost all religious ideologies are themselves rooted in the paradox between human freedom and divine will that finds expression in various forms of determinism. Here, however, the similarity ends. The religious approach to the insignificance of the person comes from a direction opposite to that of the materialistic ideologies. In the latter the deterministic pattern and the individual insignificance are part of a larger physical or material reality that eventually effaces the individual and in which the individual encounters the finality of nothingness. In the religious paradigm both what is preordained and the insignificance of the person within it are restored to a transcendental meaning that wholly transforms them. The different values and meanings assigned to the person and the individual within an ideology will have a profound influence on the way a society organizes itself for the alleviation of human suffering, on its balance of concern for present and future welfare, and on its perception of human rights and values as a whole. It will influence its social ethos, moving toward either a totalitarian value system or a more humane ideology.

Development implies choices that take a society toward or away from one or the other of these paradigms. Development requires a conscious, sensitive resolution of the conflicts between these value systems and world visions. A society that is in the process of development must have the capability to adapt to change while preserving the identity and integrity of its own collective historical experience. While renewing the life-giving sources of its own past, it has to create the freedom to move toward its future. In order to do this, it must develop a center of values out of the understanding of life that has emerged from its cultural accumulation, from which it can both be receptive to new values and respond critically and selectively to them. On this depends the capacity of a society to evolve the moral order within which it can guide and discipline the development process.

## THE NATURE OF CHOICE IN THE DILEMMA

We will now examine briefly some of the concepts implicit in the term *dilemma* and the reasons for defining the choices in the development process as dilemmas. The conceptual framework developed for the studies attempts to restore human choice to the center of the development process and human fullness to the center of the development goal. First, the approach of the study as a whole is essentially nondeterministic, although individual researchers have been free to adopt the emphasis of their choice. The study is concerned not so much with the broad direction of human well-being, as demonstrated by long-term trends in human history, as with the particular paths societies choose in this wide terrain, how they make these choices, and what their total human benefit and cost has been and is likely to be. In each historical situation the space for maneuverability and choice is always present, and the choice that is made will reflect a concern or a disregard for the central human values. In modes of social action directed at far-reaching structural changes, the available choices can take different paths, ranging from those that are highly conflict-ridden and entail heavy human costs to those that have a capacity for conflict resolution and for minimizing the short-term human costs. They can lead to sociopolitical systems that vary greatly in their framework for participation and dissent or their protection of human rights. For example, within the framework of a socialist revolution itself, there will be options for a strategy of forced industrialization, rapid structural change in the rural sector with enormous human cost, or a more humane participatory process of rural change. The collectivization process in the USSR and the development of communes in the People's Republic of China offer contrasting strategies, with a different balance sheet of gains and losses in human terms. Second, the study's approach attempted to place development processes within their cultural context and relate them to value centers that are derived from the historical experience of these societies and the visions of life that informed it.

The term *dilemma* implies a particular kind of choice, a situation in which the best alternative cannot be readily and definitely identified. Rather, the choice is between alternatives each of which is valid and important in its own right and none of which can be sacrificed indefinitely. The indivisibility of development and the need for simultaneity in its pursuit constantly confront developing societies with situations that can be called dilemmas. A dilemma offers no easy choice, no single alternative that is conclusively right. The thought frame in which the choice is always clear and the methodology for finding the right solution for every choice is always available is essentially totalitarian in its outlook; it imposes the inorganic certainty of the natural sciences on the infinitely variable living human reality. It then acquires an orthodoxy, whether religious or secular, that tries to fit this reality into its procrustean frame. It provides no outlets for communication with approaches and choices other than its own. In its inflexible search for the good it thus becomes inhumanly oppressive, transforming itself into the opposite of what it desires to be and becoming a source of evil. Both religious and secular ideologies amply illustrate this transformation throughout human history—the Inquisition and religious persecutions in feudal Europe, the revolutionary terror in France, the worst phases of the Stalinist era after the Bolshevik revolution, the holocaust in Kampuchea.

*Dilemma,* in contrast, implies humility in the presence of life's complexity, an acceptance of the limitations of human knowledge, an anguish at the possibility of losing one value when choosing another value. The unfolding of reality confounds the simple principles of logical order and sequence where every object, condition, and value has a separate identity and is not to be confused with another. In the *logical* order the good and the bad can be perceived unerringly in distinct antagonistic manifestations. The dialectical approach, on the other hand, perceives that reality cannot be contained entirely within these logical relationships; it accepts and seeks to understand the contradictory nature of reality; it recognizes that any particular condition or state of being *can,* and in the historical process *will,* become its opposite until all contradictions are resolved. The dialectical approach in its materialistic form, however, still works within the historical process; it moves from one antagonistic form to another. In this progression the contradictions in reality are successively resolved. This must be contrasted with the religious approach, which reaches further than the categories of the dialectic; in this approach the opposites are always immanent in reality; it uses the language of *paradox* and communicates through symbols to express the aspects of reality that are irreducible to a logical system. The opposites then become aspects of one overarching identity. Thus the Preserver and the Destroyer are different faces of the Infinite Being. Evil resides within good; and, as in the vision that was vouchsafed to Dame Julian, "Sin is behovely and all manner of things shall be well."[23] The diversity of the historical is contained within the unity of the transcendental.

In the religious paradigm, therefore, these fundamental paradoxes define the human condition. In Time, man has no choice but to distinguish the opposites and deal with reality through logical categories, but he needs to do so in constant awareness of the limitations of these categories, their interchangeability, and their existence in the ultimate unity. In the last frontiers of human knowledge, language folds in on itself in silence, and reality can be sought only through "the cloud of unknowing."[24] In the materialistic dialectic, on the other hand, the bourgeois relations of production are the last antagonistic forms of social production, and the communist synthesis brings "the prehistory of human society to a close,"[25] but in the latter there is no historical solution, the paradoxes are immanent in the whole of history and existence. There is no simple historical progression that fundamentally alters the paradoxical character of human existence. The synthesis always remains beyond time. The dilemmas of choice in the religious framework are best illustrated in the predicament of Arjuna in the field of battle in the Bhagavadgita.[26] What is of supreme importance is personal action and the purification of the motive in personal action. Although the correct outcome is kept in sight and the effort is directed toward that outcome, what is fundamental is the right effort, which enters the universal order—an order that may produce its own outcome, different from the original personal goal. Secular power and progress therefore must humble themselves before this knowledge. The major human choices need to be sensitive to all three categories—the logical order, the dialectical process, and the condition of paradox. To the extent that the values pertaining to all three categories are present in the ideology of a society, it is likely to have greater or lesser capacity for openness, tolerance, and humanity in the task of social engineering that it undertakes.

The intrinsic character of the dilemma also lies in the fact that the true nature of the choice itself and its outcome are never perceived in their entirety at the time the choice has to be made; neither is the full knowledge of the solution to the problems posed by the dilemma available.[27] This awareness informs the ideological approaches that perceive true knowledge as knowledge gained only through praxis or existence. *Praxis* yields the understanding of reality and the knowledge to cope with it through immersion in social action and struggle. In this sense it is essentially a methodology that belongs to the materialistic dialectic.[28] Existential knowledge, although it includes praxis, reaches out to the totality of human experience, both material and spiritual needs.[29] In a dilemma it could be said that the true choice is existential. The solution or outcome can never be anticipated in full but has to be borne out of lived experience; it has to grow, change, and mature in existential reality out of the unremitting search for genuine values. This means a continuing process of self-appraisal and renewal, and it has important implications for the ethos of a society.

## NOTES

1 Simon Kuznets, *Modern Economic Growth-Rate, Structure and Spread* (New Haven: Yale University Press, 1966), pp. 34–85; U. N. Statistical Yearbooks; World Bank Atlases. "Between 1950 and 1980, GDP of developing countries increased at an annual rate of 5.4 per cent which represents a remarkable performance...." *Trade and Development Report 1981,* U.N. Conference on Trade and Development, p. 33.

2 The discussion in this section draws on Godfrey Gunatilleke, "Re-thinking Modernisation: Asia 2000 A.D.," paper presented at the seminar on Asia 2000 A.D., organized by the Asia Society in New York, 17 July 1981, and idem, "Pluralistic Strategy of Development," *Marga Quarterly Journal* 6, no. 1 (1981).

3 Ibid.

4 "Neo-classical economists like to consider themselves 'scientific,' able to emulate in their own field the accelerating progress in the physical sciences....Such an approach has stimulated the systematic elaboration of models which reduce complex social phenomena to simple causal relationships, linking a few variables in a priori models, for which confirmatory material is then obtained." Dudley Seers, "The Congruence of

Marxism and Other Neo-classical Doctrines," in *Toward a New Strategy of Development,* A Rothko Chapel Colloquium (New York: Pergamon Press, 1979).

**5** "The danger of constructing a morality within a historiosophical vision...does not lie in the attempt to interpret one's life as a fragment of history....The danger consists in the total replacement of moral criteria by the criteria of the profit which the demiurge of history derives from our actions...." Leslek Kolakowski, *Marxism and Beyond* (London: Paladin, 1971), pp. 157–158.

**6** Bill Warren, "Imperialism and Capitalist Industrialisation," *New Left Review,* no. 81 (London, September–October 1973). See also "The Postwar Economic Experience of the Third World," in Seers, *Toward a New Strategy of Development.*

**7** Seers, "Congruence of Marxism"; F. Froebel, J. Heinrichs, and O. Kreye, *The New International Division of Labour,* trans. of *Die Neue Internationale Arbitsteilung,* Reinbek bei Hamburg: Rowohlt Taschenbuch Verlag, September 1977. English translation of introduction published in *Social Science Information,* 17, no. 1 (1978): 123–142.

**8** This is a recurrent theme in all standard texts on historical materialism; for example, see Friedrich Engels, *Socialism, Utopian and Scientific* (London: Allen & Unwin, 1892), Social Science Series.

**9** Ibid., p. 41.

**10** A. Marshall, *Principles of Economics,* 8th ed. (London: English Language Book Society and Macmillan, 1972).

**11** The writings on development in the 1950s treated development essentially as an economic phenomenon. For example, see W. A. Lewis, *Theory of Economic Growth;* W. W. Rostow, *The Take Off into Self-Sustained Growth* (New York: Macmillan, 1963); P. T. Baur, *Economic Analysis and Policies in Underdeveloped Countries* (London: Routledge, Kegan and Paul, 1965); *The Second Five-Year Plan of the Indian Government, The U.N. Strategy for the First Development Decade,* U.N. Resolution 1710 (xvi), 1961, reflected this emphasis, although a cautionary note was already present in the discussion on development in the early 1960s. "One of the greatest dangers in the development policy lies in the tendency to give the more material aspects of growth an overriding and disproportional emphasis....Human rights may be submerged and human beings seen only as instruments of production rather than as free entities for whose welfare and cultural advance, the increased production is intended." *Five Year Perspective, 1960–1964,* U.N. Publication, 1962.

**12** *International Development in the Second U.N. Development Decade,* U.N. Resolution 2626 (xxv), 24 October 1970, para. 18: "Qualitative and structural changes must go hand in hand with rapid economic growth and existing disparities should be substantially reduced"; Gunnar Myrdal, *The Challenge of World Poverty* (London: Allen Lane, 1970).

**13** *The Declaration of the New International Economic Order,* U.N. Resolution, April 1974. The ideological framework of the new international economic order (NIEO) owed a great deal to the work of Third World intellectuals and scholars, such as R. Prebisch, *The Economic Development of Latin America and Its Principal Problems,* U.N. 1950; Celso Furtado, *Development and Stagnation in Latin America—A Structural Approach,* Studies in Comparative International Development, vol. 11 (1965); Oswaldo Sunkel, *National Development Policy and External Dependence in Latin America* (1967); Samir Amin, *The Development of Capitalism in Black Africa;* Norman Girvan, ed., *Social and Economic Studies,* special issue; and *Dependence and Underdevelopment in the New World and the Old,* which appear in 22, no.1, special issue (Kingston: Institute of Social and Economic Research, March 1973).

**14** International Labor Office, *Employment Growth and Basic Needs,* Report of the Director-General, ILO, to the Tripartite World Conference on Employment, Income Distribution, and Social Progress, and the International Division of Labour (Geneva: ILO, 1976); Robert S. McNamara, *One Hundred Countries, Two Billion People—The Dimensions of Development* (New York: Praeger, 1973).

**15** *Report of the United Nations Conference on Human Environment,* June 1972 (New York: United Nations, 1973).

**16** Andrew Pearse and Matthias Stiefel, *Inquiry into Participation—A Research Approach,* and the related series of documents (Geneva: U.N. Research Institute for Social Development, 1979 and thereafter).

**17** This discussion draws on sections of Godfrey Gunatilleke, "Pluralistic Strategies of Development," *Marga Quarterly Journal* 6, no. 1 (1981). The concept of *total development* has been defined from different points of view in Dag Hammerskjold Foundation, *What Now—Another Development,* 1975; Johan Galtung, Roy Preiswek, and Monica Wemeigah, "A Concept of Development Centred on the Human Being—Some Western European Perspectives," in *Canadian Journal of Development Studies* 11 (1981); and Denis Goulet, "An Ethical Model for the Study of Values," *Harvard Educational Review* 41, no. 2 (May 1971).

**18** Organization for Economic Cooperation and Development, *Facing the Future—Mastering the Probable and Managing the Unpredictable,* 1976; M. Messaric and E. Pestel, *Mankind at the Turning Point* (London: Hutchinson, 1975); W. Leontief et al., *The Future of the World Economy* (London: Oxford University Press, 1977); Amilcar O. Herera et al., *Catastrophe or New Society* (Ottawa: International Development Research Centre, 1976); and Gerald O. Barney et al., *The Global 2000 Report to the President of the U.S.* (New York: Pergamon Press, 1980).

**19** *World Development Report 1981* (Washington, D.C.: World Bank, 1981).

**20** The critique of development that might be described as existentialist has drawn the distinction between *being* and *having*. See Denis Goulet, "Development Experts—The One-Eyed Giants," in *World Development* 8, no. 7–8 (July–August 1980); and Erich Fromm, *To Have or to Be* (New York: Jonathan Cape, 1978). This distinction is between the mode of life in which satisfaction is gained primarily through acquiring and having (characteristic of a technologically oriented society) and a mode in which the satisfaction is in the inner equilibrium achieved in relation to the world of things and the enjoyment of possessions. In the distinction between *becoming* and *being* made in this chapter, the emphasis is on the dimension of time and the equilibrium achieved in relation to wanting in the future and finding fulfillment in the present. The concepts are interlinked, as they are both concerned on the one hand with dynamic and static conditions, and on the other with moderation and excess.

**21** This discussion reiterates some of the points made in Godfrey Gunatilleke, "Commitments to Development," in *Marga Quarterly Journal* 1, no. 1 (1971).

**22** Radhakrishnan, *Eastern Religions;* Zimmer, *Philosophies of India.*

**23** Julian of Norwich, *Revelations of Divine Love* (London: Penguin, 1966), chap. 27. The methodology of the paradox is probably best illustrated in the philosophical and mystical writings of the Hindu, Buddhist, and Taoist traditions. The inadequacy of the logical category to explain reality is progressively unfolded, and the mind is finally confronted with the irreducible paradox—for example, the concept of Nirguna Brahman: "It is not this, it is not that." The Mahayana text Prajna Paramita is a magnificent example of this method. "The Mahayana Way," states Zimmer, "is to reassert the essence by means of a bold and stunning paradox." *Philosophies of India,* p. 485.

**24** *The Cloud of Unknowing,* trans. Clifton Wolters (London: Penguin, 1961).

**25** Karl Marx, "Preface to a Contribution to the Critique of Political Economy," in K. Marx and F. Engels, *Selected Works,* vol. 1 (London: Lawrence & Wishart, 1960), p. 369.

**26** Arjuna, having to do battle with his close kin, is "stricken by compassion and despair." "Before us," he says, "stand Dhritarashtra's folk, whom if we slay we shall have no wish for life." Krishna's reply in the second lesson of the Bhagavadgita expounds the metaphysic of action within the Hindu version: "Holding in indifference alike pleasure and pain, gain and loss, conquest and defeat, so make thyself ready for the fight." *Bhagavadgita* 2.38.

**27** For an economist's view of uncertainty see Kenneth Boulding, *The Image—Knowledge in Life and Society* (Ann Arbor: University of Michigan Press, 1956). "By all means let us be manifest as we can. Let us probe the secrets both of nature and her history, but let us also cultivate a sense of mystery as well as a sense of history.... The searchlight of the manifest makes more apparent the darkness of the latent beyond it. What this means in practice is that our response should be always to an uncertain image," pp. 130–131.

**28** Karl Marx, "Theses on Feuerbach," in *On Historical Materialism: A Collection* (Moscow: Progress Publishers, 1976), p. 11. Thesis II: "The question whether objective truth can be attributed to human thinking is not a question of theory but is a practical question. In practice man must prove the truth." Adolph Sancha Vasques, *The Philosophy of Praxis* (London: Merlin Press, 1977).

**29** To both the nonreligious existentialists (Heidegger, Jean-Paul Sartre, Merleau-Ponty) and the religious thinkers of this tradition (Kierkegaard, Jaspers, Marcel), human reality is free and responsible. Present choice and action *create* the future field for choice and action.

## FOR FURTHER READING

Agassi, Joseph: *Technology,* D. Reidel, Dordrecht, 1985.

Ayres, C. E.: *Science: The False Messiah,* Bobbs-Merrill, Indianapolis, 1927.

———: *The Industrial Economy,* Houghton Mifflin, Boston, 1952.

———: *The Theory of Economic Progress,* New Issues Press, Kalamazoo, Mich., 1978.

———: *Toward a Reasonable Society,* University of Texas Press, Austin, 1961.

Basalla, George: *The Evolution of Technology,* Cambridge University Press, Cambridge, 1988.

Beard, Charles: *The Industrial Revolution,* Greenwood Press, Westport, Conn., 1975.

Benjamin, Walter: *Illuminations,* Schocken Books, New York, 1968.

Berger, John: *Ways of Seeing,* Penguin Books, New York, 1977.

Billington, David P.: *The Tower and the Bridge,* Basic Books, New York, 1983.

Bugliarello, George, and Dean B. Doner: *The History and Philosophy of Technology,* University of Illinois Press, Urbana, 1973.

Burawoy, Michael: *Manufacturing Consent,* University of Chicago Press, Chicago, 1979.

Bury, J. B.: *The Idea of Progress,* 1932 rpt., Dover Publications, New York, 1955.

Chaliand, Gérard: *Revolution in the Third World,* Viking Press, New York, 1977.

Christians, Clifford G., and Jay M. Van Hook (eds.): *Jacques Ellul: Interpretive Essays,* University of Illinois Press, Urbana, 1981.

Csikszentmihalyi, Mihaly, and Eugene Rochberg-Halton: *The Meaning of Things,* Cambridge University Press, Cambridge, 1981.

De Lauretis, Teresa, Andreas Huyssen, and Kathleen Woodward (eds.): *The Technological Imagination: Theories and Fictions,* Coda Press, Madison, Wis., 1980.

Dewey, John: *Art as Experience,* Capricorn Books, New York, 1958. The critical edition of this work is published in Jo Ann Boydston (ed.), *John Dewey: The Later Works, 1925–1953,* vol. 10, 1934, Southern Illinois University Press, Carbondale, Ill., 1987.

———: *Experience and Nature,* Open Court, La Salle, Ill., 1965. The critical edition of this work is published in Jo Ann Boydston (ed.), *John Dewey: The Later Works, 1925–1953,* vol. 1, 1925, Southern Illinois University Press, Carbondale, Ill., 1981.

———: *Freedom and Culture,* G.P. Putnam's Sons, New York, 1939. The critical edition of this work is published in Jo Ann Boydston (ed.), *John Dewey: The Later Works, 1925–1953,* vol. 13, 1939, Southern Illinois University Press, Carbondale, Ill., 1988.

———: *Philosophy and Civilization,* Minton, Balch & Company, New York, 1931.

———: *Problems of Men,* Philosophical Library, New York, 1946.

———: *The Influence of Darwin on Philosophy,* 1910 rpt., Peter Smith, New York, 1951.

Dijksterhuis, E. J.: *The Mechanization of the World Picture,* The Clarendon Press, Oxford, 1964.

Durbin, Paul T. (ed.): *Research in Philosophy & Technology,* 8 vols., JAI Press Inc., Greenwich, Conn., 1978–1985.

———: *Technology and Contemporary Life,* D. Reidel, Dordrecht, 1988.

Eldridge, Seba, et al. (eds.): *Development of Collective Enterprise,* University of Kansas Press, Lawrence, 1943.

Ellul, Jacques: *Perspectives on Our Age,* William H. Vanderburg (ed.), Seabury Press, New York, 1981.

———: *The New Demons,* C. Edward Hopkin (trans.), Seabury Press, New York, 1975.

———: *The Technological Society,* John Wilkinson (trans.), Vintage Books, New York, 1964.

———: *The Technological System,* Joachim Neugroschel (trans.), Continuum, New York, 1980.
Elster, John: *Explaining Technical Change,* Cambridge University Press, Cambridge, 1983.
Faber, Roger J.: *Clockwork Garden,* University of Massachusetts Press, Amherst, 1986.
Fanon, Frantz: *The Wretched of the Earth,* Grove Press, New York, 1965.
Farrington, Benjamin: *Francis Bacon: Philosopher of Industrial Science,* Lawrence, and Wishart Ltd., London, 1951.
———: *Science and Politics in the Ancient World,* Barnes & Noble, New York, 1966.
———: *Science in Antiquity,* Oxford University Press, London, 1969.
Ferrarotti, Franco: *The Myth of Inevitable Progress,* Greenwood Press, Westport, Conn., 1985.
Ferre, Frederick: *Philosophy of Technology,* Prentice-Hall, Englewood Cliffs, N.J., 1988.
Gehlen, Arnold: *Man in the Age of Technology,* Columbia University Press, New York, 1980.
Gendron, Bernard: *Technology and the Human Condition,* St. Martin's Press, New York, 1977.
Getzels, Jacob W., and Mihaly Csikszentmihalyi: *The Creative Vision,* John Wiley & Sons, New York, 1976.
Giedion, Siegfried: *Mechanization Takes Command,* W. W. Norton & Company, New York, 1969.
Gimpel, Jean: *The Medieval Machine,* Penguin Books, New York, 1977.
Gouldner, Alvin W.: *The Dialectic of Ideology and Technology,* New York: Seabury Press, New York, 1976.
———: *The Two Marxisms,* Seabury Press, New York, 1980.
Goulet, Denis: *The Uncertain Promise,* IDOC/North America, New York, 1977.
Gramsci, Antonio: *Selections from the Prison Notebooks of Antonio Gramsci,* Quintin Hoare and Geoffrey Nowell Smith (eds.), International Publishers, New York, 1971.
Grant, George: *Technology and Empire,* House of Anansi, Toronto, 1969.
Greene, Maxine: *Landscapes of Learning,* Teachers College Press, New York, 1978.
Gunatilleke, Godfrey, *et. al,* (eds.): *Ethical Dilemmas of Development in Asia,* Lexington Books, Lexington, Mass., 1983.
Habermas, Jürgen: *Legitimation Crisis,* (trans.), Thomas McCarthy (trans.), Beacon Press, Boston, 1973.
———: *The Philosophical Discourse of Modernity,* Frederick Lawrence (trans.), The MIT Press, Cambridge, Mass., 1987.
Hacker, Andrew (ed.): *U/S: A Statistical Portrait of the American People,* Viking Press, New York, 1983.
Hickman, Larry: *John Dewey's Pragmatic Technology,* Indiana University Press, Bloomington, Ind., 1990.
Hill, Stephen: *The Tragedy of Technology,* Pluto Press, London, 1988.
Hofstadter, Richard: *Social Darwinism in American Thought,* Beacon Press, Boston, 1967.
Hood, Webster F.: "Dewey and Technology: A Phenomenological Approach," *Research in Philosophy and Technology,* V, 1982, pp. 189–207.
Horkheimer, Max, and Theodor W. Adorno: *Dialectic of Enlightenment,* John Cumming (trans.), Continuum, New York, 1982.
Horkheimer, Max: *Critical Theory,* Continuum, New York, 1986.

Hughes, Robert: *Shock of the New,* Alfred E. Knopf, New York, 1981.
Ihde, Don: *Existential Technics,* State University of New York Press, Albany, 1983.
———: *Technics and Praxis,* D. Reidel, Dordrecht, 1979.
Jonas, Hans: *The Imperative of Responsibility,* University of Chicago Press, Chicago, 1984.
Kouwenhoven, John A.: *The Arts in Modern American Civilization,* Norton Library, New York, 1948.
Kula, Witold: *Measures and Men,* R. Szreter (trans.), Princeton University Press, Princeton, 1986.
Laudan, Rachel (ed.): *The Nature of Technological Knowledge,* D. Reidel, Dordrecht, 1984.
Mannheim, Karl: *Freedom, Power, and Democratic Planning,* Oxford University Press, New York, 1950.
———: *Man and Society in an Age of Reconstruction,* Harcourt, Brace and Company, New York, 1940.
Marcuse, Herbert: "Some Social Implications of Modern Technology," *Studies in Philosophy and Social Science* vol. IX, 1941, pp. 414–439.
Mayr, Otto (ed.): *Philosophers and Machines,* Science History Publications, New York, 1976.
———: *Authority, Liberty & Automatic Machinery in Early Modern Europe,* Johns Hopkins University Press, Baltimore, 1986.
Merchant, Carolyn: *The Death of Nature,* Harper & Row, Cambridge, 1980.
Mesthene, Emmanuel G.: *How Language Makes Us Know,* Martinus Nijhoff, The Hague, 1964.
Mitcham, Carl: "Schools for Whistle Blowers," *Commonweal,* 114, April 10, 1987:201–205.
Mumford, Lewis: *The Golden Day,* 1926 rpt., Dover Publications, New York, 1968.
Pacey, Arnold: *The Maze of Ingenuity,* The MIT Press, Cambridge, Mass., 1976.
Powers, Richard: *Three Farmers on Their Way to a Dance,* McGraw-Hill, New York, 1985.
Rescher, Nicholas: *Unpopular Essays on Technological Progress,* University of Pittsburgh Press, Pittsburgh, 1980.
Rifkin, Jeremy, and Ted Howard: *The Emerging Order,* G. P. Putnam's Sons, New York, 1979.
Roszak, Theodore: *The Making of a Counter Culture,* Doubleday, Garden City, N.Y., 1969.
Slack, Jennifer Daryl: *Communication Technology and Society,* Ablex Publishing, Norwood, N.J., 1984.
Stanley, Manfred: *The Technological Conscience,* Free Press, New York, 1978.
Tichi, Cecelia: *Shifting Gears,* University of North Carolina Press, Chapel Hill, 1987.
Veblen, Thorstein: *The Engineers and the Price System,* Viking Press, New York, 1947.
———: *The Instinct of Workmanship,* Macmillan Company, New York, 1914.
———: *The Place of Science in Modern Civilization and Other Essays,* Russell & Russell, New York, 1961.
Walker, Robert H.: *Life in the Age of Enterprise,* Capricorn Books, New York, 1971.
White, Lynn, Jr.: *Dynamo and Virgin Reconsidered,* The MIT Press, Cambridge, 1976.
———: *Medieval Technology and Social Change,* Clarendon Press, Oxford, 1962.
———: *Medieval Religion and Technology: Collected Essays,* University of California Press, Berkeley, Los Angeles, and London, 1978.

Whitehead, Alfred North: *Science and the Modern World,* New American Library, New York, 1963.
Winner, Langdon: *Autonomous Technology,* The MIT Press, Cambridge, Mass., 1980.
———: *The Whale and the Reactor,* University of Chicago Press, Chicago, 1986.
Wyschogrod, Edith: "The Logic of Artifactual Existents: John Dewey and Claude Levi-Strauss," *Man and World,* 14, 1981:235–250.

## ACKNOWLEDGMENTS

Robert E. McGinn, "What Is Technology?" *Research in Philosophy and Technology,* ed. Paul Durbin. JAI Press, Greenwich, CT, Vol. 1, pp. 179–197, copyright © 1978. Used by permission.
Alan R. Drengson, "Four Philosophies of Technology," *Philosophy Today,* Summer, 1982, pp. 103–117. Used by permission.
Hans Jonas, "Toward a Philosophy of Technology," *The Hastings Center Report,* 9, no. 1 (1979), pp. 34–43. Reproduced by permission. Copyright © The Hastings Center.
Jacques Ellul, "The Technological Order," *The Technological Order: Proceedings of the Encyclopaedia Britannica Conference,* ed. Carl Stover. Wayne State University Press, Detroit, pp. 10–24. Reprinted by permission from Encyclopaedia Britannica, Inc.
D. O. Edge, "Technological Metaphor and Social Control, "*New Literary History,* Vol. 6, Autumn 1974, pp. 135–147. Used by permission.
John J. McDermott, "Glass without Feet," from *The Texas Humanist,* Jan./Feb., 1984, pp. 5–11, published by the Texas Committee for the Humanities. Used by permission.
Richard Powers, "The Cheap and Accessible Print," excerpt from *Three Farmers on Their Way to a Dance,* by Richard Powers. William Morrow & Company, Inc., New York, 1985, pp. 261–284. Copyright © 1985 by Richard Powers. Used by permission of William Morrow & Co., Inc.
Larry Hickman, "Autonomous Technology in Fiction." This essay appeared for the first time in *Philosophy, Technology and Human Affairs,* ed. Larry Hickman, College Station, TX, Ibis Press of College Station, Texas, 1985, pp. 316–324.
Don Ihde, "Technology and Human Self-Conception," *Southwestern Journal of Philosophy* (now *Philosophical Topics*), Vol. X, No. 1, Spring 1979, pp. 23–34. Used by permission.
Maurice Merleau-Ponty, from *Phenomenology of Perception,* Humanities Press, Atlantic Highlands, NJ, 1962, pp. 87–89, 142–147. Reprinted by permission of Humanities Press International, Inc., Atlantic Highlands, NJ.
Shoshana Zuboff, from *In the Age of the Smart Machine: The Future of Work and Power.* Copyright © 1988 by Basic Books, Inc. Reprinted by permission of Basic Books, Inc.
John J. McDermott, "The Aesthetic Drama of the Ordinary." Reprinted from *Streams of Experience: Reflections on the History and Philosophy of American Culture,* by John J. McDermott. University of Massachusetts Press, Amherst, 1986, pp. 129–140. Copyright © 1986 by The University of Massachusetts Press.
Douglas Browning, "Some Meanings of Automobiles." This essay first appeared in *Technology and Human Affairs,* ed. Larry Hickman and Azizah al-Hibri. The C. V. Mosby Co., St. Louis, 1981, pp. 13–17. Used by permission of the author.
Glen Jeansonne, "The Automobile and American Morality," *Journal of Popular Culture,* Vol. 8, 1974, pp. 125–131. Used by permission.
Robert Linhart, *The Assembly Line,* translated by Margaret Crosland. The University of Massachusetts Press, Amherst, 1981, pp. 13–27. Copyright © 1981 by John Calder Publishers Ltd.

George Gerbner, "Television: The New State Religion?" *Et cetera,* Vol. 34, 1977, pp. 145–150. Used by permission.

Edmund Carpenter, "The New Languages," from *Explorations in Communications,* ed. Edmund Carpenter and Marshall McLuhan. Beacon Press, Boston, Copyright © 1960, pp. 162–179. Reprinted by permission of Beacon Press.

Lewis Mumford, "The Monastery and the Clock," excerpt from *Technics and Civilization,* by Lewis Mumford, copyright 1934 by Harcourt Brace Jovanovich, Inc. Renewed 1962 by Lewis Mumford, reprinted by permission of the publisher.

Daniel Boorstin, "The Rise of the Equal Hour," *The Discoverers,* Random House, New York, 1983, pp. 36–46. Copyright © 1983 by Daniel Boorstin. Reprinted by permission of the publisher.

John J. McDermott, "Urban Time," from "Space, Time and Touch: Philosophical Dimensions of Urban Consciousness," *Soundings,* Vol. 57, 1974, pp. 268–271. Used by permission.

Paul B. Thompson, "Nuclear Weapons and Everyday Life." A somewhat different version of this essay first appeared in *Philosophy, Technology and Human Affairs,* ed. Larry Hickman, College Station, TX, Ibis Press of College Station, Texas, 1985, pp. 117–127. Used by permission of the author.

Ruth Schwartz Cowan, "Housework and Its Tools," from *More Work for Mother: The Ironies of Household Technology from the Open Hearth to the Microwave,*by Ruth Schwartz Cowen, pp. 3–15. Copyright © 1983 by Basic Books, Inc. Reprinted by permission of Basic Books, Inc., Publishers.

José Ortega y Gasset, from "Man the Technician." Reprinted from *History As a System and Other Essays Toward a Philosophy of History,* by José Ortega y Gasset, translated by Helene Weyl, by permission of W. W. Norton & Company, Inc., pp. 108–116, 139–156. Copyright © 1941, 1961 by W. W. Norton & Company, Inc.

Lynn White, Jr., "The Act of Invention," *The Technological Order: Proceedings of the Encyclopaedia Britannica Conference,* ed. Carl Stover, pp. 102–116. Wayne State University Press, Detroit, 1963. Reprinted by permission from Encyclopaedia Britannica, Inc.

Lewis Mumford, from *Technics and Civilization,* Harcourt Brace and World, Inc., New York, 1963, pp. 107–113, 151–158, 172–178, 212–218. Excerpts from *Technics and Civilization,* copyright © 1934 by Harcourt Brace Jovanovich, Inc. Renewed 1962 by Lewis Mumford. Reprinted by permission of the publisher.

Paul Levinson, "Toy, Mirror and Art: The Metamorphosis of Technological Culture," *Et cetera,* vol. 34, 1977, pp. 151–167. Used by permission.

Autumn Stanley, "Women Hold Up Two-Thirds of the Sky: Notes for a Revised History of Technology," *Machina Ex Dea: Feminist Perspectives on Technology,* ed. Joan Rothschild. Pergamon Press, Oxford, 1983, pp. 5–22. Used by permission.

Jacques Ellul, "The Autonomy of Technology," from *The Technological Society,* translated by John Wilkinson. Alfred A. Knopf, New York, 1964, pp. 133–147. Copyright © 1964 by Alfred A. Knopf, Inc. Reprinted by permission of the publisher.

Jacques Ellul, "The Present and the Future," *Perspectives On Our Own Age,* ed. William H. Vanderburg, translated by Joachim Neugroschel. Seabury Press, New York, 1981, pp. 59–84. English translation copyright © 1981 by the Canadian Broadcasting Corp. Reprinted by permission of Harper and Row, Publishers, Inc.

Herbert Marcuse, "The New Forms of Control," from *One Dimensional Man.* Beacon Press, Boston, 1964, pp. 1–18. Copyright © 1964 by Herbert Marcuse. Reprinted by permission of Beacon Press.

Arnold Pacey, from *The Culture of Technology*, The MIT Press, Cambridge, Mass., 1983, pp. 13–34. © Arnold Pacey, 1983.

Langdon Winner, "Techné and Politeia: The Technical Constitution of Society," *The Whale and the Reactor,* University of Chicago Press, Chicago, 1986, pp. 40–58. A different version of this essay was published in *Philosophy and Technology,* ed. Paul Durbin and Friedrich Rapp (Dordrecht: D. Reidel, 1983), pp. 97–111. Used by permission of University of Chicago Press and the author.

John Dewey, from "Science and Society," *John Dewey: The Later Works, 1925–1953, Vol. 6: 1931–1932,* edited by Jo Ann Boydston, pp. 53–63. Textual editor, Anne Sharpe. Copyright © the Board of Trustees, Southern Illinois University. Reprinted by permission of Southern Illinois University Press.

William H. Lowrance, "The Relation of Science and Technology to Human Values," from *Modern Science and Human Values,* Oxford University Press, New York, 1986, pp. 145–150. Copyright © 1985 by Oxford University Press, Inc. Reprinted by permission.

Herbert I. Schiller, "Paradoxes of the Information Age," from *Information and the Crisis Economy,* Ablex Publishing Corporation, Norwood, NJ, 1984, pp. 95–111. Used by permission.

C. E. Ayres, "The Path of Progress," *Southwest Review,* Vol. 27, 1943, pp. 229–244. This essay first appeared in the *Southwest Review.* Used by permission.

Jürgen Habermas, "Technological Progress and the Social Life-World," from *Toward a Rational Society,* by Jürgen Habermas translated by Jeremy J. Shapiro. Beacon Press, Boston, 1970, pp. 50–61. Copyright © 1970 by Beacon Press. Reprinted by permission of Beacon Press.

G. Gunatilleke, Neelan Tiruchelvam, and Radhika Coomaraswamy, eds., "The Ethics and Order of Change," from *Ethical Dilemmas of Development in Asia.* Lexington Books, Lexington, Massachusetts, 1983, pp. 6–27. Used by permission of the Asia Society.